MARINE BIOLOGY

MARINE BIOLOGY
AN ECOLOGICAL APPROACH

James W. Nybakken

California State University at Hayward and the Moss Landing Marine Laboratories

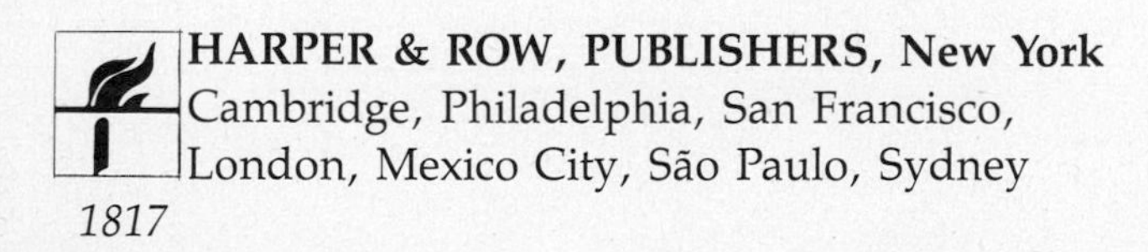

HARPER & ROW, PUBLISHERS, New York
Cambridge, Philadelphia, San Francisco,
London, Mexico City, São Paulo, Sydney

1817

Sponsoring Editor: Claudia Wilson
Project Editor: Holly Detgen
Designer: T. R. Funderburk
Production Assistant: Jacqui Brownstein
Compositor: Black Dot, Inc.
Printer and Binder: R. R. Donnelley & Sons Company
Art Studio: J & R Art Services, Inc.
Cover photograph courtesy of Dr. Lovell Langstroth

Marine Biology: An Ecological Approach

Library of Congress Cataloging in Publication Data
Nybakken, James Willard.
Marine biology.

Includes index.
1. Marine biology. 2. Marine ecology. I. Title.
QH91.N9 574.5′2636 81–7236
ISBN 0–06–044849–0 AACR2

To Bette, Kent, Scott,
and the students of the
Moss Landing Marine Laboratories

Contents

Preface

In recent years, there has been an increasing interest in the field of marine biology at the undergraduate level in our two-year community colleges as well as at the four-year colleges and universities. This interest apparently derives from the growing awareness and concern at all levels of society of the importance of the world's oceans as sources of food, as reservoirs of minerals, as major suppliers of oxygen and regulators of climates, and as the ultimate dumping ground for the mounting burden of human waste materials. This concern has been popularized and focused by the various international disputes on fishing rights, the whale problem, international law of the seas conferences, and numerous television programs and books addressed to the general public about marine life.

As a result of this heightened public awareness, marine biology courses have attracted a rather wide spectrum of students with varying backgrounds who desire to obtain a basic understanding of the biological processes that operate in the oceans.

It is important to understand that in the oceans, just as on land, there are scientific principles that govern the organization and perpetuation of organisms and associations. While these principles operate somewhat differently in the ocean than on land due to the physical properties of water, they can be readily set forth and may be understood by those with a minimum background in other sciences or mathematics.

The texts in marine biology available to serve these courses appeared to me to lack an ecological approach to the entire marine environment; they tended to emphasize specific areas, habitats, and organisms to the exclusion of broader concepts and processes. I was thus stimulated to write this book as a basic introduction to marine biology, emphasizing ecological principles governing marine life throughout the world, not as a purely taxonomic or regional approach, so that it would be useful in all parts of the world. I have, furthermore, purposely downgraded any discussion of pure oceanography, except as it bears upon organization of the associations or communities, in order to increase coverage of the biology. The sequence of topics in this book closely follows the sequence in my undergraduate course in marine ecology, which I have taught for several years at the Moss Landing Marine Laboratories.

This text is designed for the undergraduate student in marine biology. It presumes a certain minimum of background in very basic concepts of chemistry, physics, and biology, but no more than would be obtained from a high school or introductory college course in each of these fields. Familiarity with the major invertebrate phyla is helpful, but not necessary. Some acquaintance with the basic ecological concepts is also helpful, but if lacking, may be obtained from Chapter One. While generally aimed at the lower-division undergraduate student, I have tried to imbue this book with sufficient rigor and detail so that it will also be useful in upper-division undergraduate courses in marine ecology and biological oceanography. To this latter end I have included a section on additional references and text citations to the primary literature.

This text stresses ecological processes and adaptations that act to structure marine associations and permit their persistence through time. It is thus not a guide to the local fauna and flora. It is presumed that familiarity with this, if necessary, will be obtained in laboratory and field trips in which the instructor will provide the taxonomic expertise.

Depending on the time available in the course, this book includes perhaps more material than can be adequately covered in a single quarter or semester. In that case, the instructor may choose to concentrate on those chapters or areas of most concern in his or her geographical region. Thus, for example, instructors in temperate zones may wish to de-emphasize coral reefs and many may wish to spend less time on deep-sea biology and more on the intertidal, to which they have direct access. Suggestions for use of the book under varying course conditions are given in the instructor's manual. However, I hope that the comprehensive nature of this book means that whatever instructional choices are made, information will be available from the book.

As the author of this textbook, I am indebted to a large number of marine scientists whose research work is the foundation on which this book is based and who are too numerous to mention in entirety here. A few of this vast number are mentioned in the text and listed in the literature cited. I am, however, particularly appreciative of the help given me by my colleagues and students at Moss Landing Marine Laboratories, many of whom offered suggestions, pointed out errors or additional literature to be studied, and read and criticized various chapters. In this regard, I would like to thank Dr. Gregor Cailliet, Dr. George Knauer, Dr. Victor Morejohn, Dr. Michael Foster, and Dr. Michael Moser. I wish to thank also those colleagues who contributed photographs and drawings, including Dr. Paul Dayton, Dr. Joseph Connell, Dr. Fred Grassle, Dr. Charles Birkeland, Dr. Richard Young, Dr. Richard Schwartzlose,

Dr. Robert Hessler, Dr. Glen VanBlaricom, Dr. John Oliver, Dr. Lovell Langstroth, Mr. Michael Kelly, Dr. Richard Mariscal, and Dr. Clyde Roper. I am particularly indebted to Lynn McMasters, who was primarily responsible for transforming my scribbles into acceptable artwork; Rosemary Stelow, who did an outstanding job in typing, proofreading the manuscript, and picking up my errors; and Sheila Baldridge, the Moss Landing librarian, who could always find the missing reference and who put up with long absences of a significant number of books and periodicals.

I am also grateful to the following outside reviewers whose comments and suggestions were of great help to me in improving the manuscript: Dr. Peter Frank, Dr. Robert Hessler, Dr. William Jorgensen, Dr. James Lanier III, Dr. Roger Lloyd, Dr. Don Mauer, Dr. John Morrill, Dr. J. A. Musick, Dr. C. H. Peterson, Dr. E. R. Pariser, Dr. Wendell Patton, Dr. Robert Simon, and Dr. Ivan Valiela.

Because of the broad scope of the book and necessary limitations in pages, I have often been forced to treat certain complex or poorly understood processes through generalizations that some may consider serious oversimplifications. Undoubtedly some errors and omissions remain. I therefore close with a request to all who may use this book that they feel free to pass on to me any errors, omissions, mistakes in interpretations, comments, or suggestions that they may have, so that I might correct my ignorance and improve any future revisions.

James W. Nybakken

Chapter One
INTRODUCTION TO THE MARINE ENVIRONMENT

About 71 percent of the surface of this planet is covered by salt water. Beneath this surface, the water depth averages 3.8 km, giving a volume of 1370 $\times$ 10^6 km^3. Since life exists throughout this immense volume, the oceans constitute the single largest repository of organisms on the planet. These organisms include representatives of virtually all phyla and are tremendously varied. All, however, are subject to the properties of the sea water that surrounds them, and many features common to these plants and animals are the result of adaptations to the watery medium and its movements. Before we can begin a consideration of the major associations or assemblages of organisms, we need to examine briefly the physical and chemical conditions of sea water and aspects of its motion (oceanography), and some basic ecological principles and terms that will be central to an understanding of the processes discussed in the remainder of this text. Finally, we will make some comparisons between aquatic and terrestrial ecosystems to point out some fundamental differences in organization.

PROPERTIES OF WATER

Water is the substance that surrounds all marine organisms. It also composes the greater bulk of the bodies of marine plants and animals, and it is the medium in which various chemical reactions take place, both inside and outside living organisms.

CHEMICAL COMPOSITION

Pure water is a very simple chemical compound composed of two atoms of hydrogen (H) joined to one oxygen (O) atom. Expressed symbolically, it is H_2O.

The H atoms are bounded to the oxygen asymmetrically such that the two hydrogens are at one end of the molecule and the oxygen at the other. The bonding between the H atoms and the oxygen is via shared electrons, each H sharing its single electron with the oxygen. In this manner, oxygen receives the two electrons needed to complete its outer electron shell, and each hydrogen the one needed for its outer shell. However, the larger oxygen tends to draw the electrons furnished by the hydrogen closer to its nucleus. This creates a slight negative charge at the oxygen end of the molecule, while the removal of the electrons away from the hydrogens results in a slight positive charge to that end. This electrical separation creates a polar molecule.

The polar nature of the water molecule means that the hydrogen end, which is positive, will attract the negative, or oxygen, end of other water molecules. This gives rise to weak bonds, called *hydrogen bonds*, between adjacent water molecules. These bonds are only about 6 percent as strong as the bonds between the hydrogen and oxygen in the water molecule itself, and they are easily broken and re-formed (Fig. 1-1). It is this hydrogen bonding between adjacent water molecules and the polarity of the water molecule that are responsible for many of the unique chemical and physical properties of water. Indeed, if water were not polar and did not form such bonds with adjacent molecules, it would be a gas, not a liquid, at room temperature, and its freezing point would be lower than temperatures found over most of the earth's surface. Under such conditions, life as we know it would be impossible.

PHYSICAL AND CHEMICAL PROPERTIES

Because of the hydrogen bonding, water tends to stick firmly to itself, resisting external forces that would break these bonds. This is called *cohesion*. At air-water

Figure 1-1 Diagrammatic representation of a series of water molecules, indicating the polar nature of the molecules and the hydrogen bonding.

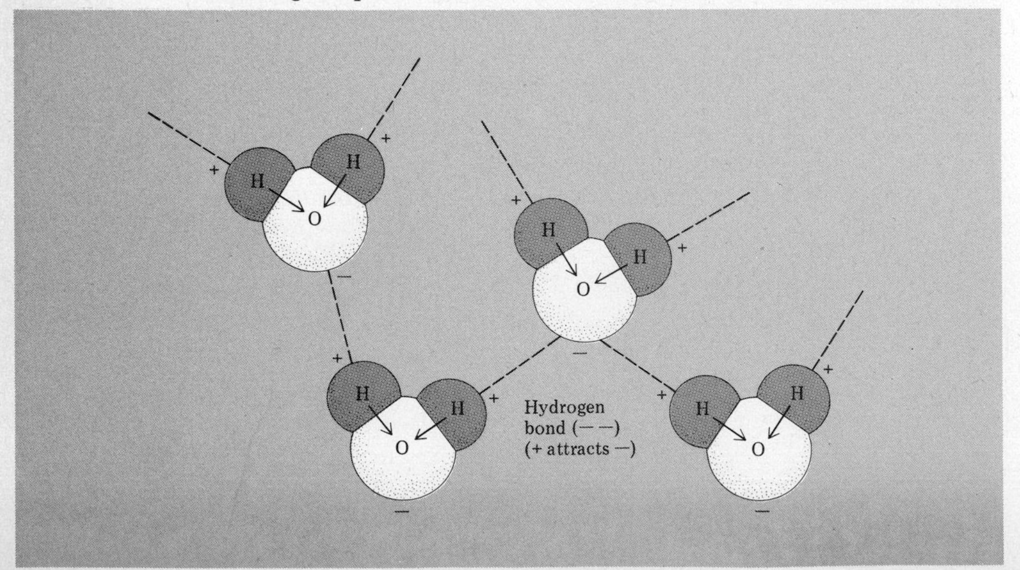

interfaces, the strength of cohesion forms a "skin" over the water surface strong enough to support small objects. This is *surface tension*. The surface tension of water is the highest of all common liquids and permits the existence of associations of organisms either suspended below it or moving over the top. Cohesion is also responsible for the *viscosity* of water. Viscosity is a property measuring the force necessary to separate the molecules and allow passage through the liquid. This resistance to flow or movement is important in the sinking rate of objects (pp. 47–50) and in problems of movement in the water by animals (see pp. 89–94). Cohesion properties are temperature dependent, increasing with decreasing temperature.

Another set of properties concerns the effects of heat. Water, when heated, evaporates slowly in comparison with other liquids of simple molecules. This means that the *heat of vaporization* is high, the highest of most common substances. This is a direct result of the strength of the hydrogen bonding between molecules, which must be broken in order to allow the escape of a molecule. Because of the high heat of vaporization, water evaporates slowly and by so doing, absorbs considerable heat, leading to cooling. Similarly, this high heat of vaporization means a high boiling point (100°C), with the result that water is a liquid on earth rather than a gas. Related to this is the *latent heat of fusion*, which is the amount of heat gained or lost per unit mass when a substance changes from a solid to a liquid or vice versa. Water, again, has the highest value of most common liquids. Ice, when melting, therefore takes up large quantities of heat and in forming ice, large quantities of heat are given off.

The high values for both the heat of vaporization and latent heat of fusion mean that it takes more heat to effect a change in temperature in a given quantity of water than in virtually any other common substance. This high *heat capacity* means that water is a strong cushion against both rising and falling temperatures and therefore ameliorates the climate; also, it means that the range of temperatures experienced within any body of water is less than that in air.

Water has a peculiar *density-temperature* relationship. Most liquids become more dense as they are cooled. If cooled until they become solid, the solid phase of such liquids is more dense than the liquid phase. This is not true for water. Water becomes more dense as it is cooled until it reaches 4°C. Cooling below this temperature decreases the density and when freezing occurs, there is a marked decrease in density. Therefore, ice is lighter than water and floats upon it. This property is of utmost importance to life in the oceans, for otherwise, major volumes of the oceans would be uninhabitable as large blocks of ice.

The chemical properties of water that are of importance are those concerned with the solvent capacities (Fig. 1-2). Water is almost a universal solvent, with the ability to dissolve more substances than any other liquid. This is because the solvent action is of two types: one depends on the polar character of the molecule, the other on the hydrogen bonding. Various nonpolar organic and inorganic compounds containing oxygen atoms or hydrogen atoms bonded to either oxygen or nitrogen atoms are held in solution by hydrogen bonding. The

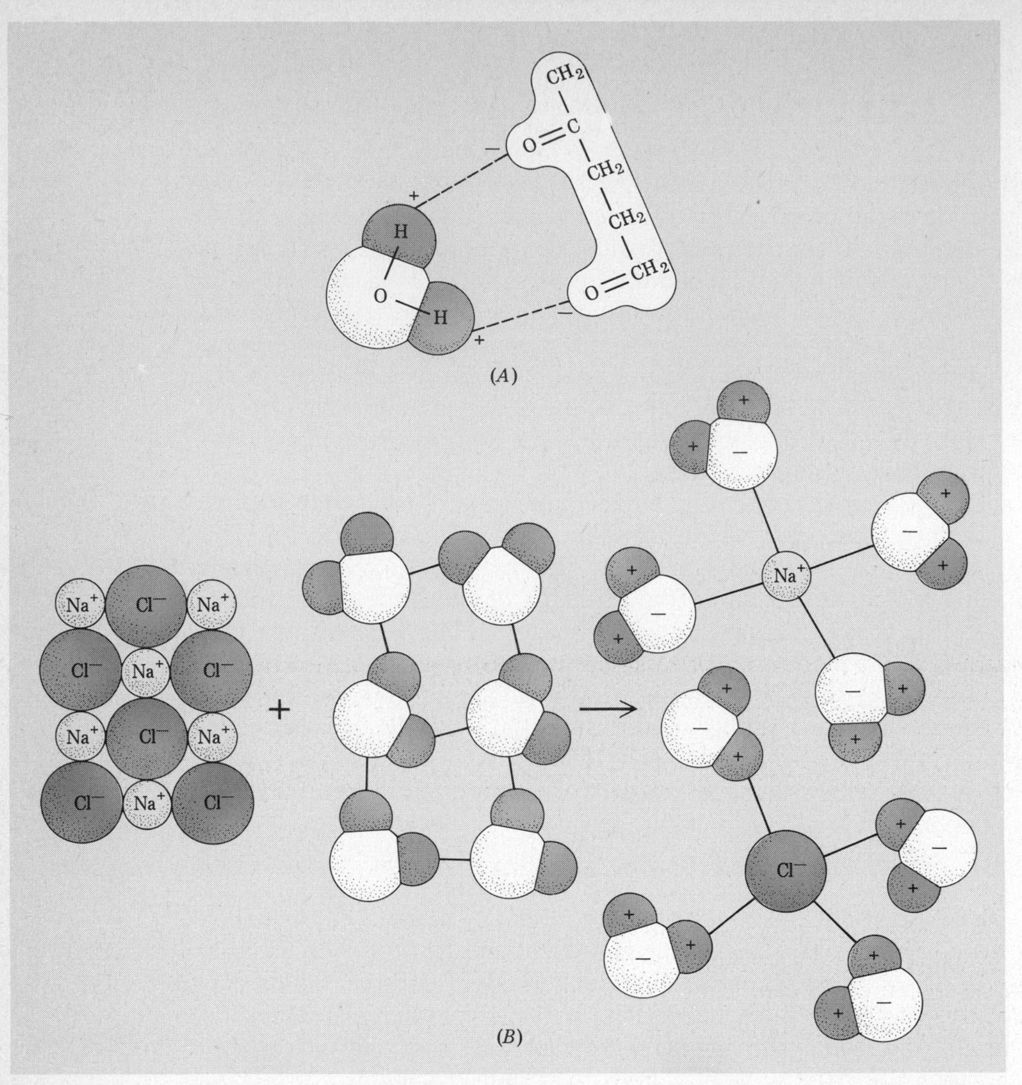

Figure 1-2 *(A)* Solvent action of water through creation of hydrogen bonds. *(B)* Solvent action of water through interaction of salt ions with charged parts of water molecules.

polar dissolving action of water for various salts depends on the interaction of the salt ions with the charges on the water molecule. When a salt is dissolved in water, it breaks into its component ions. Thus, common table salt, NaCl, when put into water, breaks into Na^+ and Cl^- ions. It is through interaction of the charges on the salt ions and the water molecules that salts are held in solution.

Still other chemical properties of water depend upon the breaking of the strong bonds between the hydrogen and the oxygen in the water molecule. In any volume of water, a few of the water molecules have separated the charges so completely that the molecule breaks into two charged parts, H^+ (hydrogen ion) and OH^- (hydroxyl ion). Here, one hydrogen atom moves away, leaving its electron with the bonding electrons on the oxygen atom, therefore making the free hydrogen positive and leaving the OH negative. Certain substances, when they dissolve in water, do so by reacting with these ions. An example is carbon dioxide (CO_2), which reacts as follows:

$$CO_2 + HOH \rightleftharpoons H^+ + HCO_3^-$$

These properties are all summarized, together with some other ones, in Table 1-1.

SEA WATER

Sea water is pure water in which are dissolved a variety of solids and gases. A 1000 g sample of sea water will contain about 35 g of dissolved compounds, collectively called salts. In other words, 96.5 percent of sea water is pure water and 3.5 percent dissolved substances. The total amount of dissolved material is termed the *salinity*. Marine biologists and oceanographers, however, generally

TABLE 1-1 Some Properties of Water

Property	Compared to Other Substances
Surface tension	The highest of all common liquids
Conduction of heat	The highest of all common liquids, except mercury
Viscosity	Relatively low viscosity for a liquid (decreases with increasing temperature)
Latent heat of vaporization; the quantity of heat gained or lost per unit mass by a substance changing from a liquid to a gas or gas to liquid phase without an increase in temperature (cal/g)	The highest of all common substances
Latent heat of fusion; the quantity of heat gained or lost per unit mass by a substance changing from a solid to a liquid or liquid to solid phase without an accompanying rise in temperature (cal/g)	The highest of all common liquids and most solids
Heat capacity: the quantity of heat required to raise the temperature of 1 g of a substance 1°C (cal/g/°C)	The highest of all common solids and liquids
Density: mass per unit volume (g/cm³ or g/ml)	Density determined by (1) temperature, (2) salinity, (3) pressure, in that order. The temperature of maximum density for pure water is 4°C. For sea water the freezing point decreases with increasing salinity.
Dissolving ability	Dissolves more substances in greater quantities than any other common liquid

After Ingmanson and Wallace, 1973.

prefer to refer to salinity in terms of parts per thousand, abbreviated as ‰. Thus, if a typical sea water sample has 35 g of dissolved compounds in 1000 g, it has a salinity of 35 ‰.

Dissolved substances include inorganic salts, organic compounds derived from living organisms, and dissolved gases. By far the greatest fraction of the dissolved material is composed of inorganic salts present as ions. Six inorganic ions comprise 99.28 percent by weight of the solid organic matter. They are chlorine, sodium, sulfur (as sulfate), magnesium, calcium, and potassium (Table 1-2). These can be considered the major ions. An additional five minor ions add an additional 0.71 percent by weight so that 11 ions together make up 99.99 percent by weight of the dissolved substances.

The salinity of various parts of the open ocean away from coastal areas varies within a narrow range, usually from 34 to 37 ‰, and averages 35 ‰. The differences in salinity are due to differences in evaporation and precipitation. Higher values occur in tropical oceans where there is a high evaporation, and lower values in temperate oceans where there is less evaporation. In inshore areas and partially enclosed seas, the salinity is more variable and may be near 0 where large rivers discharge fresh water, to near 40 ‰ in the Red Sea and Persian Gulf.

It has been found that whereas there are the above-described variations in salinity, and hence in the total amount of dissolved salts in various areas, the ratios among the most abundant ions remain virtually constant. This is important; it means that concentrations of nearly all ions in a given sample of water can be determined by measuring only one. This fundamental relationship is the

TABLE 1-2 Major and Minor Constituents of 34.8 ‰ Sea Water

Ion	Percent by Weight
A. Major	
Chloride (Cl^-)	55.04
Sodium (Na^+)	30.61
Sulfate (SO_4^{2-})	7.68
Magnesium (Mg^{2+})	3.69
Calcium (Ca^{2+})	1.16
Potassium (K^+)	1.10
Subtotal	99.28
B. Minor	
Bicarbonate (HCO_3^-)	0.41
Bromide (Br^-)	0.19
Boric acid (H_3BO_3)	0.07
Strontium (Sr^{2+})	0.04
Subtotal	0.71
Total	99.99

basis for the measurement of salinity in sea water. Salinity can be established by measuring a single parameter, which may be the chlorinity (= chlorine concentration), for example, or the electrical conductivity or refractive index, which also depend on the salt content.

Among the remaining 0.01 percent of dissolved substances in sea water are several inorganic salts that are of crucial importance to organisms in sea water. Included here are the nutrients, the phosphates and nitrates, which are required by plants to synthesize organic material in photosynthesis, and silicon dioxide, which is required by diatoms and radiolarians to construct their skeletons. In contrast to the previous ions, nitrates and phosphates do not exist in a constant ratio with other elements or ions and tend to be in short supply in surface waters, varying in abundance as a result of biological activity (see Chapter Two for discussion). Supplies of these essential nutrients may actually become limiting to plant production in some cases (see pp. 60–66).

Other substances existing as trace amounts include elements essential to life processes, such as iron, manganese, cobalt, and copper. Although present in minute amounts, they are not limiting to the existence of life. Certain organic compounds such as vitamins are also present in minute amounts, but little is known of their variation.

The salt content of sea water has a definite effect on its properties. The maximum density of pure water occurs at 4°C but with sea water, density continues to increase to the freezing point. Because of the salt content, the freezing point is also reduced from 0°C, the amount being a function of the salinity. For sea water of 35 ‰, the freezing point is −1.9°C. Upon freezing, however, the density decreases so that ice floats on the surface. The importance of this density increase below 4°C is that very cold and very dense surface water can be formed and sink to the bottom of the ocean basins.

Two gases dissolved in sea water are of metabolic importance: oxygen and carbon dioxide. The solubility of gases in sea water is a function of temperature; the lower the temperature, the greater the solubility. Therefore, the colder the water, the more oxygen it can hold. Even so, the solubility of gases in water is not great. At 0°C, 35 ‰ sea water contains about 8 ml/liter O_2, whereas air has 21 ml/liter. At 20°C, 35 ‰ sea water contains only 5.4 ml/liter. The reason that deep ocean water does not become anoxic (devoid of oxygen) through biological activity is that when water sinks from the surface, it is so cold that it has a maximum amount of oxygen, more than can be consumed by the limited populations of animals in deep water. Oxygen is not distributed uniformly with depth in the ocean. A typical vertical profile of the oxygen content shows a maximum amount in the upper 10–20 m, where photosynthetic activity by plants and diffusion from the atmosphere often leads to supersaturation (Fig. 1-3). With increasing depth, the oxygen content declines. This decline reaches a minimum somewhere between 500–1000 m in open ocean waters. This *oxygen minimum zone* may have oxygen values that approach zero in some areas. Below this zone, oxygen values increase somewhat with depth, but usually do not approach surface values except in the tropics (Fig. 1-3). The occurrence of the

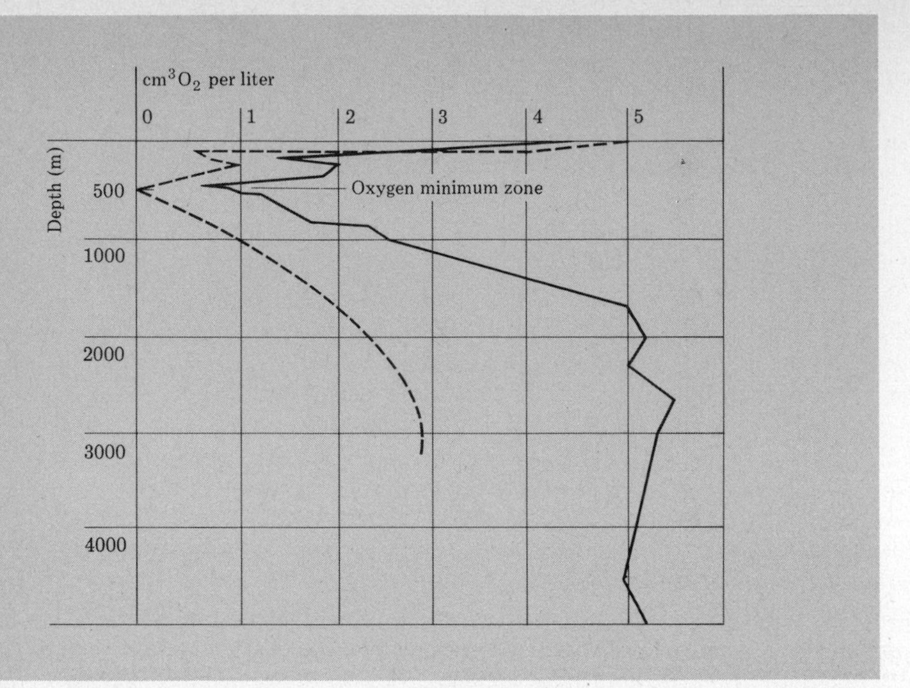

Figure 1-3 Change in dissolved oxygen with depth in the eastern tropical Pacific Ocean (dashed line) and the tropical Atlantic Ocean (solid line). (Modified from D. E. Ingmanson and W. J. Wallace, Oceanology: An introduction, © 1973 by Wadsworth Publishing Company, Inc. Reprinted by permission of Wadsworth Publishing Company, Belmont, Calif. 94002.)

oxygen minimum zone is usually attributed to biological activity depleting the oxygen, but no generally acceptable theory satisfactorily explains it.

The solubility of carbon dioxide is somewhat different from that of oxygen, since it reacts chemically in the water. Carbon dioxide is abundant in sea water, and sea water has a considerable capacity to absorb the gas. This is because carbon dioxide, upon entering sea water, reacts with the water to produce carbonic acid.

$$CO_2 + HOH \leftrightharpoons H_2CO_3$$

Carbonic acid further dissociates into a hydrogen ion and a bicarbonate ion:

$$H_2CO_3 \rightleftharpoons H^+ + HCO_3^-$$

Bicarbonate may further dissociate into another hydrogen ion and a carbonate ion:

$$HCO_3^- \rightleftharpoons H^+ + CO_3^{2-}$$

The major reservoir of carbon dioxide in the ocean is the bicarbonate ion, as can be seen from its place among the major ions in sea water (Table 1-2). In contrast to oxygen, it is more abundant in sea water than in air. Therefore, carbon dioxide is rarely limiting to plants in sea water.

The carbon dioxide–carbonic acid–bicarbonate system is a complex chemical system that tends to stay in equilibrium. Thus, if CO_2 gas is removed from sea water, the equilibrium will be disturbed and carbonic acid and bicarbonate will shift to the left in the above equations, until more CO_2 is produced and a new equilibrium set up.

The above reactions result in the production or absorption of free hydrogen ions (H^+). The abundance of hydrogen ions in solution is a measure of *acidity*. More H^+ ions mean a more acid solution, and fewer H^+ ions a more alkaline solution. *Alkaline* solutions are those that have large numbers of OH^- ions and few H^+ ions. Acidity and alkalinity are measured on a logarithmic scale of 1 to 14 units. These units are called pH units and the scale the *pH scale*. The higher the concentration of H^+ ions on this scale, the lower the pH value. Hence, low pH values indicate acid conditions. Conversely, high pH values indicate low H^+ concentrations and high OH^- concentration. The neutral point is pH 7, where equal numbers of both ions occur (Fig. 1-4).

Although pure water is neutral in pH because dissociation of the water molecule produces equal numbers of H^+ and OH^- ions, the presence of CO_2 and the strongly alkaline ions sodium, potassium, and calcium in ocean waters tend to change this, so that sea water is slightly alkaline, usually ranging from pH 7.5 to 8.4. The carbon dioxide–carbonic acid–bicarbonate system functions as a *buffer* to keep the pH of sea water within a narrow range. It does this by absorbing H^+ ions in the water when they are in excess and producing more when they are in short supply. This is accomplished by shifting the above reactions to the right when there are too few H^+, thereby producing more bicarbonate ion and carbonate ion and to the left when there are too many, producing more undissociated carbonic acid and bicarbonate ion.

BASIC OCEANOGRAPHY

This is not a book on *oceanography*, which is the discipline devoted to the study of all aspects of the physics, chemistry, geology, and biology of the sea. We are

Figure 1-4 The pH scale.

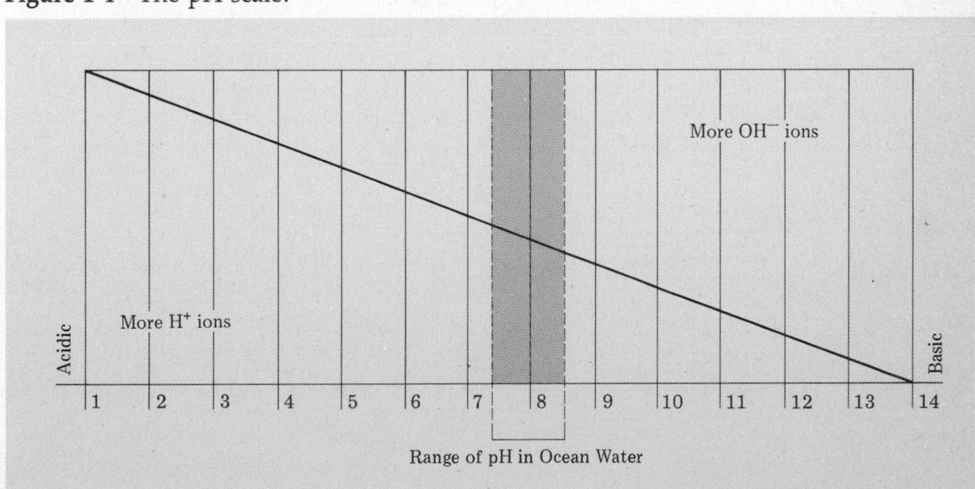

concerned here only with giving a basic understanding of life in the sea and how it is organized and persists. However, in order to understand the ecology of various marine associations, it is necessary to know something about the structure and motion of the ocean water masses.

GEOGRAPHY AND GEOMORPHOLOGY OF THE OCEANS

Whereas all the major oceans are connected to each other, the world ocean has been separated for convenience into four major divisions, Pacific, Atlantic, Indian, and Arctic, in order of decreasing size. Projecting from, or partially cut off from, these larger oceans are smaller marginal seas such as the Mediterranean, Caribbean, Baltic, Bering, South China, and Okhotsk. The locations of the major oceans and seas are given in Fig. 1-5. The Pacific, Atlantic, and Indian Oceans all converge in the area around the Antarctic continent, giving a contiguous body of water around the southern continent.

The oceans are not equally distributed over the earth. Oceans cover more than 80 percent of the Southern Hemisphere but only 61 percent of the Northern Hemisphere, where most of the earth's land masses occur.

On the margins of the major land masses, the ocean is very shallow, overlying an underwater extension of the continent called the *continental shelf*. Forming only 7–8 percent of the total ocean area, the continental shelf slopes gently from shore to a depth of 200 m. The shelf extends offshore for up to 400 km off eastern Canada, but extends only a few kilometers offshore along most of the Pacific

Figure 1-5 Major oceans and seas of the world.

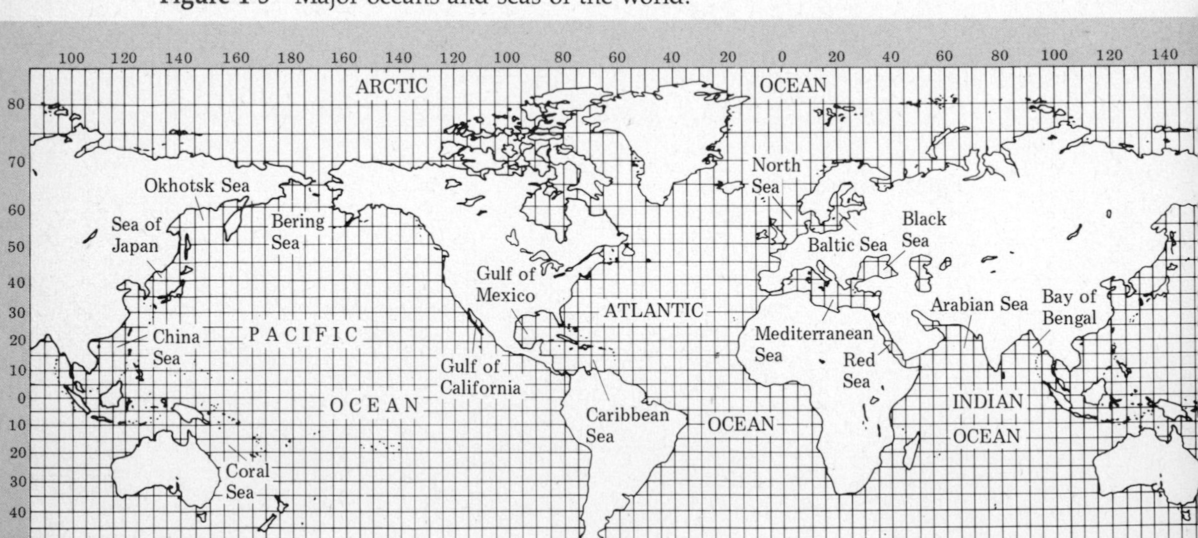

coast of North America. At the outer edge of the shelf, there is an abrupt steepening of the bottom to become the *continental slope.* The continental slope descends precipitously to depths of 3 to 5 km. At these depths, the bottom becomes the flat, extensive, sediment-covered *abyssal plain.* Such plains monotonously cover the floors of vast areas of the oceans at depths between 3 and 5 km (Fig. 1-6).

Abyssal plains are broken in several places by various *submarine ridges.* These ridges are extensive, contiguous submarine mountain chains which have been found in all oceans. The best known is the mid-Atlantic ridge which bisects the Atlantic Ocean into east and west basins and runs from Iceland into the south Atlantic, where it links with a similar ridge in the Indian Ocean. Occasionally, the ridges break the surface to form islands. The Azores, Ascension, and Tristan da Cunha are islands formed by the mid-Atlantic ridge. These extensive ridge systems mark the boundaries of the various crustal plates of the earth and are often the sites of volcanic activity.

In certain areas, the abyssal plains are cut by deep, narrow troughs called *trenches.* Most of these trenches lie in an arc bordering the islands and continents in the Pacific Ocean. The trenches have depths from 7,000 to more than 11,000 m. The deepest area known is the 11,022 m Challenger Deep in the Marianas Trench.

Finally, there may be isolated islands and submarine *sea mounts* formed by isolated volcanic action. Such mountains, as opposed to the ridges, rise individually from the abyssal plain.

TEMPERATURE AND VERTICAL STRATIFICATION

Temperature is a measure of the energy of molecular motion. In the world oceans, it varies horizontally with changes in latitude and also vertically with depth. Temperature is a singularly important factor in governing the life processes and the distribution of organisms. Vital life processes, collectively

Figure 1-6 Diagrammatic cross section of an ocean basin showing the various geographic features. (Not to scale.)

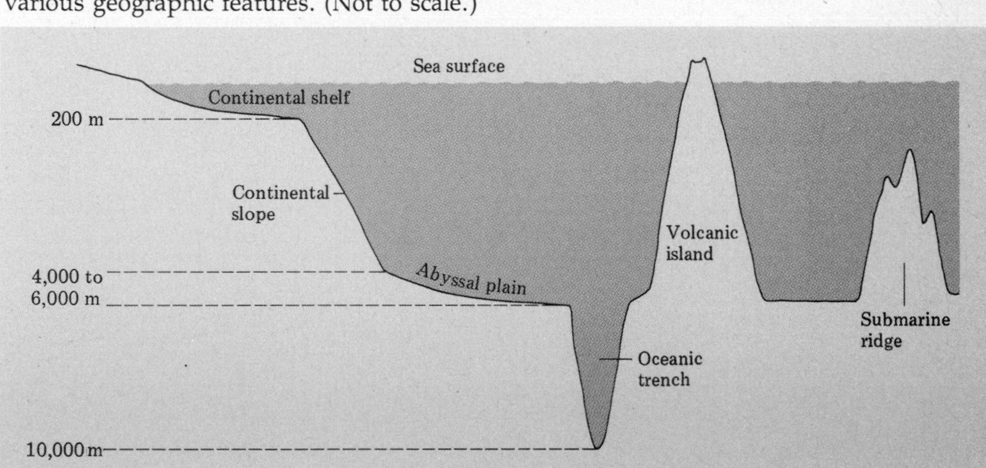

termed *metabolism,* function only within a relatively narrow range of temperatures, usually between 0° and 40°C. However, a few organisms are able to tolerate temperatures somewhat above and below these limits, such as the blue-green algae living in 85°C hot springs. Within the temperature range in which life processes operate, metabolism is temperature dependent. In general, for organisms which do not regulate their internal temperatures, metabolic processes are increased by a factor of two for each 10°C rise in temperature. With the exception of the marine birds and mammals, marine organisms are *poikilothermic* or *ectothermic*, meaning that their body temperatures vary with that of the surrounding water mass. *Homiothermic* or *endothermic* birds and mammals have the ability to regulate their own internal temperature, regardless of the temperature of the water mass. Most marine organisms are adapted to live and reproduce within more narrow temperature ranges than the total 0–40°C range. Since most are also poikilothermic and since the sea water temperatures vary latitudinally, the distribution of marine organisms follows closely the geographical differences in ocean temperatures.

On the basis of surface ocean temperatures and overall organism distribution, four major biogeographical zones may be established: *polar, tropical, warm temperate,* and *cold temperate* (Fig. 1-7). Transition zones between these areas also exist and boundaries may vary somewhat with season so that these zones are not absolute.

Figure 1-7 Major biogeographical regions of the world's oceans, based on temperature.

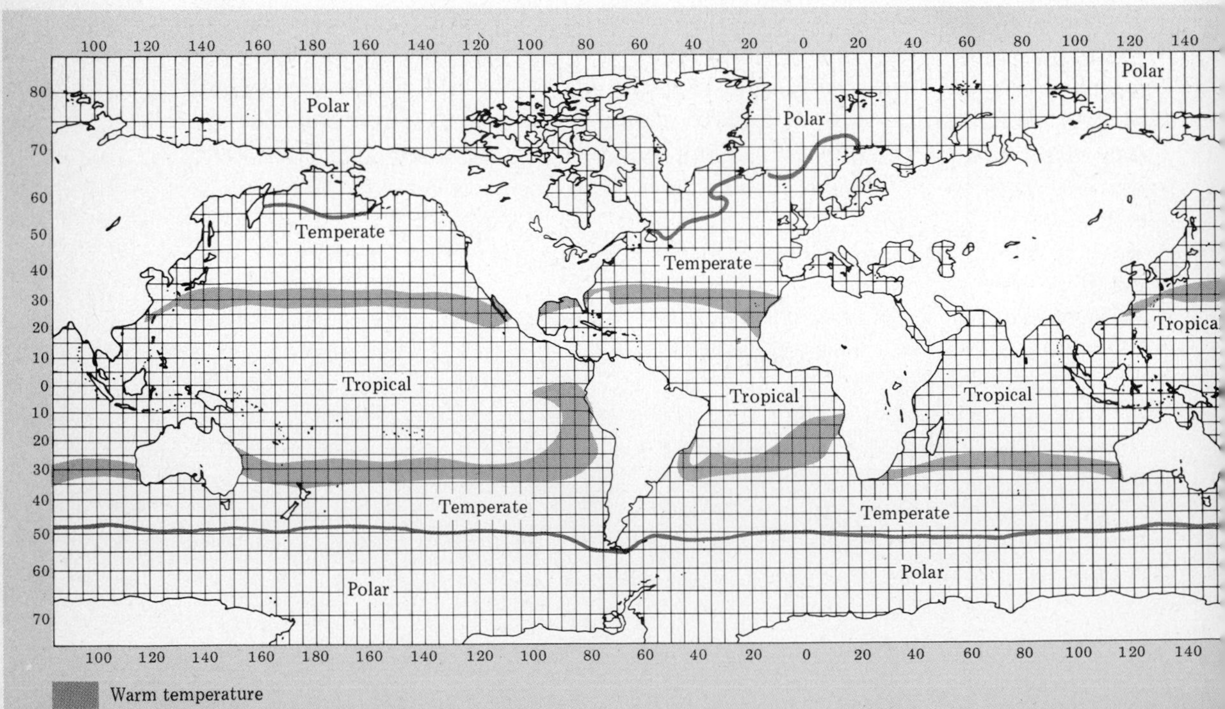

Temperature in the oceans also has a marked variation with depth. Surface waters in the tropical regions are very warm year round, 20–30°C, and temperate zone surface waters are warm in the summer.

Below the warm surface water, the temperature begins to fall, and over a narrow depth range of 50–300 m undergoes a very rapid decline. The depth zone of most rapid temperature decline is the *thermocline*. Below the thermocline, the temperature continues to fall with depth, but at a very much slower rate, so that the water mass below the thermocline is nearly isothermal all the way to the bottom. The thermocline is a persistent feature of tropical waters and occurs in temperate waters in the summer months. Thermoclines are absent in polar waters. The importance of this zone is discussed in the next chapter.

Temperature also has an effect on the density of sea water. Warm sea water is less dense than cold sea water of the same salinity. Density is also a function of salinity, increasing salinity causing increasing density. However, the range of temperatures found throughout the world oceans is greater than the range of salinities, and therefore, temperature is more important in affecting the density.

WATER MASSES AND CIRCULATION

As a result of different temperatures and salinities and their effect on density, the sea water of the world's oceans can be separated into different water masses. The *upper water mass* (surface water mass) of the oceans includes all well-mixed water above the thermocline. Below the thermocline is a *deep water mass* extending to the bottom.

The upper water mass of the oceans is in constant motion. The motion is produced primarily by the action of winds blowing across the surface of the water. These winds produce two kinds of motion, *waves* and *currents*. Waves range in size from ripples only a few centimeters in height to storm waves which may tower as high as 30 m. Other than height, waves are further characterized by *wavelength*, which is the horizontal distance between the tops or crests of successive waves. The *period* of a wave is the time required for two successive wave crests to pass a fixed point (Fig. 1-8). In addition to wind, waves may be generated by earthquakes, volcanic explosions, and underwater landslides, which create the destructive waves known as tsunamis, and by the attraction of the moon and sun, which produce the standing waves known as tides (see Chapter Six).

Wind wave height in the open ocean is dependent on the force of the wind, the distance or fetch over which the wind blows, and the length of time the wind blows.

All waves behave similarly. Once generated, the waves move outward and away from the center of origin. This horizontal progression does not result in significant horizontal transport of the individual water molecules. The water

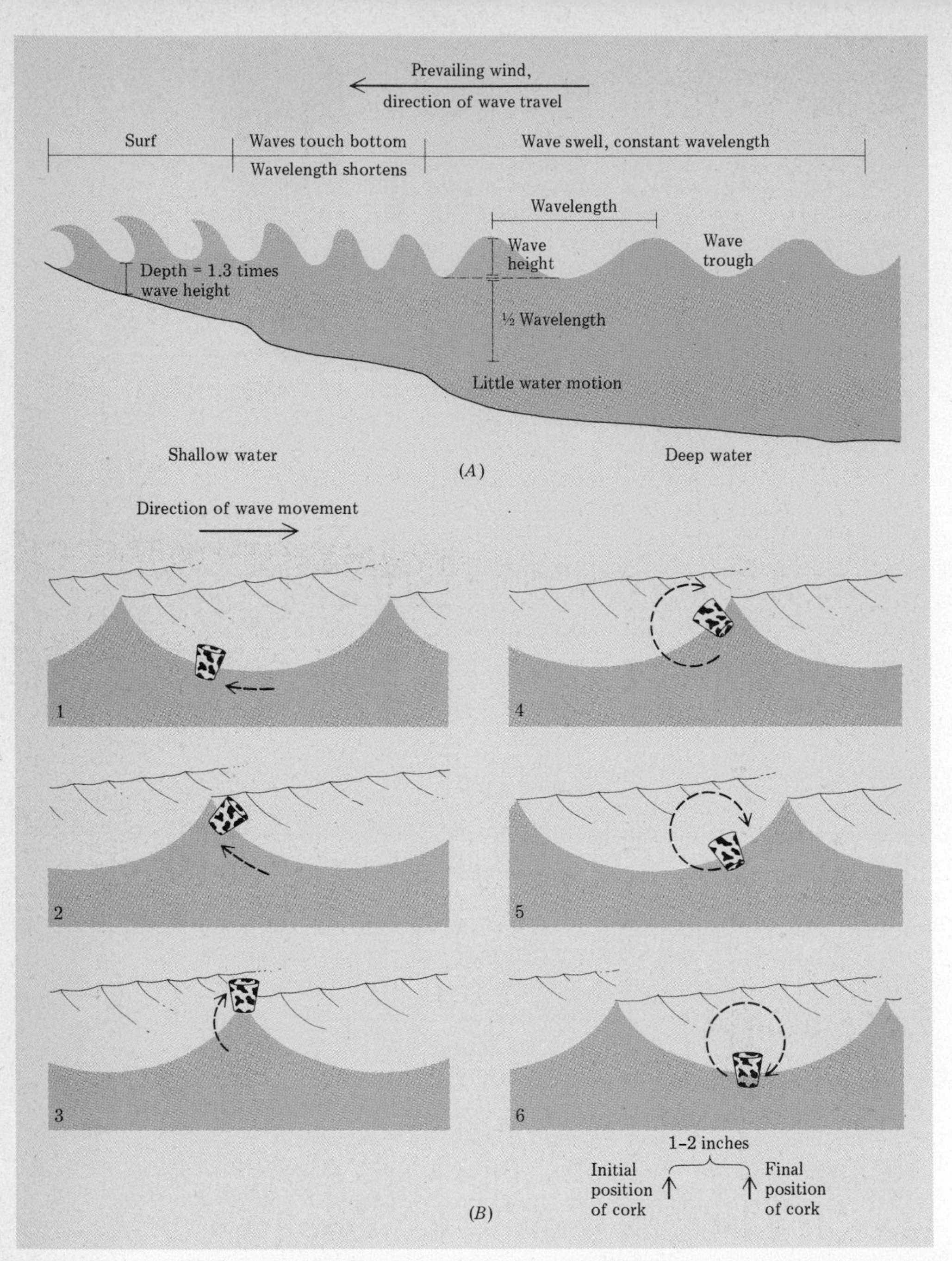

Figure 1-8 *(A)* Features of waves and change of wave form as it enters shallow water. *(B)* Cork float in the water demonstrates the passage of a wave, indicating that the water itself does not move. As the wave passes the cork transcribes an imaginary circle (dashed line).

molecules transcribe a circle, moving upward and to one side as a wave crest approaches and to the other side and down to near their original position as the wave crest passes by (Fig. 1-8). The wave form and the energy are transported horizontally. The passage of waves generates movement, not only in the surface water molecules, but also in water down to a depth approximately equal to one-half the wavelength. With each depth interval below the surface equal to one-ninth the wavelength, the orbits followed by the water particles diminish by one-half. Therefore, by the time a depth of half a wavelength is reached, the movement is almost imperceptible.

As waves enter shallow water and begin to encounter the frictional resistance of the bottom, they slow their forward motion and the wavelength decreases. As a result, they begin to increase in height and become steeper. At a point where the water depth is 1.3 times the height of the wave, it will "break," releasing the energy onto the shore. Waves are important biologically only in shallow water, as discussed in Chapter Six.

Currents are water movements which result in the horizontal transport of water masses. The major ocean current systems are produced by a few major wind belts which succeed each other latitudinally around the world and where the winds are steady and persistent in direction. The backbone of the system is the northeast trade wind blowing from northeast to southwest between the equator and 30°N latitude and the southeast trades in similar position south of the equator, moving air from southeast to northwest. Between 30° and 60°N and S latitudes, the westerlies blow from the southwest, to the northeast in the Northern Hemisphere and to the southeast in the Southern (Fig. 1-9).

These winds set the surface waters into motion, producing the slow, horizontally moving currents that are capable of transporting huge volumes of water across vast distances in the oceans. Such currents influence the distribution of marine organisms and also lead to the displacement of biogeographical zones

Figure 1-9 The main wind belts of the earth and their prevailing direction of motion (arrows).

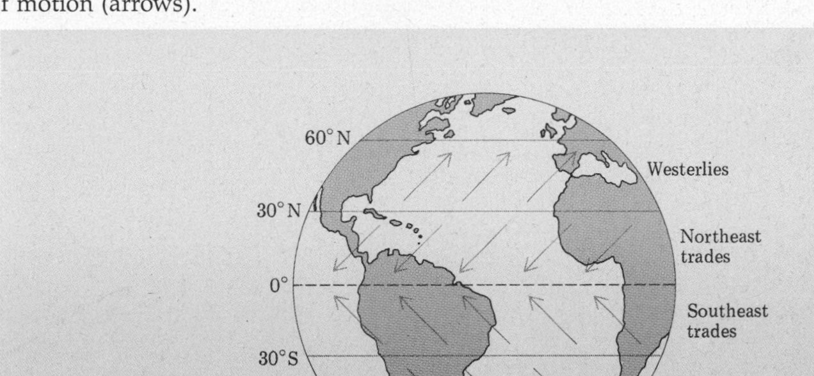

through the transport of warm water into colder regions and vice versa. The major currents of the world's oceans are shown in Fig. 1-10.

A comparison of Figs. 1-9 and 1-10 will show that the ocean currents do not flow parallel to wind direction. They are deflected into roughly circular gyres which move clockwise in the Northern Hemisphere and counterclockwise in the Southern Hemisphere. The deflections and gyres are the result of the *Coriolis force*. The Coriolis force, in turn, is the result of the rotation of the earth on its axis. The spinning of the planet imparts a deflection to moving water, displacing it to the right in the Northern Hemisphere and to the left in the Southern. Because the rotation of the earth is from west to east and because the deflection of the currents created by the trade winds imparts a water movement at the equator parallel to the equator moving from east to west, the net water movement is from east to west, piling up water on the western side of ocean basins. As water builds up on the western sides, it meets the continental or island chain land masses and is deflected north or south as continental boundary currents. These boundary currents, in turn, moving poleward, fall under the influence of the westerly winds. The westerlies impart more energy to the currents and drive them in an easterly direction, eventually crossing the ocean basins to return water to the eastern side of the basin. Continental land masses on the eastern side deflect the moving water toward the equator, which completes the pattern. These huge circular current patterns are called *gyres* and are found in all major ocean basins (Fig. 1-10).

These global currents are surface currents. How deeply do they affect the water column? This can be estimated by noting what happens to deeper layers of

Figure 1-10 The major ocean surface current systems.

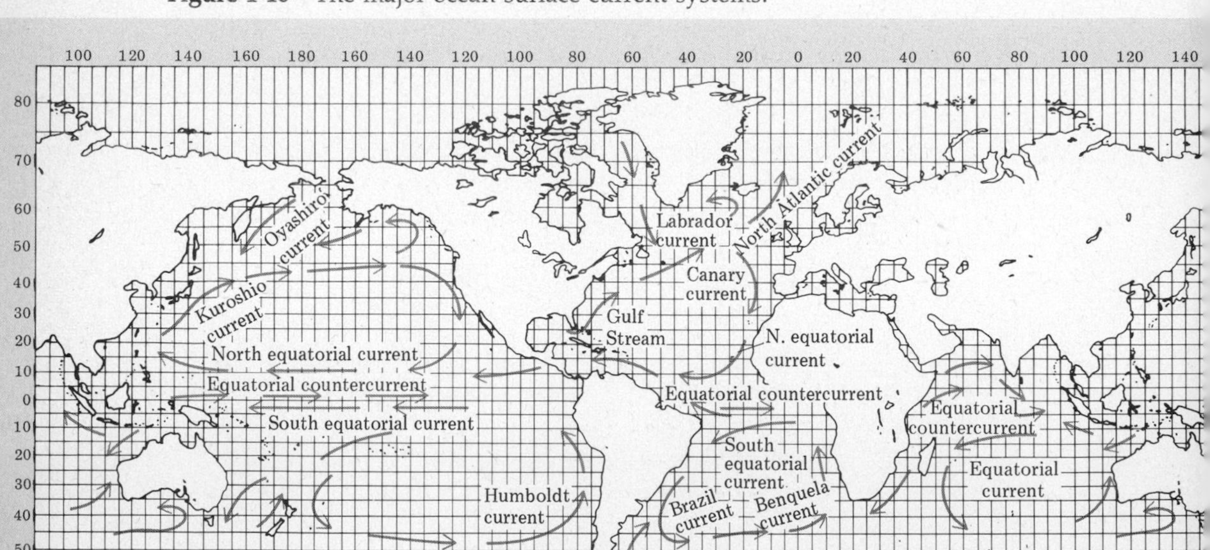

water when wind stress causes the surface water to move. As noted above, because of the Coriolis force, the direction of the water movement is deflected from that of the initiating wind. The energy of the wind is passed down through the water column and it sets each successively deeper water layer into motion. Each layer, however, receives a decreasing amount of energy, and therefore, its velocity is lower. At the same time, due to the Coriolis force, each layer set in motion is deflected with respect to the one immediately above it. The result is the *Ekman spiral* of current directions and velocities from the surface downward (Fig. 1-11). The depth at which wind stress fails to impart motion varies, but is roughly limited to the upper few hundred meters.

In certain areas and under certain conditions, the wind-induced lateral movements of the water may also bring about a vertical circulation or *upwelling* of water. Along the eastern margins of ocean basins, for example, the wind-driven surface currents along the continental margins flow toward the equator. At the same time, the Coriolis force tends to push these surface waters offshore. This water is then replaced by deeper water transported vertically to the surface. Similarly, along the equator, the two equatorial currents flowing west are

Figure 1-11 The Ekman spiral. The wind-driven surface current moves at an angle of 45° to the direction of the wind, to the right in the Northern Hemisphere, to the left in the Southern. Successively deeper water layers are deflected even further with respect to those immediately above them and move at slower speeds. Net water movement is at 90° to the wind. (Modified from Arthur Strahler, The earth sciences, 2nd ed., Harper & Row.)

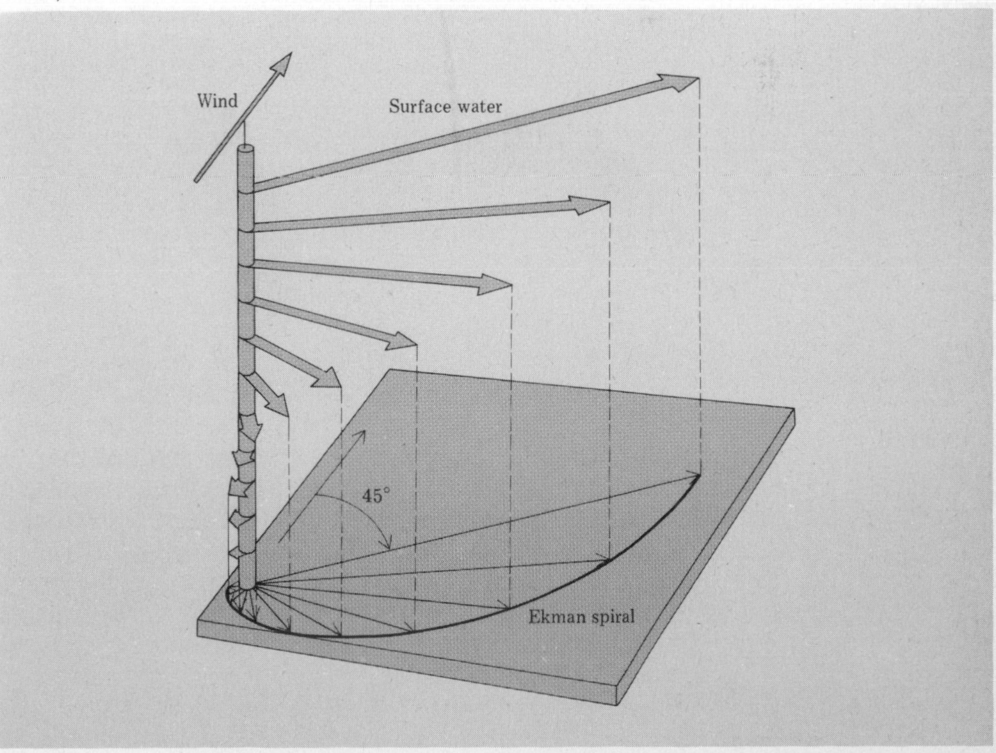

deflected away to the right north of the equator and left to the south. To replace this water, subsurface water upwells to the surface (Fig. 1-12).

Movement in the deep water mass is quite different from that in the surface mass. The deep water mass is isolated from the wind; therefore, its motion

Figure 1-12 *(A)* Coastal upwelling. Along the western margins of continents, the wind and Coriolis force act to move water offshore as indicated by the solid dark arrows. This water is replaced by water moving up from the depths (light arrows). *(B)* Equatorial upwelling. Along the equator, Coriolis force acts on the westward flowing currents, pulling water north in the Northern Hemisphere and south in the Southern Hemisphere (dark arows). This is replaced by cool water moving up from the depths (light arrows).

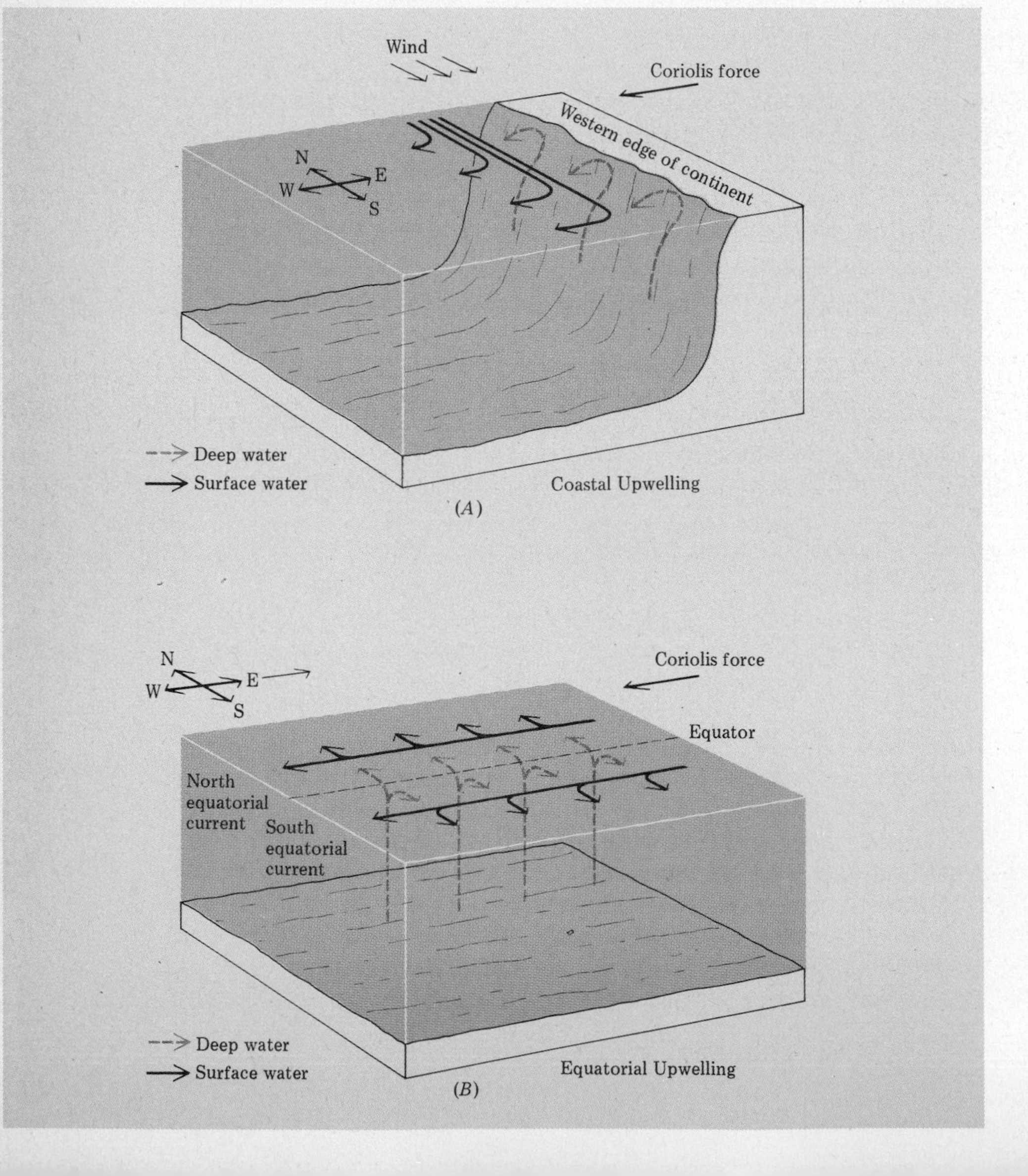

cannot be dependent upon it. Movement in the deep water mass does, however, result from changes occurring in water at the surface. Sea water increases in density with a decrease in temperature and an increase in salinity. When sea water increases in density, it sinks. Therefore, in order to move water into the deep basin of the oceans, it is necessary to increase its density at the surface. This is accomplished in two ways. Warm water from the tropics is high in salinity due to evaporation. In the northern Atlantic Ocean, such warm saline water is transported out of the tropics by the Gulf Stream. In the region of Iceland and Greenland, it meets the very cold waters of the Labrador Current moving south. Mixing then cools this highly saline water, increasing its density, and it sinks to form North Atlantic deep water. Similarly, the meeting of warm water moving south with cold water in Antarctic seas also causes the sinking of water masses. A final, very high density water mass is produced in the Weddell Sea in Antarctica, where very cold water becomes more saline when winter freezing occurs. This mass sinks to become the bottom water of most ocean basins.

Since these dense waters are all cold and produced at the surface, they have large amounts of oxygen, which are then transported to the depths. Without this oxygen, deep water would be anoxic.

These masses move very slowly north and south to form the deep water of all ocean basins. Many hundreds of years are required for these masses to move through the ocean basins. North Atlantic deep water, for example, may have been away from surface contact for hundreds of years between when it sank near Iceland and when it again surfaced in the Antarctic region. It is, however, a major contributor to the productivity of Antarctic seas, because when it surfaces, it brings with it large amounts of nutrients accumulated during the many years beneath the photosynthetic zone. A general outline of these deep-water masses and movement is given in Fig. 1-13.

SOME ECOLOGICAL PRINCIPLES

Because this is a text that emphasizes habitats and the ecology of marine organisms and because certain users of this book may lack a background in basic ecology, it seems necessary at this point to cover briefly some basic ecological concepts and terms.

TERMS AND DEFINITIONS

Ecology is the science that treats the spectrum of interrelationships existing between organisms and their environments and among groups of organisms. It is important to realize initially that living things do not exist as isolated individuals or groups of individuals. All organisms interact with others of their own species, with other species, and with the physical and chemical environments that surround them. In this interactive process, the organisms have an

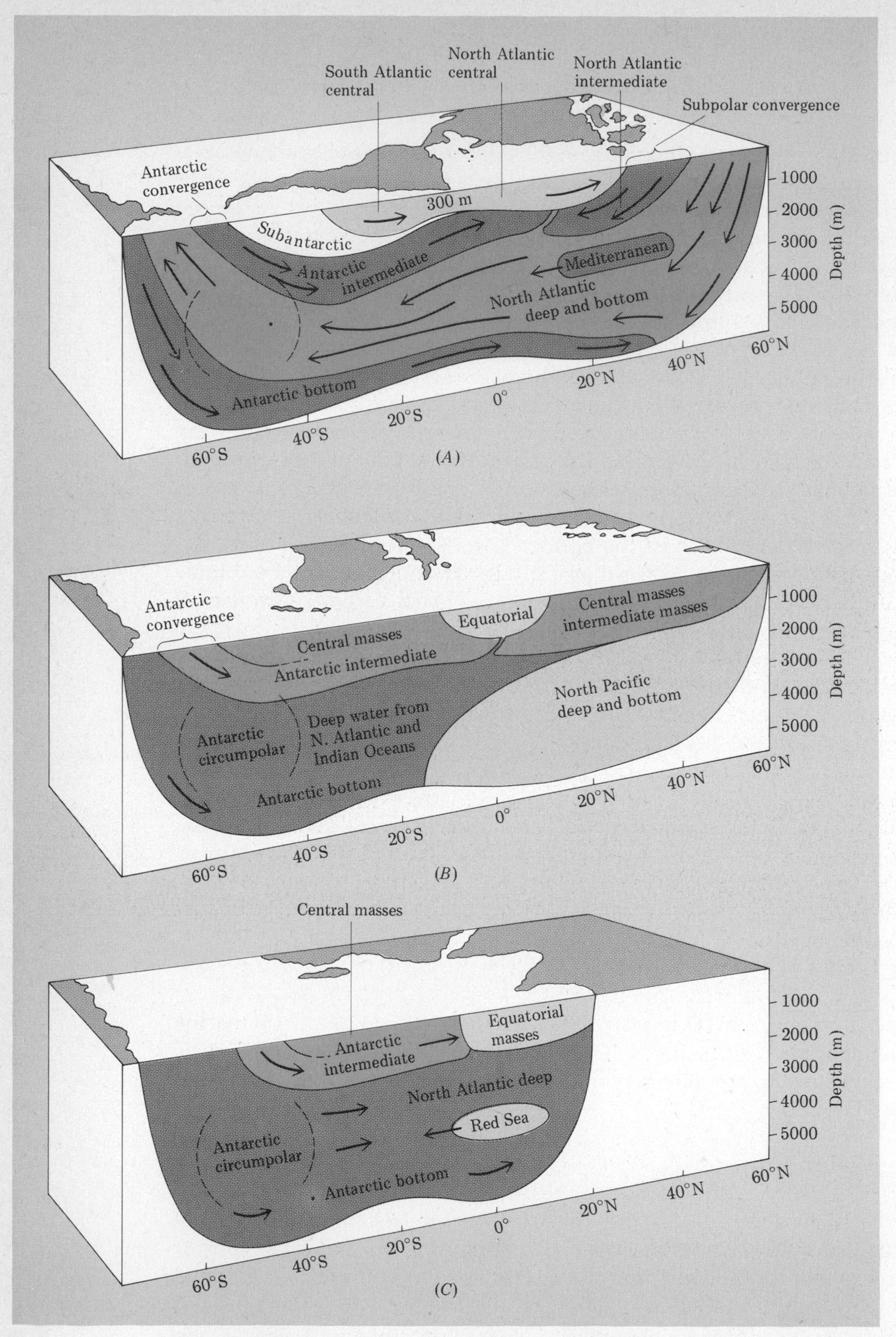

Figure 1-13 Subsurface water masses and circulation patterns in the three major oceans. *(A)* Atlantic Ocean. *(B)* Pacific Ocean. *(C)* Indian Ocean. (From D. E. Ingmanson and W. J. Wallace, Oceanology: An introduction, © 1973 by Wadsworth Publishing Company, Inc. Reprinted by permission of Wadsworth Publishing Company, Belmont, Calif. 94002.)

effect on each other and on the surrounding environment. Similarly, the various factors of the environment affect the activities of the organisms. The various organisms and the environmental parameters may be further organized into various levels, each of which is somewhat broader than the preceding. A *species* is a natural group of actually or potentially interbreeding individuals reproductively isolated from other such groups. All the individuals of a given species in an area constitute a *population*. Several species populations that tend to occur together in various geographical areas constitute an ecological *community*. A community or a series of communities and the surrounding physical and chemical environment together constitute an *ecosystem*. Ecosystems are the most complex entities in the above series and have many complex and interactive components. They are so large and complex that ecologists tend to study them by concentrating on their component parts, such as various communities or populations. Ecosystems, however, may be considered on a large or small scale, depending on the number of communities included and the dimensions of the surrounding nonliving environment. On the largest scale, it is possible to consider the earth as a single ecosystem comprising all the various terrestrial, fresh water and marine communities. At the other end of the scale, it is possible to consider a tidepool or fresh water pond as an ecosystem. Both have the biological and nonbiological components the definition requires.

The oceans of the world may also be considered a single ecosystem in which a series of communities is influenced by and, in turn, influences the physical and chemical factors of the surrounding sea water. This large ecosystem may, in turn, be subdivided into smaller sections or areas in which the physical and chemical parameters have differential effects on the populations of organisms, thus imposing changes in the composition and adaptations of organisms within the area subject to those effects. For example, the physical factors acting on a rocky seashore, waves and movements of the bottom, are not those that affect small organisms floating in the open ocean. Therefore, both areas contain different organisms and different adaptations. It is the aim of this book to give an understanding of the functioning of the total marine ecosystem through a consideration of the functioning of these major subsections, which form the bases of the subsequent chapters.

ECOSYSTEM COMPONENTS

An ecosystem is a functional unit of variable size composed of living and nonliving parts, which interact. The component parts and the whole system function through a sequence of operations involving energy and the transfer of energy. With few exceptions, the original energy source is the sun. Energy from the sun is captured by the *autotrophic* component, the green plants. Energy captured is stored in the chemical bonds of organic materials in the plants, which are the "food" which drives the *heterotrophic* component of the system. Heterotrophic organisms include all other life forms, which obtain their energy through consumption of the autotrophic plants or through consumption of organisms that have ingested plants. Such an arrangement of autotrophs and succeeding levels of heterotrophs is called a *trophic structure,* in which each successive consumer level is called a *trophic level.* Trophic structure is a characteristic feature of all ecosystems. The first trophic level is the autotrophic or producer level, where the energy is initially captured and stored in organic compounds. As the energy is passed from level to level in such a system, the majority of it is lost through heat and metabolic use by the organisms. The amount lost is variable, but substantial, ranging between 80–95 percent. The system thus becomes self-limiting in that, at a certain point, not enough energy remains to be passed on to sustain another level. This is visualized as an energy pyramid or trophic pyramid (Fig. 1-14). In such a structure, the trophic level that consumes the plants (autotrophs, first level) includes animals called *herbivores.* Herbivores, in turn, are consumed by *carnivores,* which, in turn, are eaten by still other, often larger, carnivores. All levels above the second therefore consist of carnivores or omnivores. Within each trophic level or population, the amount of living material at any instant in time is the *standing crop.*

The final component of the trophic structure of an ecosystem is the *decomposers.* These are organisms, chiefly bacteria, that break down the complex organic

Figure 1-14 Trophic pyramids. (*A*) Biomass pyramid. (*B*) Energy pyramid. The thickness of the bars indicates the relative amounts. (From R. L. Smith, Ecology and field biology, 2nd ed., Harper & Row.)

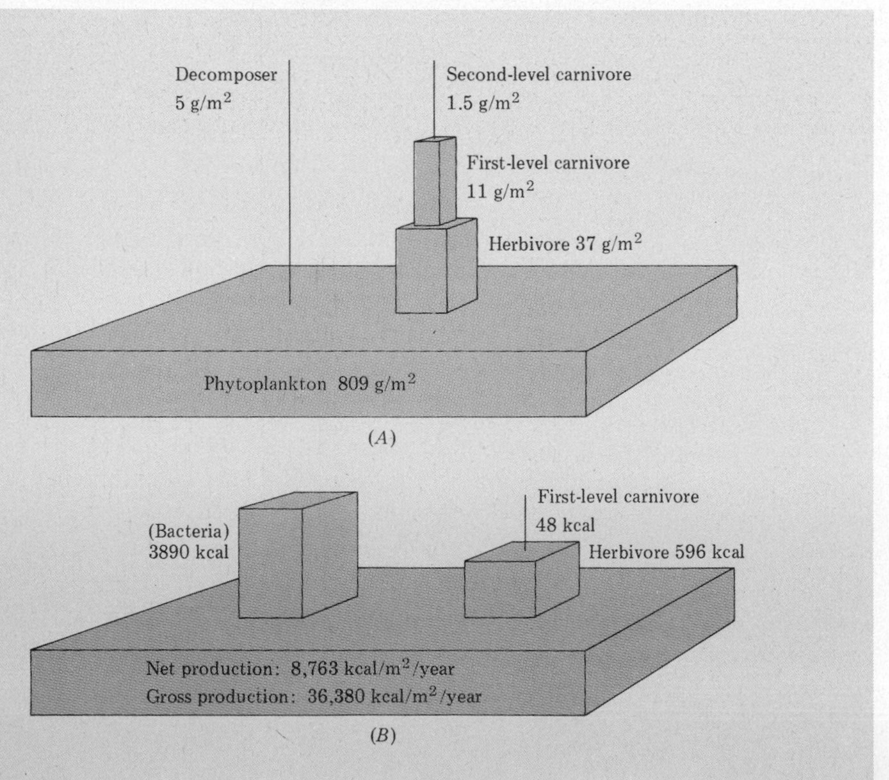

molecules of dead organisms, releasing simple molecules usable again by the autotrophs. Decomposers act on every trophic level.

The necessary abiotic components of the trophic structure of an ecosystem are an energy source and a nutrient and water source. Plants cannot fix energy and produce complex organic molecules without sunlight for energy or without a series of inorganic nutrients, of which nitrates and phosphates are the most important.

Since ecosystems are composed of from one to many communities, each, in turn, composed of many populations of producers, consumers, and decomposers, the transfer of energy through an ecosystem may follow several pathways. Each pathway that transfers energy from a given source plant or plants through a given series of consumers is called a *food chain.* The combination of all food chains in a given community or ecosystem is called a *food web.* The food web is thus a summary of all the pathways by which energy moves from one level to another through a community or ecosystem.

BIOGEOCHEMICAL CYCLES

Among many of the chemical elements and compounds in ecosystems there is a cycling back and forth between organisms and the physical environment. Such repeated transfers are called *biogeochemical cycles.* A few of these cycles involve chemical compounds vital to the continued maintenance of life in the ecosystem and are therefore extremely important. Perhaps the most significant are the cycles involving carbon, nitrogen, and phosphorus (Fig. 1-15).

In all these cycles, there is a major reservoir or pool of the element from which the element is continually moving in and out as it passes through organisms. Each cycle also contains a sink into which a certain amount of the chemical passes and from which it is not recycled in the normal course of events. Over long periods of time, the loss to the sink may become limiting, unless the sink can be tapped again. This latter situation usually develops through geological action that acts to move the sink into an area where organisms, erosion, or other factors release the elements. Biogeochemical cycles tend, finally, to have self-regulating feedback mechanisms that keep the cycles in equilibrium.

In the carbon cycle, the reservoir is in the form of CO_2, which exists in water in the carbonic acid–bicarbonate–carbonate system (Fig. 1-15). It is fixed into organic compounds by plants, transferred to animals through herbivory and predation, and returned to the reservoir via respiration and bacterial action. A nonrenewable loss occurs when heavy carbonate materials such as shells are deposited in deep oceans as calcareous sediments.

In the phosphorus cycle, the major reservoir is in phosphate rock. Here, erosion brings the chemical into the water where it cycles through animals and plants and is returned to the general circulation through decay and excretion. The loss is to deep sediments, which are unavailable to organisms.

Air is the major reservoir in the nitrogen cycle. Nitrogen gas is not used by

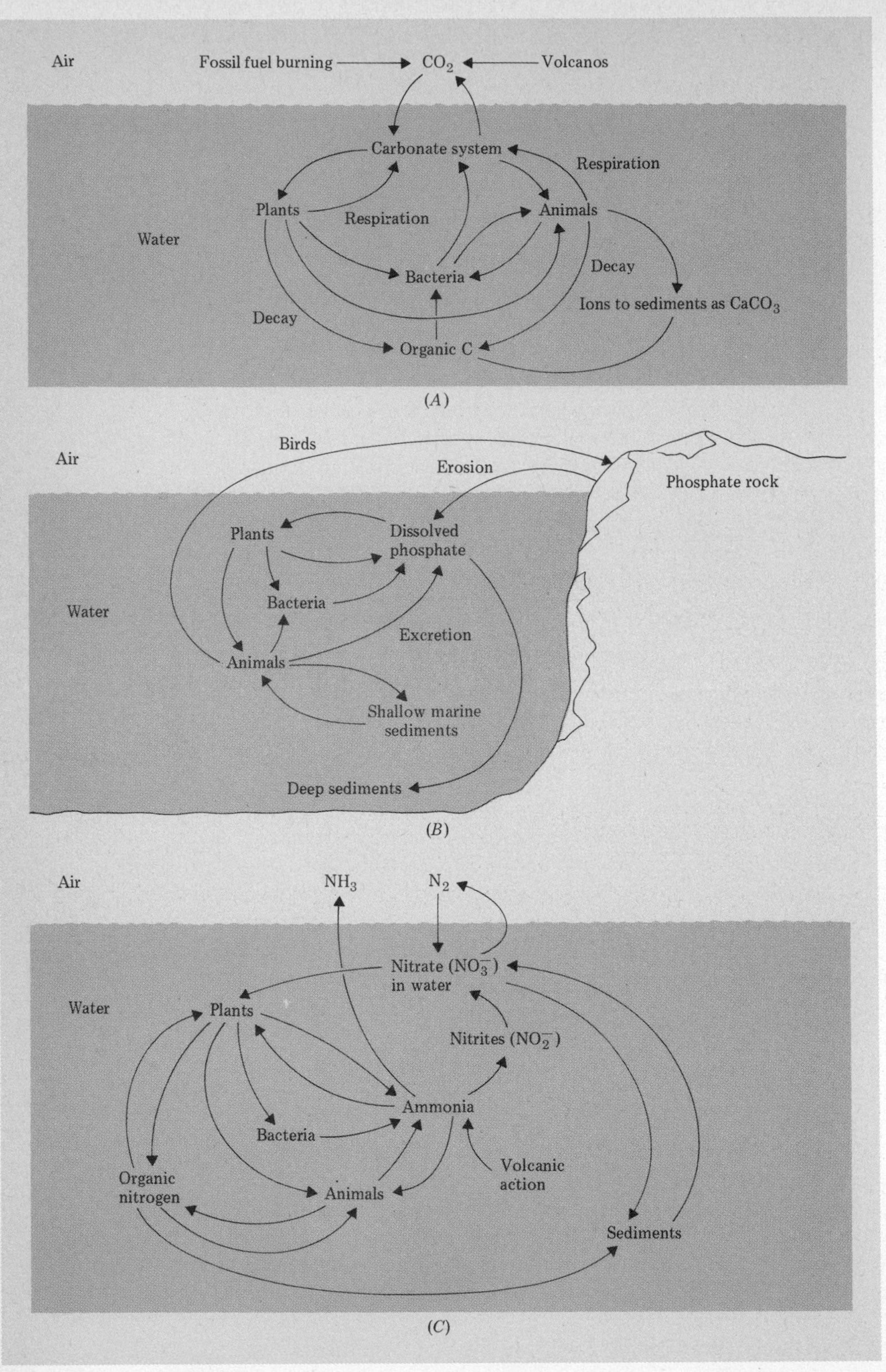

Figure 1-15 Biogeochemical cycles in the sea. *(A)* The carbon cycle. *(B)* The phosphorus cycle. *(C)* The nitrogen cycle.

most organisms and must be converted into a nitrogen compound to enter the cycle. This is normally done by bacteria and/or algae. Useful nitrogen compounds may also be produced through volcanic activity. Once in compounds, the nitrogen cycles through organisms with some loss to deep sediments (Fig. 1-15).

BIOTIC STRUCTURE OF ECOSYSTEMS

Communities and ecosystems have the same trophic levels throughout the world. However, the species constituting each differ among geographical areas and, in some areas, each level may have more or fewer species than another. This implies that ecologically equivalent species replace each other geographically and that the species structure of communities is variable, even if the trophic structure is not. For example, sea grass beds in the tropics may have several species of grasses, whereas those in the temperate zone but one. They are different species, but ecologically fulfill the same role, as autotrophs. Species that are ecological equivalents are those that perform the same function or role in a given community. They are therefore said to occupy the same *ecological niche.* Ecological niches may be broad or narrow. Narrow niches mean that the role or function in the community or ecosystem has been more finely subdivided and the species is therefore more specialized. Broad niches, on the other hand, mean the function is more generalized. Species may therefore be either *specialists* or *generalists* with reference to their niches. In our preceding example, the autotroph niche is broad in the temperate zone and therefore occupied by a single generalist grass; in the tropics, several more specialized grasses subdivide the niche.

In contrast to niche, an ecological *habitat* refers to the place where an organism is found. Therefore, the small copepod crustacean *Calanus finmarchicus* (see Fig. 2-5) has as a habitat the surface waters of the North Atlantic Ocean, but its niche is that of a herbivore feeding on the small plants in the area.

Most communities have a characteristic species structure that consists of a few species that are abundant and a larger number of species that are rare. The abundant species are usually called *dominants* and are often used to characterize a community. Thus, we can speak of a "mussel bed" on the rocky shore reflecting the dominance by the mussels *Mytilus edulis* or *Mytilus californianus.* This pattern of few common and many rare species is true at each trophic level. It is also true whether the community has many or few species in total. Species structure in ecological communities may be measured in several ways. *Species richness* is a simple listing of the total number of species in a community or trophic level. *Species diversity* is a measure that combines into a single figure both the number of species (richness) and the distribution of the total number of individuals among the species (evenness). It is expressed through various mathematical diversity indices, which need not concern us here. High species diversity is generally taken to be indicative of benign, stable environments and low diversity of stressful, fluctuating environments.

Each species in a community has certain tolerances with respect to each and every environmental factor. If the limit of tolerance is exceeded in some area by a given factor, say temperature, the species will be absent. Similarly, each species requires a certain minimum amount of various materials. If the concentrations of these necessities, such as nitrate, fall below the minimum, the species disappears. More importantly, if any *single factor* exceeds the tolerance level or any single necessary substance is reduced below the minimum, the species will be eliminated. This is true even if all the other factors and substances are favorable. This is known as *the law of the minimum.*

Since these limiting factors vary for each species and since communities are defined on the basis of recurrent groups of species, it follows that the boundaries between communities, representing changes in various environmental factors, are also not sharp. The result is areas of transition between adjacent communities where species gradually drop out and others come in. Such boundary zones are called *ecotones.*

Communities are not static units. They change in structure and composition with season and over longer periods of time. Terrestrial communities tend to change in an orderly fashion over periods of many years until they reach a stage which perpetuates itself indefinitely as long as the climate does not change or there is no disturbance. In such a sequence, each community modifies the environment, in turn making it suitable for the next community. This orderly process of community change controlled through modification of the physical environment is called *ecological succession* and the terminal, persistent community the *climax.*

ECOLOGICAL CONTROL AND REGULATION

Populations, communities, and ecosystems are all regulated by various factors. The major controlling factors for ecosystems and communities are energy, the physical factors collectively termed climate or environment, and the interaction among various species that compose the systems. Ultimately, nearly all systems on earth are limited by the amount of energy available from the sun. However, tolerance limits of various species to abiotic factors such as temperature, light, nutrients, and salinity also limit the extent of populations and communities in the oceans. The final aspects of regulation and control considered here are those dealing with the interaction among populations that act to keep populations within limits.

Biologists have long known that all species possess the reproductive potential to produce much larger populations than are observed under natural conditions, and that if population explosions occur, they are quickly reduced. What are the biological factors that exert control? They are grouped under predation, competition, and disease and parasitism.

Competition is an ecological term referring to the interaction among organisms for a necessary resource that exists in short supply. Competition may be *intra-specific* (among individuals of the same species) or *inter-specific* (among individuals of different species). Competition may be for any number of things, but usually is limited to items such as light, food, nutrients, water, and space. In a competitive interaction, either the competitors manage to share the limited resource or one excludes the other. In the first case, both individuals are hampered, which inhibits their growth, development, and reproduction, thus limiting the numbers. In the second case, one individual is eliminated, again controlling population. Interspecific competition is usually between two closely related species and has led to the *competitive exclusion principle,* which says no two species with exactly similar requirements can coexist in the same place at the same time; that is, complete competitors cannot coexist. As numbers increase, competition usually increases, because the limited resource becomes scarcer. This increased competition increases the stress on animals and plants and absorbs energy otherwise used for reproduction, thus limiting populations. Direct competitive interactions among organisms are rarely observed and are usually inferred from changes observed in population numbers and distributions in nature.

Predation can be defined as the consumption of one species by another. The animal that is the consumer is the *predator;* the victim the *prey.* A special case concerns those animals that consume plants. They are called *herbivores.* A *grazer* is an animal which feeds on plants or sessile animals. Predators and herbivores vary considerably in their ability to regulate the numbers of organisms they consume. In some cases, the predator may be the most important factor in regulating numbers of a prey species. In other cases, a predator may have little effect on a prey population. In the first case, the removal of the predator will have a marked effect on the prey population, causing it to increase dramatically. In the latter case, predator removal has little effect on the prey population. We shall see examples of both these situations in the following chapters. A special case of predation concerns those predators that have a profound effect, not only on their prey population, but also on the entire community of which they are a part. In these cases, the removal or depletion of the numbers of the predator has the effect of causing great changes in the presence and abundance of many species in the community, most of which are not the prey of the predator. As a result, the entire community structure may be changed. Such predators are called *key industry* or *keystone* species. An example of such a species is the starfish *Pisaster ochraceus* (see pp. 220–221).

Parasitism and disease are the final biological controls on populations. As with predation, they may exert strong control or they may have little effect. *Parasites* are organisms living in or upon other organisms from which they derive nourishment and shelter. Many, perhaps all, marine organisms have parasites, but we know considerably less about the role of parasites and disease in regulating populations of marine organisms than we do for terrestrial organisms. Perhaps the best documented change induced in a marine community by a disease is the loss of the eelgrass beds in the Atlantic Ocean in the 1930s (see pp. 186–188).

COMPARISON OF TERRESTRIAL AND MARINE ECOSYSTEMS

It is quite true that ecological principles apply equally whether one is dealing with terrestrial, fresh water, or marine ecosystems. However, the special conditions prevailing in salt water that are the result of the physical and chemical properties of water, have, directly or indirectly, channeled evolution and adaptation of marine organisms. The result has been some striking differences in the organization of marine communities when compared with terrestrial communities. These will be considered next.

PHYSICAL AND CHEMICAL DIFFERENCES

Sea water has several physical features that have a profound effect on the organization of marine communities. They are its greater density with respect to air and its ability to absorb light. The greater density of sea water means that relatively large organisms and particles can float around in it. This is not possible in air. One significant result of this is that the marine ecosystem has evolved a whole community of small organisms that are perpetually afloat, the *plankton*. No such comparable community exists terrestrially floating in the air. As a result, all communities in the sea are bathed in a medium that itself contains a community! The presence of this suspended community has, in turn, led to the evolution of animals that are adapted to filter organisms and particles out of sea water. These *filter feeders* are unique to aquatic systems. There is nothing really comparable on land. The closest analog would be spiders, which capture flying insects in their webs. Furthermore, since the waters of the oceans are in perpetual motion, many of these filter feeders are *sessile* (fixed in place), extending their filtering nets up into the moving water mass. Sessile animals are not possible in the terrestrial environment because the bathing medium flowing by contains no suspended food organisms as particles. Finally, marine organisms, especially the sessile species, have evolved motile larval forms. Such motile larvae, when put into the plankton, permit dispersal of the species.

As a result of the suspension of a whole community in the bathing medium, it is more difficult to have isolated communities and isolated species distributions in the sea.

In contrast to air, water strongly absorbs light. As a result, light entering water can only penetrate to a certain depth before it is completely absorbed. Although this depth varies, it is sufficiently shallow (100 m) that the great majority of the volume of water in the oceans is without light. This means that plant life and primary productivity are limited to an extremely narrow band near the surface. Whereas terrestrial communities, with the exception of caves, all possess sufficient light for plant growth, many marine communities exist without the benefit of an autotrophic component.

Not only does water absorb light, it does so differentially, depending upon the particular wavelength. The result is that certain wavelengths penetrate deeper than others. This is not true on land, where the same spectral composition of light impinges at all levels in any community. The importance of this to marine communities is discussed on pp. 57–60.

Another physical factor indirectly responsible for differences between marine and terrestrial organisms is *gravity*. Because marine plants and animals are buoyed up by water, they do not need to have a significant amount of their biomass invested in structural material such as skeletons or cellulose in order to hold themselves erect against the force of gravity. Similarly, where movement is concerned, terrestrial animals must raise their mass against the force of gravity for each step. Such movement requires significantly more energy than swimming movements do in aquatic organisms and hence, more energy storage.

This absence of large amounts of structural material and energy storage in the bodies of marine organisms is reflected in the differences in the dominant biochemical compounds found in terrestrial and marine organisms. Among terrestrial organisms, the dominant compounds are carbohydrates; among marine organisms, the predominant material is protein. Living organisms that are composed primarily of carbohydrate are long lived, slow growing, and rich in stored energy. Protein-dominated organisms, on the other hand, are rapid growing and without significant energy storage.

A final physical difference between terrestrial and marine systems concerns oxygen. In air, oxygen constitutes a nearly constant 21 percent of the volume throughout the world. However, water holds less oxygen, and furthermore, its concentration can vary with temperature and salinity (pp. 7–8). Thus, marine communities are subject to variations in oxygen not found among their terrestrial counterparts.

STRUCTURAL AND FUNCTIONAL DIFFERENCES

One striking difference between terrestrial and marine communities is the insignificance in the latter of large macroscopic plants. Terrestrial communities are universally dominated by large flowering plants that are persistent and long lived. With the exception of certain large kelp plants, the communities of the sea have no large plants and the dominant autotrophs are microscopic plants of various groups of algae. In turn, this means that the dominant herbivores of the sea are also small, often microscopic, animals, in contrast to the large-bodied herbivores common to terrestrial communities. There are, for example, few herbivores in the sea to compare in size with the antelopes of the African plains or the bison of the North American grasslands. The dominant marine herbivores are microscopic crustaceans called *copepods*.

Another characteristic of terrestrial plants is that significant portions are composed of rigid structural materials such as wood and fiber, which themselves are relatively undigestible by most herbivores. As a result, herbivore grazing in terrestrial communities rarely removes significant amounts of the plant community matrix. By contrast, the small herbivores of the seas usually consume the entire plant, and a major difference between terrestrial and marine ecosystems is the ability of the herbivores to completely remove the plants (see pp. 70–73).

Terrestrial communities are characterized by a matrix of long-lived plants and a fauna that is generally shorter lived. In the seas, at least among those communities associated with the sea bottom, the plants, if present, are short lived, and the matrix of the community is of relatively long-lived animals. Thus, one speaks of redwood forests, oak forests, tall grass prairies, on land, but of coral reefs, clam beds, oyster reefs, and mussel beds in the seas.

As a result of the microsopic nature of most marine plants and herbivores, the majority of large animals in the sea are carnivores. Large marine animals are therefore on higher trophic levels than large land animals. From the standpoint of human beings, harvestable food in the form of animal protein is available primarily from large herbivores on land (second trophic level), such as cattle and sheep, whereas harvestable food from the sea is usually in the form of carnivores from the third or fourth trophic levels, such as salmon, tuna, and halibut. A corollary of this is that most marine food chains have about five links (steps) to reach the top carnivore, whereas terrestrial food chains tend to be shorter, averaging three links to reach the top carnivore (Fig. 1-16).

There is also a question about the efficiency of energy production and transfer between terrestrial and marine ecosystems. Generally, the production of organic material is higher in terrestrial ecosystems than in marine, but the efficiency of transfer from first to second trophic levels is higher in marine ecosystems. Thus, although starting out with less energy fixed, marine systems appear to lose less with the first transfer. Attempts to further assess the efficiency of energy transfer in marine food chains are complicated because of the various feedback mechanisms which tend to "blur" the trophic level to which to assign the animal. The primary group responsible for this blurring of the trophic level is the filter feeders. These animals tend to feed upon suspended particles, and the distinction as to what is retained on their filters is made on the basis of size, not whether the particle is animal or plant or living or dead. This means they often take in both plant and animal material and feed on several trophic levels. A second reason for blurred trophic levels is the tendency of marine species to switch positions in the trophic hierarchy. Thus, juvenile fishes, for example, may feed on herbivorous copepods, but when adult change to feed upon other fishes higher in the trophic spectrum. An additional problem is the ability of

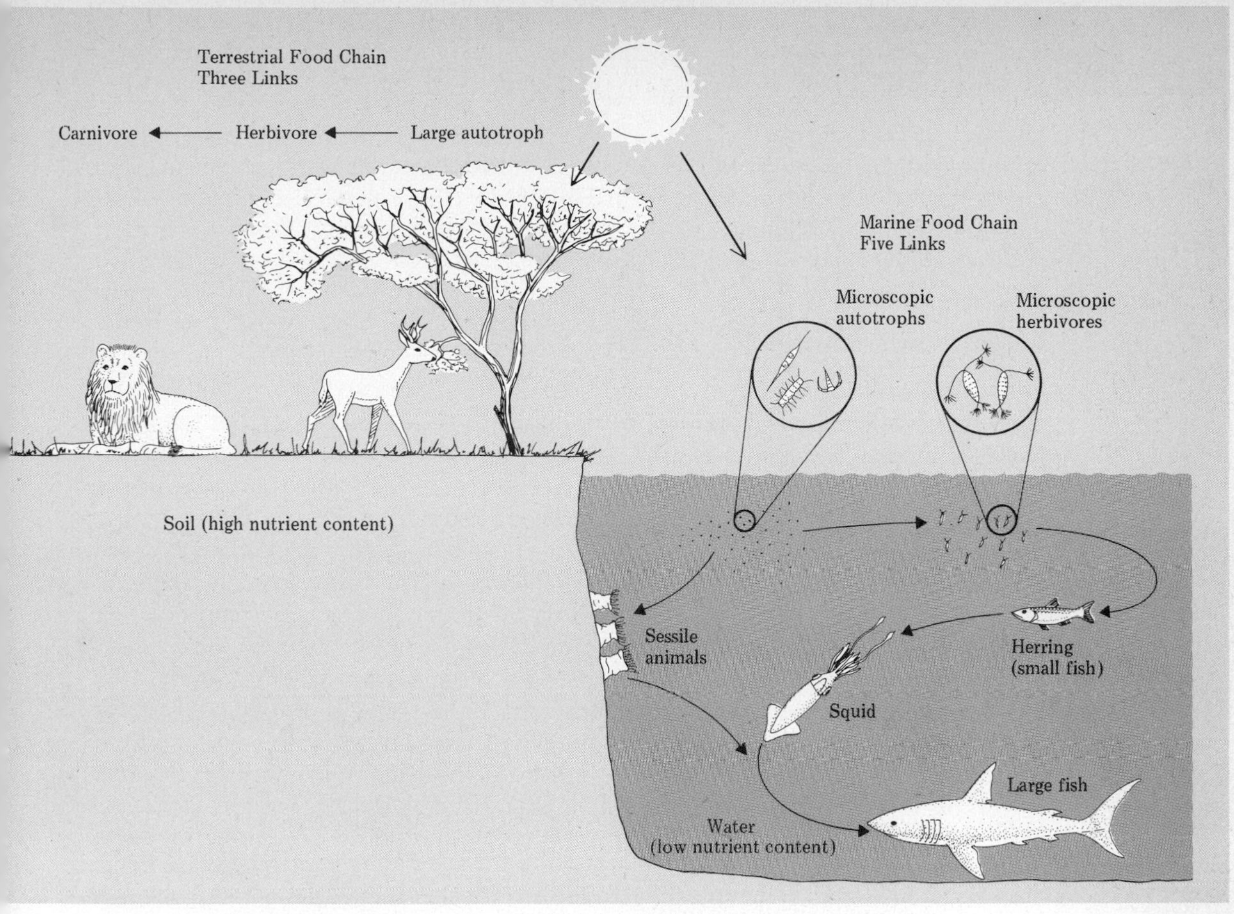

Figure 1-16 Diagrammatic representation of the differences between marine and terrestrial food chains.

various marine organisms to switch trophic levels as adults. Thus, for example, salmon may feed on squid at one time, but switch to lantern fish at another, thus shortening the food chain since lantern fishes are themselves the food of squid. Similarly, euphausiid crustaceans ("krill") may be herbivores in shallow water and carnivores in the open ocean.

The above-mentioned differences between terrestrial and marine ecosystems should be kept in mind as we explore the various habitats and communities of the sea in the remainder of this book.

DIVISION OF THE MARINE ENVIRONMENT

The marine ecosystem is the largest aquatic system on the planet. Its size and complexity make it difficult to deal with as a whole. As a result, it is convenient to divide it into more manageable subdivisions, each of which can then be discussed in terms of the ecological principles that govern the adaptations of the organisms and the organization of the communities. No universally acceptable scheme of subdivision of the marine environment has yet been proposed. The one followed here is modified from Hedgpeth (1957) and has enjoyed widespread use among biologists for 20 years. It therefore probably comes closer than any to a mutually acceptable division.

MAJOR SUBDIVISIONS OF THE WORLD OCEAN

Beginning with the waters of the open ocean, there are subdivisions that can be made in both the vertical and horizontal directions. The entire area of the open water is termed the *pelagic* realm; pelagic organisms are those that live in the open sea away from the bottom. This is in contrast to the *benthic* realm, which is a general term referring to organisms and zones of the sea bottom. Horizontally, the pelagic realm can be divided into two zones. The *neritic* zone encompasses the water mass that overlies the continental shelves. The *oceanic* zone includes all other open waters (Fig. 1-17). Progressing vertically, the pelagic realm can be further subdivided. Two schemes are possible. The first is based upon light penetration. The *photic* zone is that part of the pelagic realm which is lighted. Its lower boundary is the limit of light penetration and varies in depth with clarity of the water. Generally, the lower boundary is between 100–150 m. A synonym for this zone is the *epipelagic* zone. Because it is the zone of primary production in the ocean, it is of major importance and the subject of Chapters Two and Three. The permanently dark water mass below the photic zone is termed the *aphotic* zone.

The pelagic part of the aphotic zone can itself be subdivided into zones that succeed each other vertically. The *mesopelagic* is the uppermost of the aphotic areas. Its lower boundary in the tropics is taken to be the 10°C isotherm, which may be at 700 to 1000 m, depending on the area. Next is the *bathypelagic*, lying between 10°C and 4°C, or in depth between 700 and 1000 m and 2000–4000 m. Overlying the plains of the major ocean basins is the *abyssal pelagic*, which has its lower boundary at about 6000 m. The open water of the deep oceanic trenches between 6,000–10,000 m is called the *hadalpelagic* (Fig. 1-17).

Corresponding to the last three pelagic zones are three bottom or *benthic* zones. The *bathyal* zone is that area of bottom encompassing the continental slope and down to about 4000 m. The *abyssal* zone includes the broad abyssal plains of the ocean basins between 4000–6000 m. The *hadal* is the benthic zone of the trenches between 6,000–10,000 m. This entire aphotic area, although encompassing by far the largest volume and area of the world's oceans, is little known and forms the subject of Chapter Four.

The benthic zone underlying the neritic pelagic zone on the continental shelf is termed the *sublittoral* or *shelf* zone. It is illuminated and is generally populated with an abundance of organisms constituting several different communities, including sea grass beds, kelp forests, and coral reefs. The majority of this area is covered in Chapter Five, but the specialized coral reef communities are discussed in Chapter Nine.

Two transitional areas exist, one between the marine environment and the

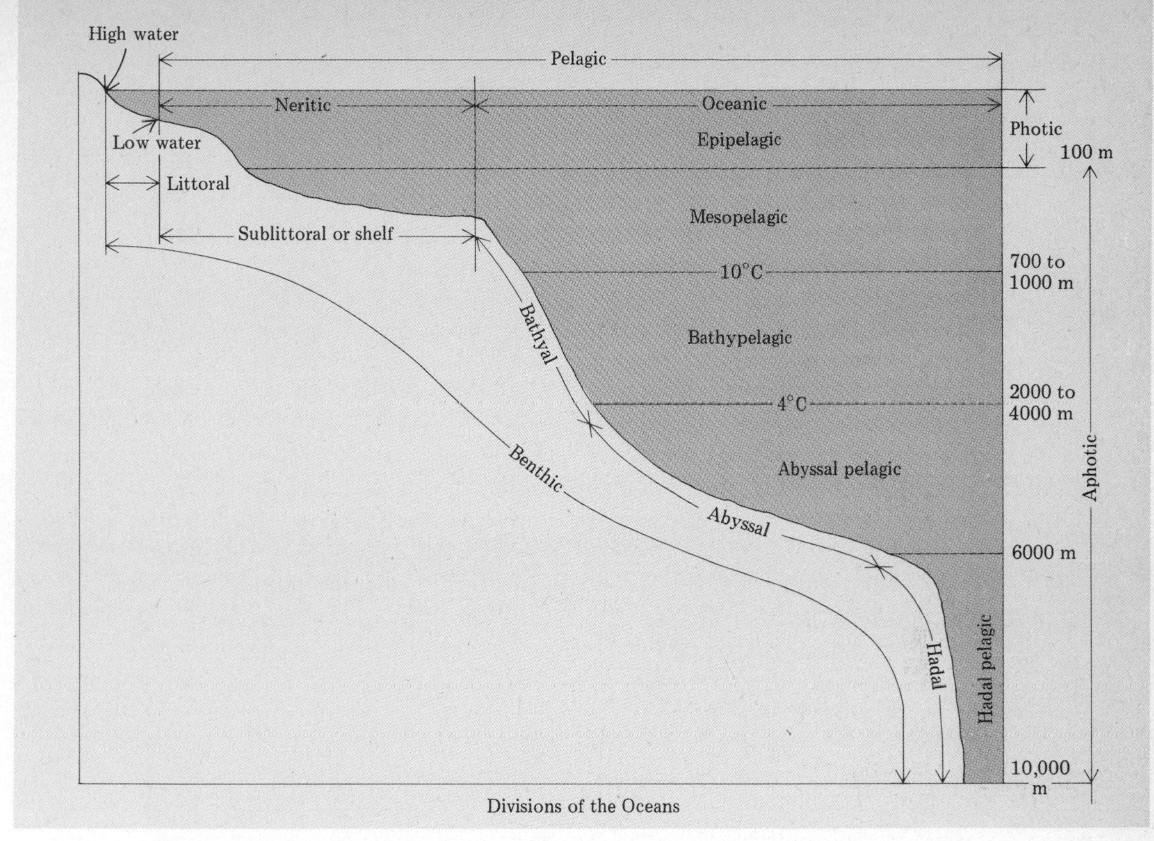

Figure 1-17 Divisions of the oceans (not to scale). (Modified from J. Hedgpeth, ed., The treatise on marine ecology and paleoecology, vol. I, Ecology, 1957, The Geological Society of America.)

terrestrial and the other between marine and fresh water. The *intertidal zone* or *littoral zone* is that shore area lying between the extremes of high and low tide; it represents the transitional area from marine to terrestrial conditions. It is a zone of abundant life and is well studied. It is the subject of Chapter Six. *Estuaries* represent the transition area where fresh and salt water meet and mix. They form the subject of Chapter Eight.

It is well to remember that the above classification scheme is not absolute, but is used for convenience, and that organisms and communities may extend into two or more zones. This is particularly true for those organisms that are powerful swimmers and divers. These organisms are discussed in Chapter Three.

References

Briggs, J. C. 1974. Marine zoogeography. McGraw-Hill, N.Y. 475 pp.

Gross, M. G. 1972. Oceanography: A view of the earth. Prentice-Hall, Englewood Cliffs, N.J. 581 pp.

Hedgpeth, J. 1957. Classification of marine environments, pp. 17–28, and Concepts of marine ecology, pp. 29–52. *In:* Hedgpeth, J. E. (ed.). The treatise on marine ecology and paleoecology. Vol. I, Ecology. Memoir 67, Geol. Soc. of Amer.

Ingmanson, D. E., and W. J. Wallace. 1973. Oceanology: An introduction. Wadsworth, Belmont, Calif. 325 pp.

Kormondy, E. J. 1969. Concepts of ecology. Concepts of modern biology series. Prentice-Hall, Englewood Cliffs, N.J. 209 pp.

Moore, J. R. (ed). 1971. Readings from *Scientific American*, oceanography. Freeman, San Francisco. 417 pp.

Odum, E. P. 1971. Fundamentals of ecology, 3rd ed. Saunders, Philadelphia. 574 pp.

Thurman, H. V. 1978. Introductory oceanography, 2nd ed. Merrill, Columbus, Ohio. 506 pp.

Whittaker, R. H. 1975. Communities and ecosystems, 2nd ed. Macmillan, N.Y. 385 pp.

Chapter Two
PLANKTON AND PLANKTON COMMUNITIES

As we have seen in the previous chapter, there are certain fundamental physical differences between terrestrial and marine environments, which contribute to the differences that we shall observe in the organization of the communities in the two areas. Perhaps nowhere are these differences more dramatic and easier to see than in a consideration of the free-floating and weakly swimming associations of organisms that we collectively call the *plankton*. It is appropriate to begin a consideration of the ecology of the oceans with the plankton because the plants of the plankton contribute by far the greatest amount of photosynthesis in the oceans. Thus, it is in the plankton that the major amount of the energy of the sun is trapped, which may subsequently be transferred to the many other communities of the ocean. This vital role of initial fixation of energy makes plankton so important in the economy of the oceans. As there could be no life on land without the energy-fixing grasses, trees, and shrubs, similarly there could be no life in the oceans without the energy-fixing minute planktonic plants.

TERMS AND DEFINITIONS

The term plankton is a general term as defined above. Planktonic organisms are animals and plants that have such limited powers of locomotion that they are at the mercy of the prevailing water movements. This is in contrast to the *nekton*, which comprise the strong swimming animals of the open sea, capable of exercising movement against the prevailing water flow. Plankton may be further subdivided. *Phytoplankton* comprise the free-floating plants of the sea that are capable of photosynthesis. *Zooplankton* are the various free-floating animals.

Because planktonic organisms are traditionally captured with the use of nets of various mesh sizes, they have also been classified on the basis of size. This size classification does not distinguish between plant and animal. Under this type of classification, a minimum of five subdivisions is recognized. *Megaplankton* are all those organisms above 2.0 mm. *Macroplankton* comprise the organisms from 0.2 to 2 mm in size. *Microplankton* are those plankton that fall between about 20 μm and 0.2 mm. These first two groups are those usually captured in standard plankton nets. The *nanoplankton* are very small organisms ranging in size from 2 to 20 μm. The smallest plankton are termed *ultraplankton* and are less than 2 μm in size. The nanoplankton and ultraplankton cannot be captured in plankton nets, because in order to filter them out of the water columns, the mesh would have to be so fine that when the nets were pulled behind a boat, the water would not pass through. As a result, these organisms can be obtained only be centrifuging samples of sea water or filtering water samples on fine filters such as millipore filters. A newer scheme of size classification includes all organisms from viruses through whales.

A final set of terms concerns the life history characteristics of the plankton organisms. *Holoplankton* are those organisms that spend their entire lives in the plankton. *Meroplankton*, on the other hand, are those species that spend but a part of their lives in the plankton. Meroplanktonic organisms include a large number of larvae of animals that as adults either live on the bottom or swim as nekton.

THE PHYTOPLANKTON

Although this text is designed to cover the principles of organization of life in the sea, not to detail the various taxonomic groups, it is worthwhile to include here a brief description of the common phytoplankton. The larger phytoplankton, that is those normally captured in nets, consist predominantly of only two groups, diatoms and dinoflagellates, which dominate the net phytoplankton throughout the world.

DIATOMS

The diatoms are easily differentiated from the dinoflagellates because they are enclosed within a unique glass "pill box," and have no visible means of locomotion. Each box is composed of two parts or valves, one valve fitting over another (Fig. 2-1). The living part of the diatom is within the box. The box is constructed of silicon dioxide, the same material that is a major constituent of glass. Each box is highly ornamented with species-specific designs, pits, and perforations, and this feature has made these organisms very popular with microscopists and, more recently, with scanning electron microscopists (Fig. 2-2).

Diatoms may occur singly, each individual occupying a single box, or they may occur in chains of various kinds, which themselves add to the ornamentation seen in the individual boxes (Figs. 2-2, 2-4).

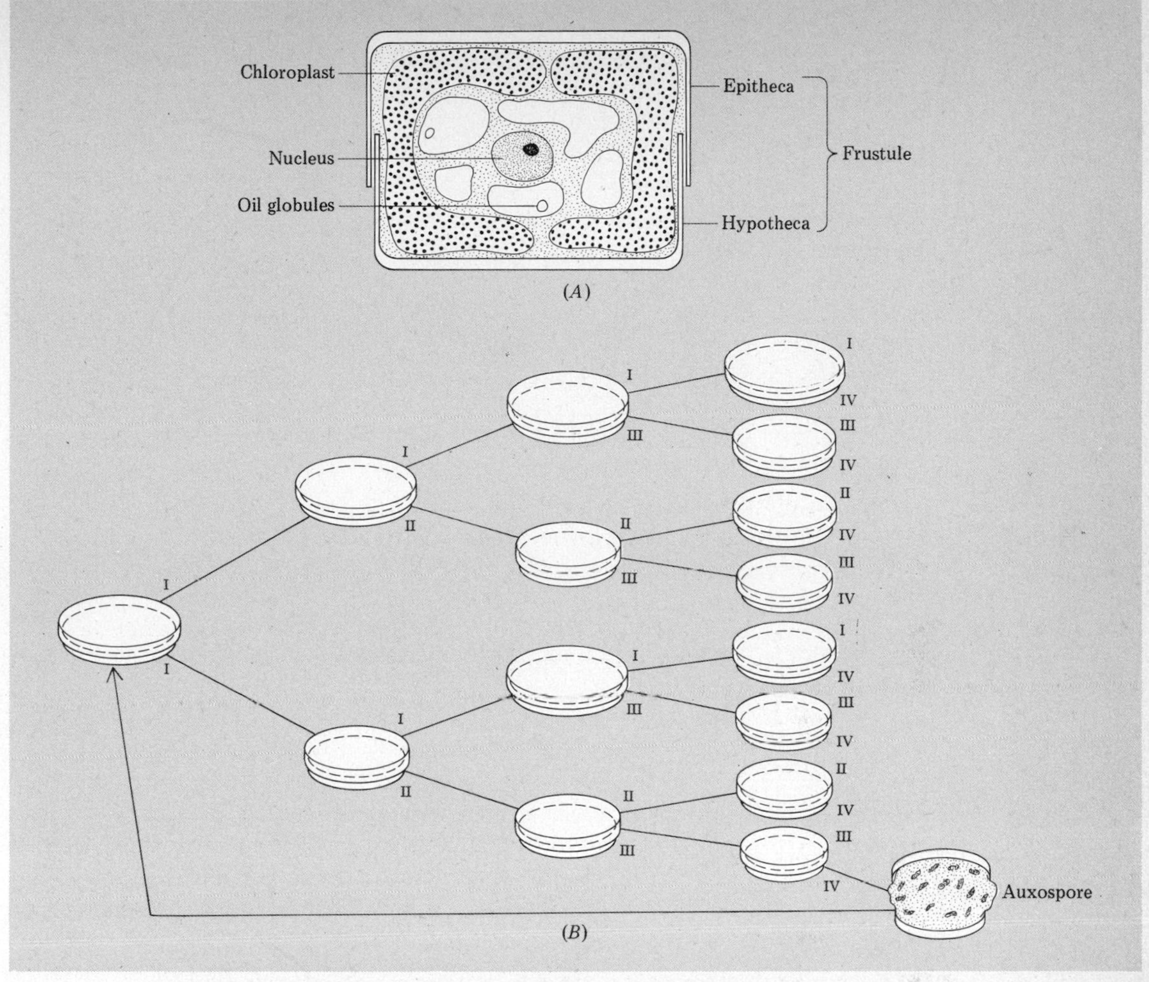

Figure 2-1 Diatom features. *(A)* Diagrammatic representation of the structure of a diatom. *(B)* Diagrammatic representation of the diminution of size in successive divisions and the restoration of size following auxospore formation.

In reproduction, each diatom divides into two halves and one half then occupies the top valve of the box, while the other takes the bottom. Each then secretes a new top or bottom valve, as the case may be, so that the typical box is recreated. Since each of these new valves is secreted *within* the old valve, as the process continues through several generations, the size of the diatom decreases. Thus, individuals of species of diatom vary in size. Obviously, there must be a limit to the process of reduction. This occurs after a certain number of generations, when the diatom casts off both valves and becomes a structure called an *auxospore*. Within this spore, new valves are secreted which reestablish the original size of the diatom species (Fig. 2-1).

Diatoms are abundant as species and individuals. Although it is not possible to illustrate all here, some of the more common planktonic genera are shown in Figs. 2-2 and 2-4.

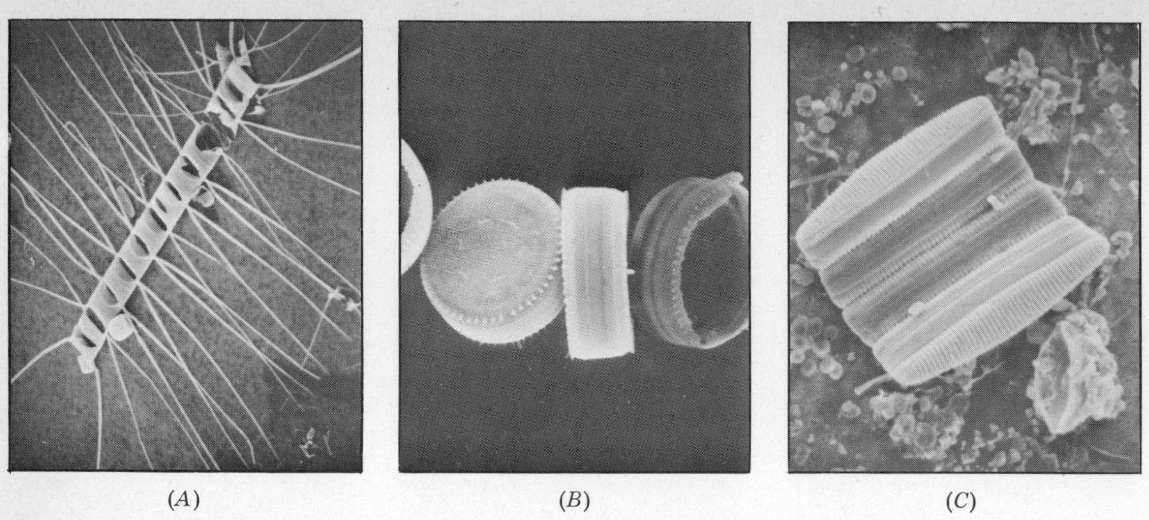
(A) (B) (C)

Figure 2-2 Scanning electron micrographs of diatoms. *(A) Chaetoceros* sp. (700×). *(B) Thalassiosira decipiens* (1,000×). *(C) Pseudoeunotia doliolus* (1,400×).

Some diatoms are not planktonic but benthic, but we shall return to these in Chapter Six.

DINOFLAGELLATES

The second major group, the *dinoflagellates*, is recognizable by possessing two flagellae, which they use to move themselves through the water. They lack an external skeleton of silicon but are often armored with plates of the carbohydrate cellulose (Figs. 2-3, 2-4). Dinoflagellates are generally fairly small organisms and are solitary, rarely forming chains. They reproduce by simple fission, as do the diatoms, but in this case, each of the daughter cells retains half of the original cellulose armor and forms a new part to replace the missing half without any

Figure 2-3 Scanning electron micrographs of dinoflagellates. *(A) Dinophysis* sp. (700×). *(B) Ceratium* sp. (500×). *(C) Dinophysis* sp. (500×).

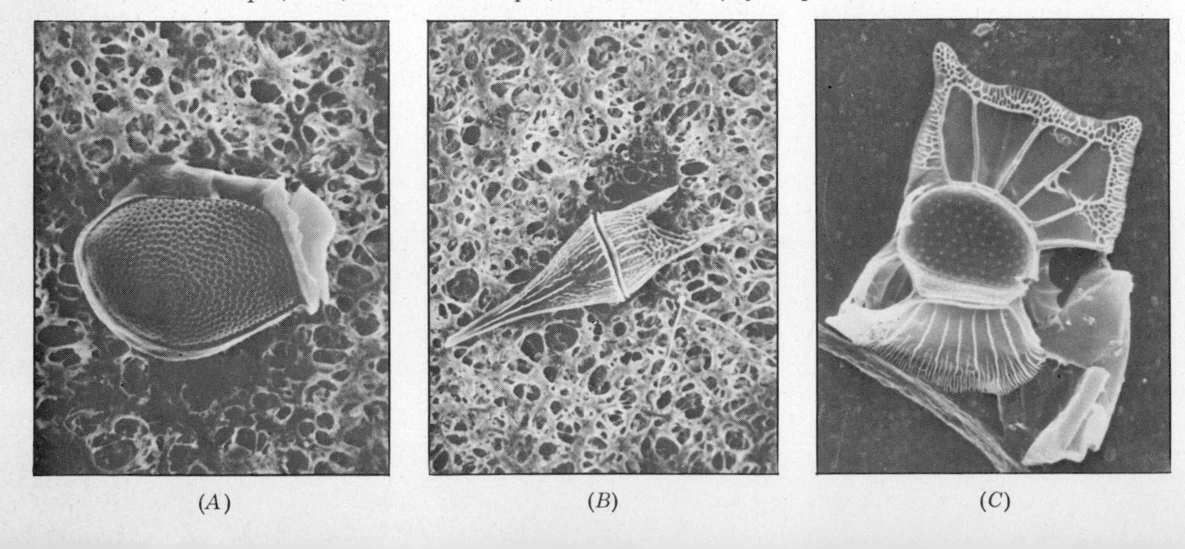
(A) (B) (C)

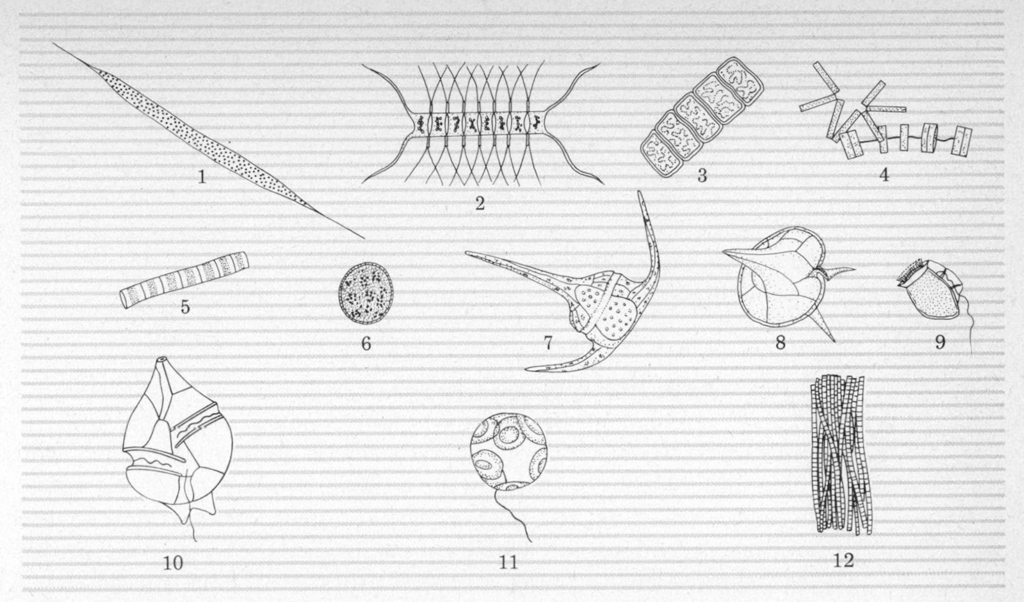

Figure 2-4 Diagrammatic representation of some characteristic genera of marine phytoplankton. Diatoms: (1) *Rhizosolenia,* (2) *Chaetoceros,* (3) *Navicula,* (4) *Thalassiosira,* (5) *Skeletonema,* (6) *Coscinodiscus.* Dinoflagellates: (7) *Ceratium,* (8) *Peridinium,* (9) *Dinophysis,* (10) *Gonyaulax.* Coccolithophores: (11) *Coccolithus.* Blue-green alga: (12) *Trichodesmium.* (Not to scale.)

diminution in size. Hence, successive generations do not change in size. Dinoflagellates also are capable of producing toxins that are released into sea water. If dinoflagellates become extremely abundant (2–8 million cells per liter), the cumulative effect of all the toxins released may affect other organisms, causing mass mortality. Such extreme concentrations, or blooms, of dinoflagellates are called "*red tides*" and are responsible for the massive localized mortality seen in fish and invertebrates in various places.

Some dinoflagellates, such as the common *Noctiluca* (see Fig. 2-10), are also highly bioluminescent (see Chapter Five for discussion) and when present in large numbers, can actually light up the wakes of boats and the breaking waves on a beach. Many dinoflagellates, such as *Noctiluca,* are not photosynthetic.

OTHER PHYTOPLANKTON

Minor constituents of the phytoplankton include the blue-green algae (Cyanophyceae), the coccolithophores (Coccolithophoridae, Haptophyceae), and the silicoflagellates (Dictyochaceae, Chrysophyceae) (Fig. 2-4). Marine blue-green algae are found mainly in the tropics, where they occasionally form dense mats of filaments and color the water. The Red Sea was named from the red color of the blue-green alga *Trichodesmium erythraeum,* which is a common member of the phytoplankton in the tropics. Blue-green algae are closely related to bacteria, with whom they share the feature of a lack of a formal nucleus in the cell; their pigments for photosynthesis also are not in chloroplasts. Coccolithophores are very small, down to 5 μm, with characteristic calcium carbonate plates ("coccoliths") embedded in the outer layer (Fig. 2-4). Coccolithophores can occur in great numbers, especially in the tropical seas, and they may contribute considerably to the primary productivity. Silicoflagellates are also small, single-celled organisms with a skeleton of glass (silicon dioxide). They are widespread in the world's oceans, but are usually not common.

THE ZOOPLANKTON

In contrast to the phytoplankton, which is dominated by two groups of plants, the zooplankton, constituting the animal members of the plankton, is extremely diverse, consisting of a host of larval and adult forms representing most of the animal phyla. From an ecological standpoint, however, one group of zooplankters stands out as being far more important than the others. The subclass Copepoda (class Crustacea, phylum Arthropoda) are small holoplanktonic crustaceans which dominate the zooplankton throughout the world's oceans (Fig. 2-5). These small animals are of vital importance in the economy of the ecosystems in the oceans, because they include the primary herbivorous animals in the sea. It is they who graze upon the aquatic pastures of phytoplankton and provide the vital link between the primary production of the plants and the numerous large and small carnivores.

COPEPODS

Free-living plankton copepods are generally small, between one and several millimeters in length. They swim weakly, using their jointed thoracic limbs, and have a characteristic jerky movement. They employ their very large antennae to slow their rate of sinking. Most free-living planktonic copepods have a characteristic body shape and hence are readily recognizable (Fig. 2-5).

Copepods graze on phytoplankton either by means of a complex filtering mechanism that employs the fine setae (hairs) covering certain of their appendages around the mouth (maxillae), or else by grasping plants with their appendages. Some copepods are carnivorous, seizing prey with their appendages. In the filtering process, the swimming movements of the thoracic legs create a water current that passes into the midventral line of the body, where it flows through the fine setae of the appendages around the mouth. Phytoplankton cells are removed from the water and passed to the mouth (Fig. 2-6).

In copepods, the sexes are separate, and sperm are transferred to the female as packaged spermatophores. After fertilization, the eggs are enclosed in a sac carried by the female, attached to her body. They hatch as *nauplius* larvae and progress through several naupliar stages and then several more *copepodite* stages before becoming adult (Fig. 2-5). As we shall see, the life history of the adult and larval forms have an effect on the phytoplankton cycles observed in the sea.

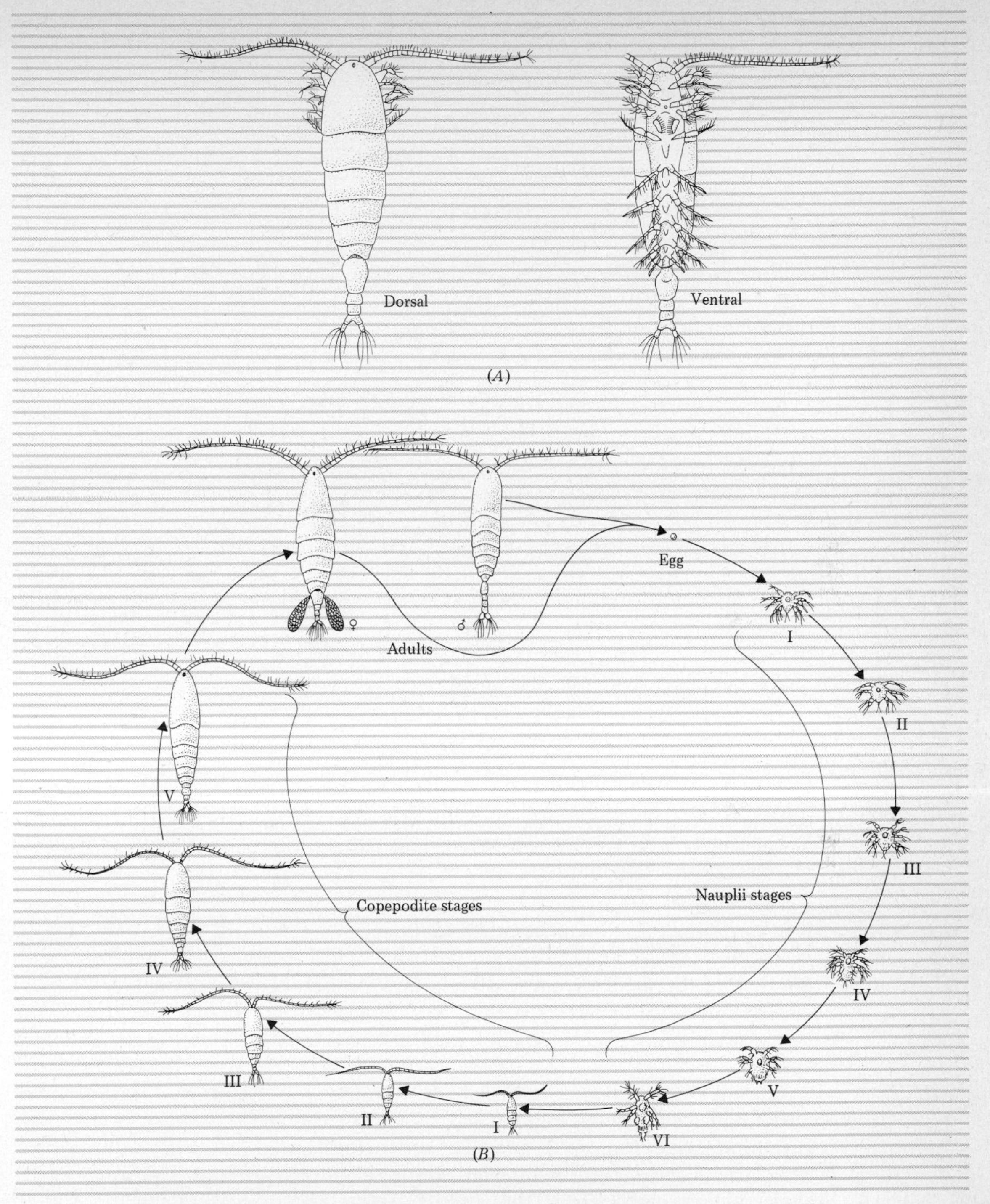

Figure 2-5 Copepods. *(A)* Typical copepod showing major anatomical features. *(B)* Outline of the typical life cycle of a copepod.

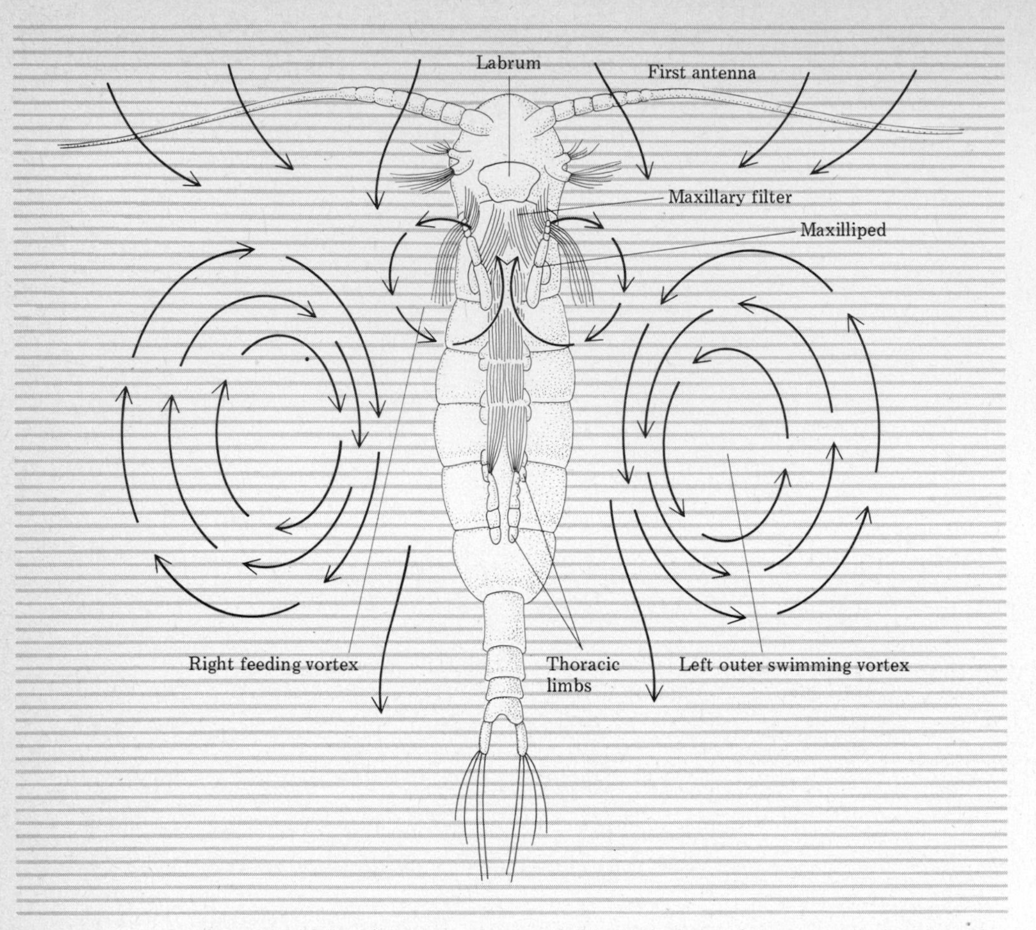

Figure 2-6 Feeding in a copepod. Ventral view of a copepod showing the currents created in swimming which bring particles in to be filtered on the setae of the maxillae. (After W. D. Russell-Hunter, Aquatic productivity, Macmillan, © 1970, W. D. Russell-Hunter. Reprinted with permission of Macmillan Publishing Co., Inc.)

OTHER ZOOPLANKTON

Because of the great diversity of zooplankton organisms, it is not practical to cover all in detail in a book such as this. Space permits us to consider only the larger taxonomic groups. Within each of these taxa are often many different species.

Among the holoplankton, the phylum Protozoa contributes an abundance of individuals to the zooplankton (Fig. 2-7*A*). The dominant groups are the order Foraminiferida (''forams'') and the order Radiolaria (see Fig. 2-11*E*). Both radiolarians and foraminiferans are single-celled organisms that produce skeletons, calcium carbonate in the case of forams and glass (SiO_2) in the case of radiolarians. Radiolarians are exclusively marine, and forams partially so. Members of these two orders are so abundant and widespread that their skeletons have formed thick layers of globigerina and radiolarian *ooze* extending over vast areas of the deep-sea floor.

Holoplanktonic members of the phylum Cnidaria include the various jellyfish-

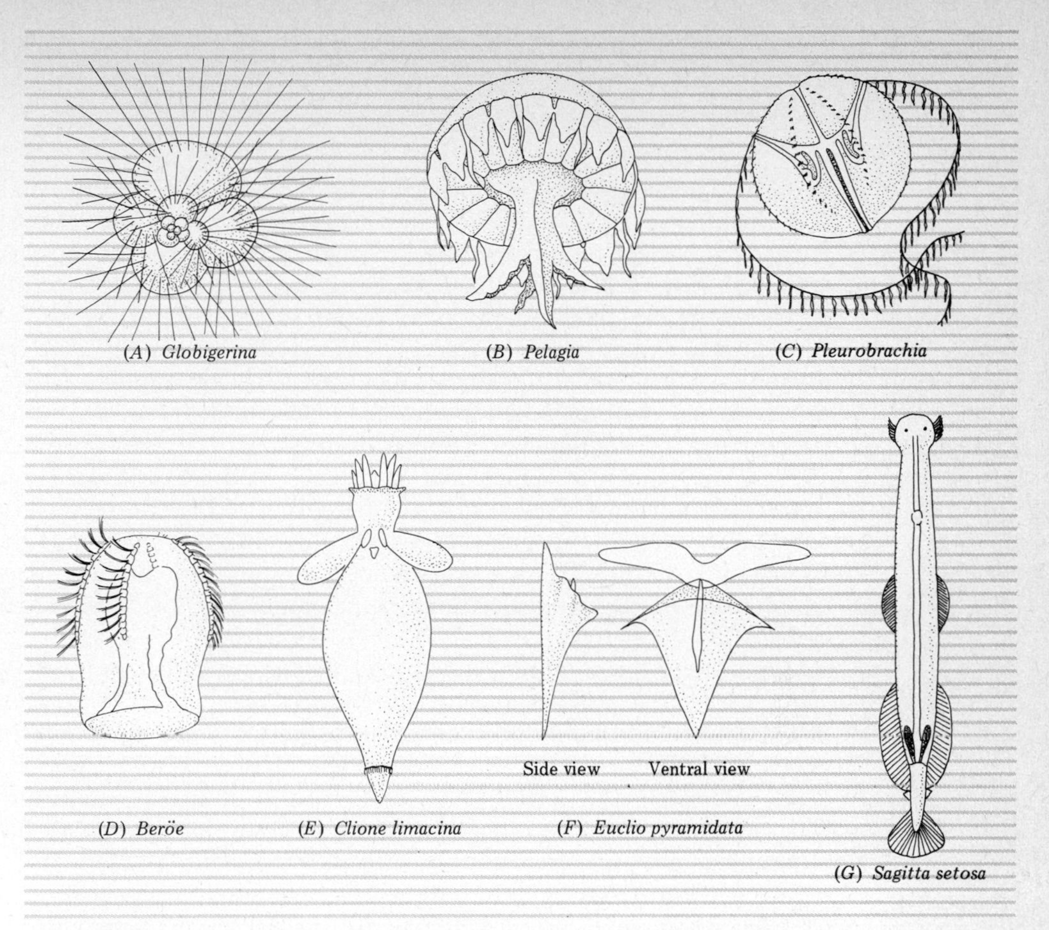

Figure 2-7 Holoplankton. Examples of some typical plankton members of the phyla Protozoa *(A)*; Cnidaria *(B)*; Ctenophora *(C, D)*; Mollusca *(E, F)*; and Chaetognatha *(G)*. (Not to scale.) (*A, E, F, G,* after E. Newell and R. C. Newell, Marine plankton, a practical guide, Hutchinson. *B, C, D,* after D. Smith, A guide to marine coastal plankton and marine invertebrate larvae, Kendall/Hunt. Copyright © 1977 by DeBoyd L. Smith. Reprinted with permission of the publisher.)

es of the classes Hydrozoa and Scyphozoa (Fig. 2-7*B*) and the curious, complex colonies known as siphonophores (see Fig. 2-11*A,B*). Scyphozoan jellyfishes are among the largest plankton organisms, and may occasionally be found in large numbers.

Closely related to the Cnidaria is the phylum Ctenophora. With few exceptions, this phylum is entirely planktonic (Figs. 2-7*C,D*, 2-12*A*). All are voracious carnivores, capturing food with sticky tentacles or engulfing them with an oversized mouth. Locomotion is via rows of fused, large cilia called *ctenes*.

The phyla Nemertinea and Annelida are represented in the holoplankton by a few highly specialized forms, which are not abundant. The annelids include polychaete worms of the families Tomopteridae and Alciopidae, the latter with the largest and best developed eyes in the Annelida. Most planktonic nemerteans live in deep water.

The phylum Mollusca is the second largest phylum in the animal kingdom. It is usually considered to be composed of slow-moving benthic animals. However, it includes a considerable variety of specially adapted holoplanktonic forms. Perhaps the most highly modified planktonic mollusks are the *pteropods* and *heteropods* (Figs. 2-7*E,F*, 2-10*D*). Both groups are closely related to snails and are classified in the class Gastropoda. Pteropods are of two types, shelled (O. *Thecosomata*) and naked (O. *Gymnosomata*). Shelled pteropods have fragile shells and swim using their winglike foot. They are herbivores and are preyed upon by the faster-swimming naked pteropods. Heteropods are large animals with transparent, jellylike bodies. They are carnivores. A special case is the pelagic shelled snail, *Janthina*, which maintains itself at the surface by clinging to the underside of its own bubble raft. A final group of planktonic mollusks are the squids. Many, perhaps most, squids are powerful, fast swimmers that must be considered nekton. There are, however, a significant number of very small squids that are not strong swimmers and must be considered planktonic (see Fig. 2-10*B*).

It is among the phylum Arthropoda that the greatest number of plankton organisms occur. In the seas of the world, virtually all belong to the class Crustacea. In addition to the dominant copepods, the holoplanktonic crustaceans include members from the order Cladocera (Fig. 2-8*A,B*), subclass Ostracoda (Fig. 2-8*D*), order Mysidacea, order Amphipoda, order Euphausiacea (Fig. 2-8*C*), and order Decapoda. Most of these taxa are small filter feeders straining plants and/or small animals out of the water.

The phylum Chaetognatha is a very small phylum consisting of about 65 species, which, with one exception, are planktonic. Chaetognaths, or "arrow worms" as they are called, are abundant members of the plankton throughout the world (Fig. 2-7*G*). All are voracious predators on copepods and other planktonic organisms.

The final phylum represented in the holoplankton is the Chordata. Planktonic chordates belong to the classes Thaliacea ("salps") and Larvacea (Fig. 2-8*E,F*). These gelatinous bodied animals are filter feeders. Larvaceans construct a "house" around themselves and pump water through a screen in the house to filter out their food. "Houses" are continually built and shed.

Compared to the holoplankton, meroplankton as a group are even more diverse. A bewildering array of larval forms constitute the meroplankton. These larvae are derived from virtually all animal phyla and from all different marine habitats. The number of larval forms is greater than the number of species in the sea that produce them because many species go through a series of larval stages before becoming adult. Thus, each of these species may be represented in the plankton by several different stages. This is particularly true in the Crustacea, where some decapods may have as many as 18 different larval stages, and most have more than one. Phyla which are not represented at all in the holoplankton, such as Bryozoa, Phoronida, Echinodermata, and Porifera, have larval forms (Fig. 2-9*A,B,C,D,E,F*). Other abundant phyla, such as Nemertinea, Mollusca, and Annelida, which have but a few highly adapted holoplankton groups, have

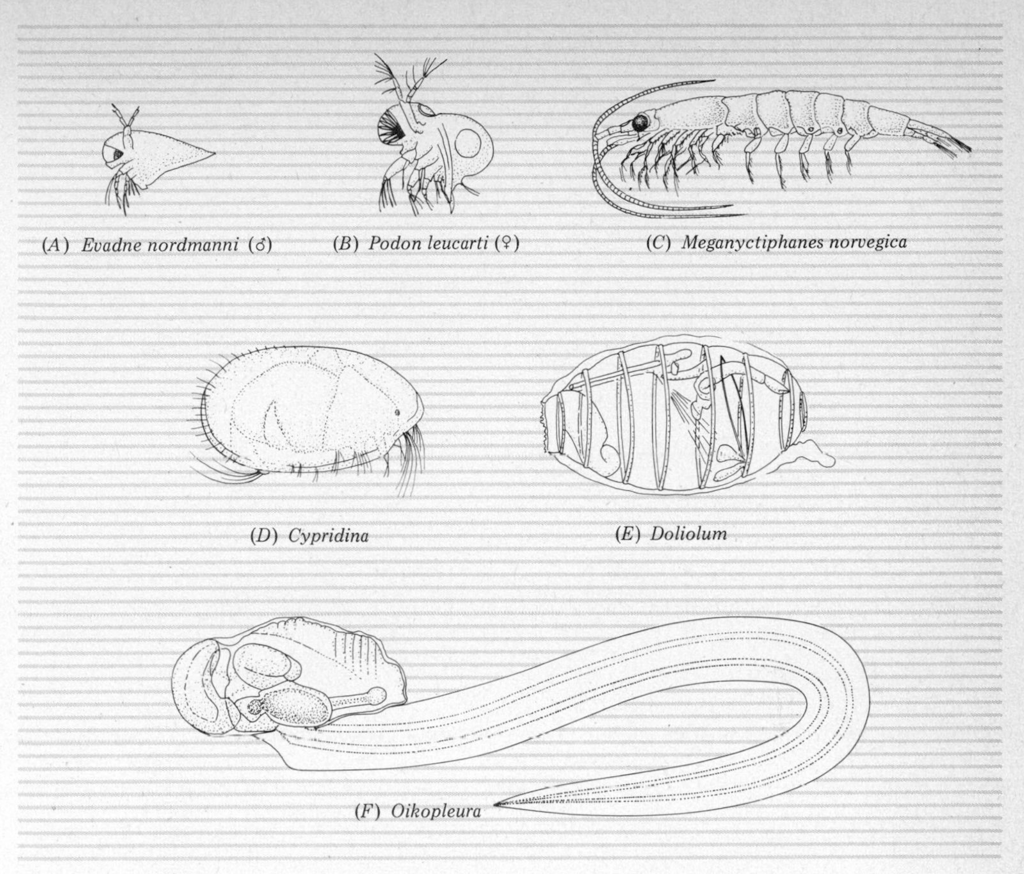

Figure 2-8 Holoplankton. Typical planktonic forms of the phyla Arthropoda *(A, B, C, D)* and Chordata *(E, F)*. *(A, B)* Order Cladocera, class Crustacea. *(C)* Order Euphausiacea, class Crustacea. *(E)* Class Thaliacea. *(F)* Class Larvacea. (Not to scale.) (*A, B, C,* from E. Newell and R. C. Newell, Marine plankton, a practical guide, Hutchinson. *D, E, F,* from D. Smith, A guide to marine coastal plankton and marine invertebrate larvae, Kendall/Hunt. Copyright © 1977 by DeBoyd L. Smith. Reprinted with permission of the publisher.)

planktonic larvae (Fig. 2-9*G,H,L*). The Crustacea have by far the greatest number of larval types, and these include larvae of the abundant holoplankton organisms also. Even the nekton may be represented here, as many fishes have eggs and larvae in the plankton. There is not space available in this book to categorize all these larval types. Some representative types are illustrated in Fig. 2-9.

FLOTATION MECHANISMS

Plankton tend to have a density (mass per unit volume) that is somewhat greater than that of sea water. This means that any plankton organism will eventually tend to sink in the water column. This would be deleterious for both phyto-

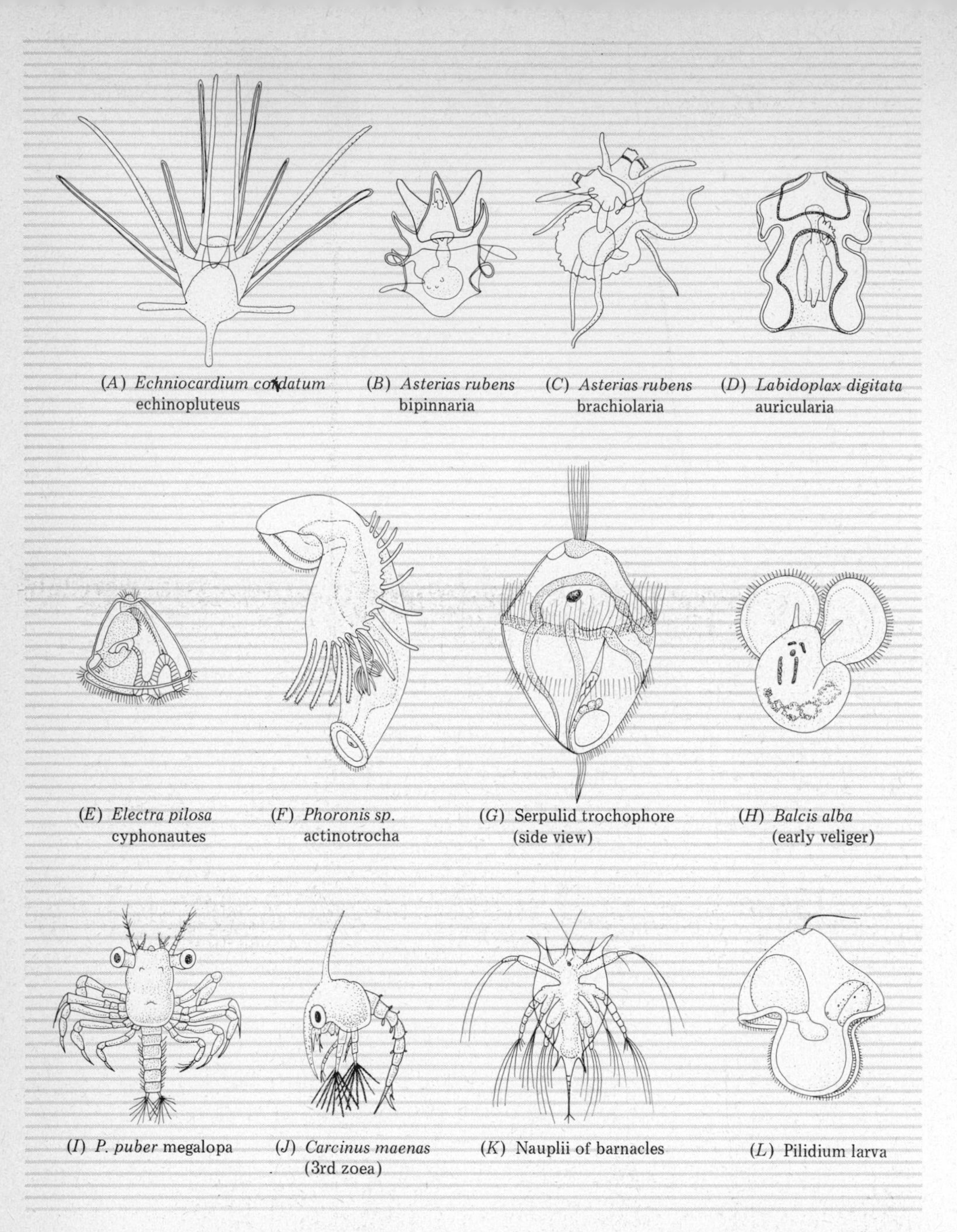

Figure 2-9 Meroplankton. Examples from several phyla. *(A)* Echinodermata, sea urchin. *(B, C)* Echinodermata, starfish. *(D)* Echinodermata, sea cucumber. *(E)* Bryozoa. *(F)* Phoronida. *(G)* Annelida, Polychaeta. *(H)* Mollusca, gastropod. *(I, J)* Arthropoda, Crustacea, Decapoda. *(K)* Crustacea, barnacle. *(L)* Nemertinea. (*A–J,* from E. Newell and R. C. Newell, Marine plankton, a practical guide, Hutchinson. *K–L,* after D. Smith, A guide to marine coastal plankton and marine invertebrate larvae, Kendall/Hunt. Copyright © 1977 by DeBoyd L. Smith. Reprinted with permission of the publisher.)

plankton and zooplankton; phytoplankton would sink below the lighted areas of the sea and hence be unable to photosynthesize, and zooplankton would sink out of the area where their phytoplankton food occurs. Since sinking is undesirable for both groups, since they are weak swimmers unable to cope with winds and currents, and since living flesh tends to sink, how then do these organisms cope with the problem of staying in the upper layers of the sea?

PRINCIPLES

In order to answer the above question, we must consider some physical and chemical characteristics that will bear on this problem. In the first place, the density of sea water is a function of two parameters: its temperature and its salinity. Sea water becomes more dense as its salinity increases and less dense as its temperature increases. Thus, the density of a given body of water may not be constant, especially in the temperate zone seas, where there are marked differences in temperature over the course of the year. A final physical characteristic is that of viscosity of sea water. Viscosity is related to temperature and salinity in that the more saline water is and the lower its temperature, the more viscous it is and the less rapidly things sink in it.

A second set of principles has to do with the effect of shapes on the rate of sinking in dense liquids. Objects of different shape but similar weights fall at differing rates in both gases, such as air, and liquids, such as water. The rate of fall is proportional to the amount of resistance the body offers to the gas or liquid through which it moves. Objects with a great amount of surface for a given weight tend to fall at slower rates than objects of similar weight with less surface area. Included here is the physical law that surface area increases as the square of the linear dimensions of the object, but the volume increases as the cube of the same dimensions. This means that the smaller the body, the greater the surface area relative to the mass.

From the above physical laws governing changes in density and viscosity of water and those governing resistance, it is possible to derive a very simple equation that will relate sinking rates of organisms to these parameters. The equation is :

$$SR = \frac{W_1 - W_2}{(R)\ (V_w)}$$

where SR = sinking rate; W_1 = density of the organism; W_2 = density of sea water; $W_1 - W_2$ = amount of overweight (overweight = the amount by which the flesh of the organism exceeds the weight of a similar volume of water); R = surface of resistance; V_w = viscosity of water.

The organisms can do nothing about the viscosity of water. Hence, all adaptations of the organism to reduce the sinking rate must be concerned with either reducing the amount of overweight or else increasing the surface of resistance.

Let us consider first those adaptations designed to reduce the amount of overweight.

REDUCTION OF OVERWEIGHT

One mechanism that may be employed to reduce overweight is simply to alter the composition of the body fluids such that they are less dense than an equal volume of sea water. It is necessary, however, to maintain the same number of solute particles in the body fluid as in the surrounding sea water in order to avoid osmotic problems. Osmosis is the passage of water across semipermeable membrane, such as cell walls, to equalize dilution on both sides of the membrane. How is this change in composition accomplished? Perhaps the most common way is to replace heavy chemical ions in the body fluids with lighter ones. This allows the animal to maintain the same osmotic condition (same number of particles) while becoming lighter with respect to sea water. An example is the dinoflagellate *Noctiluca* (Fig. 2-10), whose internal fluid contains ammonium chloride (NH_4Cl) and which is iso-osmotic with sea water but is less dense, having a specific gravity of 1.01 versus the 1.025 of sea water. Similarly, the cranchiid squid, which, for squid, have very fat, bulbous bodies, are filled with NH_4Cl, making them less dense than sea water.

In a similar fashion, certain planktonic forms such as salps, ctenophores, and heteropods (Fig. 2-10) actively exclude heavy ions such as SO_4^{2-} from their bodies and replace them with osmotically similar but lighter chloride ions.

Another mechanism is the development of special gas-filled floats. Since gas is much less dense than a similar volume of water, buoyancy is assured. Perhaps the most familiar examples of this approach are the floats of the Portuguese man-of-war (*Physalia*) and the swim bladders of fishes (Fig. 2-11). If the animal is able to regulate the gas pressure in the float or bladder, this mechanism can be used to regulate the position of the animal in the water column so that it can move up or down at will. This is what some fishes do.

Similar to the use of gas-filled floats is the employment of liquids that are less

Figure 2-10 Flotation mechanisms. Examples of organisms that exclude heavy ions, forming body fluids less dense than sea water. (Redrawn from J. H. Wickstead, An introduction to the study of tropical plankton, Hutchinson.)

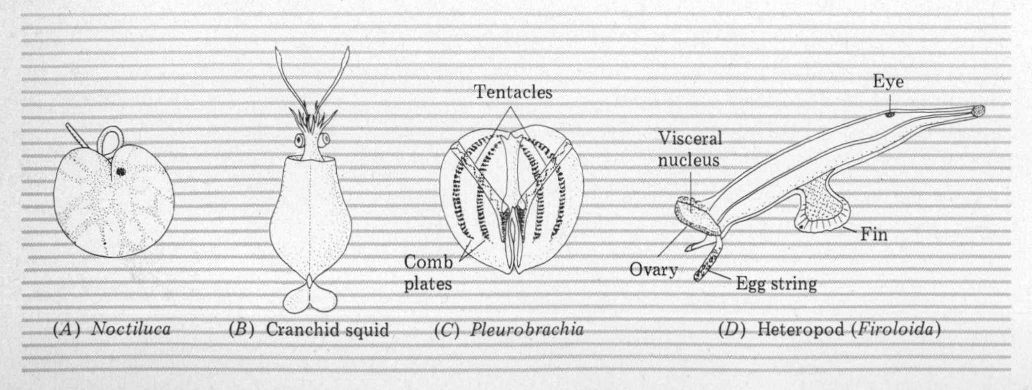

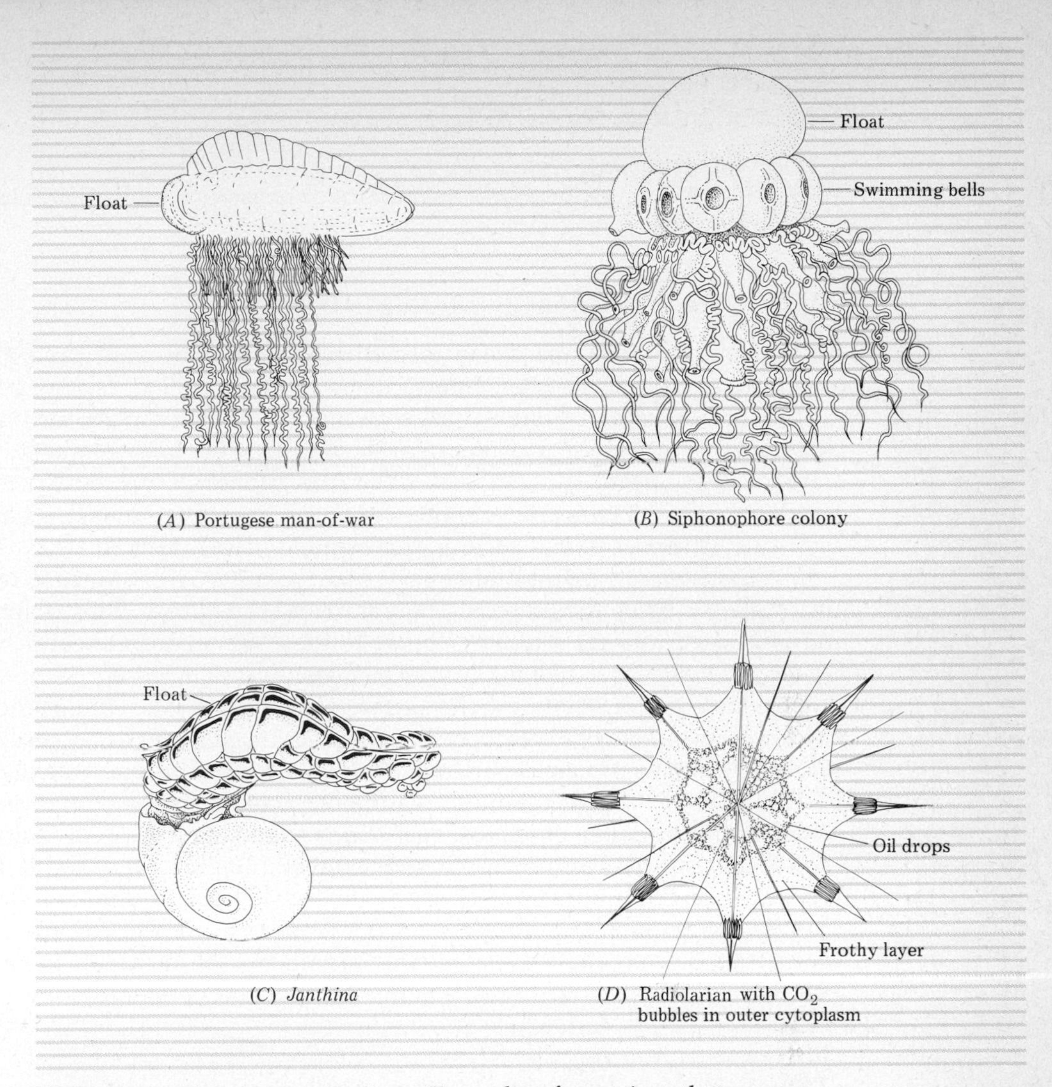

Figure 2-11 Flotation mechanisms. Examples of organisms that use gas- or fluid-filled floats. (*A*, from E. Newell and R. C. Newell, Marine plankton, a practical guide, Hutchinson. *B*, *D*, from L. Hyman, The invertebrates. Vol. I, Protozoa to Ctenophora, McGraw-Hill. Used with the permission of McGraw-Hill Book Company.)

dense than water. Prominent among these liquids are oils and fats. These can serve a dual purpose, as fats and oils also act as food reserves for the organisms. Copepods, for example, often store excess food in the form of oil droplets under the carapace, and these droplets aid in buoying the animal. Diatoms also store food as oils.

Whereas some planktonic organisms have employed the above mechanisms to alter their density and hence remain afloat, many others have not. If we return to our original equation, we can see that if overweight cannot be reduced, the only other option is to increase the surface of resistance to the water and hence, at least slow the rate of sinking.

CHANGES IN SURFACE OF RESISTANCE

Surface of resistance may be increased in a number of ways, and all may be observed in planktonic organisms.

One of the most common observations made about plankton is that the general body size is small. This is particularly true for tropical plankton. As we noted earlier, surface area increases as the square of the linear dimension, whereas volume increases as the cube. This means that the smaller the organism, the greater the surface area relative to volume. Thus, by remaining small, plankton organisms offer far more surface area of resistance to sinking per unit volume of living material than if they were large. This is particularly important for tropical plankton, because they live in warmer and hence less viscous water, where sinking would be faster.

The other way of increasing the surface of resistance is to change the shape of the body. If one were to drop a round ball and a flat coin of equal weight into water, the ball would sink more quickly because it has less surface of resistance to the fluid. In a like manner, planktonic organisms have evolved various flattened body shapes or appendages. Such examples (Fig. 2-12) include a host of different species from many different phyla.

Even more common than changes in body shape is the development of various spines and body projections. These structures add considerable resistance, but little to the weight. Such adaptations are common in the various diatoms (Fig. 2-1), radiolarians, foraminiferans, and crustaceans.

WATER MOVEMENTS

A final mechanism of buoyancy has to do not with the organisms, but with the nature of water movements in the ocean. In the ocean, the surface water heats up during the day and cools at night. This alternating heating and cooling changes the density and leads to the creation of *convection cells,* which are small units of water that are either sinking or rising according to their density. These are gentle movements, and plankton may be moved by them.

PRIMARY PRODUCTION

The basis for all life on the earth resides in the ability of green plants to use the energy of sunlight to synthesize energy-rich organic molecules from inorganic materials. This is the process of *photosynthesis.* The general equation for this process is:

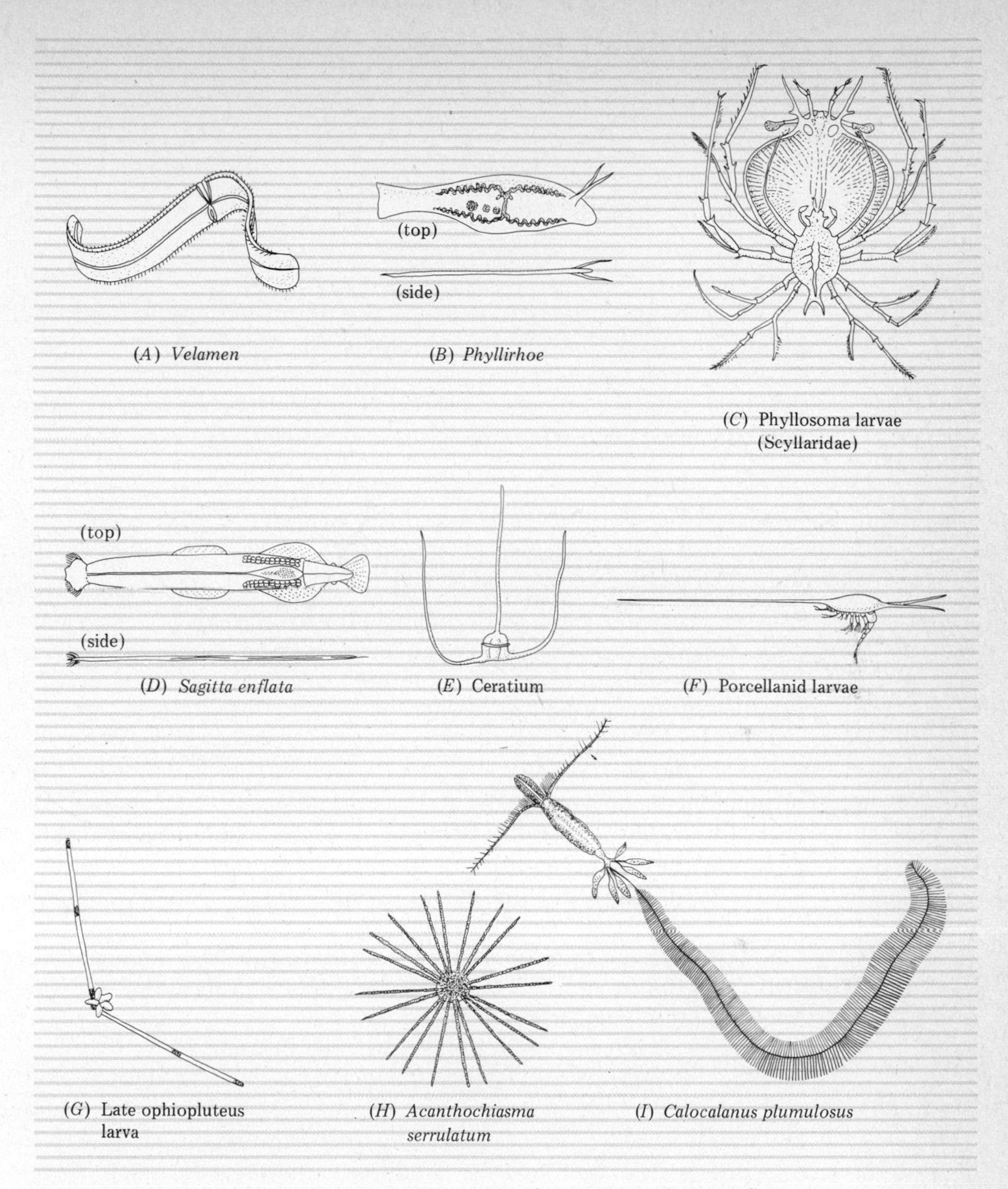

Figure 2-12 Flotation mechanisms. Examples of organisms that are flat or with spines: *(A)* Phylum Ctenophora. *(B)* Phylum Mollusca, class Gastropoda. *(C)* Phylum Arthropoda, class Crustacea. *(D)* Phylum Chaetognatha. *(E)* Order Dinoflagellata. *(F)* Phylum Arthropoda, order Decapoda. *(G)* Phylum Echinodermata, class Ophiuroidea. *(H)* Phylum Protozoa, class Sarcodina. *(I)* Phylum Arthropoda, class Crustacea. *(A–G,* from J. H. Wickstead, An introduction to the study of tropical plankton, Hutchinson. *H,* from E. Newell and R. C. Newell, Marine plankton, a practical guide, Hutchinson. *I,* from *The Cambridge Natural History.* Vol. 4, Crustacea, reprinted by Wheldon & Wesley.)

$$6CO_2 + 6H_2O \xrightarrow{\text{sunlight}} \underset{\substack{\text{energy-rich} \\ \text{organic} \\ \text{compounds}}}{C_6H_{12}O_6} + 6O_2$$

The basis for all life in the sea is the photosynthetic activity of the aquatic plants. There are, however, significant differences in the form of the plants and in the location and extent of maximum photosynthetic activity imposed by the particular physical and chemical conditions of the oceans (see pp. 56–66). These will be covered subsequently.

PRIMARY PRODUCTIVITY

Primary productivity is defined as the rate of formation of energy-rich organic compounds from inorganic materials. Primary productivity is thus usually considered synonymous with photosynthesis, but this is not quite correct, since a minor amount of primary productivity may be produced by chemosynthetic bacteria (see Chapter Six). For all intents and purposes here, however, primary productivity is confined to plants.

The total amount of organic material fixed in the primary productivity process is termed the *gross primary production*, or *total production*. Since some of this total production must be used by the plants themselves to operate their own life processes, collectively called respiration, a lesser amount is available for transfer or use by the other organisms of the sea. *Net production* is the term given to that amount of total production that is left after losses from respiration and that is available to support other trophic levels.

For either gross or net primary production, the rates are usually expressed in terms of grams of carbon fixed per unit area or volume of sea water per time interval. Thus, production may be reported in grams of carbon per square meter per day, g C/m^2/year, or any other convenient units.

A few other terms require definition at this point. The *standing crop*, as applied to plants, is the total amount of plant biomass present in a given volume of water at a given time. Primary productivity and standing crop can vary considerably on a time scale of days to a year, and this variation is the result of a large number of factors that act both directly and indirectly on the photosynthetic process within plants and on the plants themselves. These will be discussed later.

As we saw in Chapter One, water has a profound effect on the light that penetrates it. Because water absorbs light, there is less light energy available as one goes deeper, and eventually, light disappears. For plants, this means that at some depth, the light energy is just sufficient for the plant to fix energy at a rate equal to the rate at which the plant uses energy in its own metabolic processes. In other words, there is a depth at which the respiratory usage of energy by the plant equals the ability of the photosynthetic mechanism to produce energy. If the plant goes deeper, the respiratory needs continue at the same rate, but the decreasing light is insufficient to allow the photosynthetic process to keep up and so there is a net loss of energy. The depth at which the rate of respiration of a plant is just equalled by the rate of photosynthesis is called the *compensation depth*. Above this depth, photosynthesis exceeds respiration and there is a net production of carbon, or a net primary production.

The compensation depth varies over the world's oceans because it is dependent upon the clarity of the water. The clearer the water, the greater the light penetration and the deeper the compensation depth. In general, the compensation depth is deeper in clear, open ocean waters and shallower in inshore waters where large amounts of particulate matter are found in the water.

Since the compensation depth varies with depth of light penetration, it follows that it should also be possible to define the compensation depth in terms of light intensity. In fact, this is true, and one can speak of a *compensation intensity*, which is the light intensity at which photosynthesis equals respiration. Although this intensity is somewhat different for different phytoplankton species, it is approximately that depth at which the intensity has been reduced to 1 percent of the surface light intensity. The compensation depth can then be defined alternatively as the depth to which 1 percent of the incident light penetrates.

It is well to remember that the compensation depth varies, not only in different geographic areas of the world's oceans as noted previously, but also from day to day and season to season in any one geographical area, as a result of changes in light with season (see pp. 57–60) and as a result of changes in water clarity.

MEASUREMENT OF PRIMARY PRODUCTIVITY

If we return to the photosynthetic equation,

$$6CO_2 + 6H_2O \xrightarrow[\text{energy} + \text{plants}]{\text{light}} C_6H_{12}O_6 + 6O_2$$

we can see that it should theoretically be possible to measure the rate of production of organic compounds by measuring the rate of appearance or disappearance of some component of the above equation. Thus, if one could measure the rate of disappearance of CO_2 or the appearance of O_2, this would be a measure of the rate of photosynthesis. In fact, this is done, but for various reasons, is practical only for the two components mentioned.

The classical method that has been used for many years to measure primary productivity is called the *light-dark bottle method*. In this method, two identical bottles are employed. One bottle is completely transparent, while the other is made completely opaque by painting it black or covering it with something like aluminum foil. Into each is placed the same volume of sea water, taken from the body of water for which you wish to estimate the productivity. The water thus contains the naturally occurring phytoplankton and zooplankton organisms.

The oxygen content of the water to be added to the two bottles is determined on a separate sample to establish the initial concentration. The bottles are then stoppered, attached to a line, and returned to the appropriate level in the sea. In the bottle that is dark, no photosynthesis can occur, but the plants and animals within continue to respire and use up oxygen. In this bottle, then, the original oxygen content will decrease as the respiring plants and animals use up the dissolved oxygen. In the light bottle, on the other hand, photosynthesis continues in excess of respiration and hence, oxygen builds up over the initial values, since it cannot escape. After the two bottles have incubated for a period of time determined by the investigator, the bottles are brought up, unstoppered, and the water is analyzed for oxygen content. Usually, the oxygen content is determined through the Winkler method. Once the oxygen content is known for each bottle after incubation, it is possible to calculate the rate of photosynthesis from these results and the initial concentration of oxygen of the water when put into the bottles. In order to understand how this is done, it is necessary to consider what has happened in each bottle. When the original sea water was added to each bottle, it contained a certain amount of oxygen already present (initial oxygen conc. = O_I). In the light bottle, this oxygen was used by the organisms for respiration, but at the same time, the phytoplankton organisms were producing more oxygen in photosynthesis. The final oxygen concentration in the light bottle, then, is the result of oxygen added by photosynthesis plus initial oxygen concentration, less the oxygen used in respiration. Since we know the initial amount of O_2, in order to estimate the total amount of oxygen produced by photosynthesis, we need to know how much is consumed in respiration. If there were no respiration, only photosynthesis, in the light bottle, the increase in oxygen would be greater than that actually observed. This is where the dark bottle comes in. In the dark bottle, only respiration occurred and the oxygen concentration decreased. Since both bottles have the same volume and were incubated the same length of time, the decrease in oxygen in the dark bottle is a measure of the total respiration in the light bottle. To find the total amount of oxygen produced in photosynthesis, then, it is necessary to make a few simple additions and subtractions. First, the initial oxygen concentrations of the water (before incubation) are subtracted from the final oxygen concentrations in both light and dark bottles. The value of the subtraction in the case of the light bottle is the *net photosynthesis*, or the amount of photosynthesis in excess of respiration. The value from the dark bottle is simply the respiration of the organism. *Gross photosynthesis* is then obtained by adding the amount of oxygen respired to the net photosynthesis; it is a measure of the total amount of photosynthesis that occurred during incubation.

Of course, the values for both gross and net photosynthesis will differ with different incubation depths, because light values change with depth and affect the photosynthetic process.

The second method for estimating primary productivity is the *^{14}C method*. This is today the preferred method of calculating primary productivity. In this method, radioactive ^{14}C is introduced into a bottle containing a measured volume of sea water containing phytoplankton. The ^{14}C is usually introduced as $H^{14-}CO_3$ (bicarbonate), since that is the major reservoir (90 percent) for CO_2 in the ocean. A known quantity of $H^{14-}CO_3$ is added and the bottle is then incubated for a period at the appropriate depth in the sea (or in special racks on deck). At the end of the incubation period, the bottles are brought up and the water filtered onto fine membrane filters or glass fibers, which catch all phytoplankton organisms. After these filters are dried, the amount of radioactivity on them is measured with a counter. The amount of ^{14}C which appears on the membrane (e.g., in the plants) compared to the amount initially added is a measure of the rate of productivity. If it is assumed that no ^{14}C that was fixed into organic compounds in photosynthesis was lost from the tissues of the plants or animals and no ^{14}C was taken up by any other organism, process, or particle, the difference between the ^{14}C recorded on the membrane and that initially added is a measure of the gross primary productivity. In order to correct for possible nonphotosynthetic uptake of ^{14}C, a dark bottle is often run along with the light bottle. Gross photosynthesis may then be obtained by subtracting the counts from the dark bottle from those of the light bottle.

Both of the above methods have errors and problems associated with them and both rest upon certain assumptions. In the case of the light-dark bottle method, the assumption is that conditions in both bottles are similar except for light. In fact, this may not be the case and will introduce error. Bacterial growth tends to be accelerated by the introduction of surfaces and hence, the increased bacterial numbers may affect the oxygen levels in the bottles. This problem is common to both ^{14}C and light-dark bottle methods. Another problem common to both methods is that phytoplankton organisms enclosed in a bottle undoubtedly do not act normally, and this may be reflected in their photosynthetic rate. For the light-dark bottle technique, a serious problem is comparative insensitivity when concentrations of phytoplankton are low or when waters are polluted or contain high bacterial concentrations. In the ^{14}C technique, the assumption made is that the phytoplankton do discriminate between ^{14}C and ^{12}C and therefore, a correction factor of 5 percent is included in the calculation. The chief additional problems with this method are that there may be cell breakage on the filters, allowing some ^{14}C to leak out, and that the method does not account for ^{14}C fixed in photosynthesis and then respired out as $^{14}CO_2$, which then goes through the filter. Since we cannot account for this latter problem, it is not always clear whether the method measures gross or net photosynthesis or something in between.

STANDING CROP

Standing crop may be plants, animals, or both. The standing crop (at any point in time) is the result of the difference between the factors tending to increase the numbers of individuals, namely reproduction and growth, and those factors tending to decrease biomass or numbers, death and sinking or lateral transport out of the area. If reproduction and growth rates are, or have been, high and death or removal low, then standing crop will be high and vice versa.

Standing crop of phytoplankton in the oceans is a difficult factor to measure accurately. Partly, this has to do with the patchy distribution of plankton organisms and the problems of sampling (see p. 53 for further discussion of this) and partly because of the problems inherent in the methods. The usual method of measuring standing crop is to measure some component common to all plants, usually the chlorophyll *a* content of a given volume of sea water. Since all plants must have chlorophyll to photosynthesize, the total amount of chlorophyll in a given volume of sea water should be a direct measure of the total biomass of plants present. Chlorophyll can be measured by taking advantage of its ability to fluoresce when excited by an appropriate wavelength of light or by directly extracting the chlorophyll from the plants with a chemical such as acetone and then measuring the amount of color in a colorimeter. The problems with this method are numerous. The method assumes, for example, that chlorophyll content is constant, but it is not; it varies among the phytoplankton species as well as among cells of the same species, depending on their condition. It also varies with time of day and light intensity. Finally, in the method that extracts chlorophyll before measurement, the very method of extraction may alter the chlorophyll.

Despite all the above problems, it is still the method of choice because the only other way of estimating standing crop of plants would be to actually remove the plants and measure their dry weight or amount of organic carbon; so far, no rapid and accurate method is available to make these determinations.

FACTORS AFFECTING PRIMARY PRODUCTIVITY

The functioning of the entire ecosystem of the world's oceans is dependent for energy almost exclusively on the photosynthetic activity of marine plants. Of these plants, by far the greatest amount of energy are fixed by the small phytoplankton confined to the thin layer of lighted surface waters of the oceans. Because of the great importance of these pelagic plants to all habitats in the seas, it is important to understand the conditions under which their productivity is either enhanced or inhibited; this is the first step toward reaching an understanding of how our oceans function.

PHYSICAL AND CHEMICAL FACTORS

If we consider terrestrial plants, there is a series of physical and chemical factors that affect their growth, survival, and productivity. Critically important factors would include light, temperature, nutrient concentration, soil, and water. Considering the freely floating phytoplankton, we find that the number of factors from the aforementioned list that are important is considerably reduced. Obviously, for plants suspended in water, neither water nor soil are of any importance. Temperature, which can vary in the terrestrial biosphere to such an extent as to influence productivity, has a considerably narrower range in the marine environment. Because this range almost never encompasses lethal limits for life, and because the changes are always gradual due to the inherent physical properties of water, temperature is of lesser importance to productivity in the ocean. Thus, of all the physicochemical factors affecting terrestrial plant production, only two, *light* and *nutrient concentration*, are of significance in limiting productivity in the ocean. Since, however, the phytoplankton are suspended in water and therefore affected by forces that act to move water around, and because both light and nutrients are also affected by the water masses, a new, very important factor not seen on land enters into consideration. This factor is a composite one, which we may call *hydrography*, and comprises all those factors that act to move water masses around in the oceans, such as currents, upwelling, and diffusion. It is the interplay of these three factors, light, nutrients, and hydrography, that provides the limits to phytoplankton productivity in the oceans and the geographical differences that we observe. It is now necessary to consider how each factor acts separately to limit or enhance production and then to consider how all three act in concert to produce the observed productivity patterns of the oceans.

Light. Photosynthesis is possible only when the light reaching the algal cell is above a certain intensity. This means that the phytoplankton are limited to the uppermost layers of the ocean where light intensity is sufficient for photosynthesis to occur. The depth to which light will penetrate into the ocean, and hence, the depth at which production can occur, is dependent upon a number of factors; these include absorption of light by the water, the wavelength of light, transparency of the water, reflection from the surface of the water, latitude, and season of the year.

A number of meteorological features influence light before it even impinges on the surface of the water (Fig. 2-13). Factors such as clouds and dust act to interfere with light in such a way that lesser amounts survive the passage through the atmosphere to impinge on the surface of the water. This reduces the available light initially, without reference to water conditions.

When light strikes the surface of the water, a certain amount of light is reflected back; the amount depends upon the angle at which the light strikes the surface of the water. If the angle from the horizontal is low, a large amount will be reflected. Conversely, the nearer the angle is to 90° (that is, perpendicular to the horizontal surface of the water), the greater will be the penetration and the lesser will be reflection. Light that is reflected is lost to the system, so from the standpoint of phytoplankton productivity, maximum penetration is the most desirable.

The angle at which the light strikes the surface of the water is directly related

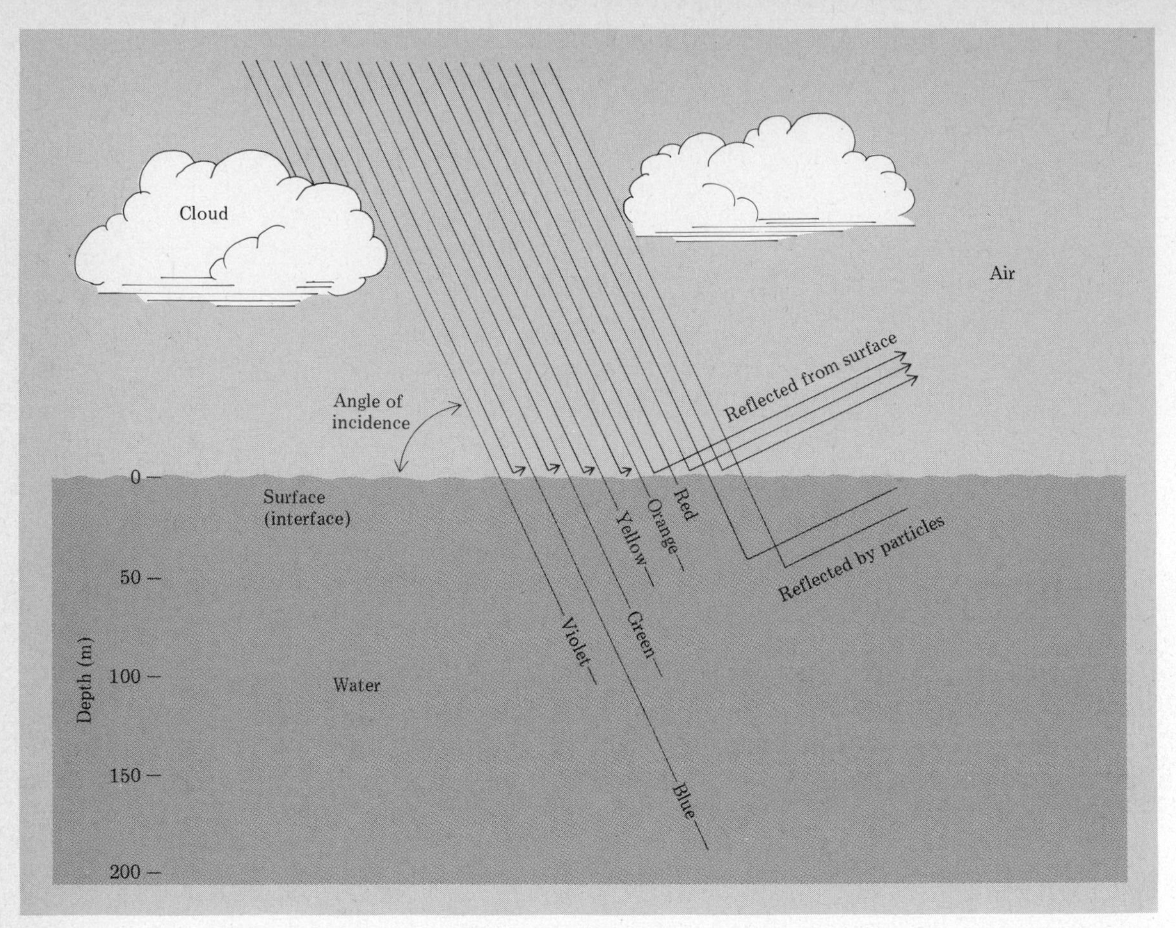

Figure 2-13 The fate of incident light in the ocean.

to the maximum height of the sun above the horizon. In the tropical regions of the earth, the sun is directly overhead at midday, or virtually perpendicular to the sea surface, giving an angle for maximum penetration of light into the water column. Furthermore, this position changes little with seasons of the year, so that light conditions are maximal all year long. As one progresses toward the poles from the equator, the sun may be directly overhead during the summer months, but may be far from this position at other times of the year (Fig. 2-14). The closer one approaches the poles, the greater becomes the difference in the height of the sun above the horizon among the seasons of the year. This reaches its maximum in the Arctic and Antarctic regions, where the sun is absent during the winter or is so low to the horizon that no light can penetrate the water (Fig. 2-14). The presence of ice in these areas also reduces light. This means that as one moves away from the equator, north or south, the amount of light penetrating the surface of the ocean, and hence available for use, changes significantly with season. As a result, the amount of photosynthesis that can occur also changes, being maximal in the summer and minimal in the winter. Only in the tropics is the light optimal at all seasons.

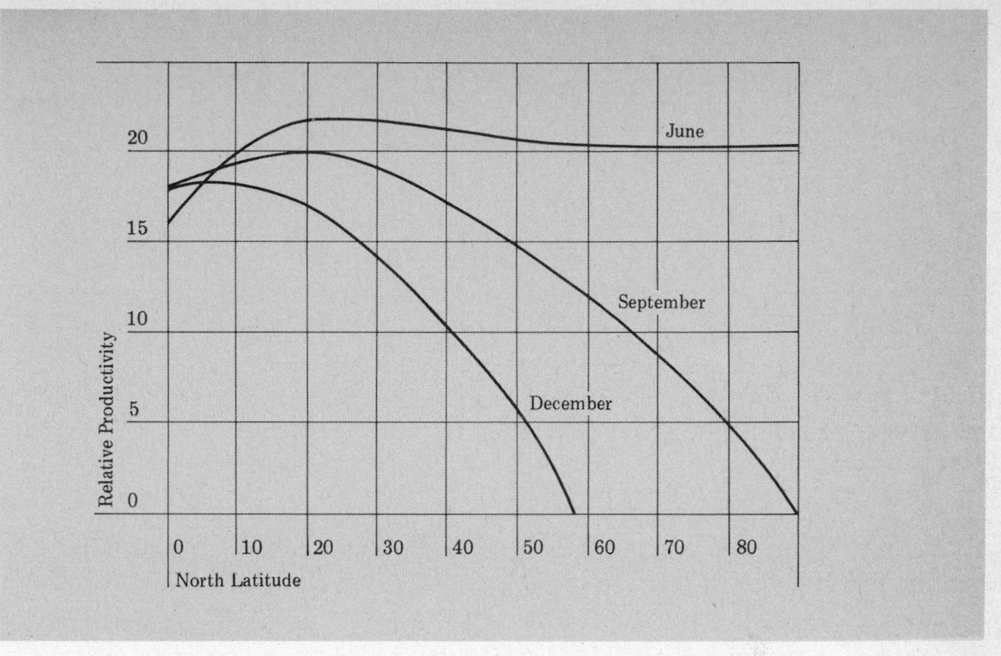

Figure 2-14 Relative amount of photosynthesis as a function of latitude for three seasons of the year. (From J. Ryther, Geographical variations in productivity. *In:* The seas, vol. II, edited by M. N. Hill, Wiley.)

The portion of light that enters the water column is subject to further reduction from two additional processes acting on it within the water. The first is reflection from various suspended particles in the water column. Suspended living or dead particles intercept the light and either absorb it or reflect it back to the surface. This light is unavailable for use and further reduces available light. Secondly, water itself absorbs light, making it unavailable for the plants. The amount absorbed is a function of wavelength and depth in perfectly clear, clean water. This absorption of light by water is the reason that the vast majority of the water masses of the ocean are dark below a certain level. Because of this absorption of light in the water, photosynthesis is automatically restricted to the thin, uppermost lighted layer. Water, however, does not absorb all wavelengths of light equally. Sunlight, as it arrives at the surface of the ocean, is composed of radiation in a spectrum of wavelengths measured in nanometers (1 nanometer = 10^{-6} mm). This spectrum includes all the visible colors ranging from violet to red (Fig. 2-13), or wavelengths from about 400 to 700 nm. As these wavelengths enter into sea water, the violet and red components are very quickly absorbed by the water. The green and blue components are absorbed less rapidly and hence penetrate most deeply. Eventually, they too are absorbed by the water. All the red and violet light is absorbed within the first few meters of even the clearest sea water, whereas 10 percent of the blue light may penetrate to more than 100 m under similar circumstances. It is important here to realize that for any given wavelength of light, a certain fraction of its remaining intensity is lost by absorption with each additional increment of water depth. Thus, even though blue and green light penetrate deeper into the water column, the intensity decreases with depth, and it is intensity that is needed by plants. Intensity is measured by the *extinction coefficient*, which is the ratio between the intensity at a given depth and intensity at surface. For pure water, it is 0.035.

The depth to which a given intensity of light penetrates is thus a function of the transparency of the water and the differential absorption by the water. Since differential absorption by water is constant, the changes in depth of effective light penetration are due primarily to particle concentration. Where there are large numbers of particles in the water, such as in coastal waters, the depth of light penetration may be severely reduced and the amount of light insufficient for photosynthesis below a few meters. On the other hand, in the clearest tropical water, where few interfering particles exist, light intensity may be sufficient for photosynthesis down to 100–120 m.

Phytoplankton photosynthesis is light dependent. The rate of photosynthesis is high in high light levels and decreases as the light intensity decreases. On the other hand, the rate of respiration of the phytoplankton cells is essentially constant at all depths. This means that, as the algal cells go deeper in the water column, the rate of photosynthesis declines as the light intensity decreases, until at some point, the photosynthetic rate equals the respiration rate. At this point, there is no net production of organic material. As defined earlier, this depth is called the compensation depth, and it is the depth to which 1 percent of the incident radiation penetrates. The compensation depth marks the lower limit of the euphotic zone and varies geographically from a few meters in very turbid inshore waters to depths of 120 m or more in the open waters of tropical oceans. It also varies on a seasonal basis in temperate areas where high turbidity during certain seasons (the plankton bloom) reduce it to a few meters (see p. 62), whereas at other times, the sparse populations of organisms increase it.

The compensation depth also changes with season, due to the change in the position of the sun, and it may be virtually absent during the winter months in high latitudes.

For most phytoplankton, the phytosynthetic rate is an almost linear function of light intensity over a range of moderate light intensities. Near the surface of the water column, however, where the light intensities are the highest, most species show a leveling off or a decrease in photosynthesis (Fig. 2-15). This is due either to an inhibition at high light levels or else to saturation of the photosynthetic apparatus such that it is not possible to increase rates. Different species have different curves of photosynthetic rate plotted against light intensity, giving different optimallight intensities for maximum photosynthesis. This may be of considerable significance in seasonal succession (see p. 78).

Nutrients. The major inorganic nutrients that are required by phytoplankton for growth and reproduction are nitrogen (as nitrate, NO_3^-) and phosphorus (as phosphate, PO_4^{2-}). Other inorganic and organic nutrients may be required in small or minute amounts but none have the profound effect on productivity that nitrogen and phosphorus do. These elements are of great importance partly

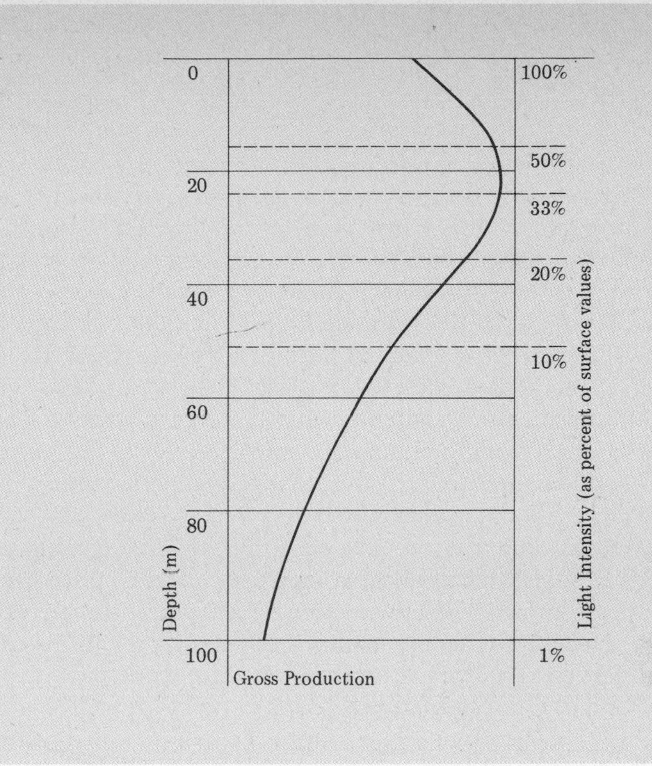

Figure 2-15 Inhibition of photosynthesis in the surface layers of water as measured by gross production at different levels in the water column. (After E. Steeman-Nielsen, Productivity, definition and measurement. Chap. 7. *In:* The seas, vol. II, edited by M. N. Hill, Wiley.)

because they occur in such small amounts in sea water. They are thus the limiting factors for phytoplankton productivity under most conditions, and the oceans of the world can be considered as nutrient-poor deserts when compared to terrestrial counterparts. For example, fairly rich agricultural land contains about 0.5 percent nitrogen in the upper meter of soil. In one cubic meter of soil, this amount of nitrogen is sufficient to permit the production of 50 kg of dry organic matter. Terrestrial plants, under ideal conditions, can produce several kilograms of dry organic matter in excess of their own need per year per square meter. Thus, the reservoir of nitrogen in that one cubic meter is sufficient to allow a plant to grow for many years. As a result, it is possible in terrestrial systems to have long-lived plants that continue to grow for many years. Forests are a good example, for they represent the accumulation of years of organic production. Of course, eventually, unless the nitrogen is renewed in the soil, nutrient exhaustion will limit growth.

On the other hand, the richest ocean water contains only about 0.00005 percent nitrogen, or 1/10,000 of the amount in soil. This means that a cubic meter of such water could permit a production of only 5 g of dry organic matter. In contrast to soil, however, where plant roots may penetrate only 1 m, oceanic phytoplankton should, theoretically, have access to the nitrogen in a column of water that extends as deep as the plants can exist. What is this depth and does it compensate for the greatly reduced nitrogen concentration? As we learned in the previous section, phytoplankton are limited in depth by light. If we assume the most ideal light conditions, we might have plant production extending down to as much as 100–120 m. This would mean that the plants would potentially have access to a 100 m^3 (1 m^2 × 100 m deep) volume and hence a potential production of 500 g of dry weight (5 × 100), or about 1/100 of the amount on dry land. In fact, production levels of that magnitude never occur and the maximum amount of organic production that can accrue under a square meter of sea water under ideal conditions is only about 25 g. Why the discrepancy?

Several factors act in concert to reduce the maximum productivity. In the first place, the above example assumed that the nitrogen content of sea water was constant throughout the water column. Such is usually not the case. The upper layers of water usually have a reduced concentration compared to lower waters, as will be shown subsequently. Also, due only to light absorption by water, the production at 100 m would be less than at 10 m. More importantly, the increasing numbers of phytoplankton cells have a profound effect. As the phytoplankton population grows in the upper 100 m of water, the plants themselves absorb more and more of the light. As a result, there is less and less light penetrating to the deeper levels. Less light means that the compensation depth begins to move upward and becomes shallower (Fig. 2-16). Thus, the original 100-m reservoir of nutrients is reduced and, as plants increase in numbers, more and more of the water column and nutrient supply become inaccessible to the phytoplankton, thus reducing total potential productivity. Finally, as the plants grow, they absorb the nutrients, and absorbed nutrients are not available to other plants. The result of all these factors acting together is to reduce the theoretical production. The reduction in the compensation depth and the absorption of nutrients is so great that it has been estimated that by the time a phytoplankton population reaches a density of 2 g/m^3, an original 100-m compensation depth would have been decreased to as little as 3.5 m, and all nitrogen would be transferred into plant bodies. Hence, in terms of production and of standing crop of plant material, the oceans appear to be deserts in comparison with fertile land. The rate of production between land and sea may not be significantly different, but the differences in nutrient concentration preclude that rate from continuing in the sea for long periods of time.

If, however, the nutrients are used up in the upper, lighted zone, an untapped reservoir remains in the water mass below the photic zone. Since this mass is several orders of magnitude larger in volume than that of the lighted (photic) zone, it represents a considerable reservoir of nutrients that could greatly enhance production. If this vast reservoir could be tapped, then high rates of production and large standing crops could be sustained on a long-term basis.

Unfortunately, certain physical factors prevent general access to this reservoir over most of the world's oceans. Water, as we learned in Chapter One, has different densities, depending upon its temperature and its salinity. Cold, saline

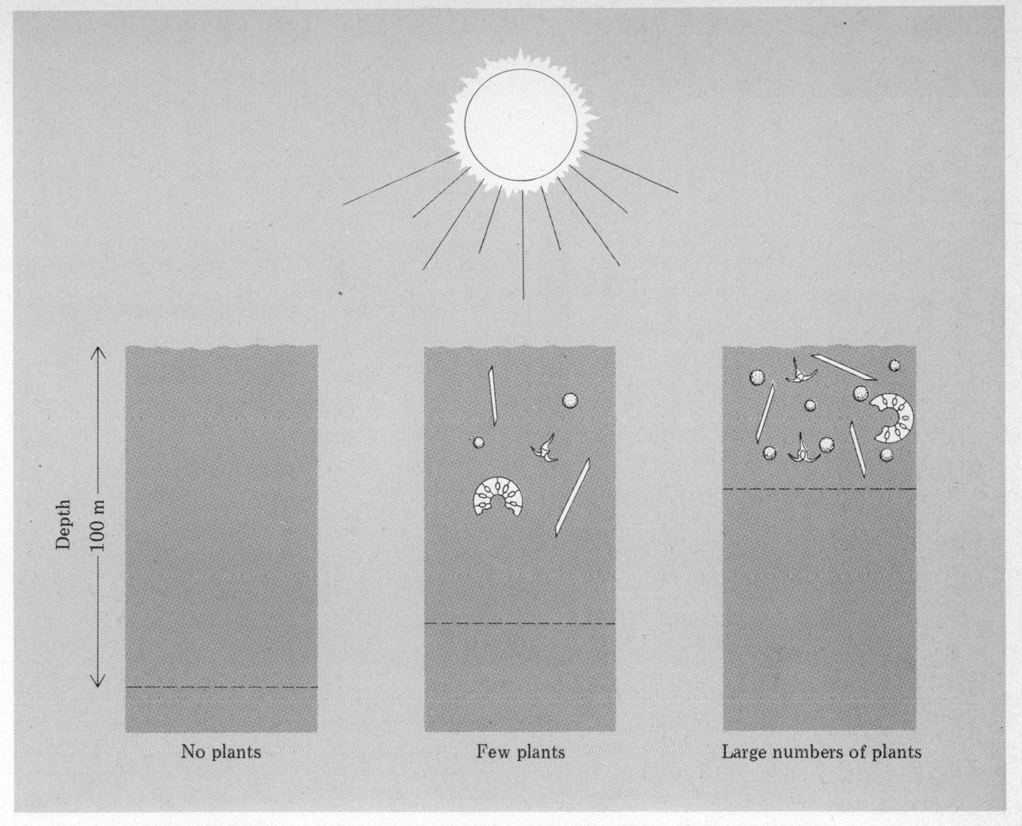

Figure 2-16 Change of compensation depth with increasing phytoplankton numbers. Dashed line indicates compensation depth.

water is more dense than warm, less saline water. Below the euphotic zone throughout the world's oceans, the water is cold, therefore dense. In the euphotic zone, the water varies in temperature. In the tropics, it is warm, therefore less dense, all year round. In the temperate zone, this water is warm in summer and cold in winter. In polar regions, it is cold and dense year round. In the tropics, the difference in density between the warmer upper and cold lower layers is of such magnitude that the two do not mix. Hence, the nutrients cannot reach the euphotic zone. In the temperate seas, the same situation prevails in the summer, but in the winter, the temperature of the two water masses becomes similar and mixing can occur. In polar regions, there are no significant differences in temperature and mixing can occur year round.

Another component is required to effect the tapping of the reservoir, and that is a mechanism to effect mixing. Even though the water masses approach the same density, some force must be exerted to mix the masses together. The force that is available to effect this over most of the world's oceans is the wind. Strong winds blowing over the surface of the water create the mixing force (Fig. 2-17). The wind is of sufficient strength to cause mixing of the water masses and to transport nutrient-rich water into the euphotic zone, when the two masses are similar in density and temperature. In polar seas this is possible at all times, and

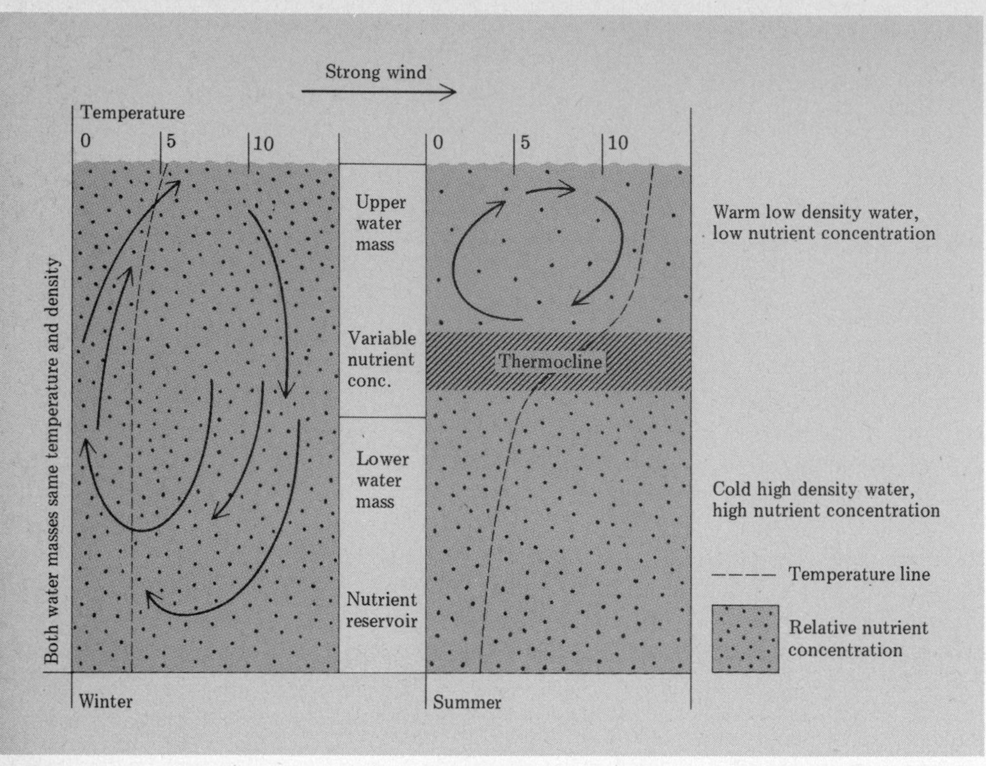

Figure 2-17 Thermal structure of temperate ocean water masses in two seasons, winter and summer, showing the relationships between nutrient concentration, temperature, density, and wind-induced mixing.

in temperate seas only in the winter. The great differences in density between the euphotic and aphotic zones all year round in the tropics and in the temperate zones in the summer are of such magnitude, however, that no wind, even from storms such as hurricanes, is of sufficient strength to mix the layers. This means that the upper layer remains nutrient-poor in the tropics at all times and in the temperate zone in summer, unless certain specific hydrographic conditions act to overcome the stability induced by density differences.

The special hydrographic conditions that act to bring nutrient-rich deep water up into the euphotic zone are upwelling, divergence of currents, and special currents. Upwelling occurs where the surface water moves away from shore and is replaced by nutrient-rich deep water brought to the surface. Persistent upwelling along the west coast of North America is responsible for high productivity in that region. Divergences are the result of transverse ocean currents flowing away from each other, bringing up deeper nutrient-rich water. Such a divergence occurs between the North Equatorial Current and the Equatorial Countercurrent in the Pacific; this is one area in the open tropical ocean where productivity is high, due to upwelled nutrients (Fig. 1-12).

Turbulence and Critical Depth. While vertical mixing brings up nutrients, it is also responsible for carrying phytoplankton cells down into the depths. As long as vertical mixing is confined only to the upper illuminated zone, the plant cells can be carried downward only a short distance and will remain where there is sufficient light for photosynthesis. When mixing includes the lower water mass, however, it is possible for the plant cells to be carried well below the compensation depth. If the mixing is especially vigorous, the phytoplankton may spend most of their time below the compensation depth and there will be no net production, because the time spent in the lighted zone in active photosynthesis is insufficient to fix as much organic matter as is used when they are below the compensation depth. This effect of turbulence has led to the development of the critical depth concept (Fig. 2-18). The *critical depth* is that depth at which total photosynthesis of the phytoplankton in the water column equals total respiration. It is different from the compensation depth, which is the depth at which incident light is reduced to 1 percent, or the depth at which the rate of photosynthesis equals the respiration rate. The critical depth is always deeper than the compensation depth, because we are dealing here with a vertical mixing process in which the algal population is circulated between lighted and unlighted areas. When it is in the light, it can photosynthesize in excess of respiration and build up organic matter. The critical point is the length of time

Figure 2-18 The relationship between the compensation depth and the critical depth. Critical depth is the depth to which the total phytoplankton biomass may be circulated and still spend enough time above the compensation depth to have a total production equal to its total respiration during the same time period. (Modified from D. H. Cushing, Productivity of the sea, Oxford biology reader no. 78, Oxford University Press.)

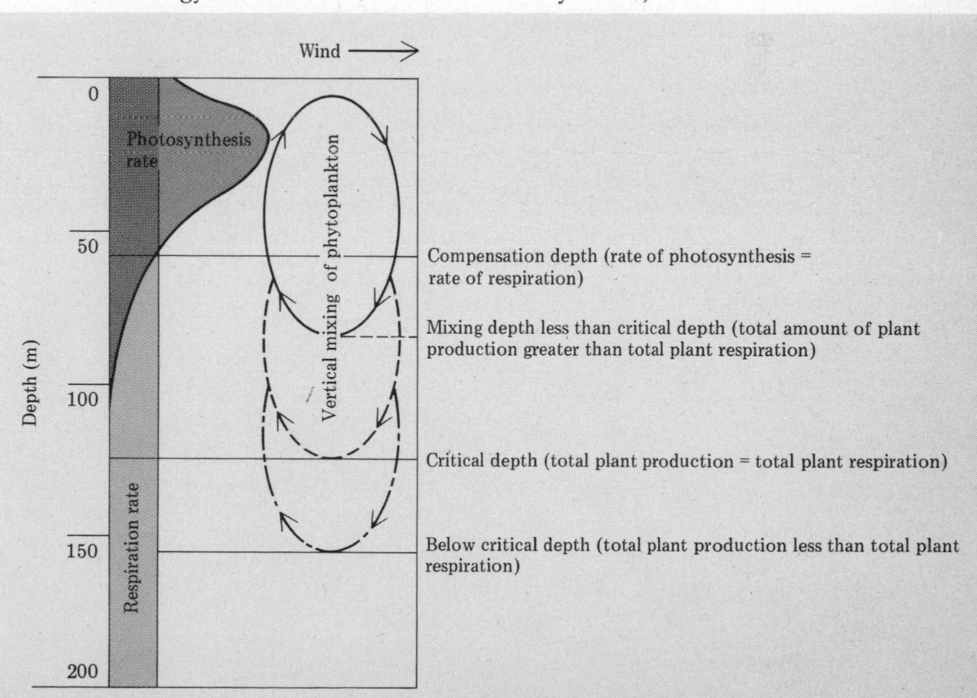

spent in each area; this in turn depends on how deep the mixing occurs. Deeper mixing means more time spent below the photic zone, where respiration uses up organic material more rapidly than the short time in the lighted zone can compensate for. Whenever the wind-driven vertical mixing is less than the critical depth, photosynthesis is greater than respiration and a net production occurs.

GEOGRAPHICAL VARIATIONS IN PRODUCTIVITY

Having now considered the role of light, transparency, and nutrients separately, it is possible to consider how these factors interact with each other to produce the geographical or latitudinal variations in productivity that we observe in the oceans.

Tropical Seas. In the tropical seas, the upper waters are well lighted throughout the year because the sun does not show marked changes in height above the horizon (Fig. 2-14). Light conditions are, therefore, optimal for phytoplankton production. At the same time, the continual input of energy from the sun maintains the surface layers of water at temperatures much higher than those in deeper waters. This means that there is a great difference in density between surface and deep waters and because of that, mixing does not occur. We say that such waters are *thermally stratified.* This thermal stratification extends throughout the year (Fig. 2-19). In the tropical seas, the sun, and hence light conditions, are optimal for high productivity, but because the sun's energy creates a thermal stratification in the water column that prevents mixing and the upward transport of nutrients, the productivity is low but constant throughout the year (Fig. 2-20). Tropical seas are very clear and have the deepest compensation depths, but they are that way because there are few phytoplankton in the water column due to the low nutrient content.

Temperate Seas. In the temperate zone seas, the amount of light varies seasonally. As a result, the amount of solar energy entering the water varies, which in turn alters the temperature in the upper water layers. The thermal structure of the water column thus changes seasonally (Fig. 2-19). In the summer months, the sun is high, days are long, and the upper layers heat up and become less dense than underlying layers. In other words, the water column is thermally stratified and no mixing occurs. In the fall, the amount of solar energy entering the water column decreases, days become shorter, upper layers cool and thermal stratification decreases. Finally, a point is reached where the temperature of the surface layers has been reduced to such an extent that the density of the layer is little different from that of the underlying mass. At this point, mixing can occur whenever sufficient wind is available. In winter, usually the storm season in the temperate zone, the sun is lowest on the horizon, solar energy input to the water is at a minimum, thermal stratification is at a

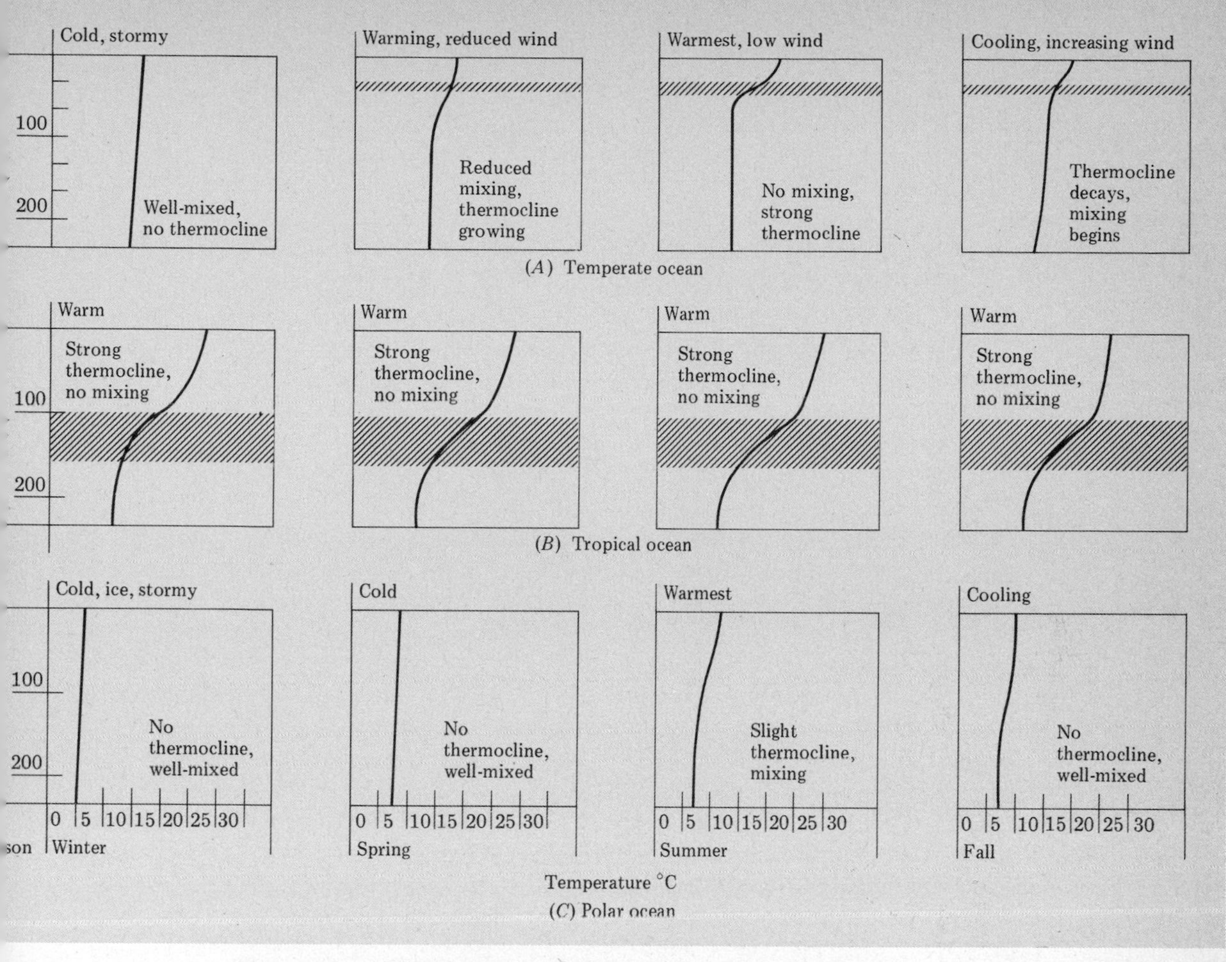

Figure 2-19 The thermal structure and extent of mixing in temperate, tropical, and polar seas during the four seasons of the year.

minimum or absent, and mixing occurs. With the onset of spring, the days become longer, the solar energy increases and the upper layers begin to rise in temperature, and the system moves toward reestablishment of thermal stratification.

In contrast to the tropics, all the major factors that affect productivity change seasonally in temperate seas. This is reflected in the change in production over the year (Fig. 2-20), with a major peak in spring, a lesser peak in the fall, and low productivity in winter and summer. We may explain this as follows: The low winter productivity is because of low light levels, due to the low position of the sun on the horizon and because the winter storms mix the isothermal water column and plant cells below the critical depth. In the spring, the light levels are improving and the increased light and solar energy increase the temperature of the upper layers. With increasing temperature come increasing differences in density between upper and lower layers. Under such conditions, the wind

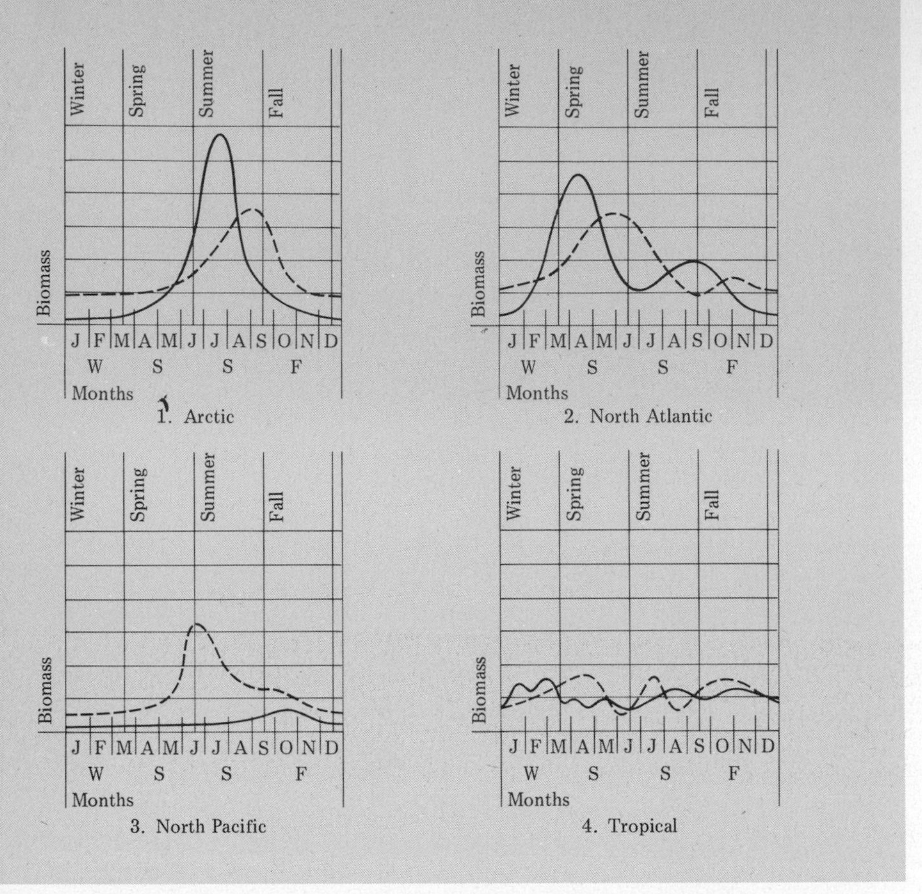

Figure 2-20 Summary of the seasonal cycles of phytoplankton and zooplankton in four different geographical areas. The solid line represents phytoplankton, the dotted one zooplankton. (Modified from T. R. Parsons and M. Takahashi, Biological oceanographic processes, Pergamon Press.)

cannot mix the water to as great a depth as in winter, and at some point, algal cells are no longer carried below the critical depth. Since nutrients in upper layers have been replenished during the winter mixing, conditions are good for phytoplankton growth, and we observe the *spring bloom* (Fig. 2-20). As spring passes into summer, the water column becomes more thermally stratified, mixing with lower levels ceases, and light conditions reach optimal levels. Because mixing ceases due to stratification, nutrient replenishment ceases and production falls, even though light levels are optimal. With the advent of fall, the thermal stratification begins to break up and nutrients are returned to upper levels. If, in the fall, the mixing alternates with calm weather such that the plants spend more of their time in the upper layers and are not carried below the critical depth, a small bloom will occur, because of the increased nutrients. This bloom declines in late fall, due to decreasing light and increased mixing. In the winter, low light levels and deep mixing of the water column keep productivity low.

Polar Seas. In polar areas, productivity is restricted to a single short period in the polar summer, usually July or August in the arctic (Fig. 2-20). At this time, the snow cover on the ice has disappeared, allowing sufficient light to enter the water through the ice to permit phytoplankton growth. In areas outside of the permanent ice pack, breakup of the ice at this time opens the leads, allowing sufficient light to enter the water, permitting phytoplankton growth. Following the single burst of production, the production quickly declines. Nutrients are not limiting and the water column is never strongly stratified (Fig. 2-19). The reason for the lack of production at other times is due primarily to light. Light intensity is insufficient for a fall bloom, and during the long winter, light is either absent or else prevented from reaching the water column by a layer of snow over the ice pack.

PRODUCTIVITY IN INSHORE AND COASTAL WATERS

The previously discussed latitudinal variations in phytoplankton productivity apply to open ocean areas away from the influence of land masses. The situation in the water masses adjacent to land is somewhat different. There are several factors that contribute to this difference.

In the first place, inshore waters tend to receive a considerable input of the critical nutrients, PO_4 and NO_3, due to runoff from the adjacent land (where, as we have seen, the nutrients are far more abundant). Because of this input, inshore waters do not show nutrient depletion. A second factor contributing to the difference is the water depth. In most inshore waters, water depth is shallow. In most cases, this depth is shallower than the aforementioned "critical depth." Thus, the phytoplankton cannot be carried below the critical depth in any kind of weather. Given sufficient light, production can occur at any time, even in the winter. A third factor is that shallow inshore waters rarely have a persistent thermocline, and hence no nutrients are locked up in bottom waters. A final influencing factor is the presence of large amounts of terrigenous debris in the water, which may act to restrict depth of the photic zone and thus counteract the high nutrient concentration and shallow depth.

Interaction of these factors on a latitudinal basis produces changes, both in the cycle of productivity and in the total production when compared to offshore areas (Table 2-1). In the temperate regions, instead of a bimodal production cycle, as seen offshore, production remains high all through the summer, because nutrients are not limiting due to runoff from land and lack of a permanent thermocline. Average production in inshore temperate waters on a yearly basis is higher than in offshore waters, due to the greater nutrient concentrations and lack of critical depth problems. That production is not even higher inshore is probably due to the presence of large amounts of light-absorbing debris in shallow water, and the fact that in offshore waters, production can occur to a greater depth. In other words, in shallow waters,

TABLE 2-1 Production for Several Different Geographical Areas

Location	Productivity in g C/m^2/year
Long Island Sound (temperate inshore)	380
Continental shelf	100–160
Tropical oceans	18–50
Temperate oceans	70–120
Antarctic oceans	100
Arctic ocean	<1

After Raymont, 1966.

production is limited to the upper 5–10 m, whereas offshore, it may go as deep as 50 m.

In tropical waters, the difference between inshore and offshore waters is particularly dramatic. Inshore tropical waters have a productivity as much as ten times that of offshore waters. This must be attributed in large part to the increased nutrient concentration inshore compared with offshore areas.

Inshore production is further enhanced through contributions of benthic plants, a component not present in offshore areas. The production of benthic algae and sea grasses is considered in Chapter Five, but is mentioned here for completeness.

GRAZING

As on land, the plant production is transferred into the food chain of the pelagic community through the grazing activity of herbivores. Because the phytoplankton are small, the herbivores are also small. A large number of invertebrate planktonic species are herbivores, but by far the dominant herbivores in all oceans are various species of copepods (Fig. 2-7). The dominance of the herbivorous zooplankton by copepods is so great (70–90 percent of biomass) that it is possible to consider only the copepods in discussing the effects of grazing on phytoplankton populations. Furthermore, in most seas, only from one to three species of copepod are numerically dominant.

Thus far, we have considered the changes in phytoplankton production as a function only of interacting physical factors. In the oceans, however, phytoplankton exist together with herbivorous copepods. What effect do they have on the phytoplankton populations? Evidence suggests a great effect.

Careful inspection of the numbers of phytoplankton cells per unit volume of water has shown that for temperate seas the rate of decline of phytoplankton numbers following the spring bloom is steeper than would be predicted based upon the nutrient decrease. Furthermore, nutrients never disappear completely from the upper waters, as would be expected if they were limiting. It can also be shown that the curve of copepod abundance in temperate seas is similar in form to that of the phytoplankton, only delayed slightly in time (Fig. 2-20). All of the above have suggested to some workers (Riley, 1946; Cushing, 1959; Harvey et al., 1935) that the copepods are responsible for the regulation of the phytoplankton populations.

The rapidity with which copepods can remove phytoplankton cells is really quite remarkable (Table 2-2). In experimental conditions, Fleming (1939) established a diatom population that initially had 1,000,000 cells per liter and divided once each day. It was grazed by a copepod population adjusted to remove 1,000,000 cells per day. Under such conditions, the diatom population would remain in a steady state of 1,000,000 cells per liter. When the copepod density was doubled (double the grazing rate), the effect was to reduce the diatom population to 27,000 cells per liter in five days. If the copepod numbers and hence grazing rate were increased five times, the diatoms were eliminated in five days. Since the rate of diatom reproduction remained unchanged in the above scenarios, it means that any increase in copepod numbers in the plankton could have a potentially significant effect on the numbers of diatom individuals. It is important here to understand that this decrease in numbers is independent of any change in the photosynthetic rate and, hence, the rate of primary production. It is quite possible to have a diatom population decreasing in numbers due to grazing, while actually increasing the rate of primary production. Of course, with fewer plant cells, the total amount of carbon fixed would decline with declining algal cell populations, but it would still be possible for the *rate* of fixation of carbon in each cell to increase. It is for this reason that the population density or standing crop of phytoplankton may be a poor or misleading measure of primary productivity.

It is thus possible to observe a situation in which the standing crop of phytoplankton is declining due to grazing, while the rate of primary production is increasing or remaining steady. Under such circumstances, the diatom population is dividing rapidly because the photosynthetic rate allows the buildup of protoplasm for cell division, but as fast as the diatoms reproduce,

TABLE 2-2 Changes in the Numbers of Individuals of a Phytoplankton Population with a Constant Rate of Reproduction When Subjected to Two Different Grazing Intensities

	Phytoplankton Population Density (cells/liter)		
Time in days	Initial grazing intensity	Grazing intensity doubled	Grazing intensity increased fivefold
0	1,000,000	1,000,000	1,000,000
1	1,000,000	487,000	62,000
2	1,000,000	237,000	3,900
3	1,000,000	106,000	240
4	1,000,000	56,000	15
5	1,000,000	27,000	1

Adapted from Fleming, 1939.

they are grazed off by the copepods and other herbivores. In this case, the major portion of the carbon fixed in photosynthesis appears not in the standing crop of phytoplankton, but in the standing crop of zooplankton. Does such a situation occur in the oceans and if so, what is the magnitude of the carbon fixed that appears in the zooplankton versus phytoplankton? Plankton ecologists have used various mathematical equations employing the concentrations of nutrient, water transparency, light availability, and photosynthetic rates to calculate what the productivity of a given body of water should be and then comparing it with what is observed. The results for many areas have been remarkable. For example, Hart (1942) reported that, in the Antarctic seas around South Georgia Island and in the English Channel, the standing crop of phytoplankton was 2 percent of calculated production. In oceanic areas of the Antarctic seas where grazing was more intense, Mare (1940) found that the standing crop of algae was only 0.5 percent of calculated production. In other words, the great majority of primary productivity resides not in the bodies of the plant cells, but in the zooplankton. These data also show that copepods are extremely efficient grazing animals and are, in fact, capable of reducing the populations of phytoplankton drastically in patches of the ocean. This is in contrast to the situation in natural terrestrial communities, where the grazers rarely remove all the vegetation and the standing crop of plants is high.

The ability of copepods to decrease phytoplankton populations to very low levels means that it is likely that the sudden decrease in the phytoplankton curve after the spring bloom could be caused by the grazing of increased numbers of copepods rather than as a result of nutrient depletion. This concept is supported by the finding that, in the Antarctic seas, where nutrients are not limiting due to upwelling, the phytoplankton population also drops precipitously after the spring bloom. Additional evidence comes from the fact that grazing, and hence breakup of phytoplankton cells during passage through the copepod guts, releases the nutrients, which are then excreted back into the water. Under these circumstances, nutrients would be constantly regenerated in the upper layers and productivity never halted. Additional support for grazing as a control mechanism for phytoplankton is furnished by the patchiness of plankton. In the oceans, there are dense patches of phytoplankton and clear patches with copepods. One explanation is that copepods overgraze the phytoplankton.

If the copepods are so efficient at reducing the plant population, how does the spring bloom ever get started? The answer to this is that the copepods reproduce more slowly than the phytoplankton. Thus, in early spring, following the winter low period, the phytoplankton are able to respond more rapidly to the improving light conditions and lowered turbulence. They reproduce rapidly and build up to high levels in absence of significant grazing. The copepods, also at low levels, also begin to breed, but their cycle is longer and they cannot build up large enough populations until after the phytoplankton have bloomed (see subsequent section on copepod cycles). After that delay, however, when their populations are high, they quickly graze down the plant population to the low summer level. Copepod levels also then decline because of less food, predation, and because copepod egg production is dependent on food available to the breeding adult.

The above situation prevails in temperate and subpolar seas. In the tropics, where nutrients limit population levels of plants to low levels at all times, the situation is different. In these areas, there are no real pulses in either the plant or copepod cycles and instead, there is a steady but inconspicuous consumption by copepods of the small phytoplankton crop.

Copepod Cycles. Because the copepods dominate the zooplankton in all world oceans, because they are the major herbivore, and because they can determine the form of the phytoplankton population curve, it is worthwhile to review the major aspects of their life cycle, which contributes to our understanding of the aforementioned grazing patterns. This section will concentrate on copepod cycles in temperate and polar waters.

Relatively few of the numerous copepod species have been thoroughly studied, but in most temperate and polar waters, only a few species predominate and for some of these, notably *Calanus finmarchicus* of the Atlantic Ocean, the life cycle is well known. Since other species of *Calanus* also dominate Arctic and North Pacific waters and have similar life cycles, a general scheme may be established.

All copepods have a similar pattern of development. The eggs hatch as naupliar larvae and pass through a series of six naupliar stages. They then enter another larval stage called the *copepodite*, passing through five of these before becoming adult.

For large copepods such as *Calanus*, the early winter months are spent in the last or fifth copepodite stage in the deeper water of the open ocean. Here they persist without feeding, living on fat or oil reserves in their bodies. They do not migrate at this time (see p. 75). Those copepodites that survive the winter begin their final molt to the adult stage in late winter or early spring. At this time, the total numbers of individuals are at a minimum for the year. After molting to the adult condition, the copepods migrate to the surface waters, and begin feeding and spawning. The time usually coincides with the vernal phytoplankton bloom. This first brood of eggs then goes through the naupliar and copepodite stages and itself matures and reproduces. This cycle of generations continues through the spring and summer and the number of these generations per year is dependent on the environmental conditions. Off Scotland and Maine, for example, there may be three complete generations. Further north, the number is reduced to two off Norway, and in Arctic seas, such as off Greenland, to one. Further south, off the middle Atlantic states and English Channel, four generations may be produced. Each cycle from egg production through adult to egg production again takes about two months under favorable conditions. The life span of these large calanoid copepods is about two months for those individuals born at this time (Fig. 2-21), which means that adults do not live long after breeding.

The last brood produced in late summer does not complete the developmental cycle. Instead, when they reach Stage V copepodites, these animals migrate down into deeper water and remain there through the winter, subsisting on stored food reserves, to repeat the process. For this particular generation, the life span is then much longer, perhaps five to seven months, depending on the area.

Egg production is influenced by the abundance of phytoplankton available as food. Thus, successive generations of copepods build up large numbers in the spring, feeding on the spring bloom of phytoplankton. Similarly, the decrease in copepod numbers in the summer reflects the decrease in phytoplankton due to excessive grazing by the large numbers built up in the spring.

Summarizing, the life cycle of *Calanus* copepods is about two months from egg through larval stages to adult and successive generations are produced through the spring and summer in temperate and polar waters. The number of generations decreases as one moves poleward, due to the increasing shortness of the season. Copepod numbers reflect phytoplankton densities.

While the above is the situation for the dominant large *Calanus* type copepods, there are other smaller copepods such as *Acartia*, *Pseudocalanus*, and *Microcalanus*, which co-occur with the larger *Calanus* and which differ slightly in life cycle. In these forms, the generation time is shorter and they tend to have more generations per year than the larger ones. Although they build up Stage V copepodites in the fall, there is no conclusive evidence of overwintering and it may be that these small forms breed throughout the year, albeit reduced in the winter and at other periods of low phytoplankton availability.

Figure 2-21 The life cycle of *Calanus finmarchicus* in the North Atlantic Ocean.

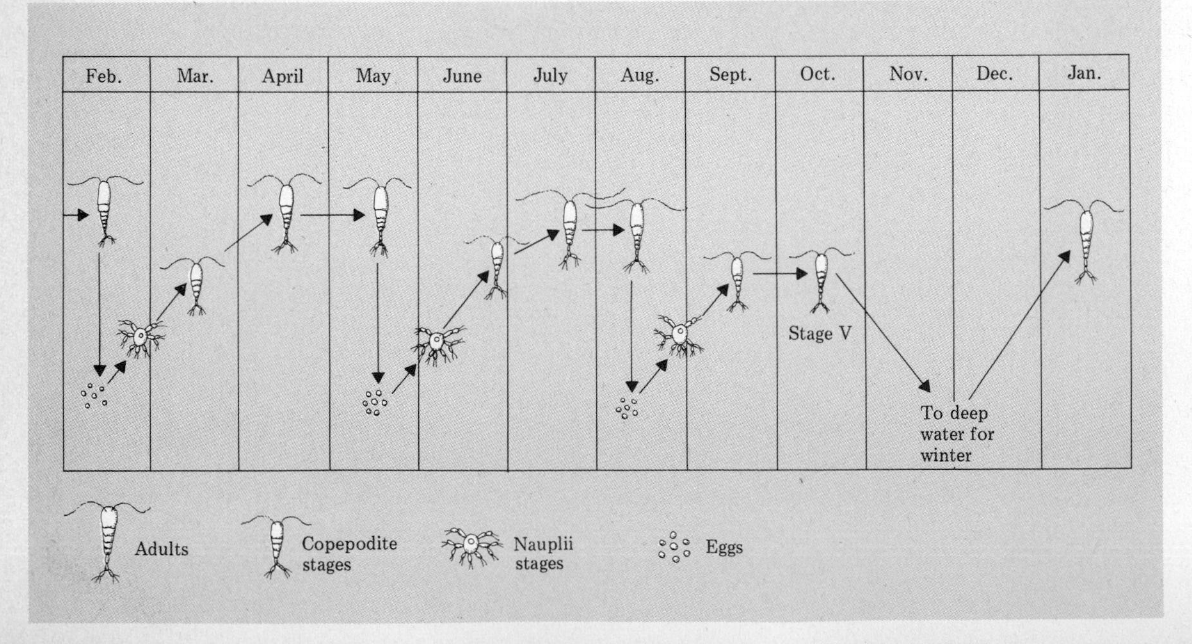

This latter type of life cycle with continuous breeding year round is found in many tropical copepods. Because of their small size, the constancy of the conditions and the continuous phytoplankton growth, development is rapid and many broods are produced per year. This spreads out the grazing pressure more or less equally over the year to coincide with unfluctuating phytoplankton populations.

In general then, the copepod cycles are maintained such that there is maximum utilization of almost all phytoplankton production, with little unused. However, a certain, but as yet undetermined, amount of energy is lost through fecal pellets and exoskeletons sinking out of the photic zone.

Vertical Migration. One of the more puzzling phenomena in the sea is that of the vertical migration of zooplankton. *Vertical migration* is the name given to the daily migration of certain zooplankton organisms down into the depths during daylight hours and up into the surface waters at night. It is entirely separate from the seasonal migration noted above, where copepods descended in the fall to deeper waters to overwinter.

Daily (diel) vertical migrations have been known for more than 100 years. What makes this phenomenon so puzzling is that it is difficult to offer a satisfactory explanation as to why such small organisms should expend significant amounts of their limited energy resources for this purpose. The distances traversed in these daily movements may be from 100–400 m. Since we are considering animals of only a few millimeters in size, this is the equivalent of humans walking 25 miles to work and then 25 miles back at night.

Vertical migration has been observed in all zooplankton taxa, but not all zooplankton migrate. With respect to taxonomic group, size, or feeding habits, there are few generalities that can be drawn from a comparison between those that do migrate and those that do not. This inability to find correlations make its occurrence even more difficult to understand.

It is now generally agreed that the major stimulus that initiates and controls vertical diel migrations is light. Vertical migrators respond negatively to light, moving themselves deeper as the surface light intensity increases. Conversely, they move toward the surface as the surface light intensity decreases. In a typical pattern, the migrators are found at the surface during the night; with the approach of dawn and advent of light, they begin their descent. As the light intensity increases through the morning, the animals move deeper, usually keeping themselves at a certain light intensity. At noon or whenever the intensity is maximal, they will be at the deepest position. Then, as the sun sinks through the afternoon and light intensity decreases, they begin their journey toward the surface again, arriving there after sunset and remaining until just before dawn. Since the advent of sonar or echo sounding gear in World War II, it has been possible to actually observe the above phenomenon by watching the trace formed by sound reflected from the bodies of these animals move up and down on the recorder. Their bodies form a layer in the water column (Fig. 2-22).

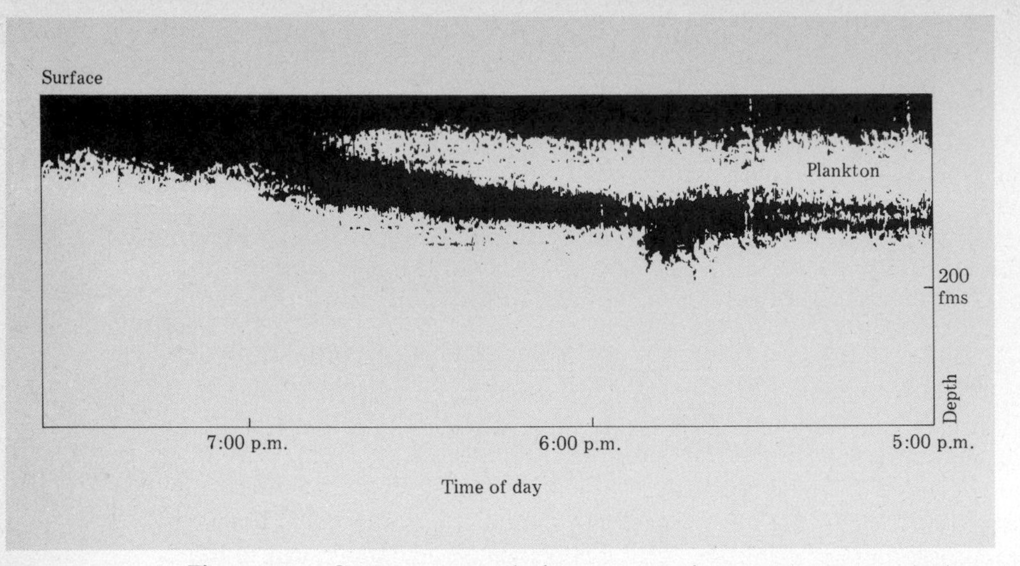

Figure 2-22 Sonogram record of movement of vertical migrating plankton. (After G. B. Farquhar, ed., Proceedings of an international symposium on biological sound scattering in the ocean, 1970, U.S. Government MC Report 005.)

Since the ocean is not uniform in its physical parameters in a vertical direction, other factors may modify the above light-regulated scheme. For example, temperature may also influence migration. Some migrators are limited by temperature change in the depth to which they go. In such cases, the maximum depth to which they descend is set by temperature, which overrides the stimulus of light. Yet another constraint is the presence of the bottom. Hence, in open ocean areas where submarine ridges or mountains rise to within daytime depths of the vertical migrators, the migrators may congregate at the bottom. Such concentrations of animals are a prime food source and may be responsible for the observation that banks and mountains are often densely populated with predators.

Although light is the main stimulus to which the animals respond and the guide to their positioning in the water column, it does not explain why the animals make the journey. What advantage can there be for undertaking such energy-demanding movements? Over the years, a number of explanations have been offered by marine ecologists. At the present time, three seem plausible, but there is no universal agreement among ecologists as to which, if any, of the explanations is correct.

The first explanation, which cannot be attributed to any single person, but has been reviewed by McLaren (1963), is that the zooplankton move from the lighted waters into dimly lit, deeper waters to avoid predation by visual, diurnal predators such as various fishes, cephalopods, and birds that forage at the surface. The problems with this explanation are that (1) the daytime depths of residence for many zooplankters still have sufficient light for visual predators to operate; (2) other zooplankton migrate deeper than would be required by this hypothesis; (3) many of the migrators possess and use bioluminescent organs which would seem to counteract the effect of the darkness; and (4) many potential predators are also migratory.

A second hypothesis for the occurrence of vertical migration, suggested by Hardy (1953), is that it allows the weakly swimming zooplankton to effectively change their horizontal position. In order to understand how this is possible, it is necessary to return to the previously described principle known as the *Ekman spiral.* As noted in Chapter One, as one moves downward from the surface waters, the speed and the direction of a surface current also change. The speed decreases and the direction of its movement changes to the right of the surface-most current (Fig. 1-11). This means that, as a zooplankton organism descends, it will encounter currents that not only move more slowly, but in slightly different directions. Once the zooplankter has descended into the deeper, slower-moving water, the surface currents will bring different water across above the animal. If the zooplankter then ascends again, it will find itself in a different water mass than that from which it descended. What is the value of this? Since zooplankton organisms lack the ability to swim against currents, they would automatically be confined to a given water mass in the ocean. Given their great ability to graze out the phytoplankton from an area, they would soon be without food and starve. However, by simply sinking into the depths and then rising again, they would come up in a different surface water mass, potentially ungrazed. This explanation is supported by the universal observation that phytoplankton and zooplankton are very patchy in occurrence. This mechanism might also be employed by zooplankton to avoid or leave very dense phytoplankton patches. Hardy and Gunther (1935) have suggested that a toxic substance is produced by the phytoplankton, which has deleterious effects on zooplankton.

A final reason for diel migration, suggested by McLaren (1963) and McAllister (1969), concerns production and energetics. This two-part hypothesis suggests that: (1) the rate of phytoplankton production is greater when subjected to discontinuous nocturnal grazing as opposed to a population subjected to continuous grazing; and (2) the zooplankton obtain an energetic advantage by spending a portion of their time in cold, deep waters rather than maintaining themselves constantly in the warm, upper waters. The second part of this hypothesis rests on the ability to demonstrate that more energy is saved by remaining in deeper, cold waters during the day than is expended in making the migration. It has been demonstrated that a neutrally buoyant zooplankter requires remarkably small amounts of energy to migrate, but it remains to be proven that the amount saved by remaining at depth is in excess of this amount. If this can be demonstrated, then vertical migration can be shown to be energetically profitable. Implicit in this energetics theory is that feeding in warm water increases the efficiency of feeding and of assimilation of food, while residence at depth improves efficiency of growth.

This hypothesis is supported by the observation that, in polar regions where the temperature differential between deeper and surface waters is lowest, vertical migration is also least often observed. Conversely, the strongest vertical migration patterns are observed in the tropics, where the greatest differential exists between surface water and depth.

The above argument may also be used to explain the overwintering of copepods at depth when phytoplankton is scarce. Such overwintering at low temperatures conserves energy.

In summary, it can be stated that vertical migration is a worldwide phenomenon of certain zooplankton. The stimulus is primarily light, although it may be modified by other factors such as temperature. It occurs most commonly in strongly thermally stratified seas and becomes suppressed or disappears where seas approach isothermal conditions. It is also absent in temperate seas during winter months. The hypotheses for its occurrence include avoidance of predation by visual predators, change of position in the water column, and as a mechanism to increase production and conserve energy. It is probable no single explanation for occurrence will be found true and that some combination of the above hypotheses is likely.

SEASONAL SUCCESSION IN PHYTOPLANKTON

In addition to the marked changes in abundance observed in the phytoplankton over the course of a year, there is also a marked change in species composition. This change in the dominant species from season to season is called *seasonal succession*. Under seasonal succession, one or more species of diatom or dinoflagellate dominate the plankton for a shorter or longer period of time and then are replaced by another set of species. This pattern is repeated yearly. This "succession" is thus different from the typical terrestrial ecological succession in which various plant associations replace each other until finally, a so-called "climax" community develops, which persists through time.

Figure 2-23 displays the dominant phytoplankton species at several geographic locations to emphasize the widespread occurrence of this successional phenomenon. Some successional changes occur even in polar (East Greenland) and tropical areas (Bermuda).

What are the causative factors of this phenomenon? Margalef (1963) suggests several, and they are herein reviewed. Considering that the seasonal succession is most often and clearly seen in temperate seas, which have a marked change in temperature over the course of a year, temperature has been suggested as a cause. Certainly, this may be one of the factors, but it is unlikely to be solely responsible, because certain dominant species recur at different temperatures. Furthermore, temperature changes rather slowly in sea water, but the replacement of dominant species often is very much more rapid.

Another suggested reason is the change in nutrient level over the year, differing concentrations favoring different phytoplankton species. While this factor may also contribute, observations suggest population changes are not closely correlated, that phytoplankton populations rise and fall much more quickly than nutrient concentrations change.

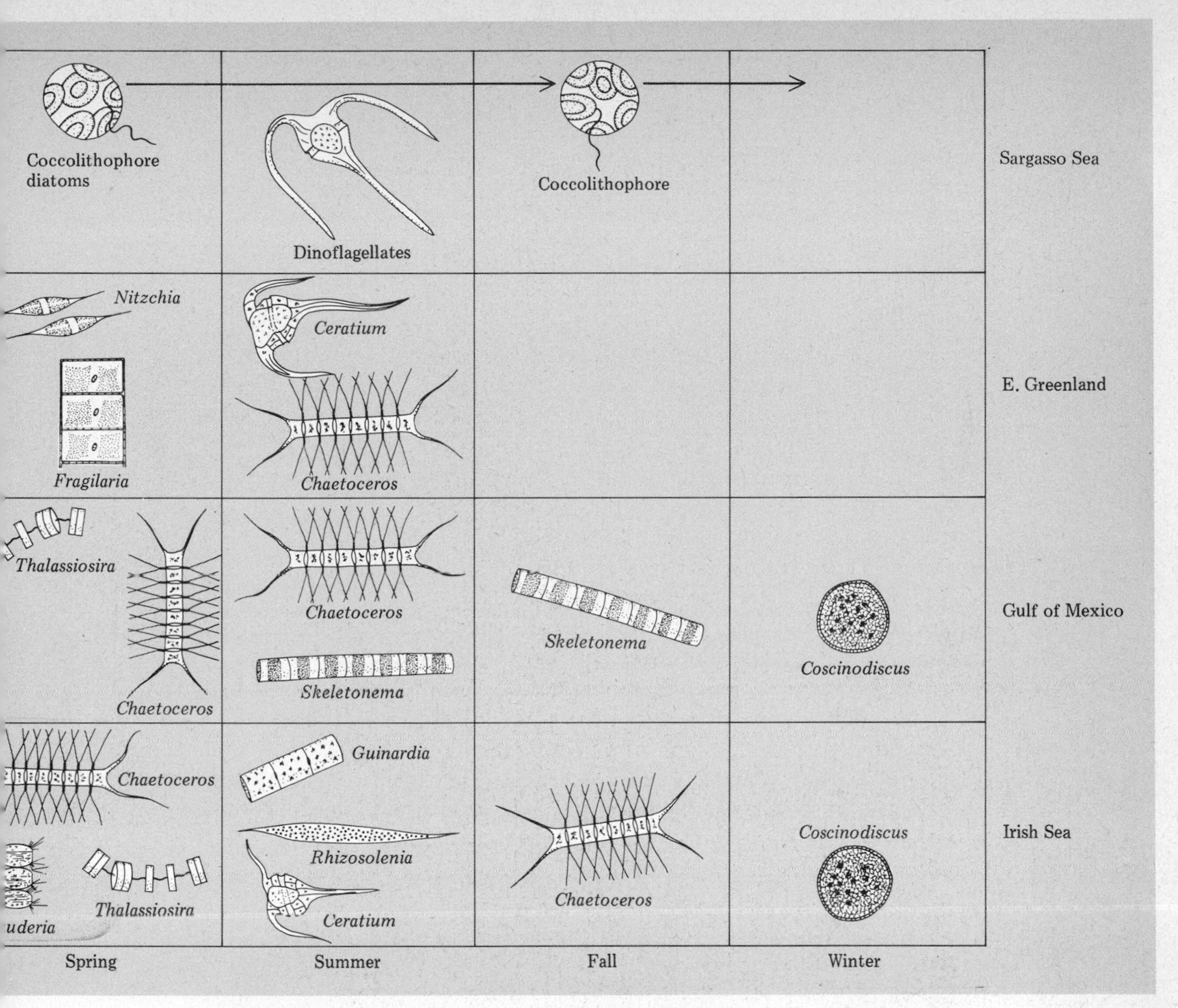

Figure 2-23 Seasonal succession of phytoplankton in four different geographical areas.

A final explanation for succession is that it is the result of *biological conditioning* of the sea water by the organisms in it. By this we mean that one group of organisms, when dominant in the water column, secretes or excretes very small quantities of organic compounds that have a definite effect on other organisms, either inhibiting or promoting their growth. At the same time, these organic compounds or *metabolites* could affect the very organisms producing them, making the water either more or less suitable for their own existence.

These organic metabolites could, and probably do, include a number of different classes of organic compounds. Some are likely toxins such as those released by the dinoflagellates during "*red tides*," which inhibit growth of other plants. In such cases, the population explosion of dinoflagellates is so great that the water becomes brownish red in color from the billions of dinoflagellate cells. At the same time, although each secretes only a very minute amount of a toxin, the massive dinoflagellate numbers cause the toxin to reach concentrations where many creatures are killed. It is this same toxin that can be concentrated in certain filter-feeding organisms such as clams and mussels and that renders them toxic to humans at certain times and places.

Yet another class of metabolite is the *vitamins*. It is now known that certain phytoplankton species have requirements for certain vitamins and that there are considerable differences between species as to requirements. The B vitamins, especially vitamin B_{12}, thiamine, and biotin, seem to be the most generally required. It is possible that some species may be unable to thrive until a particular vitamin, or group of vitamins, is present in the water, and these are produced only by another species. Hence, a succession could occur whereby first the vitamin-producing phytoplankter is present and then is followed by the vitamin-requiring species.

Other organic compounds that may act as inhibitors or promoters of various species include amino acids, carbohydrates, and fatty acids.

Although it is currently suspected that these organic metabolites may have an important role in the phenomenon of species succession and it has been demonstrated in the laboratory that phytoplankton species vary both in their ability to produce necessary vitamins and in their requirements for such in order to grow, evidence is still sketchy as to their real role in the sea.

A general pattern of succession for temperate waters has small-celled, rapidly dividing diatom species initiating the spring bloom. These forms are then succeeded by larger-sized diatoms in the spring and early summer and are themselves followed by an increasing number of dinoflagellates in late summer and early fall. Diatoms predominate again in the winter. In tropical regions, dinoflagellates tend to dominate throughout the year.

References

Bougis, P. 1976. Marine plankton ecology. Elsevier, N.Y. 355 pp.

Cushing, D. H. 1975. The productivity of the sea. Oxford biology reader no. 78. Oxford Univ. Press, London. 16 pp.

Cushing, D. H. and J. J. Welsh (eds.). 1976. The ecology of the seas. Saunders, Philadelphia. 467 pp.

Denton, E. J. 1974. Buoyancy in marine animals. Oxford biology reader no. 54. Oxford Univ. Press, London. 16 pp.

Fleming, R. H. 1939. The control of diatom populations by grazing. Journal du Conseil Permanent International pour l'Exploration de la Mer. 14:210–227.

Fraser, J. 1962. Nature adrift, the story of marine plankton. G. T. Foulis and Co., Ltd., London. 178 pp.

Hardy, A. 1956. The open sea: Its natural history. Vol. I, The world of plankton. Houghton Mifflin, Boston. 334 pp.

Hill, M. N. (ed.). 1963. The sea. Vol. 2, The composition of sea water, comparative and descriptive oceanography, IV, Biological oceanography. Wiley, N.Y. pp. 347–485.

Newell, G. E., and R. C. Newell. 1973. Marine plankton, a practical guide. Hutchinson Educational, Ltd., London. 244 pp.

Parsons, T. R., M. Takahashi, and B. Hargrave. 1977. Biological oceanographic processes, 2nd ed. Pergamon Press, N.Y. 332 pp.

Raymont, J. E. G. 1963. Plankton and productivity in the oceans. Macmillan, N.Y. 660 pp.

Raymont, J. E. G. 1966. The production of marine plankton. Advances in Ecological Research 3:117–205.

Russell-Hunter, W. D. 1970. Aquatic productivity. Macmillan, N.Y. 306 pp.

Smith, DeB. L. 1977. A guide to marine coastal plankton and marine invertebrate larvae. Kendall-Hunt Publ. Co., Dubuque, Iowa. 161 pp.

Wickstead, J. H. 1965. An introduction to the study of tropical plankton. Hutchinson and Co., London. 160 pp.

Wimpenny, R. S. 1966. The plankton of the sea. Elsevier, N.Y. 426 pp.

Chapter Three
OCEANIC NEKTON

In contrast to the plankton, the *nekton* comprises those organisms that have developed powers of locomotion so that they are not at the mercy of the prevailing ocean currents or wind-induced water motion. They are, in fact, capable of moving at will through the water. Most are large animals and include the largest and fastest-moving organisms in the sea. Whereas the plankton is dominated by a host of invertebrate animals, the nekton are predominantly vertebrates; among these the fishes are the most numerous, both in species and individuals, but representatives of every vertebrate class except amphibians are found here.

Whereas nekton in the larger sense includes all those organisms capable of sustained locomotion against the water motion without consideration as to habitat, in this chapter we will consider only those nektonic animals that are distributed in the epipelagic zone of the open ocean, the *oceanic nekton*. The ecology and adaptations of nekton associated with inshore waters, such as kelp beds and coral reefs, and with the deep sea will be considered in subsequent chapters. The special adaptations of the oceanic nekton are different from those of nekton living in deeper waters and inshore waters and hence, we justify this division of the nekton.

Because these animals are fast swimming and often wide ranging in the immense open seas of the world, they are difficult to study at sea and are virtually impossible to keep captive under natural conditions. We know very little about most aspects of the ecology or life history of these forms. In the dearth of such field data and virtual absence of laboratory information, we have been forced to infer much of our understanding of their ecology through study of anatomical and physiological characteristics of captured specimens.

Most students will have little opportunity for direct contact with living oceanic nekton forms except for fishes and occasional seal, sea lion, or porpoise sightings or strandings; however, the immensity of the ocean area inhabited by this fauna and its important role in the ecology of the upper layers of the seas of the world make it important to study. Some nektonic fishes, such as the tunas, sustain a major world fishery. This fauna also includes perhaps the most popular group of marine organisms in the public eye, namely, the whales, over which considerable controversy has raged in recent years with respect to their continued survival. To perpetuate the existence of the whales requires some knowledge of the habitat in which they live.

COMPOSITION OF THE OCEANIC NEKTON

The oceanic nekton is composed of a wide variety of various bony fishes, sharks, and rays as well as lesser numbers of marine mammals, reptiles, and birds. The only invertebrates that can be considered nekton are the various cephalopod mollusks.

Several different groups of fishes may be recognized in the nekton. These are, first of all, those fishes which spend their entire lives in the epipelagic. These fishes may be termed *holoepipelagic*. Included here are certain sharks (thresher shark, mackerel shark, blue shark), most flying fishes, tunas, marlins, swordfish, saury, oarfish, and others (Fig. 3-1). These fishes often lay floating eggs and have epipelagic larvae; they are most abundant in the surface waters of the tropics and subtropics.

A second group of oceanic fishes is termed *meroepipelagic*. These fishes spend only part of their life cycle in the epipelagic. This is a more diverse group and includes those fishes that spend their adult lives in the epipelagic but spawn in inshore waters (herring, whale shark, dolphin, halfbeaks) or in fresh water (salmon) (Fig. 3-2). Still other fishes enter the epipelagic only at certain times, such as at night, when certain of the deeper water fishes such as lantern fish migrate to the surface layers to feed. This latter group is, however, considered in the next chapter.

Most fishes spend their early life in the epipelagic but their adult lives in other areas. These juveniles form a persistent part of the epipelagic fauna, but can best be considered meroplankton, since they are restricted in their ability to move. They will not be considered further here.

The second major component of the oceanic nekton is the marine mammals. Nektonic marine mammals include the whales (order Cetacea) and the seals and sea lions (order Pinnipedia). Other marine mammals exist, such as manatees and dugongs (order Sirenia) and sea otters (order Carnivora), but these animals are not pelagic because they occupy inshore waters at all times. They will not be considered here. The role of sea otters in the economy of inshore waters is discussed in Chapter Five.

Nektonic reptiles are almost exclusively turtles and sea snakes. Marine iguanas exist in the Galapagos Islands, and salt water crocodiles inhabit many island areas of the Indo-Pacific, but these again are littoral animals, only rarely venturing out of sight of land. The fossil record indicates that, during the Cretaceous period some 60 million years ago, marine reptiles were much more

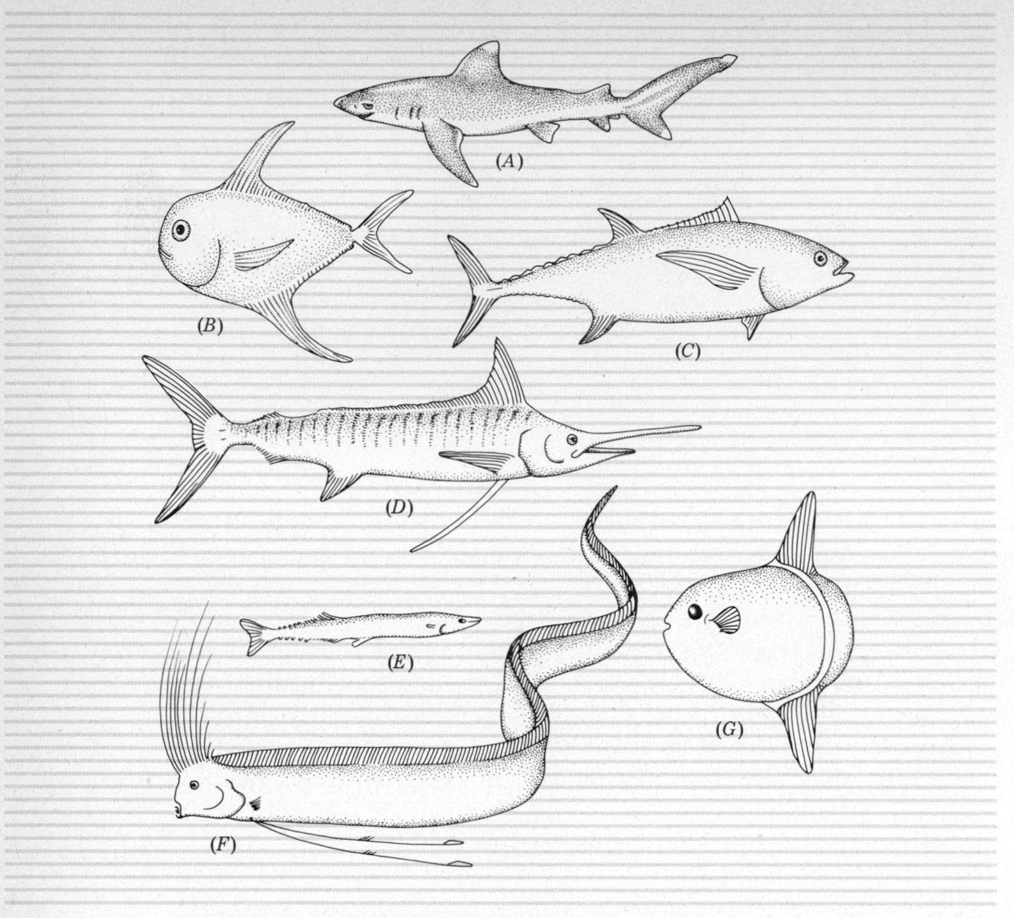

Figure 3-1 Some characteristic holoepipelagic fishes. *(A)* White tip shark, *Pterolamiops longimanus.* *(B)* Spiny marine brean, *Taractichthys longipinnus.* *(C)* Yellowfin tuna, *Thunnus albacares.* *(D)* Striped marlin, *Tetrapterus audax.* *(E)* Saury, *Cololabis saira.* *(F)* Oarfish, *Regalecus glesne.* *(G)* Ocean sunfish, *Mola mola.* (Redrawn from N. V. Parin, Ichthyofauna of the epipelagic zone, trans. from the Russian by the Israel Program for Scientific Translations.)

common and varied than today. At that time, large plesiosaurs, ichthyosaurs, and mosasaurs roamed the warm seas.

Technically, most seabirds are not nektonic, since they fly over the open ocean and not through it, but they do enter into the economy of these waters and are considered here. Perhaps the only group of truly nektonic birds are the flightless penguins of the Southern Hemisphere. But cormorants and other seabirds do dive for food and spend a good deal of time as swimmers.

ENVIRONMENTAL CONDITIONS

The environmental factors acting in the epipelagic zone inhabited by the nekton are, of course, the same as those reported in the preceding chapter, for the plankton (see pp. 55–66) and include light, temperature, density, and currents.

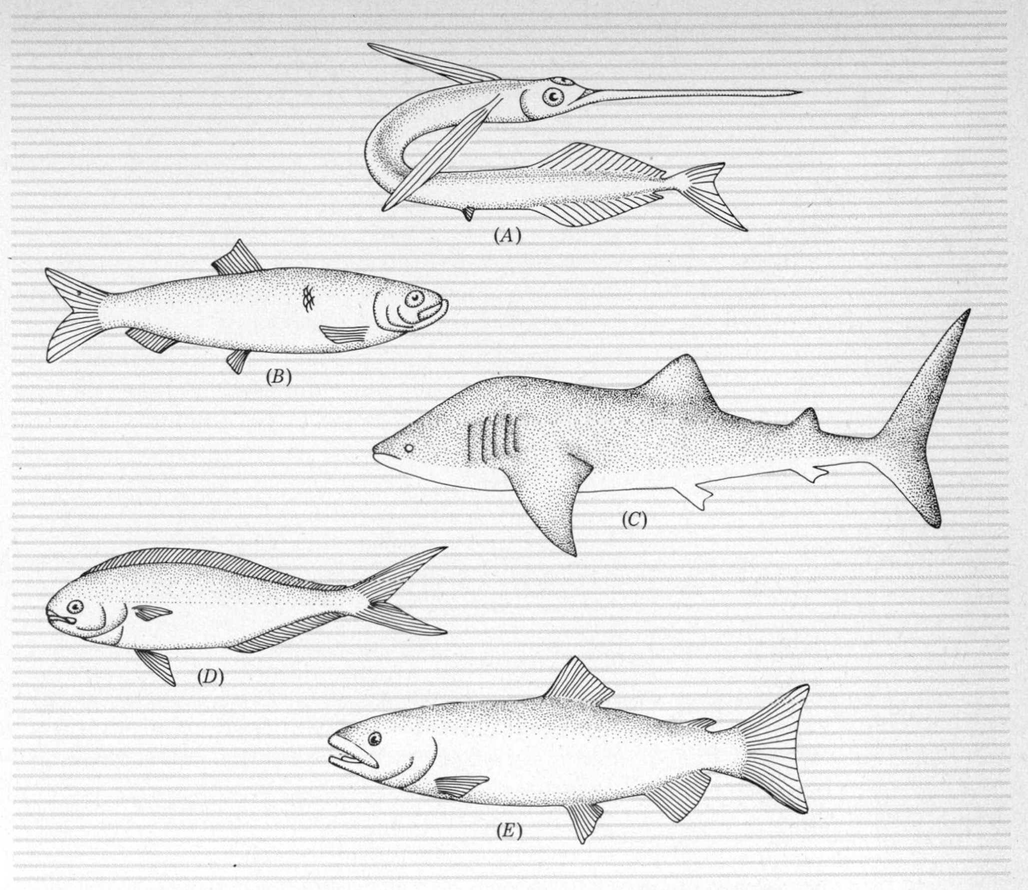

Figure 3-2 Some characteristic meroepipelagic fishes. *(A)* Ribbon halfbeak, *Euleptorhamphus viridis.* *(B)* Herring, *Clupea harengus.* *(C)* Whale shark, *Rhincodon typus.* *(D)* Dolphin, *Coryphaena hippurus.* *(E)* Salmon, *Oncorhynchus keta.* (Redrawn from N. V. Parin, Ichthyofauna of the epipelagic zone, trans. from the Russian by the Israel Program for Scientific Translations.)

However, the relative importance of different factors in selecting adaptations and life strategies of the nekton may well be different. It is important to note that the perception of this environment is very different for a large, fast-swimming fish or mammal than it is for a small copepod.

Several environmental conditions should be emphasized as being of considerable significance to the nekton and for which adaptations have evolved. In the first place, this is an area with a profound "three-dimensionality." Secondly, there is no solid substrate anywhere, and thus, the animals are always suspended in a transparent medium with no shelter that may be employed as a refuge from potential predators. Because of the general lack of structures in this area, there is nothing visible for animals to take bearings on when moving from place to place horizontally. Finally, the lack of a substrate means that there is no firm support for the animals, most of which have flesh somewhat denser than the surrounding sea water.

Three-dimensionality, combined with the lack of any obstacles, facilitates the evolution of adaptations for great mobility. Great mobility and ability to cover large distances in turn lead to development of nervous and sensory systems that will provide and process the information necessary to navigate in the area, to find and capture food, and to avoid predation.

Similarly, the lack of any shelter, coupled with the large size of most nekton forms, leads to development of faster swimming speeds as one of the few options left to escape predators and, in turn, to enable them to capture food. This lack of shelter also leads to development of camouflage as another option. Continuous suspension of the more dense body of nektonic animals in the less dense water leads to the progressive development of various adaptations toward keeping afloat.

ADAPTATIONS OF OCEANIC NEKTON

BUOYANCY

Perhaps the most significant adaptations of nektonic animals are those that serve to keep the animals suspended in the water and propel them through the water at a fast rate of speed. It is these adaptations that have facilitated the strong convergence that we observe in morphology of otherwise very different animals.

The first order of business in living in the epipelagic is to stay afloat. This, as we saw in the last chapter, is also the primary concern of the plankton. Most nektonic animals have densities very close to that of sea water. Since living tissues are generally heavier than sea water, the fact that many of these large animals are nearly neutrally buoyant means that there must be lower density areas in their bodies that counteract the higher density of most tissues.

Most fishes have a *gas* or *swim bladder* in their bodies. This structure, which forms perhaps 5–10 percent of the fish by volume, serves to counteract the denser flesh of the fish and thus give the fish a neutral buoyancy (Fig. 3-3). Most fishes can regulate the amount of gas in the gas bladder and thus change their buoyancy state. Two types of gas bladder systems are known: the *physostome,* in which there is an open duct between the gas bladder and the esophagus, and the *physoclist,* in which there is no duct. Physostome fishes move gases in and out of the bladder via the duct by gulping air at the surface, but usually fill the swim bladder through the gas gland and a *rete mirabile* system. The rete mirabile is a network of small blood vessels that branch off a large vessel. Physoclist fishes also secrete gas into the swim bladder through the gas gland and rete mirabile, but in order to void the gas, they must do so through a special gas absorptive organ called the *oval.*

In very fast-moving fishes (*Sarda, Scomber*) that also move vertically in the water column, the gas bladder cannot adjust quickly enough to compensate for pressure changes and maintain neutral buoyancy, and it thus becomes more of a

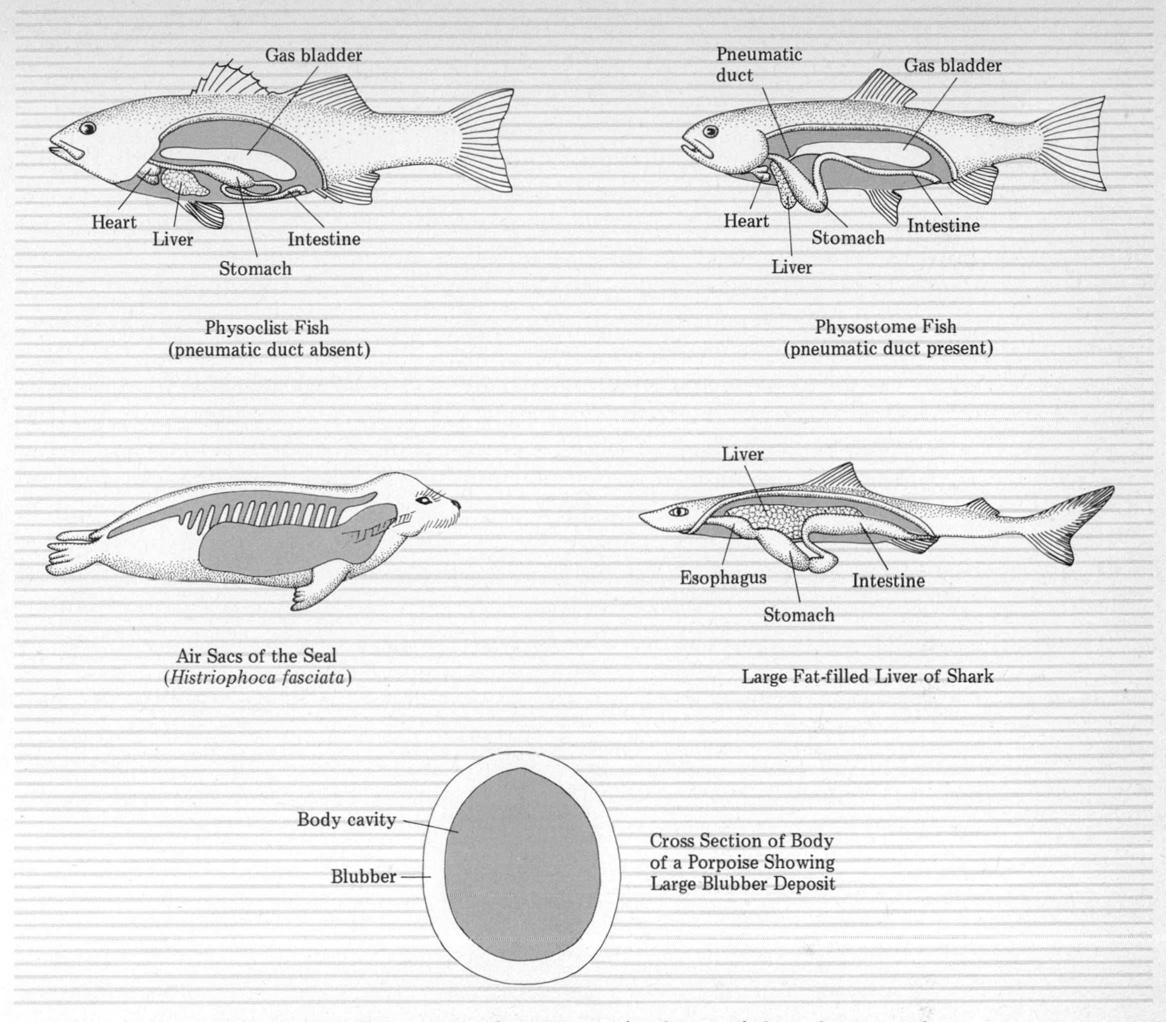

Figure 3-3 Buoyancy adaptations of nektonic fish and mammals.

liability. Hence, Denton and Marshall (1958) note that the fastest fishes tend to lack gas bladders and maintain buoyancy by other means.

Gas-filled cavities in the form of lungs also help sustain the neutral buoyancy of all the air-breathing nektonic animals. Some marine mammals have accessory air sacs (Fig. 3-3). In the case of these animals, they can regulate their buoyancy through the amount of air they hold in their lungs. Birds also have additional air sacs. In most diving seabirds (excluding penguins), air trapped under the feathers provides the greatest amount of buoyancy. Among marine mammals, the sea otter and fur seals also use air trapped in the dense wool undercoat for buoyancy.

Another mechanism of ensuring neutral buoyancy is replacement of the heavy chemical ions in the body fluids with lighter ones. We observed this in plankton as well (see pp. 48–49). The only nektonic animals in which this occurs are the squids. Squids tend to have body cavities in which heavy sodium ions are replaced with lighter ammonium. As a result, an equal volume of body fluid is less dense than the same volume of sea water. Whereas this is a common and often used mechanism among plankton, it is rare among nekton, simply because for it to be really effective, the amount of ammonium-dominated fluid must be large. Large, fluid-filled spaces give the animal a rotund appearance and reduce the mantle cavity, which thus markedly decreases its ability to move rapidly.

Increasing buoyancy through reduction of the amount of bone or other hard parts is not a viable option for these animals, since strong and rigid skeletal frameworks are needed for the muscle system to operate effectively to propel the animals through the water. This is, of course, a marked contrast to the plankton.

Still another mechanism to increase buoyancy is to lay down lipid (fat or oil) in the body. Lipid is less dense than sea water and thus can be important in contributing to buoyancy. Large amounts of lipid are present in many nektonic fishes, primarily those that lack swim bladders, such as sharks, mackerels (*Scomber*), bluefish (*Pomatomus*), and bonito (*Sarda*). Presumably, the lipid, at least partially, makes up for the absence of the swim bladder. Lipid may be deposited in various body parts such as muscles, internal organs, and body cavity, or else be localized in one organ. In pelagic sharks, for example, the lipid is localized in the much enlarged liver, and in many shark species fat deposition in the liver is part of their development; thus, some young sharks start out negatively buoyant and gradually, during growth, become neutral or positively buoyant, as fat stores build up in the liver. In marine mammals, the lipid usually is deposited as a layer of fat just below the skin, where it serves not only to aid buoyancy, but to insulate against heat loss.

In addition to these static means of maintaining or increasing buoyancy, certain nekton animals also show some hydrodynamic mechanisms for producing additional buoyancy during movement. Perhaps the most common are the formation of lifting surfaces in the anterior region, usually represented by pectoral fins or flippers, and the presence of an heterocercal tail. A heterocercal tail is one in which the upper lobe is the larger and better developed. In this system, the fin or flippers act as moveable ailerons, just as on an airplane, and when canted at the appropriate positive angle, cause the individual to rise in the water column as the tail provides the propulsive thrust (Fig. 3-4). If the tail is epicercal, its motion also provides a portion of the upward thrust as well. In some forms, the lift provided by the fins or flippers is aided by the whole anterior portion of the body, which can also be inclined at an angle to provide lift. The best development of these dynamic buoyancy forces occurs in those forms which are negatively buoyant.

In general, there is a tendency among the more primitive fishes to have hydrodynamic (water movement) adaptations to create lift, whereas the more advanced forms seem to evolve static or passive means to achieve neutral buoyancy. This is because less energy is expended to obtain neutral buoyancy than to constantly have to move to achieve the lift necessary to keep a body

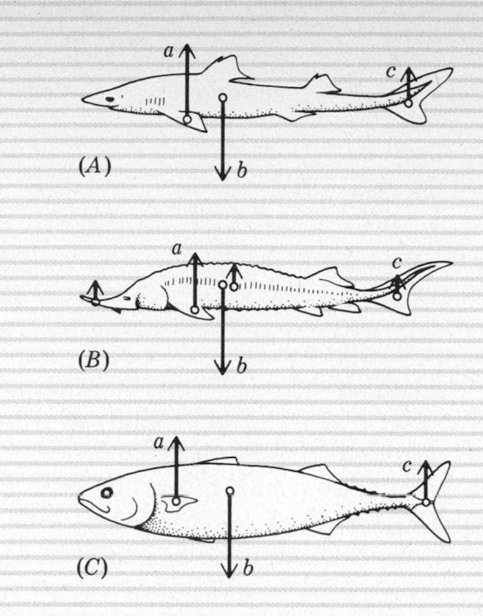

Figure 3-4 Various tail and fin shapes in fishes showing the lift provided. *(A)* Shark *(Squalus acanthias)* with slightly epicercal tail. *(B)* Sturgeon *(Acipenser stellatus)* with epicercal tail. *(C)* Mackerel *(Scomber scombrus)* with homocercal tail. *a,* Force created by the pectoral fins or analogs; *b,* force due to residual weight; *c,* vertical component of propulsive force furnished by the caudal fin. (Modified from Yu. G. Aleyev, Nekton, Dr. W. Junk b.v., 1977. Reproduced by permission of Dr. W. Junk BV.)

suspended in the water. Because of air-filled lungs, air-breathing mammals tend to be nearly neutrally buoyant.

LOCOMOTION

A second, and related, group of adaptations in nektonic animals are those having to do with movement of the animals through the water. These adaptations can be divided into two groups: those necessary to create the propulsive force, and those that reduce the resistance of the body to passage through the water.

The force necessary to propel a nektonic animal through the dense water is created by some part or parts of the animal's body. The most common means of producing forward movement is by undulatory motions of the body or of the fins. Virtually all nektonic fishes show these types of motion. In the undulatory mechanism, the animal moves itself forward by sweeping the posterior part of the body and fins from side to side. This throws the body into a series of short curves that start at the head and move down the body. This side-to-side motion is created by alternate contractions of the body musculature, first on one side and then on the other. When one analyzes the forces that result from such movements in water, it is found that the forward component is the strongest, and hence the animal moves in that direction (Fig. 3-5). A similar series of movements is used by the whales, but the flexural motions are up and down rather than side to side. The result is the same forward motion. The up-and-

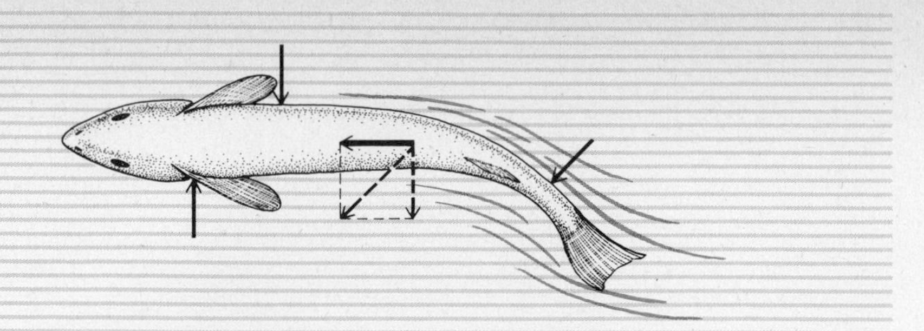

Figure 3-5 Mechanism of swimming in a fish. As the tail moves back and forth, the water exerts a force (diagonal arrows) and the forward component (heavy arrow) moves the fish ahead. The lateral component (dashed arrow) tends to move the fish to the side, but this is prevented by the force exerted by the water (vertical solid arrows). (Modified from K. F. Lagler, J. E. Bardach, and R. R. Miller, Ichthyology, Wiley.)

down motion of the forward part of the body is a compensation for the undulatory motions of the laterally displaced tail flukes. The shorter and stubbier a fish is, the less flexibility it has in its body musculature. The result is that the propulsive wave is less apparent, is confined to the posterior part of the body, and progresses most quickly along the body. The speed of a fish is related to the speed with which the muscle contraction waves pass down the body, but also depends on other aspects of the body shape; in general, short, stubby fishes are faster than long, narrow fishes (Fig. 3-6).

The seals and sea lions, when swimming, are not able to reach the speeds seen in whales. Eared seals (sea lions) swim using their front flippers as paddles, but earless seals use their webbed hind feet, spread vertically like the double caudal fin of fishes.

In the fastest of the nektonic fishes, however, the undulatory wave has been completely suppressed. These are the tuna and allies. In these fishes, the propulsive force is generated exclusively by the lunate caudal fin. This fin moves rapidly from side to side propelled by alternate contractions of the strong body muscles acting through tendons, which, in turn, run like pulleys across the bones of the narrow caudal peduncle to insert on the caudal fin base.

Another mode of propulsion is to employ undulatory movements of fins. In this mode of locomotion, the body remains stationary and the fins are moved in one of several ways to cause forward motion. This form of locomotion is slower than the previous ones. Examples of this type are found in the rays such as the manta ray (*Manta hamiltoni*) (see Plate 7), certain squids (*Todarodes*), and ocean sunfish (*Mola mola*) (Fig. 3-7). In most fishes, however, the lateral fins are used for maneuvering or lift, as in sharks with a heterocercal tail.

Except for cetaceans and sea snakes, the common form of propulsion among marine air-breathing vertebrates is through paddling movements of either the fore or hind limbs or both. The limbs of marine turtles, seals and sea lions, and penguins are all modified into flat, paddle-shaped appendages, which the animals use to move themselves through the water, as we would employ oars (Fig. 3-8). The speed of movement through the water using paddles depends on the frequency of the propulsive stroke. In organisms that employ few strokes,

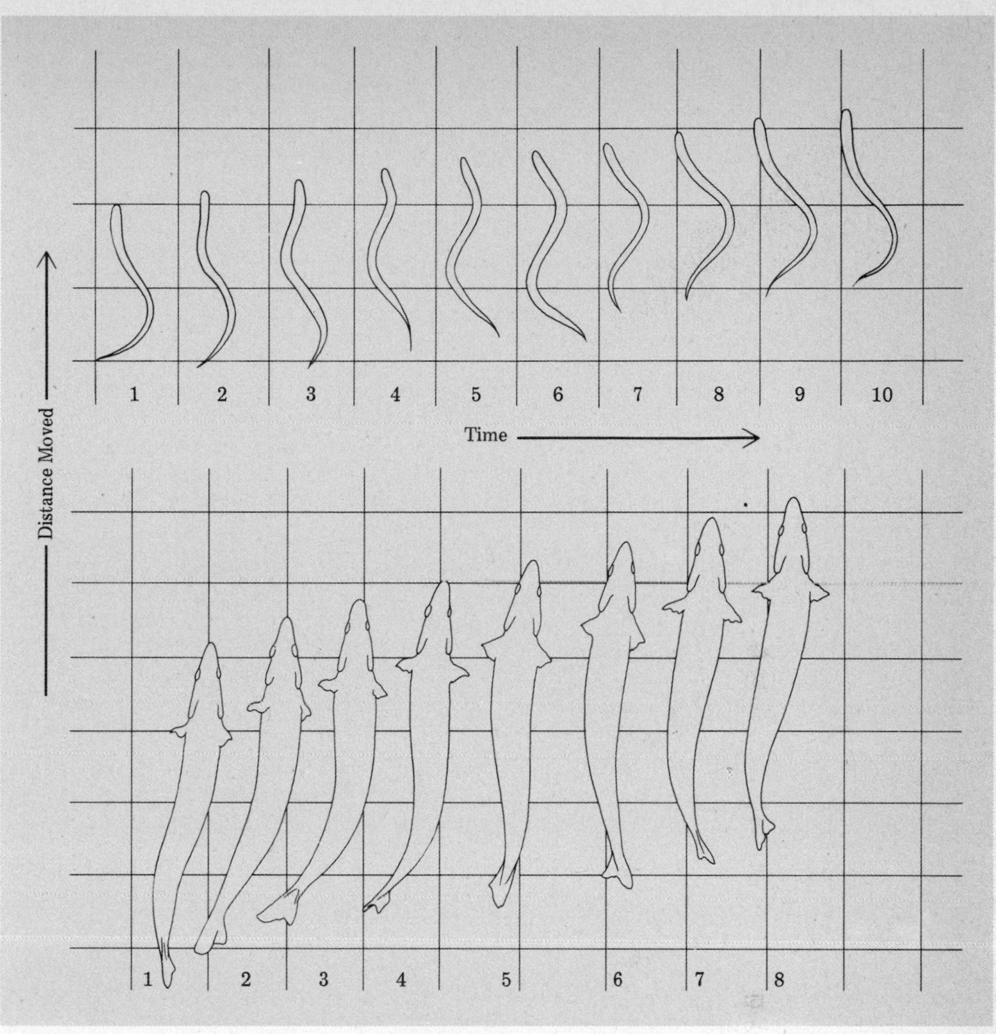

Figure 3-6 Propulsion in an elongated fish (top) and a stubby fish (bottom). (Adapted from Yu. G. Aleyev, Nekton, Dr. W. Junk b.v., 1977. Reproduced by permission of Dr. W. Junk BV.)

Figure 3-7 Fishes using undulatory fins for locomotion. *(A)* Ocean sunfish *(Mola mola)*. *(B)* Manta ray *(Manta hamiltoni)*. (Not to scale). (From Yu. G. Aleyev, Nekton, Dr. W. Junk b.v., 1977. Reproduced by permission of Dr. W. Junk BV).

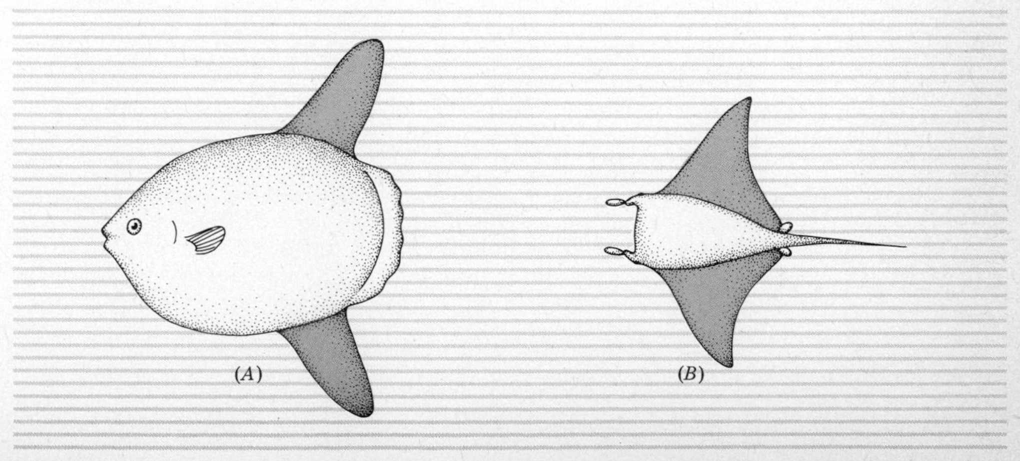

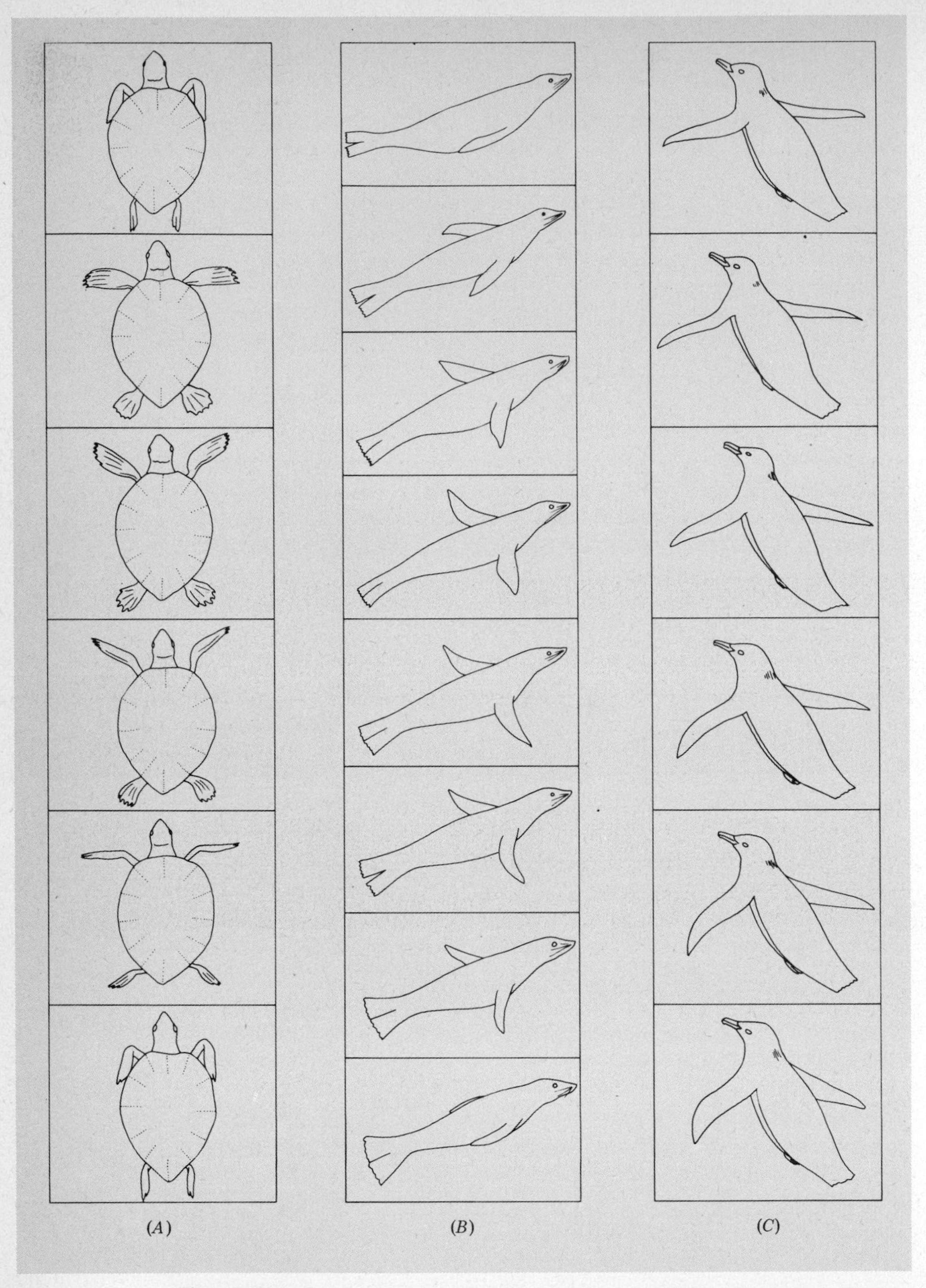

Figure 3-8 Swimming by paddle motions in three classes of marine vertebrates. *(A)* The green turtle *(Chelonia mydas)*. *(B)* The sea lion *(Arctocephalus pusillus)*. *(C)* The penguin *(Pygoscelis papua)*. (Redrawn from Yu. G. Aleyev, Nekton, Dr. W. Junk b.v., 1977. Reproduced by permission of Dr. W. Junk BV.)

such as turtles, speed is low; in others, such as penguins—where up to 200 strokes per minute have been recorded by Brooks (1917) in *Pygoscelis papua*—the rate of movement may be very fast (10 m/sec for the 200 strokes/minute).

The final type of propulsion is "jet propulsion" using water. This form of propulsion is the domain of the oceanic squids. It is capable of producing very rapid movement.

SURFACE OF RESISTANCE AND BODY SHAPE

Because water is a very dense medium, it is difficult to move an object through it, and it is even more difficult to move an object at a fast rate of speed. To propel a body through water requires considerably more energy than to move the same body through air. It requires less energy to move an object through water if it has a shape that reduces the surface of resistance to the water to a minimum. Since nektonic animals must move and since they have limited, at least finite, energy resources, it follows that a major set of adaptations must be made toward reducing the surface of resistance and hence, the drag of the body.

There are several types of drag or resistance to movement which must be overcome. *Frictional resistance* is proportional to the amount of surface area in contact with the water. Minimum frictional resistance is produced by a spherical object, which, of geometric shapes, has the minimum surface area for a given volume. If, however, one wishes to move a nektonic object through water, another type of resistance to movement becomes of great importance. This is *form resistance*, where the drag is proportional to the cross-sectional area of the object in contact with the water. In such a case, the spherical object presents a very large area and hence is unsuitable as a shape for nektonic animals. To reduce form resistance to a minimum, the shape should be relatively long and thin, such as a thin cylinder or wire. The final type of resistance that must be overcome is that of *turbulence*. Turbulence results when the smooth layered flow of a liquid over the surface of a body is disrupted and thrown into vortices or eddies, which in turn increase resistance (Fig. 3-9). This type of resistance is

Figure 3-9 Frictional forces on various shaped objects moving through water. *(A)* A flat disc. *(B)* A cylinder. *(C)* A teardrop-shaped object. All have the same cross-sectional shape and area but differ in their three-dimensional form. The teardrop shape is best for a fast-moving fish and has about one-fourteenth the resistance of that on disc *A*.

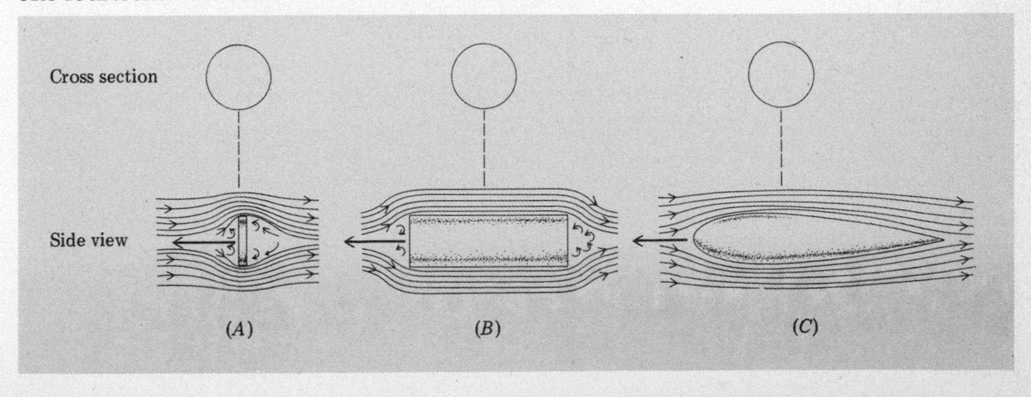

reduced in bodies having the shape of a teardrop, somewhat blunt in front and tapered to a point posteriorly. This shape is also the best compromise to minimize the frictional and form resistance as well, so we find that most of the fast-moving nektonic animals have bodies that approximate this shape (Fig. 3-10).

In addition to body shape nektonic animals have other adaptations to reduce drag. These adaptations generally fall under the category of streamlining the external surface of the body so that there are no protuberances that may break up the smooth flow of water over the body and hence, increase drag. To achieve this, species that move rapidly, such as tunas, are structured so that almost all normally protruding body structures are recessed into depressions or grooves from which they may be elevated only when needed. Thus, in fast-moving fishes, eyes, although large, do not protrude beyond the sides of the body. Pectoral and pelvic fins fit into grooves except when in use, and body scales are reduced or absent (Fig. 3-10). Similarly, in marine mammals, the hair is lost or reduced in length, for it produces more drag than bare skin. Mammary glands are flattened, and genitalia of the males do not protrude beyond the skin, except when in use.

DEFENSE AND CAMOUFLAGE

By far the strongest adaptations seen in nektonic animals are those that relate to the ability to achieve rapid forward movement in the water column. These adaptations are of such importance that they take precedence over, or preclude, any adaptations with respect to defense against potential predators, if such adaptations would necessitate modifications requiring decreased locomotory

Figure 3-10 Three views of a tuna showing the adaptations necessary for fast movement. *(A)* Front view. *(B)* Side view. *(C)* Top view.

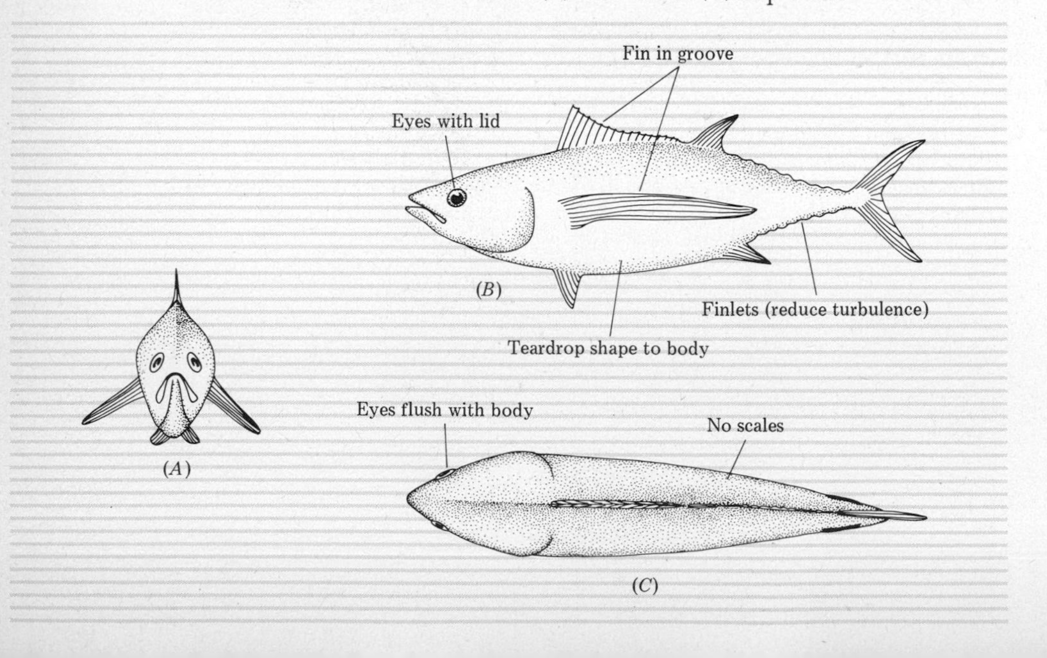

ability. This immediately limits the number of possible defensive adaptations. At the same time, most of these nektonic animals are large, which means most have few potential predators; the very largest (whales) oftentimes have few predators other than humans and killer whales. Thus, the need for elaborate defensive mechanisms is somewhat reduced.

Certain defensive mechanisms remain possible within the framework of fast locomotion. Of these, the most common and widely used is that of *camouflage*. Since we are now dealing with an environment that has no physical hiding places and is visible in all three dimensions, camouflage mechanisms could conceivably take one of three directions: transparency of the body, cryptic coloring, and alteration of body shape (= cryptic body shape).

If the body of an organism is transparent and suspended in the transparent surface water of the ocean, the animal will become invisible when viewed in the water. As we saw in the plankton, such transparency is a common defensive adaptation of many species. It is, however, not found among the nekton. The reason is that, as the size and thickness of animals increase, as with nekton, it becomes much more difficult to keep a body transparent, particularly when the body is highly muscular for propulsion. Thus, camouflage by transparency is not an option for nekton.

Camouflage, through alteration of body shape, is possible so long as the shape does not interfere with that necessary for fast locomotion. Among nektonic vertebrates, the commonest manifestation is to develop a ventral keel to the body to eliminate a conspicuous shadow on the belly of the animal when viewed from below. What is the origin of this shadow and how does body shape help to reduce it? To answer this, we must know something about how light appears in the surface waters of the ocean. When light enters the water, the rays penetrate down in a conelike path. At the same time, some of the light is reflected or scattered back in all directions by the particles in the water. This scattered light is available to illuminate objects in the water from different angles other than the surface, but its intensity is very much less than the downwelling surface light. If an animal is suspended in this water column, it is illuminated most intensely from above, while the scattered light illuminates from the sides and below. Because light intensity from the side and below is so low, a shadow appears under the animal where its body has cut out the strongly downwelling surface light (Fig. 3-11). If the body is now extended ventrally as a keel to form a sharp ventral edge rather than a rounded edge, the shadow will be eliminated when viewed from below. This is because all surfaces on the body now are oriented so that none are exclusively illuminated only by the diffuse scattered light, but also by at least some component of the intense surface light (Fig. 3-11). Without this conspicuous shadow, an animal viewed from below becomes virtually invisible in the downwelling light and hence, camouflaged from deeper-dwelling predators. The elimination of the shadow is enhanced if the keel has reflecting surfaces on it in the form of white pigment and/or scales.

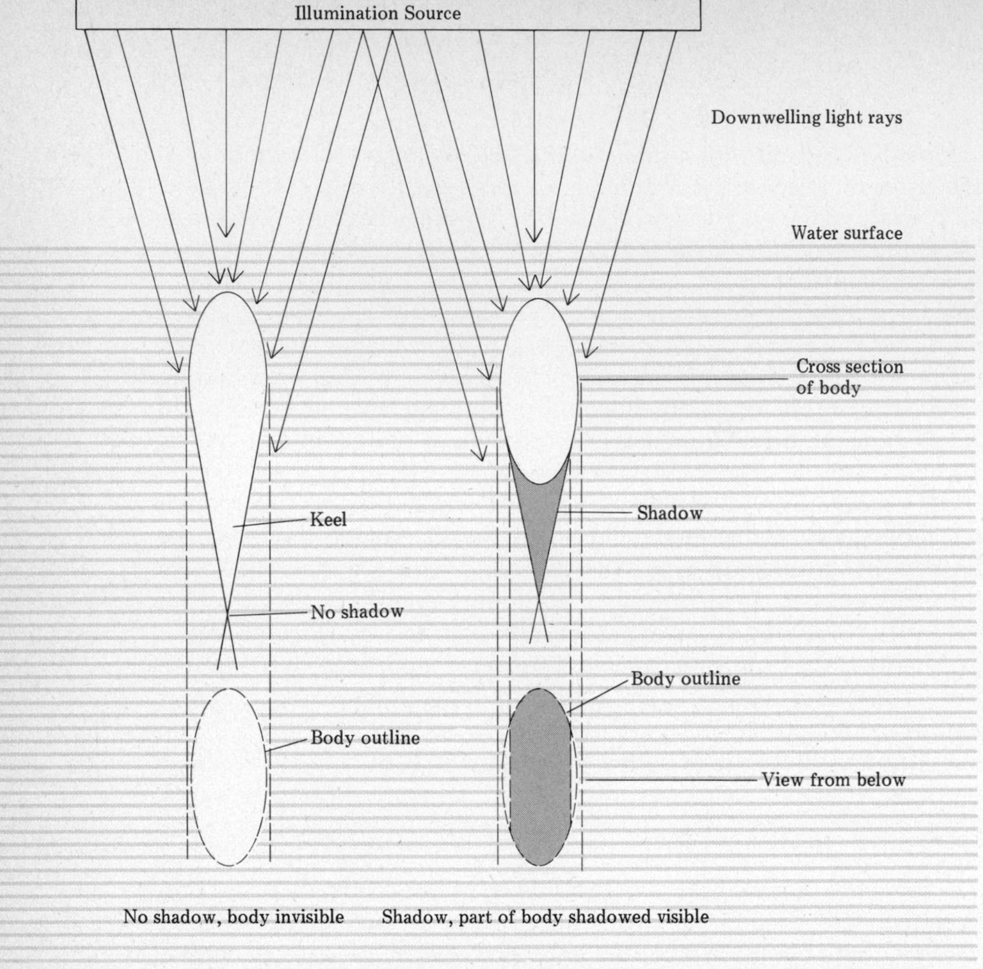

Figure 3-11 Diagram showing how a keel on the ventral surface of an animal eliminates the dark shadow normally cast downward by an unkeeled animal. The presence of the shadow means that an animal living deeper and looking upward would see the unkeeled nektonic animal due to the shadow, but would not see the keeled animal, which would blend into the lighted background. (Modified from Yu. G. Aleyev, Nekton, Dr. W. Junk b.v., 1977. Reproduced by permission of Dr. W. Junk BV.)

Cryptic coloration is also characteristic of most nektonic animals. In the lighted upper waters of the ocean, the dominant spectral colors are blues and greens. When water is viewed from above the surface or at the surface, it appears greenish or bluish looking down into the depths. It should not be surprising then, that many nektonic forms have dark blue or green on their dorsal surfaces so that potential predators looking down would be hard-pressed to make out their form in the general bluish green background. At the same time, when viewed from below, the water looks brighter or whiter toward the surface. Any dark organism swimming in this area would be conspicuous if viewed from below, even if keeled to prevent a shadow. However, if the ventral side is colored white or silver to maximize light reflectance or to blend with the downwelled light, the animal will tend to become invisible. Hence, we observe that many nekton are bicolored, dark green or blue above and white or silver below (Fig. 3-12). For certain of the very surface-dwelling vertebrates such as porpoises, the color pattern is more complex, with irregular bands of light and dark which, in fact, mimic the pattern of wave-roughened surface waters themselves (Fig. 3-13). Some porpoises that live within tuna schools are countershaded gray above and white beneath, with white spots and speckles in the gray and black spots and speckles in the white. Colored in this manner, these animals are difficult to perceive from the side view, underwater, amidst a school of tuna.

Among the abundant flying fishes there is an adaptation to produce large fins. These fishes escape predators by propelling themselves out of the water and gliding for long distances on these winglike fins.

Other than the above adaptations, there are few specialized morphological

Figure 3-12 Contrasting color patterns on various nekton. *(A)* Dall porpoise *(Phocoenoides dalli)*. *(B)* Manta ray *(Manta hamiltoni)*. *(C)* Albacore *(Thunnus alalunga)*.

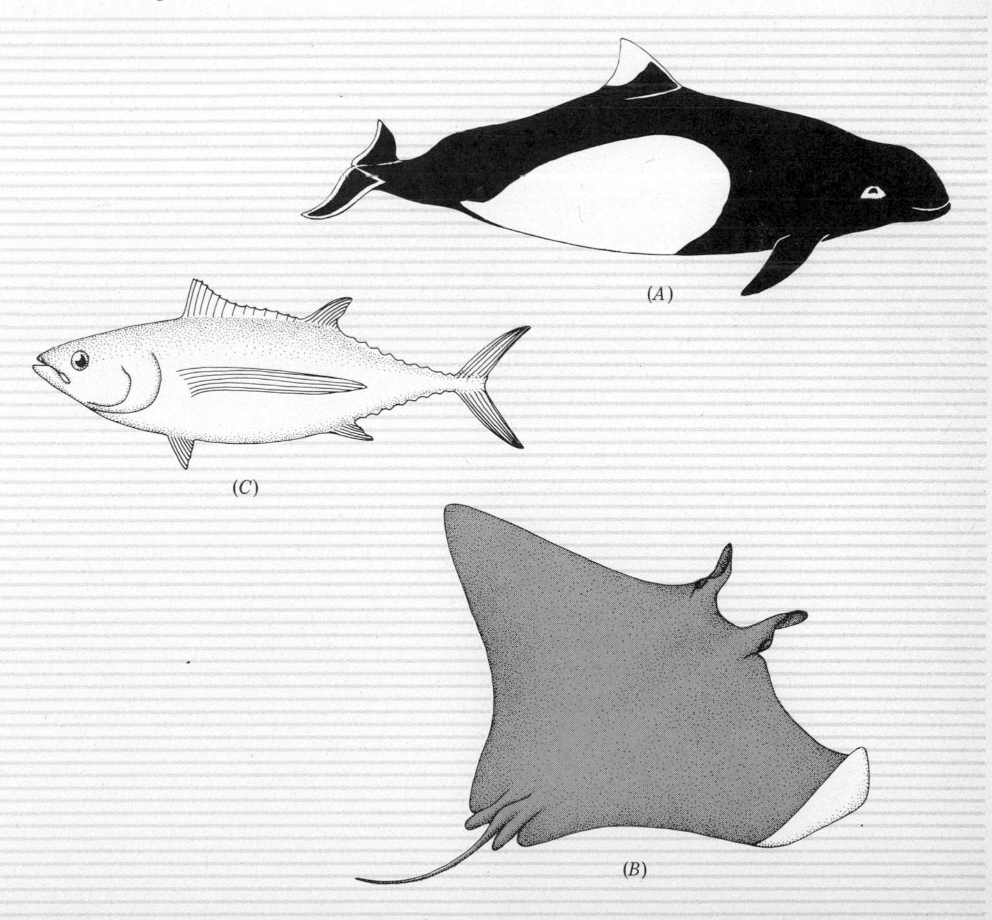

Figure 3-13 Cryptic coloring on the sides of a Pacific white-sided porpoise, *Lagenorhynchus obliquidens,* mimicking the wave-roughened surface of the water.

structures developed as defense against predators. Most likely, this is because the development of various spines and armor would interfere with the ability to move quickly by increasing resistance. However, it may also be partly due to the fact that the larger nekton are the top predators in the system and have few or no other nekton consuming them.

SENSE ORGANS

Since nekton animals are large, fast moving, and primarily predaceous, they might be expected to have well-developed sense organs. This is generally the case, but with few exceptions, such as lateral lines in fishes, the senses are no different from those possessed by other vertebrates in different habitats. Lateral lines are rows of small tubes, open to the water containing sensory pits which are sensitive to pressure changes in the water.

Most sensory information received by nekton comes via vision or hearing. Eyes tend to be well developed and complex in all forms, but the size of the eye relative to size of body varies greatly. The eyes are usually set laterally on the head in such a way that the fields of view for each do not overlap, yet encompass large areas on either side. The lack of overlap of fields of view means that binocular vision is small or absent among most nekton except for pinnipeds.

ECHOLOCATION

Among the mammals of the nekton, it is the sense of hearing that has generated the greatest number of specialized adaptations, which, in turn, suggests its great significance to these animals. The importance of sound to nektonic mammals resides in the fact that sound travels about five times faster in water than in air and has a much greater communication range than does sight. As a result, most nektonic animals show strong development of sound-receiving structures.

In the terrestrial environment, enhanced sound reception is usually indicated in external morphology through the enlarged external ears or pinna of mammals. However, such structures would create an excessive drag on aquatic vertebrates and hence, are reduced or absent. To make up for their absence, there has been a tendency among aquatic mammals to develop other structures of the head to receive sound waves.

Sound reception and production are most highly developed in those cetaceans that use them for *echolocation* in much the same way we use sonar to determine depth. In echolocation or sonar, sound waves are sent out from a source in a particular direction. These sound waves pass uninterruptedly through the water until they impinge on a solid object. When they strike an object, they are reflected and return to the source. The time interval between the initial production of the sound and its movement to a target and subsequent return after reflection is a measure of the distance between the source and the object (Fig. 3-14). As the distance changes, so will the time necessary for the sound "echo" to return. Continual production of sound waves and sensory evaluation of the reflected waves during swimming give a nektonic animal a constant check on all objects in its path. Knowing the distance to objects makes it possible for the echolocating animal to either avoid them (predators) or else to close in upon them (food source).

Low-frequency sound is used by echolocating animals to orient themselves in the water column with respect to objects around them. Low-frequency sound, however, does not produce information as to the fine structure of objects. In order to obtain that information, higher frequency sound waves must be produced and reflected from the object, which will, in turn, discriminate considerable detail. Thus, most nekton animals that have highly developed echolocation also have the ability to vary the frequency of the sound produced. Discriminatory ability of high-frequency sound is really quite remarkable.

Figure 3-14 Echolocation in a sperm whale. Sounds are sent out (solid lines) focused by the melon (spermaceti organ), and the returning echoes (broken lines) are received by the lower jaw.

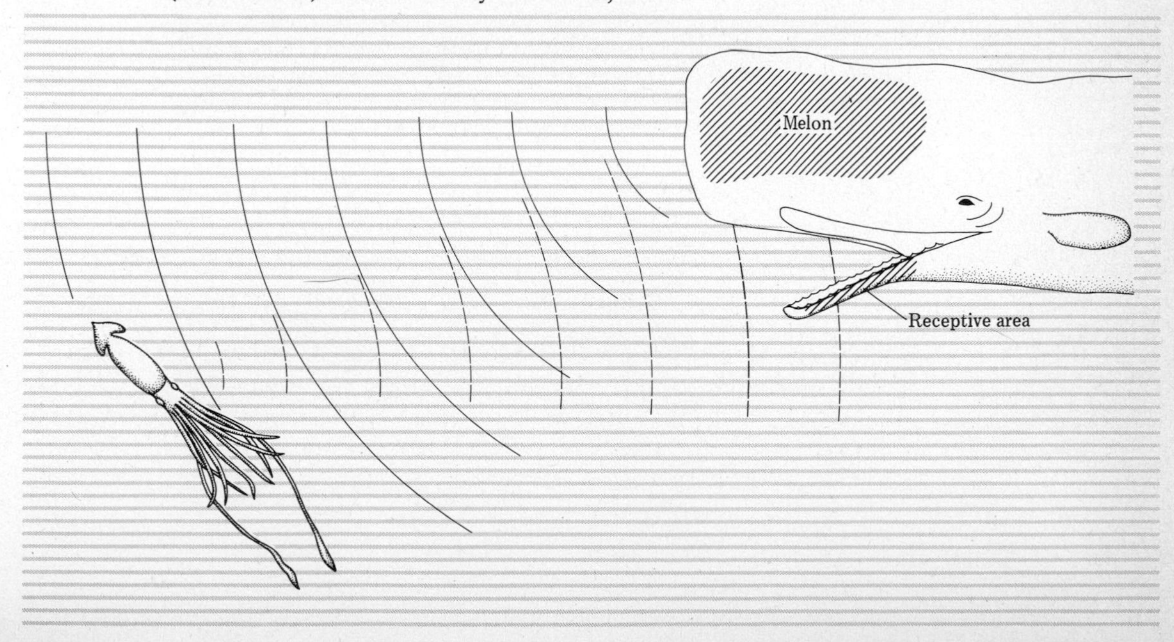

Porpoises, for example, are reported by Kellogg (1958) to be able to distinguish between two fish species of similar size and shape; Norris et al. (1961) report porpoise able to distinguish even between similar objects differing only in thickness.

It is in the toothed whales that echolocation reaches its zenith. These animals possess elaborate morphological modifications of the head and respiratory systems that permit them to send and receive sound waves varying over a wide range of frequencies.

Toothed whales have a peculiarly bulging, rounded forehead and, associated with it, a dorsally placed external nasal opening or blow hole. Internally, a complex series of air sacs is associated with the nasal passages leading from the blow hole to the lungs. The rounded forehead is caused by the presence underneath of a large, fat-filled structure called the *melon*. This fatty organ reaches its greatest development in the sperm whales, where it is called the spermaceti organ; it constitutes as much as 40 percent of the entire animal's length. The relationship of these structures is diagrammed in Fig. 3-15.

Although we do not as yet completely understand how the above elaborate system operates in sound production and reception, enough is understood to permit a description of how this apparatus likely functions. Sounds are produced by the toothed whales by movement of air through the nasal passage and associated air sacs. This air movement may be associated with breathing at the surface, but sound is also produced by recirculating internal air during diving. Special muscles in and around the nasal passages and air sacs allow these channels to change shape and volume and thus change the frequency of the sound produced. The fatty melon is apparently used as an acoustical lens to

Figure 3-15 The anatomy of the sound-producing apparatus in toothed whales. *(A)* Dorsal view of the head of a sperm whale. *(B)* Side view of the head of a sperm whale *(Physeter catodon)*. (1) Nostril, (2) bony nasal duct, (3) sound-focusing surface of cranium, (4) left nasal passage, (5) right nasal passage, (6) frontal air sac, (7) distal air cavity, (8) and (9) upper and lower spermaceti sac (acoustic lens). (From Yu. G. Aleyev, Nekton, Dr. W. Junk b.v., 1977. Reproduced by permission of Dr. W. Junk BV.)

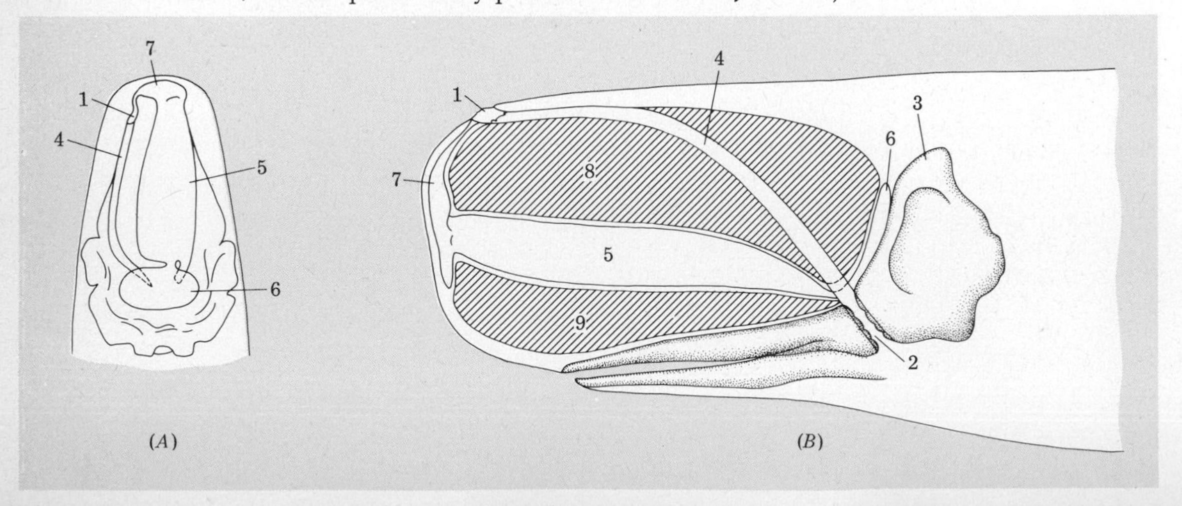

focus and direct the train of outgoing sound waves and this directionality increases with increasing frequency of the sound, so that these animals can pinpoint objects with high-frequency sound. Directionality is further enhanced by the bones of the peculiarly shaped skull of these toothed whales. Reception of the reflected waves is centered in the bone and fat deposits of the lower jaw and in the middle ear. In contrast to most mammals, where the middle ear resides in bone attached to the skull, the toothed whale's middle ear is loosely attached to the skull by ligaments and provided with special air- and fat-filled spaces.

Thus, the elaborate morphological modifications of the head region of toothed whales is primarily directed to producing and receiving a wide range of sound frequencies, which, in turn, allow the animals both to navigate without bumping into objects and to pinpoint potential food organisms for subsequent capture. Toothed whales also possess very large brains relative to their body size, brains which are second only to humans' in development of the cerebral hemispheres. It seems likely that these large brains are necessary to allow them to process rapidly the acoustical information received.

REPRODUCTION AND LIFE CYCLE

Among epipelagic fishes, no special reproduction mechanisms are apparent that would sharply set them off from their benthic or shallow water relatives. Characteristically, however, holonektonic bony fishes such as tuna and marlin spawn eggs that float and undergo development in the open-ocean waters. Some even have threadlike structures associated with them that permit them to adhere to various kinds of floating debris. Because these floating eggs are planktonic, they are subject to tremendous losses due to predation. As a result, the fishes produce tremendous numbers to offset the equally huge losses. Parin (1970) reports that skipjack and albacore tuna, for example, produce 2 and 2.6 million eggs, respectively, while striped marlin spawn over 13 million and ocean sunfish an incredible 300 million eggs. Spawning is often intermittent and is often extended over a period of many months.

Among the pelagic sharks, a different reproductive strategy is observed. These fishes produce only a few eggs or embryos. Parin (1970) notes that the thresher shark (*Alopias*), for example, has but two embryos and the blue shark (*Prionace glauca*) up to 54. Obviously, if so few young were to undergo development in the plankton, the chances that any would survive the tremendous predation are very poor. Thus, these sharks enhance the chances for survival of their few offspring by retaining the eggs in the female for a much longer period of time so that at birth or hatching, the young are large in size and immune to most potential predators.

We know relatively little about the growth of pelagic fishes, but available knowledge suggests that the growth rate is very rapid. Tunas, for example, seem to increase their weights by 2–6 kg per year and length by 20–40 cm (Table 3-1).

TABLE 3-1 Age and Length in Three Species of Tuna, Albacore *(Thunnus alalunga)*, Big-Eye Tuna *(Thunnus obesus)*, and Yellowfin Tuna *(Thunnus albacares)*

Species	Age years							
	1	2	3	4	5	6	7	8
	Length, cm							
T. alalunga	18–66	32–84	54–89	73–94	70–95	82–100	98–105	—
T. obesus	45–50	62–70	74–94	85–116	97–138	115–155	—	—
T. albacares	54–103	70–136	85–155	99–154	109–134	110–130	127–145	135–160

After N. V. Parin, 1970.

Correlated with this rapid growth, most nektonic fish seem to be short-lived; even the large tunas appear to live only five to ten years. By contrast, the pelagic sharks may live from 20–30 years.

Marine birds and turtles retain the reproductive characteristics of their terrestrial relatives. All produce shelled eggs which are laid on land. Marine birds often congregate in large groups to nest on islands or cliffs, which are inaccessible to terrestrial predators. This ensures that the usually helpless young (*altricial*) will survive until old enough to fly. However, it also makes such birds extremely vulnerable to human predation or pollution, since a large fraction of the existing population of a species may be present in one small area and can be wiped out easily. For example, the Laysan albatross nests on Midway Island, where the Navy has airport installations, and many albatrosses are killed by collision with planes. Most marine birds have definite breeding seasons and may migrate thousands of miles from their feeding grounds to their breeding grounds.

Marine turtles all lay eggs in depressions that they excavate in the sand above the high-tide level on beaches in various parts of the tropics. This is the only time that these animals normally return to land. Immediately upon hatching, the young turtles instinctively head for the ocean, where their continued development occurs, but about which we know virtually nothing. As with birds, turtles tend to migrate thousands of miles and congregate off certain beaches for breeding. Females haul out to lay eggs only on certain beaches, and since both eggs and adults are considered excellent food by humans, many marine turtles have been drastically reduced in numbers in recent years in all areas of the world. Some sea snakes give birth to living young in the water; others lay eggs on beaches.

With respect to reproduction in marine mammals, there are two groups: those that give birth on land and those that give birth in the water. We know considerably more about reproduction in the group that breeds on land because they may be observed by humans at that time. Knowledge of the reproductive patterns of those giving birth in the water is generally limited to observations made on captive animals in aquaria.

Seals, sea lions, and walruses all give birth to live young on land or on floating ice. Their young are usually unable to swim at birth and require some time before they are capable of venturing into the water. During this terrestrial period, the pups grow very rapidly and gain the strength and insulating layers of fat and fur they need in order to survive in the cold, open water. Many sea lions and seals such as fur seals, Steller's sea lions, and elephant seals are polygamous and territorial on the breeding grounds (see Plate 9). The largest and most aggressive males (harem bulls) tend to collect together varying numbers of females into a harem, occupying a small area of beach, which they then protect from other bulls (Fig 3-16). Should another bull attempt to steal a female, take over a harem, or encroach upon the territory occupied by the bull and his harem, the resident "harem master" will fight. Though such fights may be mostly bluff and noise, serious fights occur in which one or the other bull may be seriously injured or killed. In the seals and sea lions where such territorial behavior occurs, the males are usually much larger in size than the females.

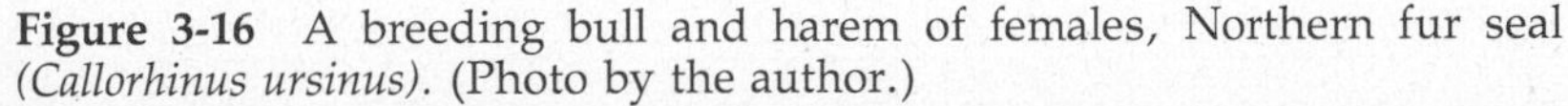

Figure 3-16 A breeding bull and harem of females, Northern fur seal *(Callorhinus ursinus)*. (Photo by the author.)

This territorial behavior and aggressiveness on the breeding ground do not extend to their life when away from the breeding grounds. It suggests that, while pelagic food and space are not limiting, breeding space is, because it is such a small area. This method of breeding also means that relatively few males, the harem masters, actually do the breeding, and all others are excluded (Fig 3-16). As with the marine birds, many of these pinnipeds migrate long distances to their breeding grounds; for example, the northern fur seal (*Callorhinus ursinus*) is pelagic all over the north Pacific Ocean, but most migrate back each summer to two small islands making up the Pribilofs in the Bering Sea, to breed.

In contrast to the pinnipeds, the cetaceans give birth to their offspring in the water. The young of whales must therefore be able to swim at birth and instinctively know how to surface for air. They also remain closely associated with the mother. Whereas young pups of sea lions may be left on the breeding grounds for days while the female forages for food in the open ocean, the juvenile whales always remain close to their mothers, where they are protected from potential predators. As with the pinnipeds, certain cetaceans may also undergo migration for breeding purposes. Often, this is a migration of thousands of miles from the feeding areas in cold water to the calving grounds in warmer waters (Figs. 3-17, 3-18). The reason for moving to warm waters for birth of the young is that the newly born young do not have the insulating blubber of

Figure 3-17 Migratory routes of whales. *(A)* Migratory routes of humpback whales *(Megaptera novaeangliae)* in the Southern Hemisphere. *(B)* Migratory route of the gray whale *(Eschricthius robustus)*. (*A*, after E. J. Slijper, Whales and dolphins, University of Michigan Press, 1962; used by permission of Springer-Verlag, Heidelberg. *B*, modified from V. B. Scheffer, A natural history of marine mammals, Scribner.)

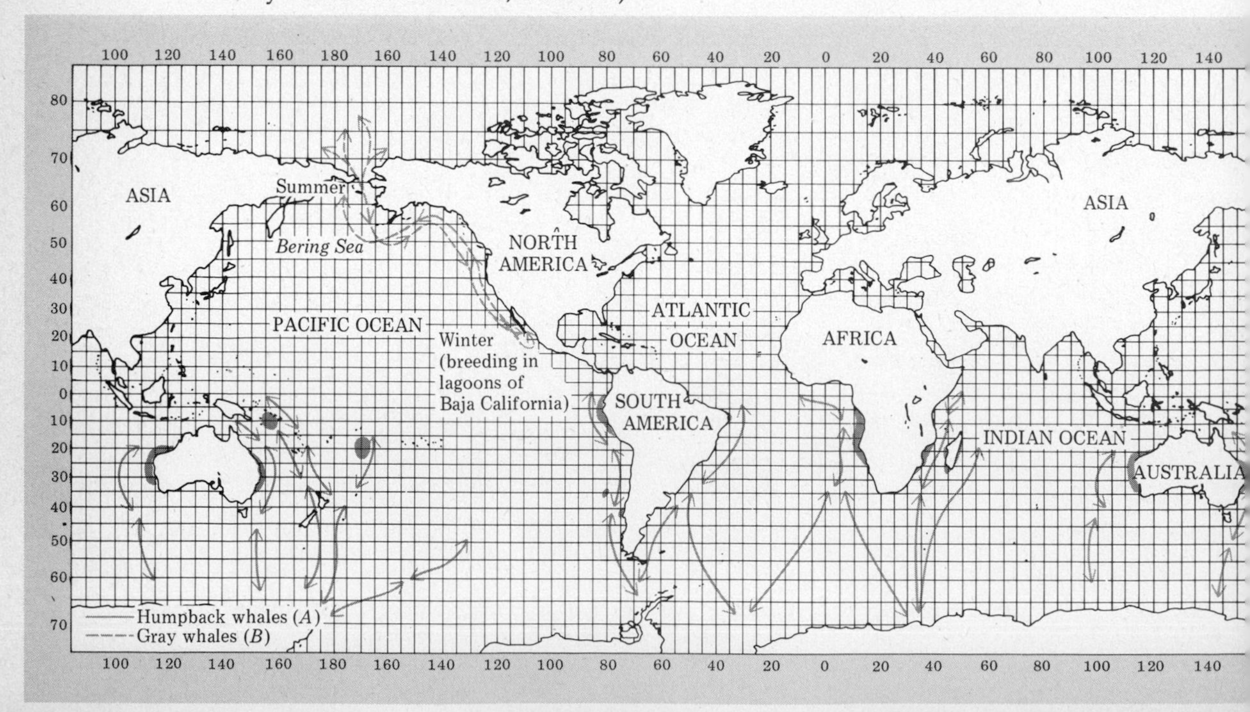

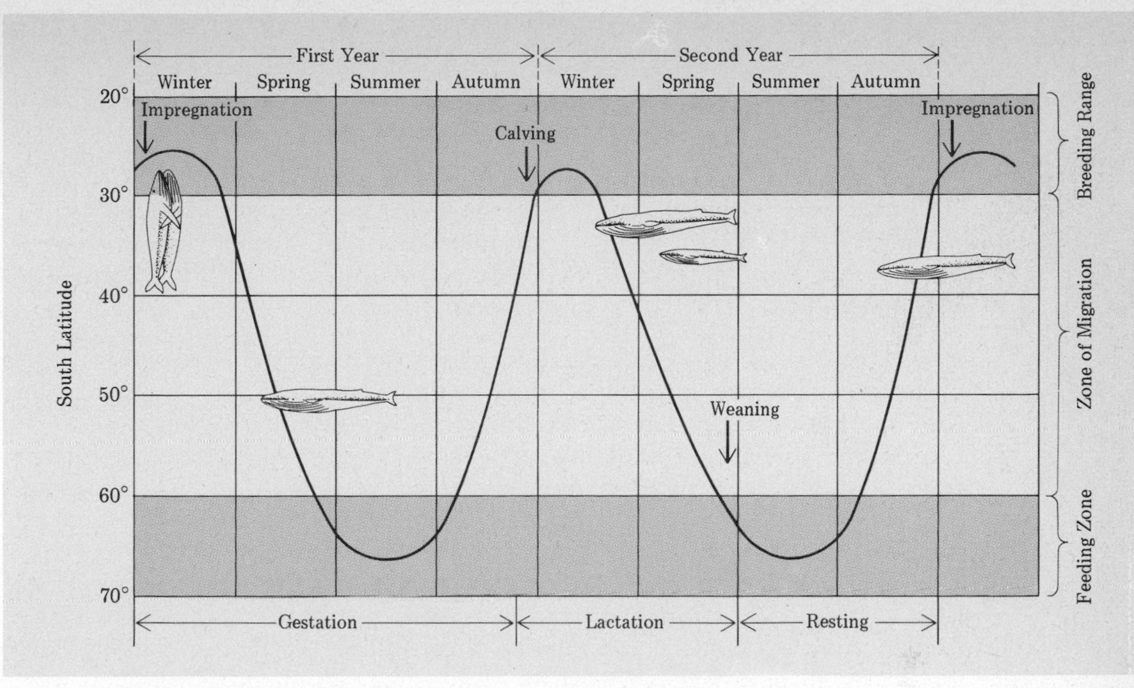

Figure 3-18 The reproductive cycle of the fin whale *(Balenoptera physalis)* in the Southern Hemisphere. (Modified from N. A. Mackintosh, The stocks of whales, Fishing News Books.)

the adults and hence will have a better chance of survival in warm water until such time as the additional insulating layers can be added.

The young of both pinnipeds and whales grow very rapidly, adding many kilograms per day. Blue whales, for example, can grow from 3 tons at birth to 23 tons at weaning, seven months later. The reason for such rapid growth is that the milk of pinnipeds and cetaceans is extraordinarily rich in fats (as much as ten times as much fat as in cow's milk) and is produced in large quantities.

Pinniped pups are nursed throughout the time they are on the rookeries. At the end of the season, in some species, they are usually abandoned by the females and must fend for themselves in the open ocean. Cetacean calves are nursed for up to a year before weaning, although nursing lasts 15 months in sperm whales and 18 months in pilot whales.

Because of the enormous amount of energy that must be put into milk production by these marine mammals to sustain a single offspring, only one young at a time is usually produced. The reproductive cycle is also such that young are produced annually (most pinnipeds) or at even longer intervals (walrus, certain whales) (Fig. 3-18). This means that stocks of these animals can be easily reduced and take a long time to recover their numbers.

Most marine mammals are long lived. Gray seals have been recorded to live 46 years and harbor seals 36 years, while small whales such as the bottlenose porpoise 32 years and large whales such as the sperm whale and fin whale up to 77 and 80 years, respectively. Correlated with the long life span is the delayed onset of sexual maturity and reproduction. Male fur seals, for example, do not become harem masters until 9 or 10 years of age and sperm whales apparently do not breed until about age 20.

MIGRATIONS

As we noted in the previous section, many marine mammals, birds, and reptiles undertake extensive migrations for breeding purposes. This migration for breeding is a common characteristic of air-breathing marine vertebrates.

Nektonic fishes also undertake extensive horizontal migrations that are equivalent in distance to those taken by the air breathers. These migrations are of great importance but, unfortunately, have been little analyzed. Migrating holonektonic fishes include the various tunas and the saury.

Salmon, which spend most of their lives dispersed in the open ocean, migrate back to fresh water streams in which they were spawned to reproduce. In this respect, they are similar to the marine air-breathing vertebrates. Salmon have the ability to return to breed in the very same stream in which they were hatched. This requires considerable navigational abilities on the part of the fish. From recent studies, it appears that part of the key to this navigational ability resides in the fishes' sense of "smell"; that is, these fish follow various water-borne "scents" to locate their home streams. The suggested mechanism is that when the young salmon migrate down the streams to enter the sea, they are imprinted with the "odors" of the various streams in succession. This sequential memory is then reversed on their return to allow them to find their way back to the same stream.

Equally remarkable migrations are made by the green turtles (*Chelonia mydas*). As with the salmon, these animals migrate from distant feeding grounds to congregate on one or a few beaches to lay their eggs. How do they locate these beaches? In the case of beaches along the shores of continents, this is not as difficult to understand. Here, a turtle could simply follow along the shore line until the beach with the right "smell" came along. However, many turtles along the Atlantic coast of South America regularly nest on tiny Ascension Island, some 1,400 miles out in the middle of the Atlantic Ocean. How do they find such a small target? Since the turtles must navigate most of the distance to a mere pinpoint of an island out of sight of land and without any landmarks in the ocean, the most reasonable assumption is that they are navigating as we humans do, using information from celestial bodies! If this is true, we as yet do not know how it is done. One way would be to use the height of the sun to establish their latitude. Ascension lies due east of the bulge of Brazil. Thus, turtles could move along the coast to the tip of the bulge, then move off directly east, correcting for deviations through height of noonday sun. This should bring them within range of the island, where it either would be visible on the horizon or could be detected and homed in on through some chemical tasted in the water.

Other extensive and complex migrations are those undertaken by tunas and their relatives. Tunas are primarily tropical fishes that make extensive migrations across oceans within the tropical zone and also move into the temperate waters during the warmer seasons. Tunas tagged in Florida have been recovered across the Atlantic in the Bay of Biscay, and Pacific tunas tagged off California have been captured in Japanese waters. It is not entirely clear why these fishes should make such monumental journeys, but it appears that such migrations permit the fishes to exploit their food resources more completely and reduce the possibility of destroying or eating out their food in any one area. Such migrations enable these fast-moving fishes to take advantage of the rich food areas of the temperate zone. One of the most important factors that guide these migrations is water temperature. As primarily tropical fishes, tunas venture into the temperate waters when the water temperature rises about 20°C. Thus, off California, the cold upwelled water is very rich in food organisms, but the tunas usually enter these waters in the summer when surface temperatures reach 20–21°C. If the year is cold and temperatures of the surface waters do not rise to this point, the tunas are absent. However, the Atlantic bluefin tuna regularly summers off Newfoundland in water below 20°C, so not all tunas are restricted to warm water. Tunas apparently always return to tropical waters to spawn and to spend the early parts of their lives (Fig. 3-19).

SPECIAL ADAPTATIONS OF MARINE BIRDS AND MAMMALS

The warm-blooded marine mammals and birds require some special adaptations to permit them to persist in ocean waters. These special adaptations are primarily concerned with maintaining temperature, diving, and osmotic regulations.

Water has a higher thermal conductivity than does air, which means that it is quick to extract heat from a warm body. Humans experience this when, as swimmers, they become chilled after a short time in water of even 80°F; in air of the same temperature, they would be comfortable indefinitely. Thus, marine mammals, which maintain elevated body temperatures with respect to the surrounding water, must somehow have adaptations to prevent their body heat from being drained away.

One means to slow the rate of heat loss is to have a large body. As you may recall from the plankton section (pp. 50–51), the ratio of surface area to volume of any body is lower for a large one than for a small one. The larger the body then, the smaller the surface area and correspondingly, the smaller the surface area in contact with the environment through which heat may escape. All nektonic marine mammals are of large size, and it may well be that the reason there are no mouse-sized marine mammals is that they would simply die from chilling. There are small marine birds (petrels, auklets), but these animals are never completely immersed in the water. Only a fraction of their

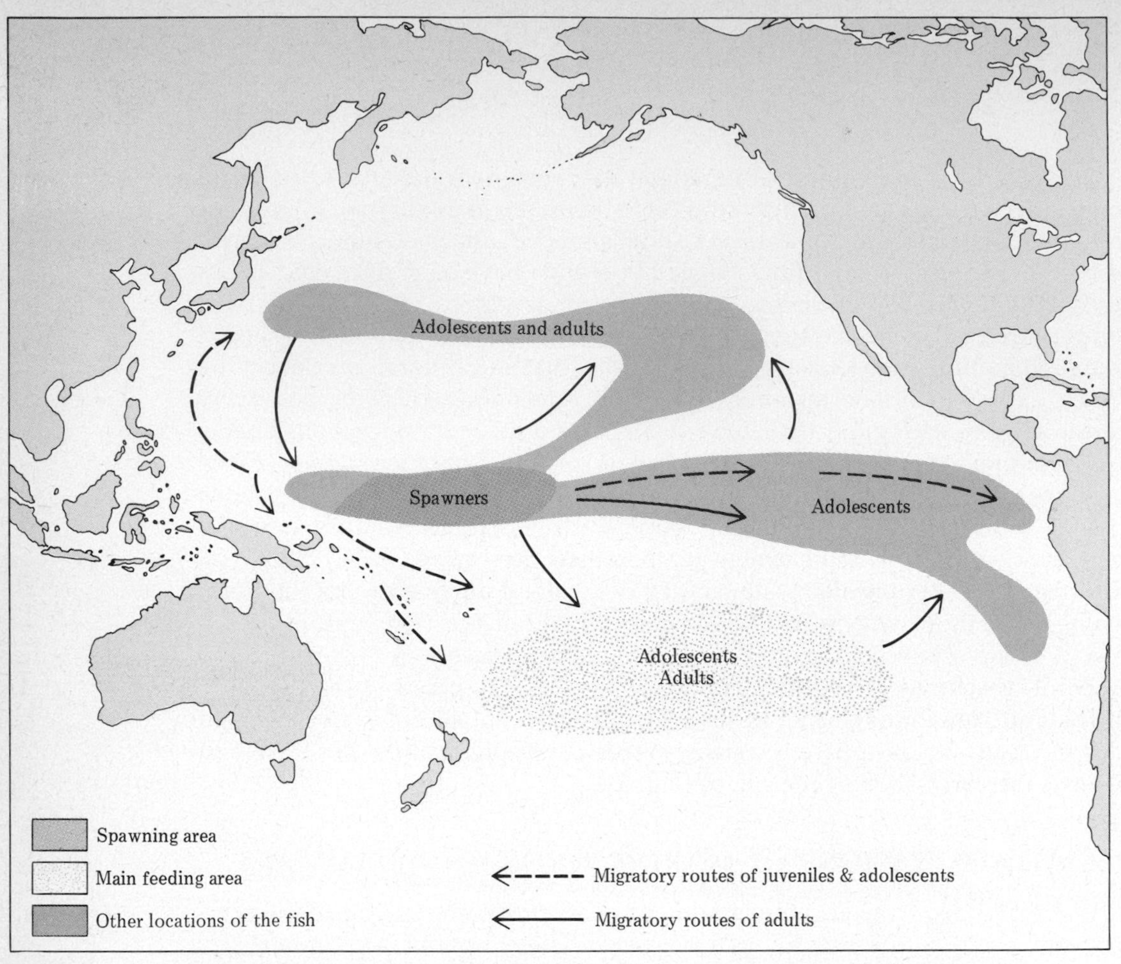

Figure 3-19 Centers of distribution and migratory routes of the big-eye tuna *(Thunnus obesus)* in the Pacific Ocean. (Modified from The Rand McNally atlas of the oceans, Rand McNally.)

body is in contact with the water at any one time (except during intermittent diving).

A second adaptation that prevents or reduces heat loss is a thick insulating layer of blubber or fat just beneath the skin. This layer reaches its greatest thickness in whales, where it may be 2 ft thick. In pinnipeds such as the walrus and elephant seal, subcutaneous fat may constitute as much as 33 percent of the weight. Blubber and fat are poor conductors of heat and hence protect the animal from losing internal heat. The thicker the blubber or fat layer, the less the heat loss. Those marine mammals inhabiting polar waters therefore have thicker layers than do temperate and tropical species.

A final adaptation concerns the circulatory system. The areas of a marine mammal that offer the greatest surface area to the water and hence greatest heat loss, and also lack the protective layer of lipid, are the fins and flippers. What adaptations prevent massive heat loss through these extremities? In cetaceans, the answer is that the arteries that bring the warm blood out to these extremities are surrounded by a number of smaller veins that bring blood back to the central core of the mammal. Because of this arrangement, the heat of the blood in the arteries can be absorbed by the cooler blood returning in the veins before it is lost to the external water through the thin flesh of the outer extremities (Fig. 3-20). This is a countercurrent system of circulation designed to save heat.

Because most of their adaptations are designed to keep the body heat in, marine mammals (in particular, pinnipeds) may, on occasion, become too warm. Hot, still days may mean considerable stress from overheating. On these uncommon occasions, the animals must act to dissipate heat. They do this by waving their flippers in the air while increasing the blood flow out to the extremities and restricting the flow back to the core through the veins. The result is greater heat loss and subsequent cooling. Seals and sea lions may also open their mouths and pant like a dog.

Most nektonic marine mammals, particularly the pinnipeds and the whales, regularly dive, or are capable of diving, to depths far greater than that of humans. Whereas exceptional humans may be capable of free dives as deep as 60 m with a breath-hold duration of six minutes, various seals and porpoises are known to dive to depths of from 160 to 600 m and hold their breaths for 6 to 40 minutes. The champion is the sperm whale, which can dive as deep as 2,250 m and hold its breath for 80 minutes. How is it possible to do this? The answer rests with a number of anatomical and physiological adaptations.

One of the first questions to be asked regarding these prolonged deep dives is

Figure 3-20 Section of the tail fluke of a bottle-nosed dolphin, *Tursiops truncatus*, showing the arrangement of veins surrounding an artery (25x). (From R. Elsner, The biology of marine mammals, edited by H. Andersen, Academic Press.)

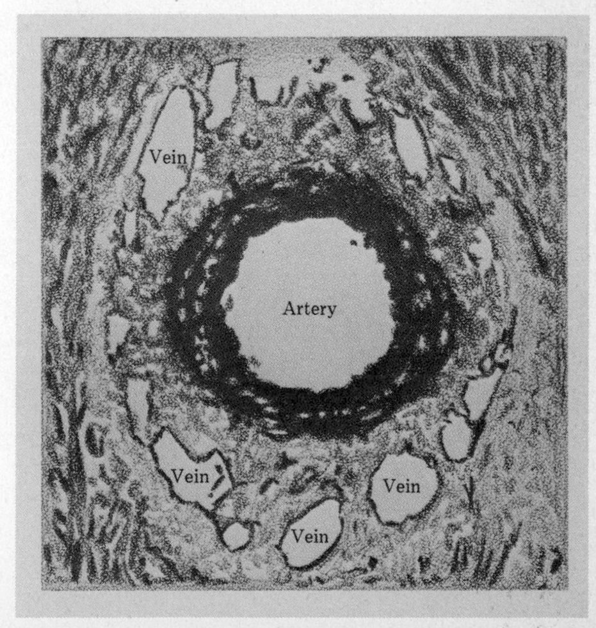

how do these animals avoid "the bends"? This is a serious affliction of human divers in which the nitrogen breathed under pressure at depth as part of their air supply bubbles out of solution in the blood when a diver ascends too quickly. These bubbles then lodge in the joints, causing paralysis and even death. The key here is that humans get the bends due to breathing gas *under pressure,* as from a scuba tank. Since marine mammals do not inhale pressurized gas but rather depend on gas inhaled at the surface at regular atmospheric pressure, this disease has less chance of occurring. The chance of its occurrence is further reduced by certain other adaptations, the most important of which is the collapse of the lungs. In a deep dive, the pressure of the outside water causes the gas exchange tissues of the lungs of the diving mammal to collapse. This collapse forces the residual gas in the absorptive area of the lung into the nonabsorptive cartilaginous air passageways, and thus diffusion of gas into the blood stops. With no gas entering the blood, there can be no nitrogen to come back out when the animal surfaces. This collapse of the lungs is aided by the fact that diving marine mammals have few ribs that are attached to the sternum, and the sternum is shortened. Therefore, the rib cage is easily pushed in. Finally, it has also been suggested that the peculiar "foam" of emulsified fat droplets and mucus that many whales have in their respiratory passages serves as an absorptive material, absorbing the nitrogen gas so that it cannot enter the bloodstream where it could cause "bends."

A second problem with deep diving is to explain how the animals manage to survive so long without access to a supply of oxygen. A consideration of this problem will suggest that the only mechanisms for surviving are for the animals to store more oxygen than nondiving forms do and to conserve carefully the amount they have. Since the major organ system involved in oxygen storage and transport is the circulatory system, it is there we may look first for adaptations.

Many diving birds and mammals have a larger blood volume than their terrestrial relatives. The elephant seal, according to Elsner (1969), has a blood volume of 12 percent of its body weight, whereas humans have a blood volume of only 7 percent of body weight and domestic dogs only 9 percent. This larger volume thus allows more oxygen to be held in the body. This oxygen capacity is increased further, as Scholander (1940) notes, because marine mammals also possess a higher oxygen capacity per unit volume of blood than terrestrial mammals (40 ml O_2/100 ml blood in elephant seals vs. 16–24 volume percent in humans), but it is not as great as might be expected. Neither the increased blood volume nor the increased O_2 capacity, however, is sufficient to account for the ability of pinnipeds and cetaceans to remain under water for extended periods. For example, harbor seals have about the same weight as humans and possess about twice the amount of oxygen in the form of increased blood volume and oxygen capacity. However, rather than being able to submerge twice as long as humans, they can stay under five to ten times as long (see Plate 10). Other adaptations must also be working.

One of these additional adaptations is that of the marked slowing of the heart beat during the period of submersion. This *bradycardia* is common to all diving, air-breathing vertebrates. The decrease in heart beat is quite dramatic. For example, in Pacific bottle-nosed porpoises, *Tursiops truncatus*, the heart beat under experimental conditions drops from about 90 beats/minute at the surface to about 20 beats/minute during a five-minute dive (Fig. 3-21).

Of more importance than slowing of the heart beat, however, are two other adaptations. The first is that during a dive the circulatory system cuts off the blood supply to various organs and organ systems, including the muscles, digestive system, and kidneys. This cutoff has the effect of conserving the limited oxygen supply in the blood for those more sensitive and vital tissues such as the brain and central nervous system. Thus, a limited supply of oxygen is made to last longer by allowing it to be used only by selected organs. In this way, the animal may remain submerged much longer than if it had to supply oxygen to all its tissues. The second adaptation is related to the first: It is that the muscular system and other organs are extremely tolerant of anaerobic conditions and continue to function when the blood flow is cut off. This results in the buildup of large amounts of lactic acid in the muscles during a dive. Lactic acid is the end product of anaerobic metabolism. The muscular system is able to continue functioning in the absence of a blood supply not only because it is very tolerant of lactic acid and anaerobic conditions, but also because of yet another adaptation. The muscular system of marine mammals is rich in an oxygen-containing compound called *myoglobin*. This compound has a structure very similar to hemoglobin, but it is better at storing oxygen. Thus, when the animal dives, the muscle blood supply is shut off. The muscles initially have a large supply of oxygen in the myoglobin, but as the dive continues, the oxygen of the myoglobin is depleted and the muscles continue to operate anaerobically, building up the product lactic acid.

Figure 3-21 Change in heart rate in the bottle-nosed porpoise, *Tursiops truncatus*, during diving. Arrows indicate the beginning and end of the dive. (From R. Elsner, The biology of marine mammals, edited by H. Andersen, Academic Press.)

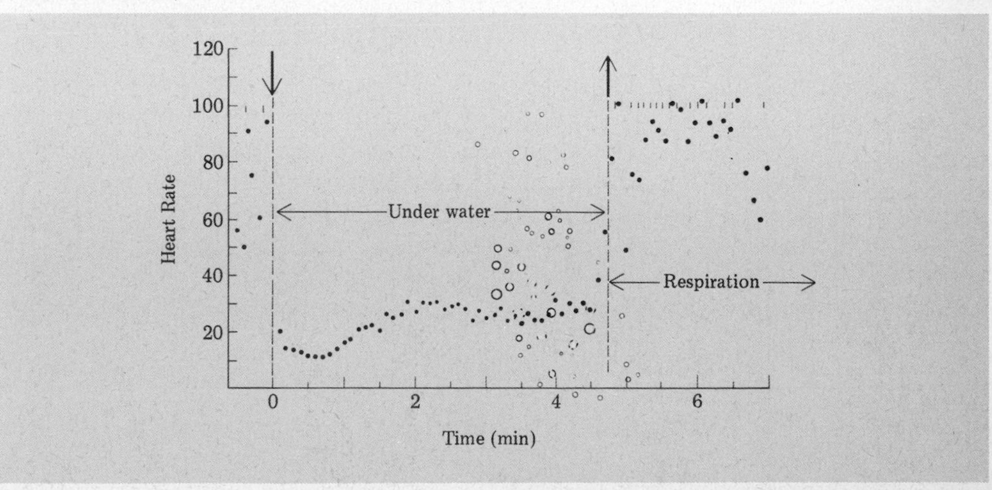

When a dive has been completed and the animal returns to the surface and begins breathing again, the blood is recharged with oxygen, the blood supply is restored to the muscles and other organs, and oxygen replenishes the myoglobin. The blood picks up the accumulated lactic acid, where it is eliminated by oxidation. At this point, the animal is again ready to undertake another dive.

Marine mammals and birds generally have internal salt concentrations in their blood and body fluids that are lower than concentrations of surrounding sea water. This means that they face a potential osmoregulatory problem in that water has a tendency to pass from their bodies to the outside in order to equalize the salt concentrations between the inside and outside of the animal. The marine mammals and birds must compensate for this water loss either by drinking sea water or by obtaining water from their food. If they drink sea water, they gain an unwanted quantity of salt, which they then must get rid of in some way. Marine birds eliminate this excess salt through special salt-secreting glands in the region of the orbit of the eye. Marine mammals, however, do not have such glands, nor do they have sweat glands as do terrestrial mammals. The only organ left to them for elimination of salt is the kidney. Excess salt must be removed via the kidneys and washed out of the body with large amounts of water. Unfortunately, we know very little about urine composition and production in marine mammals, particularly whales. It seems likely that the kidneys do remove the excess salt, since they are much larger in whales than in terrestrial mammals of similar size. Also, in those whales that live in fresh water, the kidneys are not enlarged. However, we do not at present know if whales remove salt via the kidneys and urine.

ECOLOGY OF NEKTON

As we noted in the introduction, because of the difficulties of observation and experiment with natural populations of nekton, we know very little about the ecology of these forms. Most of our information concerns the feeding relationships and probable trophic links.

FEEDING ECOLOGY AND FOOD WEBS

Basically, all adult nekton are carnivores preying upon either the smaller plankton or other nekton. By far the largest number of nekton are predators on other nekton. The plankton feeders consume the larger zooplankton and include the fishes such as flying fish and sardines (Clupeidae), but the best known are the large baleen whales. Baleen whales include all of the larger whales, such as the blue whale, the largest animal that has existed on this planet. These tremendous animals have large mouths that are devoid of teeth; they have instead sheets of "baleen" or "whalebone," which hang down like a curtain from the roof of the mouth cavity. Baleen consists of a series of closely spaced parallel plates fringed on the free end. These plates form a sieve mechanism. When the whale opens its mouth, a large amount of water and plankton rushes in. The animal then partially closes its mouth so that the only passage out is through the narrow spaces between adjacent baleen plates. The whale next raises its tongue and forces water out past the baleen, which then strains out the larger zooplankton (Fig. 3-22). The trapped zooplankton are then swallowed. The dominant zooplankton organisms fed upon in this way are euphausiids (see Fig. 2-8). These organisms exist in untold numbers in the oceans, particularly the Antarctic and Arctic seas, where they are called "krill" by whale fishermen. Baleen whales are not the only marine mammals to feed on krill; the crab-eater seal of the Antarctic does so also, but it filters out the krill on the cusps of its peculiarly shaped teeth (Fig. 3-23).

Plankton-feeding fishes are varied, and include such common forms as flying fish, salmon, and the largest fishes, the basking sharks. As with the baleen whales, the dominant food organisms appear to be various crustaceans of the zooplankton, including euphausiids, copepods, and amphipods. The type of zooplankton taken varies among different areas of the ocean as well as with different seasons. In the case of the Pacific Ocean salmon, the diet varies considerably, not only with season, but also among different years. There is some evidence that this change in diet with years is related to the competition that exists for food among the several salmon species. Thus, when there are years of high abundance of certain of the salmon, the intensified competition leads to an expanded range of food items taken and to changes in the composition of the diet. For example, Parin (1970) notes that in 1956, when pink salmon (*Oncorhynchus gorbuscha*) were low in numbers, 73.2 percent of the diet of chum salmon (*O. keta*) consisted of euphausiids, but in 1957, when pink salmon were abundant, chum consumed ctenophores and jellyfish (72.6 percent of the diet). It does not appear that planktivores are specialized on any one kind of plankton; rather, they are size selective.

Whereas few of the vertebrates of the epipelagic zone of the oceans are plankton feeders, there are a large number of small fishes living in the mesopelagic zone (see Chapter Four) that are planktivores and that migrate into the epipelagic at night to feed. Thus, the competition for plankton may be greater than first anticipated, but our understanding of these relationships is poor at present. These same small fishes also form a food source for some of the carnivorous fishes of the epipelagic.

The nekton-feeding fishes, birds, and mammals dominate the open ocean nekton. Their diet generally includes fishes, squids, or large crustaceans. The size of prey taken generally depends on the size of the predator, with the larger species taking progressively larger prey species. The largest carnivore in the oceans is the sperm whale (*Physeter catodon*), which preys upon the largest squid, diving deeply to obtain them. Apparently, the sperm whales are the only

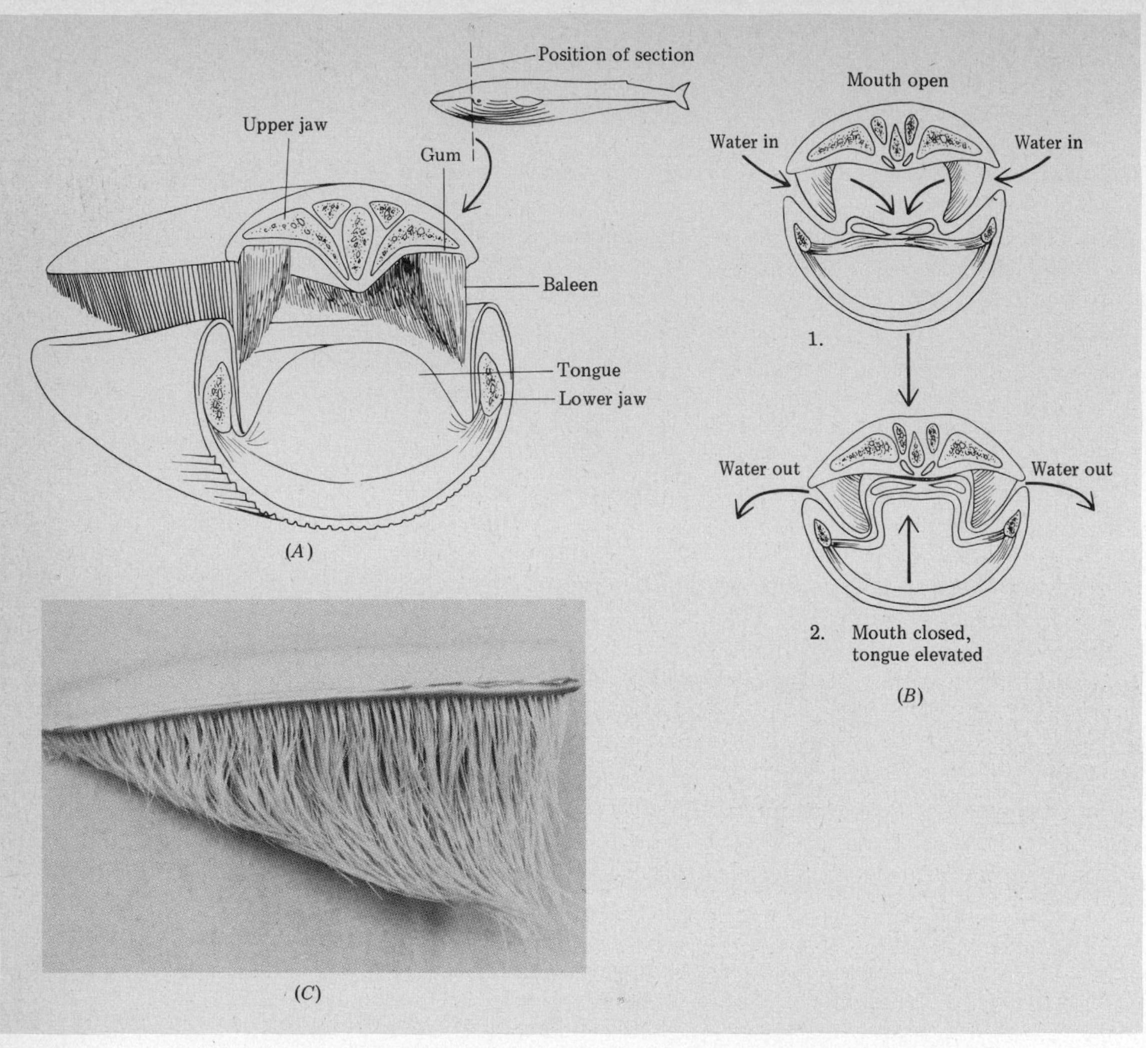

Figure 3-22 *(A)* Cross section of a baleen whale head to indicate the functional anatomy of the feeding apparatus. *(B)* The feeding process in a baleen whale: (1) mouth open and tongue depressed, water and plankton flow in; (2) mouth closed, tongue raised, water passes out through the baleen which strains out the plankton. *(C)* Photograph of the baleen from a gray whale *(Eschricthius robustus)*. (Modified from E. J. Slijper, Whales, Cornell University Press, Hutchinson. Photograph by the author.)

animals preying upon the giant squid, *Architeuthis*. (The stomach of one sperm whale yielded a giant squid 10.5 m long and weighing 184 kg.) The second largest carnivore of the nekton is the killer whale (*Orcinus orca*), which preys upon fishes, penguins, porpoises, seals, and sea lions (see Plate 6). These animals have also been known to attack the much larger baleen whales, which they do in packs of 3 to 40. In such attacks, the killers bite off pieces of the larger prey, whereas when feeding on smaller porpoises, seals, and birds, they

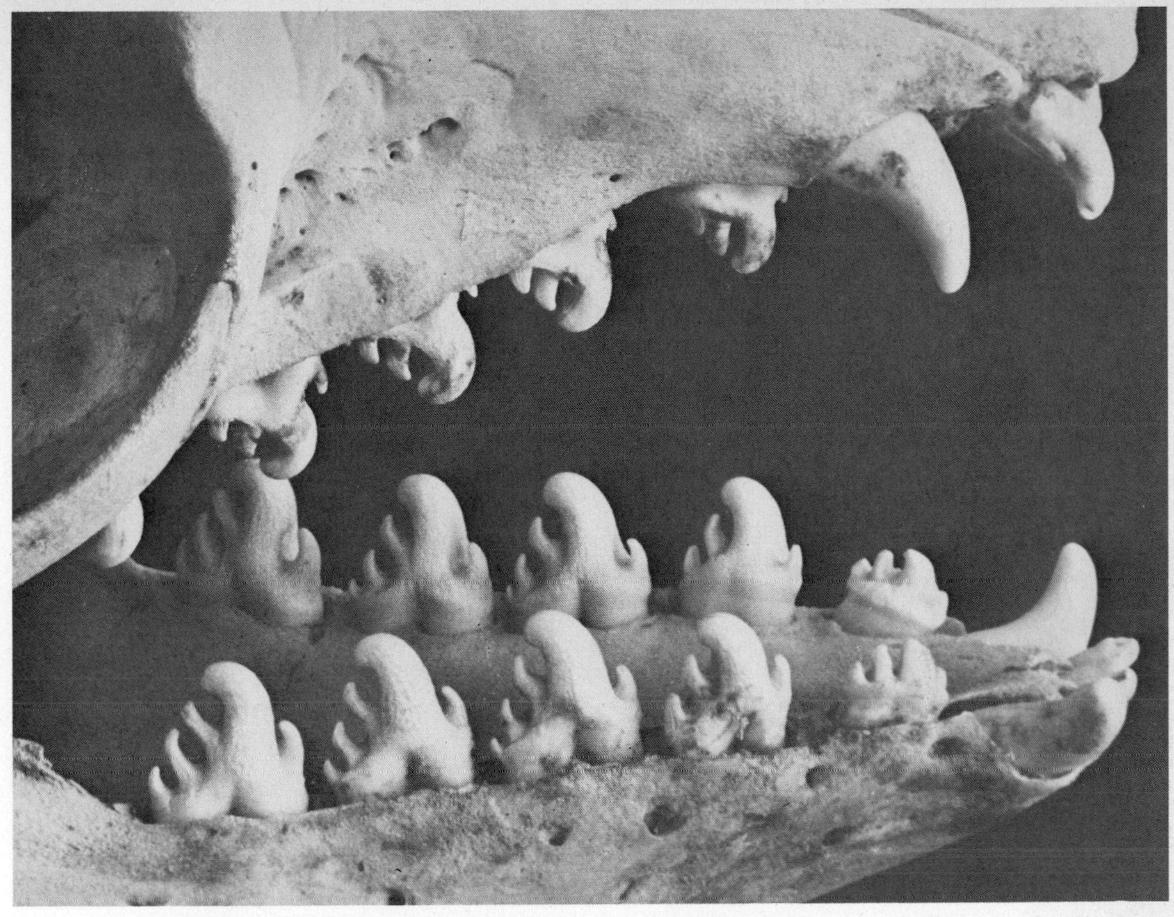

Figure 3-23 The teeth of the crabeater seal *(Lobodon carcinophagus)*. (Photo courtesy of AIBS from September 1964 cover of *BioScience*. Used by permission of Dr. V. B. Scheffer.)

swallow their prey whole. The capacity of these animals is truly enormous; one specimen, reported by Slijper (1976), was captured with parts of 13 porpoises and 14 seals in its stomach (Fig. 3-24)!

The smaller toothed whales, seals, and sea lions tend to feed on various squids and fishes, which they capture and swallow whole. Both groups have been attacked by humans because of the belief that they feed heavily on commercially important fishes such as salmon and therefore compete with humans. Although they take commercially important fishes, it remains to be demonstrated if such fishes constitute any significant part of the diet. The truly fish-eating whales include dolphins of the genus *Delphinus*. Examination of the stomachs of five species of delphinids revealed a diet composed 90 percent of mesopelagic lantern fish. These porpoises have a short, slender rostrum with many teeth. Those porpoises feeding on squid tend to have reduced numbers of teeth, but increased ridges in the palate to hold slippery squid. Included with squid eaters are the rare beaked whales (Ziphiidae) and the pilot whales.

Figure 3-24 Contents of the stomach of a killer whale *(Orcinus orca)*. (From E. J. Slijper, Whales and dolphins, University of Michigan Press, 1962; used by permission of Springer-Verlag, Heidelberg.)

Pelagic birds are generally more restricted in their feeding areas as they cannot dive as deeply as the marine mammals. They appear to feed mainly in the surface waters. They consume various small fishes and squids. Since certain species often occur in tremendous numbers, it may be that they have a very significant effect on the fish and squid populations of the upper layers, but we are ignorant of the details of this at present. They may also well be competing with various fishes and smaller marine mammals for the same food resource.

We know most about the food of various nektonic fishes and of these, the commercially important species such as tuna, albacore, and marlin are best known. They consume fishes, squids, and crustaceans.

Tunas have been intensively studied and their food is extremely varied, with as many as 180 different food items reported for striped tuna (*Katsuwonus pelamis*). Part of this varied diet is due to the great areas over which they move, thus encountering different arrays of food items. Small tunas tend to feed on species in the surface layers of the epipelagic zone, whereas larger tunas obtain food from depths where they capture mesopelagic organisms. As with planktivores, the composition of the diet varies with season and location. Marlins and swordfish have diets similar to tunas.

Nektonic sharks have a significant role in the feeding relationships of nekton, feeding on several different species of fishes, including tunas, marine birds, mammals, and squids. The largest carnivorous shark, the white shark, has the most varied diet, which includes not only fishes, birds, and mammals, but also other sharks. One was even found with remains of a basking shark in its stomach.

One of the most consistent features of feeding in nektonic fishes is the general lack of selectivity or specialization. Most fishes seem to feed on any food present of the appropriate size.

Figure 3-25 Food webs and trophic structure of the pelagic community in three different geographical areas. *(A)* Cold temperate waters. Algae: (1) diatoms. Herbivores: (2) euphausiids, (3) copepods, (4) pteropods. Filterers: (5) juvenile fish, (6) anchovy, (7) hyperiid amphipods, (8) lantern fish. Predators: (9) salmon, (10) squid, (11) lancetfish. Top predators: (12) mackerel shark, (13) porpoise, (14) seals and sea lions, (15) killer whale, (16) marine birds.

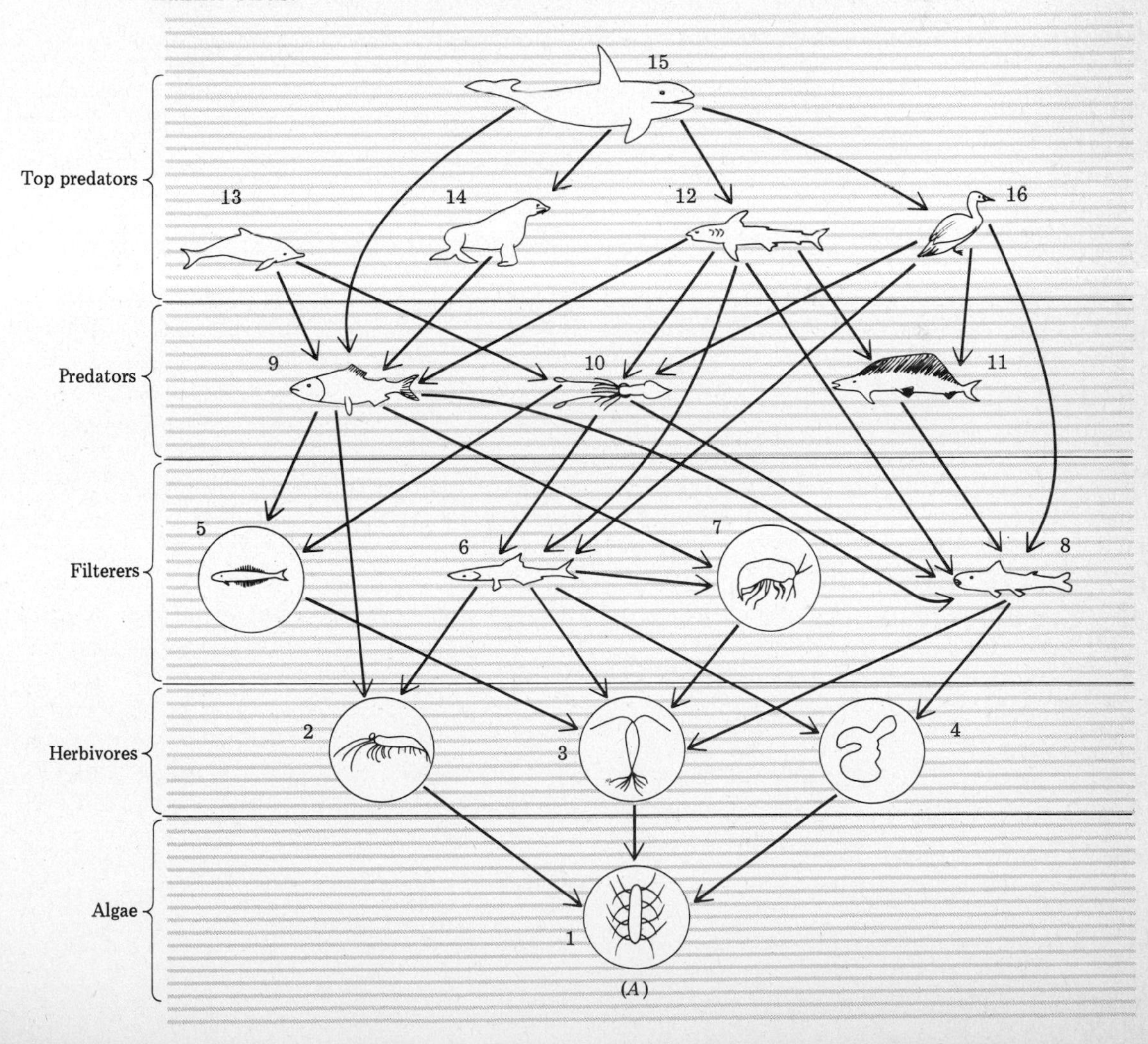

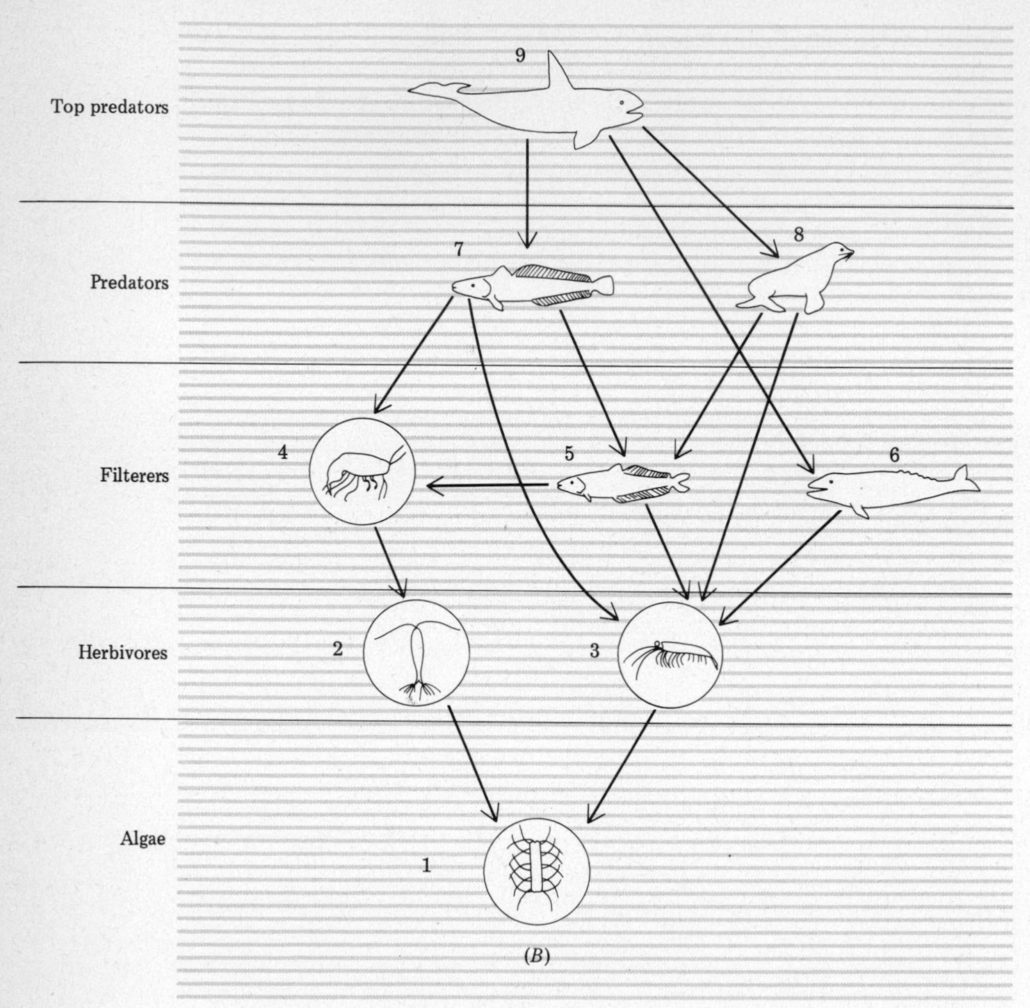

Figure 3-25 *(Continued)* *(B)* Antarctic seas. Algae: (1) diatoms. Herbivores: (2) copepods, (3) euphasiids. Filterers: (4) hyperiid amphipods, (5) planktivorous fish, (6) baleen whales. Predators: (7) predatory fish, (8) seals and sea lions. Top predators: (9) killer whale.

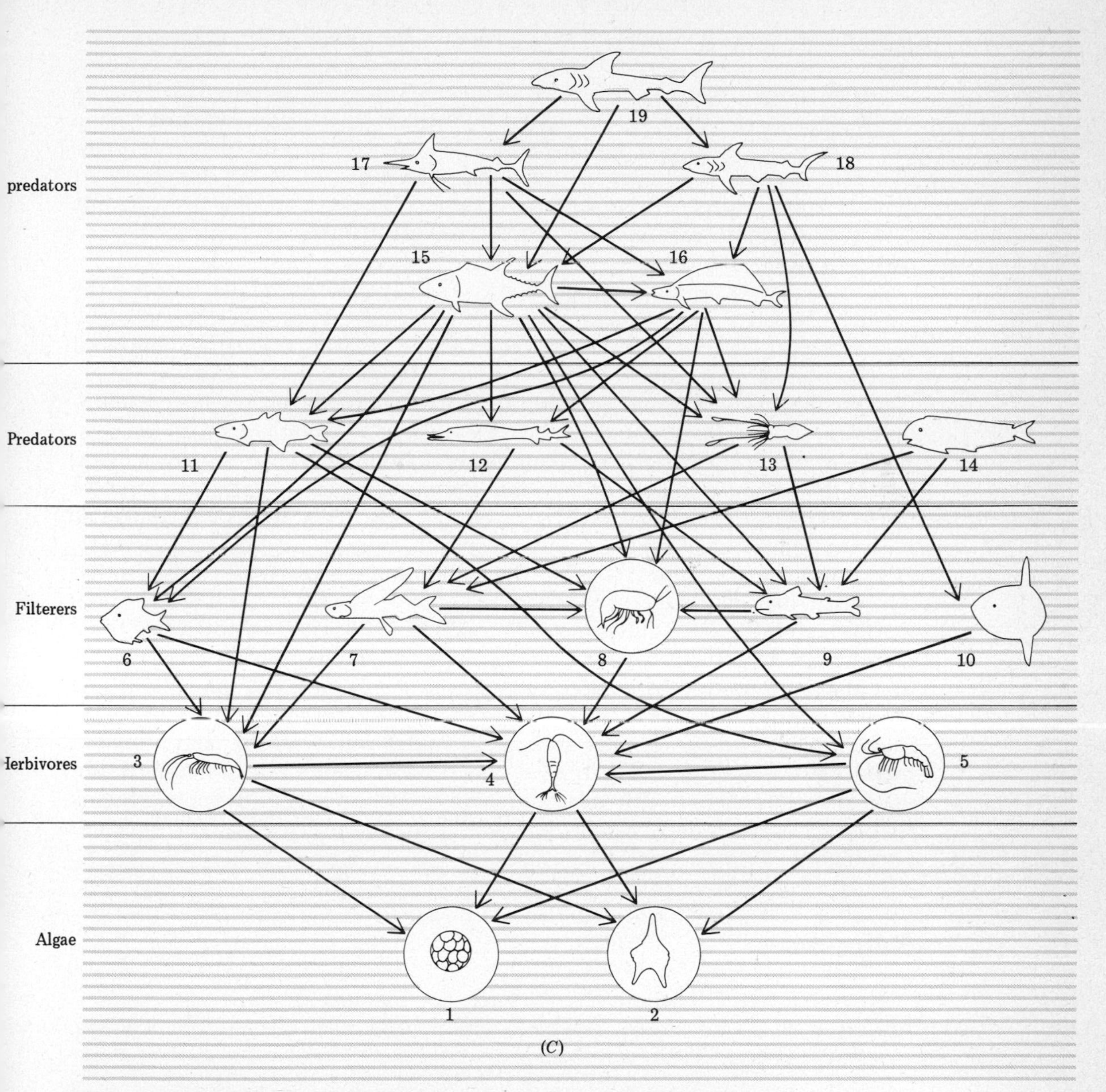

Figure 3-25 *(Continued)* *(C)* Tropical seas. Algae: (1) coccolithophores, (2) dinoflagellates. Herbivores: (3) euphausiids, (4) copepods, (5) shrimp. Filterers: (6) vertically migrating mesopelagic fishes, (7) flying fishes, (8) hyperiid amphipods, (9) lanternfish, (10) ocean sunfish. Predators: (11) mesopelagic fishes, (12) snake mackerel, (13) squid, (14) dolphin *(Coryphaena)*. Top predators: (15) tuna, (16) lancetfish, (17) marlin, (18) medium-sized sharks, (19) large sharks. (*A, B, C,* Modified from N. V. Parin, Ichthyofauna of the epipelagic zone, trans. from the Russian by the Israel Program for Scientific Translations.)

With the limited information now available, it is possible to outline the food webs that exist in the open epipelagic zones in the oceans of the world, but only in a very simplified and general way, as much more study is needed to define all of the links and their relative importance. These food webs are different among the tropical and polar oceans (see Fig. 3-25 on pp. 117–119). In the north Pacific Ocean, the main producers are diatoms, which are, in turn, consumed by various zooplankton. These serve as food for plankton feeding fish, mainly salmon and migrating mesopelagic forms. Top carnivores include the sharks and marine birds and mammals.

In the Antarctic, primary production is also centered in the diatoms and the foremost herbivores are copepods and euphausiids. These are consumed by baleen whales and certain fishes. The top-level carnivores are the penguins and various marine mammals.

In the tropical waters, the primary producers are dinoflagellates and coccolithophores. The herbivores include a wide variety of zooplankton organisms. These, in turn, provide food for a number of planktivorous fishes such as flying fish at the surface and lantern fish from the mesopelagic. All serve as food for the larger, first-level predatory fishes and squids. This level of predators is preyed upon by larger predators such as marlin, swordfish, and sharks. Finally, at the top are the largest sharks, the white and mako, which attack tuna, swordfish, and marlin.

Although these are highly simplified food webs, it should be noted that there are marked differences between the tropics and temperate-polar areas. In the first place, the food webs are more complex in the tropics, with more links and more trophic levels. This is partly due to the greater number of species present in the tropics and the general absence in colder waters of the larger, swifter predaceous fishes (tuna, marlin). The second observation is that marine mammals and birds seem to play a larger role in the food webs of the polar regions than the tropics. Indeed, larger numbers of marine mammals occur in colder seas than in warmer seas.

References

Aleyev, Y. G. 1977. Nekton. Dr. W. Junk, The Hague. 435 pp.

Andersen, H. T. (ed.). 1969. The biology of marine mammals. Academic Press, N.Y. 511 pp.

Carey, F. G. 1973. Fishes with warm bodies. Sci. Amer. 228(2):36–44.

Carr, A. 1965. The navigation of the green turtle. Sci. Amer. 212(5):79–86.

Clarke, M. R. 1979. The head of the sperm whale. Sci. Amer. 240(1):128–141.

Dunson, W. A. (ed.). 1975. The biology of sea snakes. University Park Press, Baltimore, Md. 530 pp.

Gray, J. 1957. How fishes swim. Sci. Amer. 197(2):48–54.

Hardy, A. 1965. The open sea: Its natural history. Vol. II, Fish and fisheries. Houghton Mifflin, Boston. 322 pp.

Kooyman, G. L., and R. J. Harrison. 1971. Diving in marine mammals. Oxford biology reader no. 6.

Parin, N. V. 1970. Ichthyofauna of the epipelagic zone. Jerusalem, Israel Program for Scientific Translations. 206 pp.

Scheffer, V. B. 1976. A natural history of marine mammals. Scribner, N.Y. 157 pp.

Slijper, E. J. 1976. Whales and dolphins. Univ. of Michigan Press, Ann Arbor. 170 pp.

Würsig, B. 1979. Dolphins. Sci. Amer. 240(3):136–148.

Chapter Four
DEEP-SEA BIOLOGY

By far the most extensive habitat on the planet inhabited by living organisms is that which comprises the permanently cold, dark waters of the deep oceans and the associated bottom. The shallow waters fringing the continents and islands of the world make up less than ten percent of the total area of the world's oceans, and the lighted upper waters of all oceans are an even smaller fraction of the total volume of ocean living space. Thus, of the 70 percent of the planet's surface covered with water, perhaps 85 percent of the area and 90 percent of the volume constitute the dark, cold area we call the deep sea.

Although this area is the largest habitat on earth, its biology is the least known and explored. This is primarily due to the difficulty of sampling this remote area and to the fact that, until recently, it was virtually inaccessible to humans. The recent advent of deep submersible vehicles has now allowed at least a portion of this vast area to be observed by scientists.

Although this huge area is unlikely to be sampled, much less observed, by most users of this book, the very fact that it is such a large part of the ocean, which is still a frontier where new and exciting discoveries are being made, and because it may become increasingly important as a source of materials and as a dumping site for humans, it is well to consider it.

The object of this chapter is to outline our present understanding of the basic principles concerning adaptations and organization of life forms in this vast and intriguing area.

ZONATION

When we refer to the "deep sea" in this chapter, we will be referring to that part of the marine environment that lies below the level of light penetration in the open ocean and deeper than the depth of the continental shelves (> 200 m). The entire area is permanently dark. One term for this area is the *aphotic* zone as opposed to the lighted, or *euphotic*, zone, where all primary production occurs. In tropical waters, the start of the aphotic zone is deeper (~ 600 m), whereas in temperate waters, it is shallower (~ 100 m).

Several schemes of zonation for the deep sea have been proposed over the years and reviewed by Menzies, George, and Rowe (1973), but none has been universally acceptable to all scientists working in the area. There are a number of reasons for this, but the primary one is the lack of sufficient ecological information. Hence, most schemes of zonation have been produced by simply dividing the water column into discrete zones based upon depth, or temperature changes, or both. Another approach has been to divide the deep ocean on the basis of species abundance, distribution, or associations. A problem here is that certain species are distributed at different depths in different ocean basins, depending upon special hydrological and/or ecological characteristics of the region. Similarly, the changes in temperature occur at different depths in different areas. On the other hand, kinds of species may be better at indicating "zones" than the associated environmental parameters. We shall follow the zonation scheme outlined in Chapter One and illustrated in Fig. 1-17. The relevant deep-sea zones are listed in Table 4-1 and briefly reviewed here.

A primary division that can be made in the deep sea is into benthic (associated with the bottom) and pelagic (associated with the open water) zones. Since the physical environment differs between them, these two areas are inhabited by very different associations of organisms. We know perhaps more about the benthic deep-sea fauna than about the pelagic fauna. The benthic fauna, as noted in Chapter One, can be subdivided into those that occupy a *bathyal* zone on the continental slope and those living in a large *abyssal* zone constituting most of the deep-sea floor. The animals of the trenches live in a separate *hadal* (ultra-abyssal) zone, but little is known of their biology.

In the pelagic division, there is an upper zone immediately below the euphotic zone, which contains animals, many of which migrate vertically into the

TABLE 4-1 Deep Sea Faunal Zones

Light	Pelagic Zones	Depth Range	Benthic Zones	Depth Range
Present (photic)	Epipelagic or euphotic	0–200 m	Continental Shelf or sublittoral	0–200 m
Absent (aphotic)	Mesopelagic	200–1,000 m (?)	Bathyal	200–4,000 m (?)
	Bathypelagic	(?) 1,000–4,000 m (?)		
	Abyssal pelagic	(?) 4,000–6,000 m (?)	Abyssal	4,000–6,000 m (?)
	Hadal pelagic	6,000–10,000 m	Hadal	6,000–10,000 m

After Hedgepeth, 1957.
Note: (?) = boundary uncertain.

euphotic zone at night. This is variously called the "*twilight*" or *mesopelagic* zone. This zone is inhabited by many species of animals with well-developed eyes and a great variety of light organs. The dominant fishes here are black, and the crustaceans are red. Because this is the most accessible of the deep-sea pelagic zones, we know most about it. The numbers of organisms appear to be the highest of all the pelagic zones. This zone extends down to about 700 to 1,000 m, the depth varying with location, clarity of water, and other factors.

Below the mesopelagic, it is much more difficult to establish any pelagic zones that are universally acceptable. The region between the lower limit of the mesopelagic and the upper limit of the deep trenches (taken to be 6,000 m) has been divided into two zones by Hedgpeth (1957): an upper *bathypelagic* and a lower *abyssal pelagic*, a division continued here. The boundaries of these zones are very uncertain and vary from one author to another. These areas are characterized by low numbers of species and individuals, considerably fewer than observed in the mesopelagic. The organisms tend to be white or colorless, with reduced eyes and bioluminescent organs.

The open water of the trenches is often called the *hadal pelagic* zone. It appears set off from the other areas and we are so ignorant of it that we do not know if a separate pelagic community exists in these areas.

SAMPLING THE DEEP SEA

In the preceding section, we alluded to the problems of sampling and obtaining information about the deep sea. Before continuing this discussion, it is useful to indicate briefly how our information is obtained from the deep sea and why it is difficult to obtain and often biased. An analogy has been made that characterizing the deep sea and its communities using the gear that we have traditionally used is like characterizing the terrestrial communities using samples taken with a butterfly net towed behind an airplane! Exaggerated perhaps, but close enough to have a ring of truth.

The types of gear used to sample the deep-sea benthos include various types of nets, grabs, and dredges of similar design to those used in shallow water. Representative types are illustrated in Figure 4-1. Sampling of pelagic animals is done with several types of midwater trawls (Fig. 4-2). Because the density of organisms is so low in pelagic areas below the mesopelagic, these midwater nets must have a large gape and sieve a lot of water to obtain any number of animals and to reduce the problem of simple avoidance of nets by the animals.

The main problem in sampling the deep sea, pelagic or benthic, is that the deeper the sampling, the more cable necessary to get down to depth and the longer the time needed to make a single haul. Furthermore, as more and more cable is played out in a deep haul, the greater the weight and the greater the chance for snarling the cable or otherwise fouling the gear. In order that most nets towed on the bottom fish correctly, an amount of line equalling two to three times the depth must be played out. If one were trawling at only medium depths

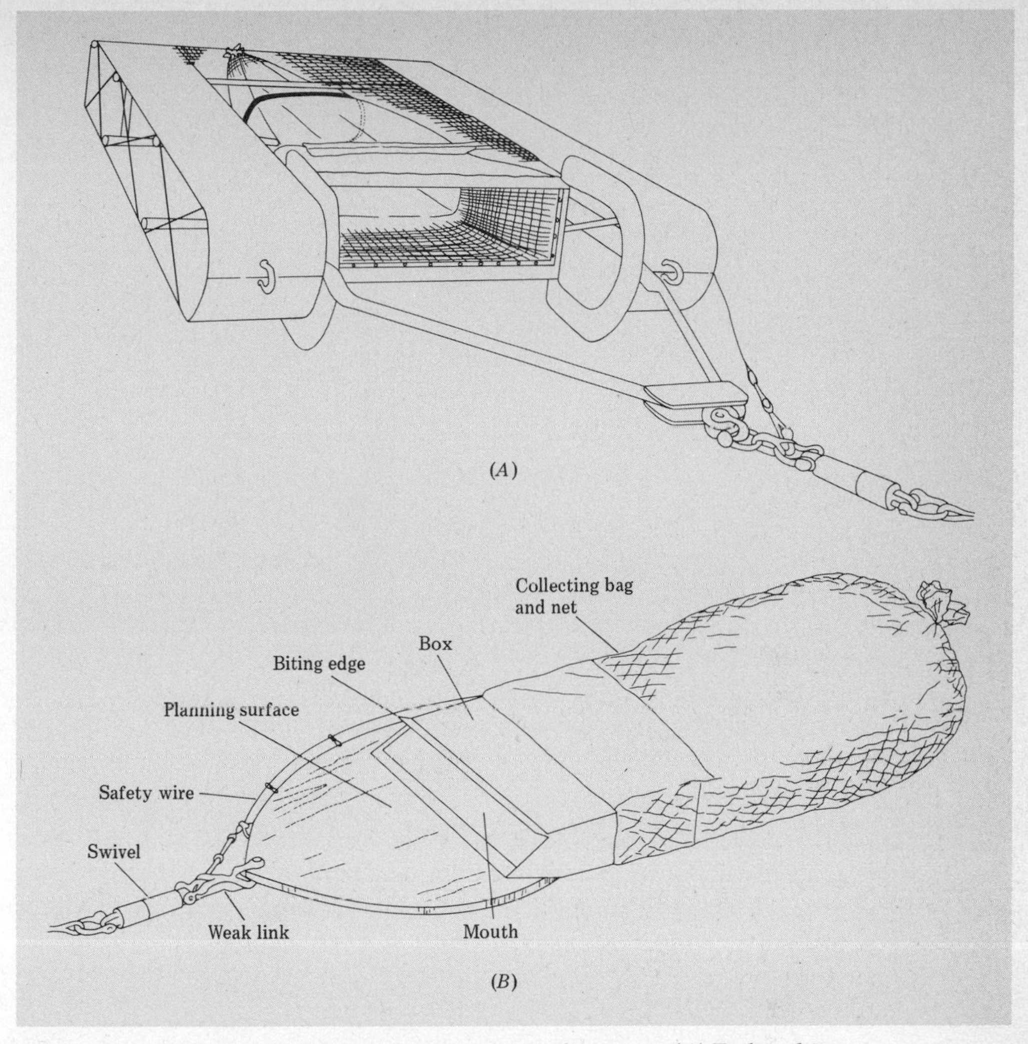

Figure 4-1 Some types of deep-sea bottom sampling gear. *(A)* Epibenthic sled. *(B)* Anchor dredge. (*A,* from R. J. Menzies, R. Y. George, and G. T. Rowe, Abyssal environment and ecology of the world oceans, 1973, Wiley. *B,* from J. Nybakken, Readings in marine ecology, Harper & Row.)

of, say, 4,000 m, this would be about 8,000–12,000 m of cable, or 8–12 km (= 5–7 miles). Not only do few vessels have the capability of carrying such amounts of cable, but it takes a considerable amount of time to feed that amount out and then retrieve it. Twenty-four hours may be required to make one trawl in the trenches. One can thus appreciate why the number of samples from deep water is not great.

Still other problems plague the deep-sea biologist. In shallow waters of the continental shelf, it is possible to ascertain if the trawl or dredge is fishing on the bottom by "feeling" the vibrations on the cable. In the deep sea, so much cable is out that the weight of it far exceeds the drag of dredge or trawl, and vibration in the cable does not reflect action of the sampling device. Failure of the gear to

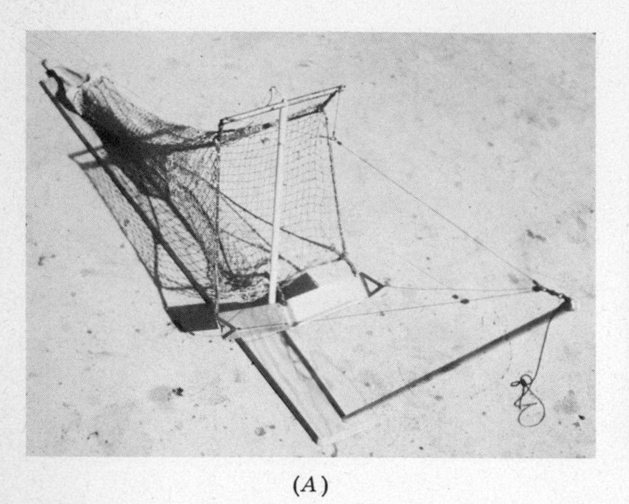

(A) (B)

Figure 4-2 *(A)* A small Issacs-Kidd midwater trawl, which is used to capture pelagic organisms. *(B)* Rigging a large midwater trawl for fishing. (*A,B,* courtesy of Dr. Greg Cailliet.)

contact the bottom or to fish correctly has been suggested by Menzies (1964) as responsible for up to 50 percent of deep-sea sample failures. Finally, with so much wire played out, the chance for snarling, knotting, or wrapping it up in the gear is considerably increased. Best results are obtained where depth-sensing devices with shipboard readout are used with the gear so that the gear can be controlled.

A major problem in sampling the deep-sea benthos is the basic qualitative nature of most dredging devices. They sample an unknown area of the bottom. Because densities of species and individuals are difficult or impossible to estimate from these samples, comparative studies of deep and shallow water have been difficult. Fortunately, recent investigations in the Pacific Ocean have employed large box corers, which largely overcome this problem by sampling a known bottom area.

Finally, in sampling the midwater animals, there is the problem of net avoidance. Trawling in deep water usually means that the ship must travel relatively slowly, because of the great weight of the cable it drags. This means that many fast-swimming pelagic animals may simply avoid the net by swimming out of its path. If this happens to any great extent, then characterizing the communities from such net hauls gives a misleading idea of what is in fact there. In recent years, deep-sea cameras of various designs have done much to increase our understanding of the real state of deep-sea communities, particularly bottom communities, and the advent of deep submersibles has further advanced our understanding of these areas (Figs. 4-3, 4-4).

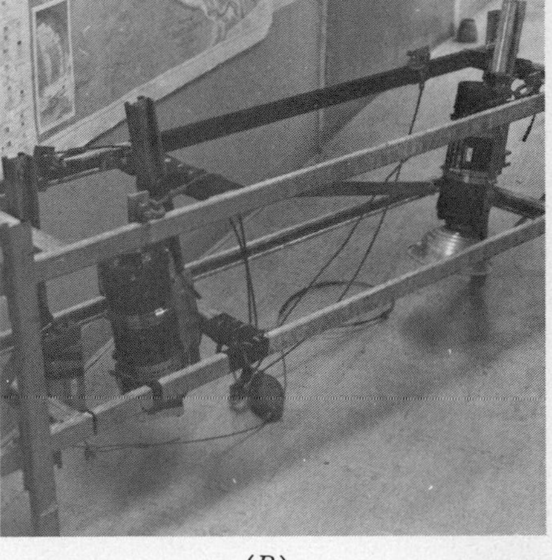

(A) (B)

Figure 4-3 *(A)* The deep submersible *Alvin*. *(B)*. A deep sea camera. *(A,* courtesy of Dr. H. Mullins. *B,* photo by the author.)

ENVIRONMENTAL CHARACTERISTICS

Before considering the life in the deep ocean, it is necessary to establish the particular physical and chemical conditions under which the organisms must function. Perhaps the major point to be emphasized in considering these factors is that, at any given level or position in the deep ocean, these factors remain remarkably constant throughout long periods of time.

LIGHT

In the deep ocean, the entire area is without light, except at the very upper parts of the mesopelagic, where there may be some light that penetrates in certain areas and under certain conditions. Since this area is either dark at all times or has such extremely low light levels, photosynthesis is not possible, precluding any plant-based primary productivity. What light is present in this dark world is usually produced by the animals themselves (see pp. 141–144). This lack of light means that the organisms have to rely on other senses to find food and mates and to maintain various inter- and intraspecific associations.

PRESSURE

Of all the environmental factors acting on deep-sea organisms, pressure shows the greatest range. Pressure increases 1 atmosphere (14.7 lb/in^2) for each 10 meters in depth. Since the deep sea varies from a few hundred meters down to the bottom of the trenches at more than 10,000 m, the range of pressure is from 20 to more than 1,000 atm. Most of the deep sea is under pressure between 200 to 600 atm. In no other marine environment does pressure exhibit such great range or play such a potential role in the distribution of organisms.

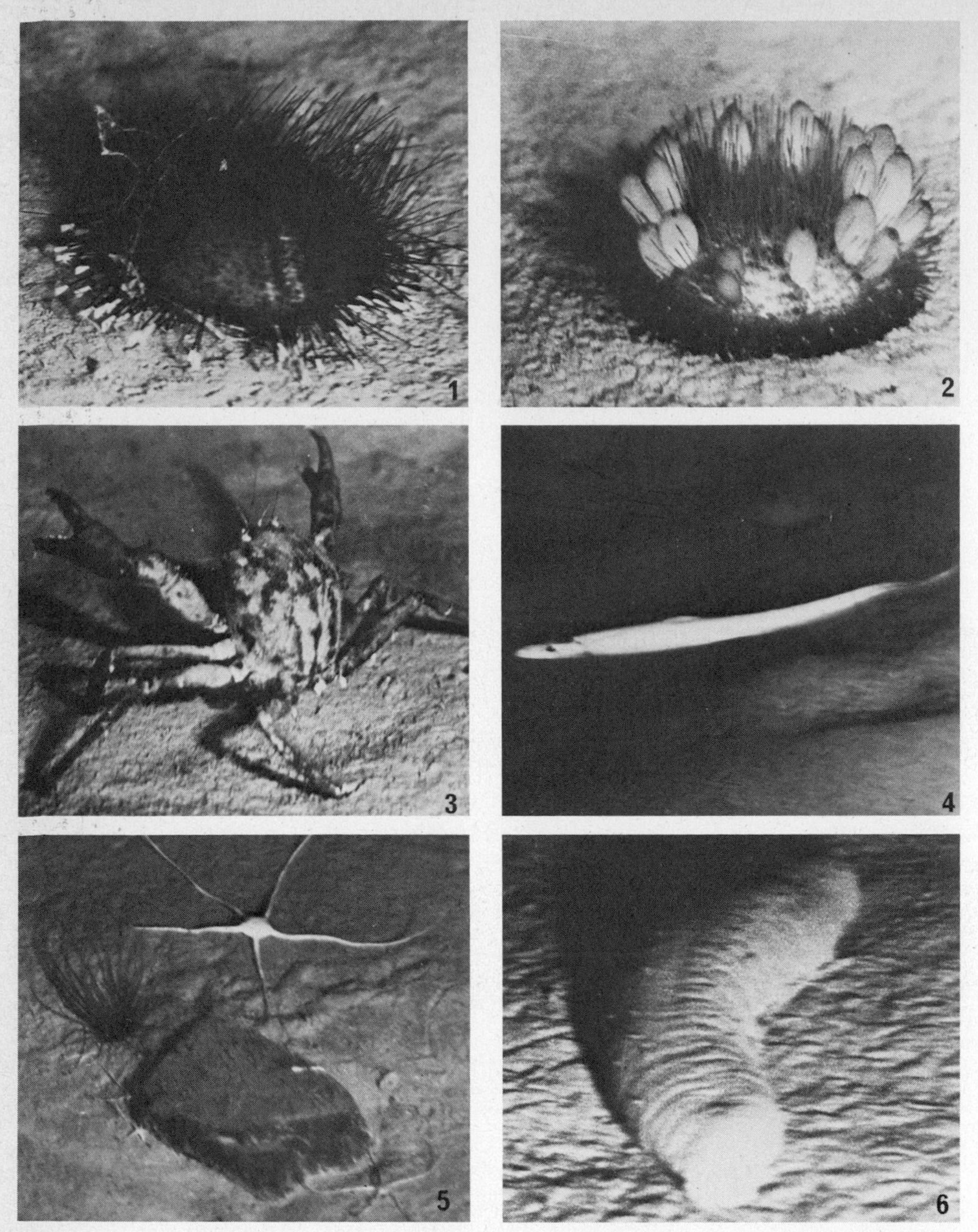

Figure 4-4 Some characteristic deep water benthic animals from the bathyal zone of the Gay Head–Bermuda transect in the Atlantic Ocean. (1) *Hygrosoma petersii* (echinoid). (2) *Phormosoma placenta* (echinoid). (3) *Geryon quinquedens* (decapod crustacean). (4) *Aldrovandia affinis* (fish). (5) *Ophiomusium lymani* (brittlestar). (6) *Mesothuria lactea* (sea cucumber). (Photos courtesy of Dr. Fred Grassle; reprinted with permission from J. F. Grassle et al., *Deep Sea Research,* vol. 22, copyright 1975, Pergamon Press Ltd.)

Unfortunately, we lack detailed information on the direct effects of pressure on most deep-sea organisms. This is primarily due to the fact that virtually all organisms trawled up from the deep sea arrive at the surface in a dead or dying state. Only in recent years, with the advent of traps that incorporate a special pressure-maintaining chamber, have undamaged larger metazoan animals been retrieved from the deep sea in good condition. Some of these have been maintained for experimental purposes, and we are obtaining more knowledge of the real biological effects of pressure.

That pressure must have a significant effect on deep-sea organisms can be extrapolated from experiments done on one group of organisms successfully retrieved alive from the deep sea. Bacteria have been retrieved alive and cultured from the deepest areas of the ocean. Figure 4-5 shows the results of culturing a deep-sea bacterium under different pressure regimes. The fact that bacteria from the deeper areas virtually cease growth and reproduction at lower pressures, but grow and reproduce actively at higher pressures corresponding to their natural environment, strongly suggests that there are special adaptations to pressure and that they can profoundly affect organisms. Since metazoan animals are anatomically and physiologically more complex than bacteria, it seems highly probable that multicellular animals will evince an even stricter pressure-dependent physiology.

Recent studies have been done by Siebenaller and Somero (1978) on the enzyme systems of two closely related fishes living at different depths. These studies have demonstrated that differences in hydrostatic pressure as small as 100 atm or less are sufficient to alter markedly the functional properties of enzymes, namely, their ability to bind to the appropriate substrate and the rate at which the reaction proceeds. Such changes are comparable to the changes in enzyme activity caused by temperature differences. These results strongly

Figure 4-5 Graph of the amount of growth of a deep sea bacterium at different pressures. (From A. A. Yayanos, Dietz, and van Boxtel, *Science*, vol. 205, no. 4408, p. 809, August, 1979. Copyright 1979 by the American Association for the Advancement of Science.)

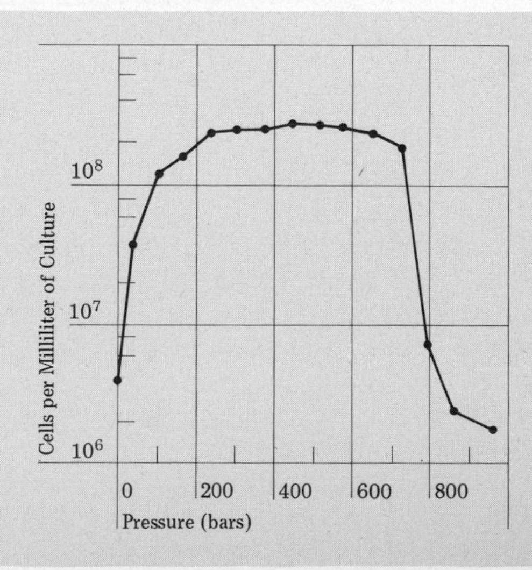

suggest that hydrostatic pressure plays a major role in adaptation to the deep-sea environment.

A great number of laboratory studies on various kinds of cells from protozoa to mammalian have demonstrated that pressure markedly affects the morphology of the cells, including ability to form mitotic spindles and to undergo mitosis. Zimmerman and Zimmerman (1972) report that amoebas tend to lose their pseudopods and ball up, and *Tetrahymena* loses its ability to locomote with cilia. Pressure also has a dramatic effect on certain physiological and biochemical processes such as typical muscle physiology. Perhaps the most dramatic effect of pressure is its effect on macromolecules such as proteins. Protein synthesis and function are markedly affected by pressure in laboratory situations, usually adversely. Since proteins are vital in living systems as enzymes and as major building blocks, it would seem that deep water animals must have some special adaptation with respect to protein structure and function in order to survive.

The body of laboratory studies of the observed definite effect of pressure on shallow water animals, coupled with the known effects of pressure on cells and macromolecules (= protein), and the recent experimental evidence from deep-sea animals themselves are strong indications that the physiology of deep-sea animals is profoundly affected by pressure and that pressure may be of considerable significance in zonal patterns of distribution.

SALINITY

Below the first few hundred meters in the world's oceans, the salinity is found to be remarkably constant throughout the depths. There are some minor differences in salinity, but none that can be considered ecologically significant.

TEMPERATURE

The area of greatest and most rapid temperature change with depth in the oceans is the transition zone between the surface waters and the deep waters, the area known as the thermocline. Thermoclines vary in thickness from a few hundred meters to nearly a thousand. Below the thermocline, the water mass of the deep ocean is cold and far more homogeneous. The temperature continues to decrease toward the bottom, but the rate of change is much slower (Fig. 4-6). Below 3,000–4,000 m, the water is essentially isothermal. What is ecologically significant is that at any given depth, the temperature is practically unvarying over long periods of time. There are no seasonal temperature changes, nor are there any annual changes. Perhaps nowhere on earth is there another habitat with such a constant temperature.

OXYGEN

Despite the fact that the deep ocean waters lie far removed from a source of oxygen replenishment, either by interaction with the atmosphere or through

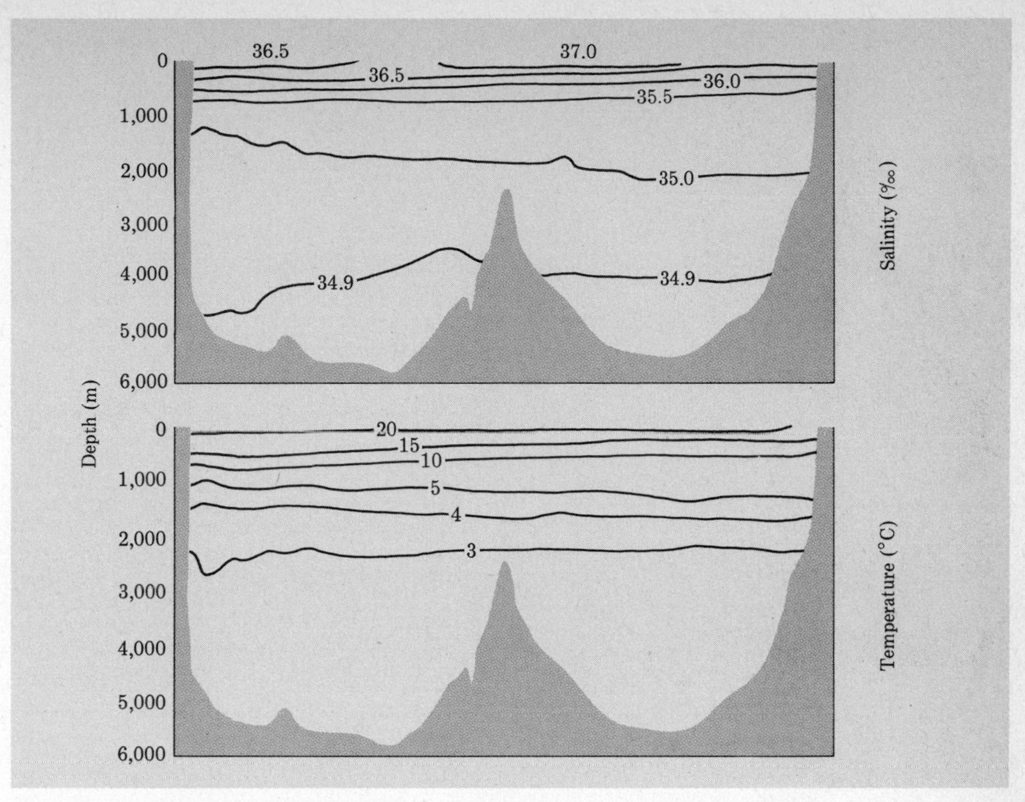

Figure 4-6 Salinity and temperature changes across the North Atlantic Ocean with respect to depth. (After D. A. Ross, Introduction to oceanography, Prentice-Hall; from F. C. Fuglister, Atlantic Ocean atlas.)

production by plants, there are essentially no abyssal or hadal areas that lack sufficient oxygen to support life. The few exceptions include the Carioca Trench off Venezuela in the Caribbean and the Santa Barbara basin off California. They are anaerobic at the bottom. The oxygen present in the deep water masses entered the water when this now deep water mass was at the surface. Virtually all of the water of the deep sea has its origin at the surface in the Arctic or Antarctic areas (see Chapter One, pp. 18–19). Here, the oxygen-rich cold water then sinks and flows north or south to make up the deep water of the world's oceans. Since it no longer gains oxygen after sinking, the reason that these deep masses are not depleted of oxygen from respiration of deep water organisms is likely due to the fact that the abundance of deep water organisms is so low that it is insufficient to deplete the resident oxygen supply. That this situation is true is supported by the fact that the oxygen concentration declines in the 20 meters or so just above the bottom in the deep sea. It is also near the bottom that the most dense concentrations of organisms occur in the deep sea.

Another peculiar feature of oxygen concentration in the deep sea is the presence of the aforementioned *oxygen minimum* zone between 500–1,000 m (Fig. 1-3). Oxygen values are higher, both above and below this zone. In the oxygen minimum zone, oxygen values may fall to less than 0.5 ml/liter. The occurrence of this zone is mainly the result of respiration of organisms, coupled with the lack of interchange of the water mass with more oxygen-rich water. The reason that an oxygen minimum zone occurs at 500–1,000 m and not deeper is that the numbers of animals in deeper waters are so low that they never deplete the available oxygen, whereas at 500–1,000 m, organisms are abundant. Above 500 m, oxygen depletion does not occur, even though the animal biomass is high, because there is constant replenishment from air and by plants.

FOOD

The deep sea is removed from the photosynthetic zone and has no primary production except for certain areas, where chemosynthetic bacteria are found (see pp. 151–152). All organisms living in the deep sea are thus dependent on food that ultimately is produced elsewhere where photosynthesis is possible and subsequently transported into the deep sea. The deep sea is therefore unique among the world ecosystems in that it has no indigenous primary productivity.

Potential food must come into the deep sea by sinking from the surface waters. Since the populations of organisms are greatest in the upper layers of the oceans, the chance that any food particle will be able to sink through all these voracious animals without being consumed or decaying is small. As a result, few food particles are available to the animals of the deep sea. This paucity of food is probably the reason that the density of deep-sea animals is very low. Without sufficient energy in the form of food, large numbers of organisms cannot be sustained.

The probability that a particle of food will decay or be consumed during sinking increases with increasing depth, since there are more time and more organisms to consume it. Thus, the deeper the organisms, the less the food available to them.

Perhaps a major portion of the particles sinking from the euphotic zone are particles of material that are not directly suitable as food. An example would be fecal pellets and shed chitinous exoskeletons of crustaceans. Most organisms cannot digest chitin. These materials, however, are acted upon by bacteria during descent into the deep sea and, after reaching the bottom, are converted into suitable food in the form of bacterial protoplasm. Because the residence time for a food particle is longer on the bottom than in the water column, there is an increase in bacteria in the bottom oozes of deep sea, which thus allows for increased numbers of larger organisms. Densities may exceed that of pelagic organisms at the same depth. This also explains the reduction in oxygen in the near-bottom water.

The amount of food available to the deep sea is correlated with the amount of primary productivity in the surface waters above or else with the proximity to a secondary source such as organic debris from terrestrial habitats. Deep-sea areas under productive surface waters and near islands or continents have more food than deep waters under low surface productivity or far from land.

Several types of food sources are available in the deep sea. Directly available foods include those deep-sea organisms that spend their early or larval stages in the lighted upper waters and then migrate into the depths, where they furnish food to the predators. Another directly available food source is the large bodies of marine mammals and fishes that may sink quickly before all readily available flesh is consumed (Fig. 4-7). This would seem to be a particularly unpredictable type of food, but Isaacs and Hessler of Scripps Institution of Oceanography have shown, by placing baited cans with cameras attached into the deep sea, that food is quickly detected and brings in consumers from great distances (Fig. 4-8).

Indirect food sources include undigestible animal and plant remains, which first must be acted on by bacteria. Examples are chitin and wood and cellulose of land plants. Other potential sources are colloidal or dissolved organic material in the water and gelatinous plankton matter (= "marine snow"). At the present time, the relative importance of these sources cannot be evaluated.

We can summarize by saying that food is very scarce in the deep sea, when compared to other areas of the oceans, and that it decreases with depth. This food scarcity is responsible for the low densities of deep-sea organisms.

ADAPTATIONS OF DEEP-SEA ORGANISMS

Correlated with the conditions of the environment in the deep sea, the organisms that live there possess a number of adaptations. Because we have been unable to experiment with living metazoan animals from the deep sea

Figure 4-7 Food sources to the deep sea.

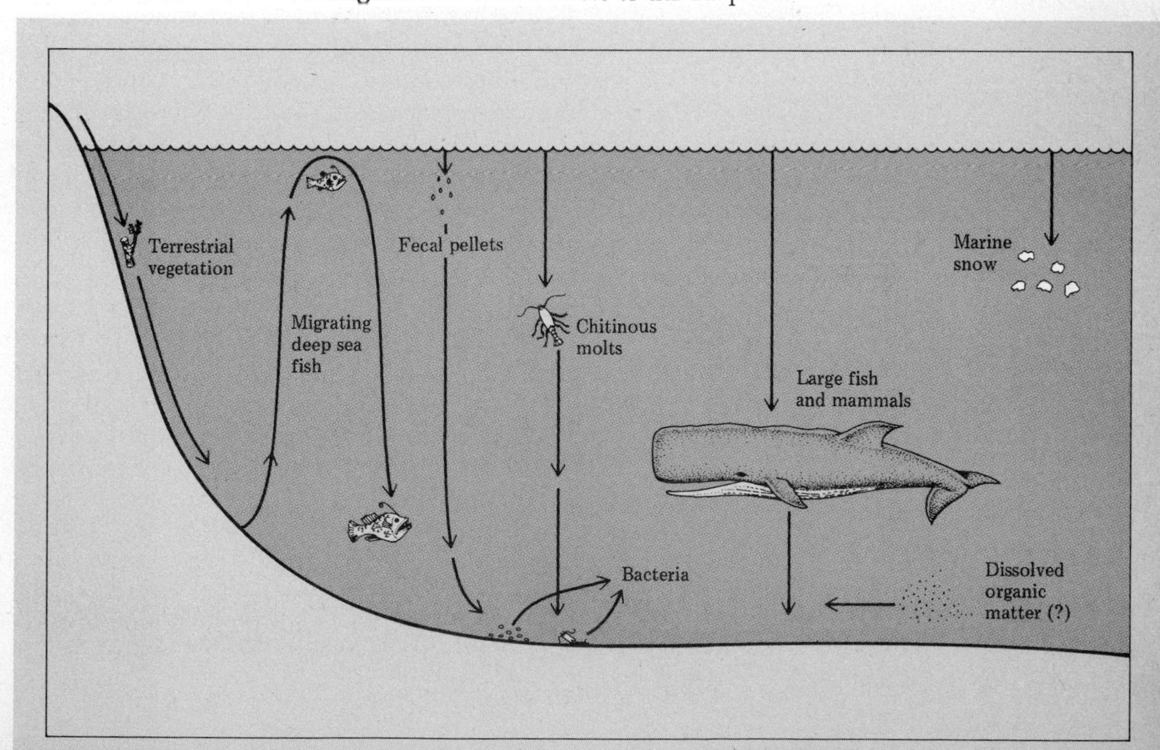

(A) (B)

Figure 4-8 Two photographs showing the rapid attraction of deep-sea animals to bait and its subsequent consumption. *(A)* At the time of reaching the bottom. *(B)* Six hours and 40 minutes after reaching the bottom, showing consumption of bait by the amphipod *Hirondella gigas* (at 9,605 m in the Philippine trench). (Photographs courtesy of Dr. Robert Hessler, Scripps Institution of Oceanography.)

except in a few cases (see pp. 127–130). we have little direct evidence of the function of many adaptations and must offer educated guesses based upon the conditions under which we know they live. We also know little about physiological and biochemical adaptions at the present time. As we perfect our ability to work with these animals, adaptations to life in theses areas will also be revealed.

One adaptation that can be observed, even in those organisms from the upper mesopelagic, is that of *color*. Mesopelagic fishes in particular tend to be either silvery gray in color or a deep black. They are not countershaded as are epipelagic fishes. The mesopelagic invertebrates, on the other hand, may be either purple or bright red in color. Mesopelagic jellyfish are often a deep, dark purple, whereas the crustaceans such as copepods, mysids, and shrimps are brilliant red (see Plate 5). Since these organisms live in waters that are virtually unlighted, organisms that are black will be invisible. But why the red color? Since red light is the first to be absorbed in sea water, organisms that are red also appear black at depth.

Organisms living even deeper, in the abyssal and bathyal zones, are often colorless or a dirty white. They seem to lack pigment. This is particularly true for animals dwelling on the bottom. Fishes, however, may be black at all depths.

Another adaptation seen particularly in mesopelagic and upper bathypelagic fishes is the presence of *large eyes*. The eyes in these fishes are much larger relative to body size than those in fishes of the lighted epipelagic zone (Fig. 4-9). Usually, the presence of large eyes is correlated with the presence of light organs (see pp. 141–144). These fishes dwell in the uppermost parts of the deep sea, where some small amounts of light may penetrate. Many also migrate vertically into the epipelagic zone at night. Thus, the large eyes give them maximum light-collecting abilities in an area in which light intensities are very low. The

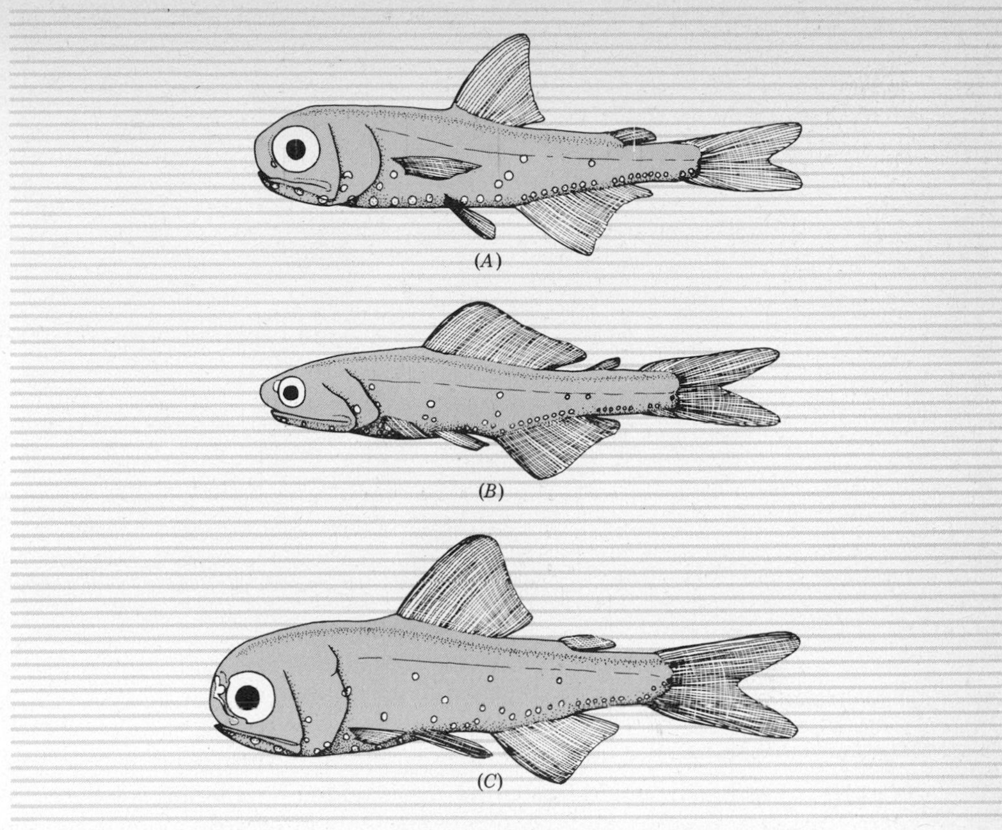

Figure 4-9 Large eyes in three mesopelagic fishes of the family Myctophidae. *(A) Myctophum punctatum. (B) Lampanyctus elongatus. (C) Diaphus metopoclampus.* (After N. B. Marshall, Aspects of deep sea biology, N. Y. Philosophical Library.)

large eyes are also presumably needed to detect the low light intensities of the light organs. Increased surface area is only one visual adaptation. These fishes also have enhanced "twilight vision" derived from the pigment rhodopsin and an increase in the density of rods in the retinal area.

Fishes dwelling in the deepest parts of the ocean (abyssal pelagic, hadal pelagic) show a different trend. Many of these have very small eyes or lack eyes entirely (Fig. 4-10). Because they dwell in permanent darkness, eyes are not necessary. Generally, fishes in depths of 2,000 m and above have eyes, often large; below 2,000 m, eyes are small, degenerate, or lost. Often, bottom-dwellers have no eyes.

Yet another adaptation in fishes is *tubular eyes* (Fig. 4-11; Plate 3). Fishes with peculiar tubular eyes are found in several families. They give the fishes an extremely bizarre appearance (Fig. 4-11). In these fishes, the eye is a short black cylinder topped with a hemispherical, translucent lens. Each eye has two retinas, one at the base of the cylinder and one on the wall of the cylinder. Apparently, these eyes function such that the main retina at the base of the tube focuses on nearby objects and that on the wall on distance objects. However, just why they should be developed in the deep sea is not clear.

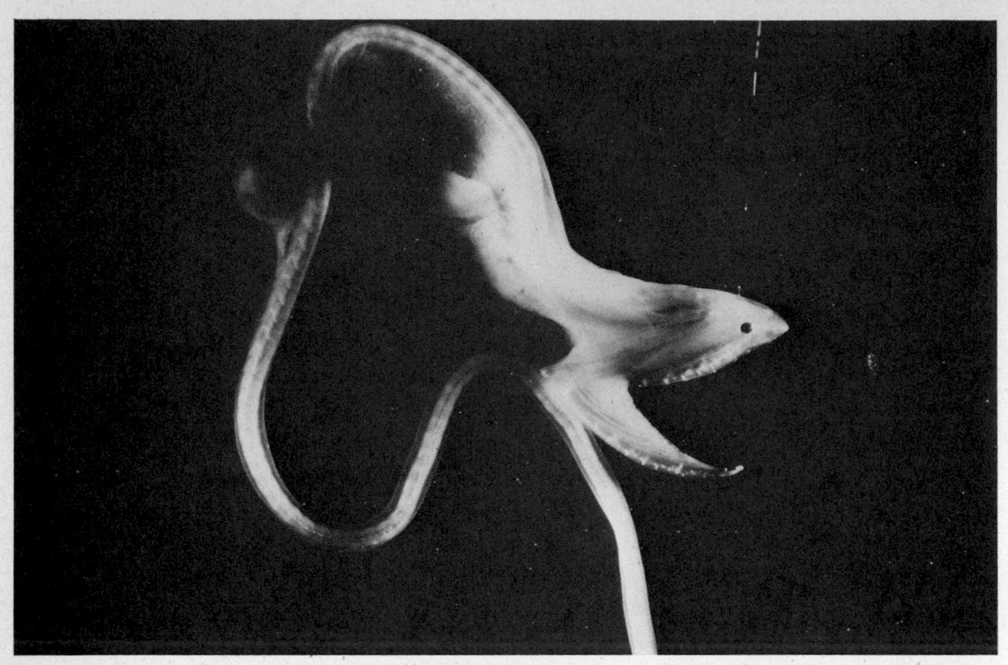

Figure 4-10 Fish with reduced eyes, a gulper eel of the genus *Saccopharynx*. (Photo courtesy of Michael Kelly.)

Among the invertebrates, certain squids of the family Histioteuthidae exhibit another peculiar adaptation in that they have one eye much larger than the other (Fig. 4-12). Such squids also display numerous light-producing organs called photophores. Recent work by Young (1975) has demonstrated that these squids,

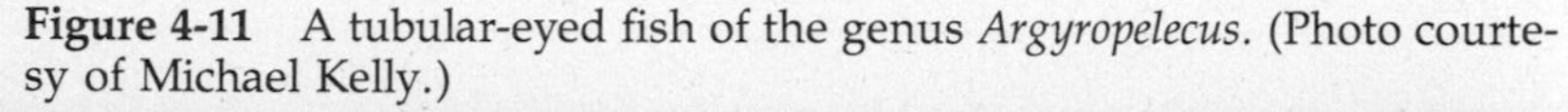

Figure 4-11 A tubular-eyed fish of the genus *Argyropelecus*. (Photo courtesy of Michael Kelly.)

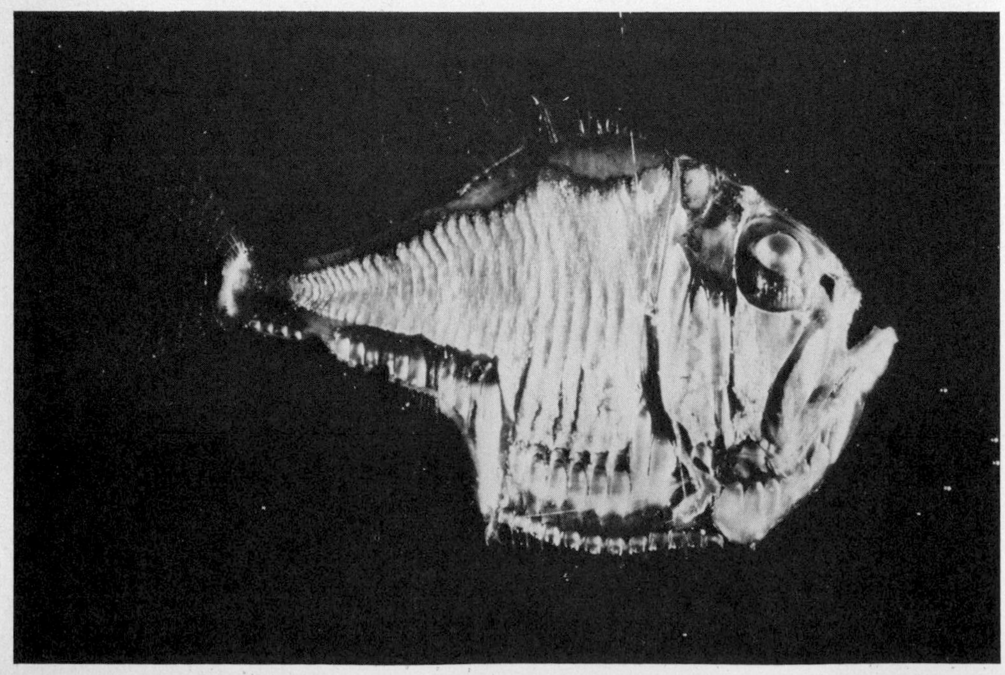

Figure 4-12 The dimorphic eyes of the squid *Histioteuthis.* (Photo courtesy of Dr. C. Recksiek.)

which live in the mesopelagic at depths between 500–700 m, usually orient with the arms down and with the large eye directed vertically upward, while the small one is oriented downward. In this position, presumably the large eye gathers the faint light downwelling from the surface, while the smaller eye responds to light from photophores. It has been further demonstrated with another midwater squid, *Abraliopsis*, that the squid responds to downwelling light by turning on its photophores just enough to match the downwelling light and hence becomes invisible when viewed from below (i.e., prevents being silhouetted).

Food scarcity in the deep sea seems to be the reason for another series of adaptations. Most deep-sea fishes have large mouths, larger relative to their size than fishes of other areas (Fig. 4-13; Plate 2). Furthermore, the mouth is often beset with long teeth recurved to the throat, an adaptation to ensure that whatever is caught does not escape. Even more bizarre is that, in some fishes, the mouth and skull are so hinged that the animals can open the mouths much wider than their own bodies (Fig. 4-14), enabling them to swallow prey larger than themselves. In fact, this ability to engulf and swallow food larger than themselves is yet another adaptation to the scarce food supply (Fig. 4-15). Food is not passed up simply because of size! Still other fishes, such as the angler fish (Ceratoidea), have responded to the scarce food by establishing themselves as traps, using a highly modified part of the dorsal fin as a lure. In this case, the lure is a luminescent organ. Other lures are found in many other fishes, such as the stomiatoids, and occur on barbels attached to the chin (Fig. 4-16).

Because food is scarce and the resulting density of organisms is very low, there is a potential problem of finding a mate for reproductive purposes in this

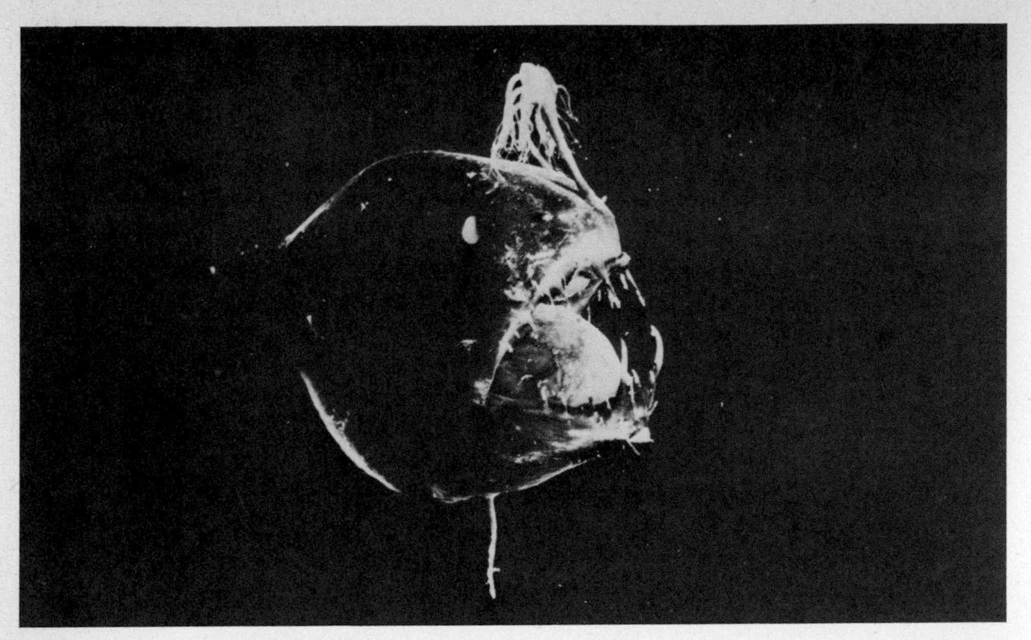

Figure 4-13 *Linophryne*, a genus of fish with a very large mouth. (Photo courtesy of Michael Kelly.)

Figure 4-14 The hinged jaw in *Chauloidus*. *(A)* Anatomical construction. *(B)* Operation during swallowing. (*A*, after A. Hardy, The open sea, 1956, Houghton Mifflin. Reprinted by permission of Houghton Mifflin Company. *B*, after N. B. Marshall, Aspects of deep sea biology, Philosophical Library.)

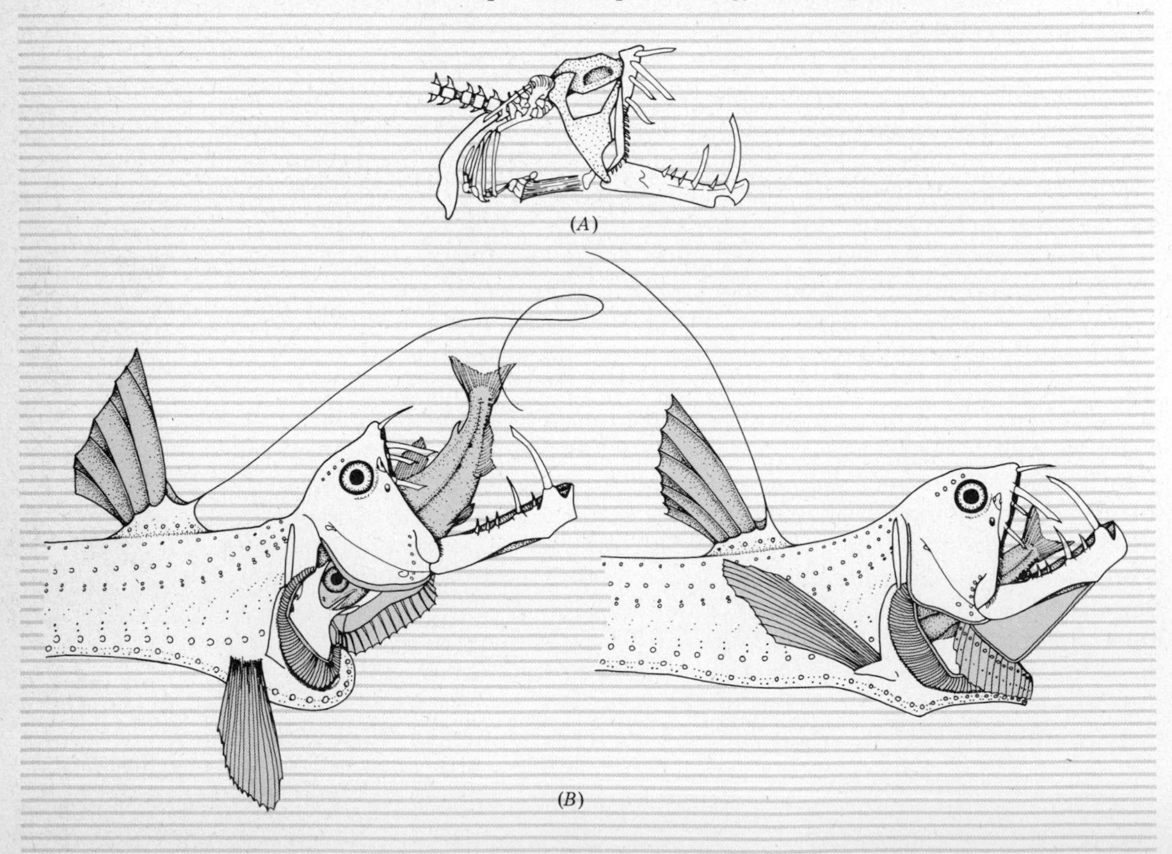

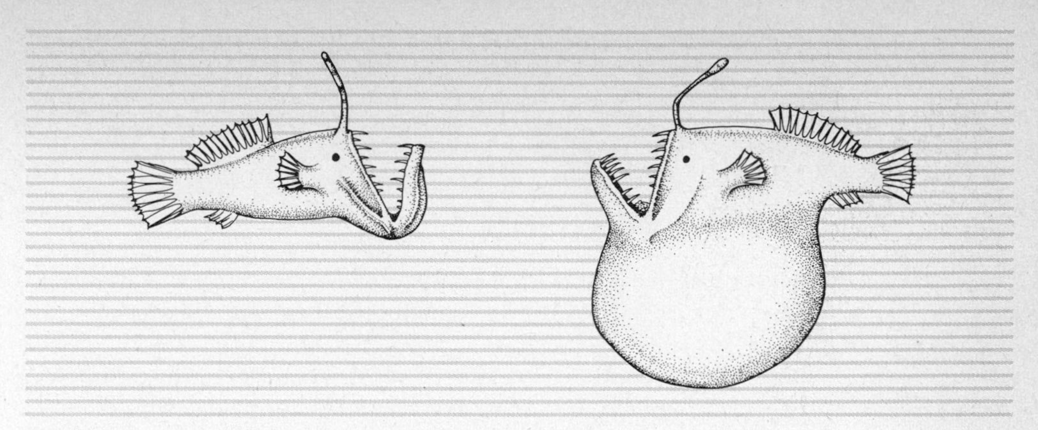

Figure 4-15 The deep water angler fish, *Melanocetus johnsoni*, before and after swallowing a larger fish. (From A. Hardy, The open sea, 1956, Houghton Mifflin. Reprinted by permission of Houghton Mifflin Company.)

Figure 4-16 Luminescent barbel of the stomiatoid fish *Idiacanthus* sp. (Photo courtesy of Michael Kelly.)

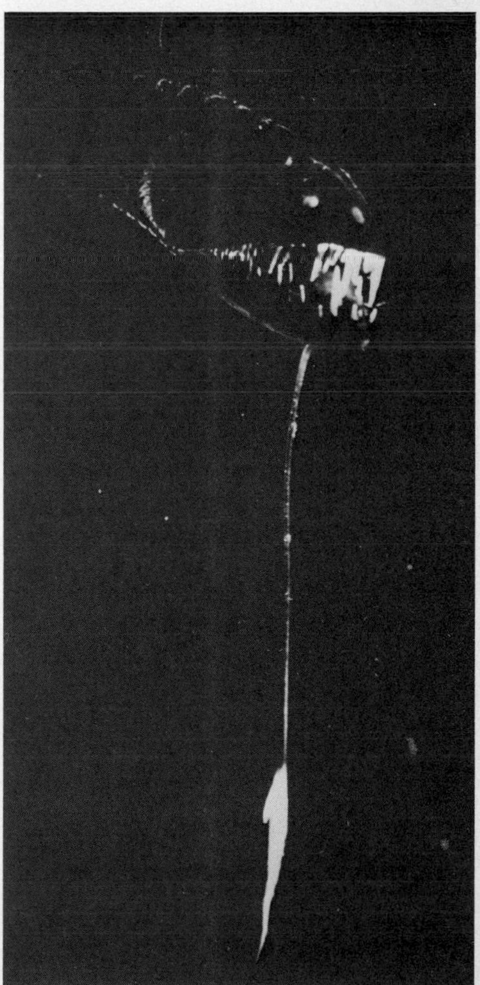

vast, dark area. One adaptation to this problem is found in angler fish, *Ceratias*. In angler fish, the large individuals are all female. The males are very tiny and are parasitic, attached to the body of the female (Fig. 4-17). Since the males are always present to provide the sperm, the search for a mate is negated. Of course, the male has first to find the female, and this is presumably accomplished via olfaction.

Body size of deep-sea organisms represents a peculiar paradox. Since food is scarce, it might be anticipated that most organisms would be small in size. Indeed, when considering fish, this appears to be the case. Most fishes captured in the deep sea are quite small, smaller than corresponding relatives in shallow water. As far as we now know, the deep sea contains few large fishes, though a few large ones have been photographed by the aforementioned baited "monster camera." On the other hand, certain invertebrate groups, particularly amphipods, isopods, ostracods, mysids, and copepods, attain a much greater size in deep waters than their relatives in shallow water. This phenomenon of larger size with increasing depth is called *abyssal gigantism*. The size that certain of these "abyssal giants" attain is really quite remarkable. The isopod *Bathynomus giganteus* reaches 15 inches (42 cm) and the amphipod *Alicella gigantea* 6 inches (15 cm). *Gigantocypcis*, the giant mesopelagic ostracod, has a diameter of several centimeters. The copepod *Gausia princeps* reaches a size of 10 mm, nearly ten times the size of most calanoid copepods.

Scientists currently disagree as to the reasons for abyssal gigantism, but there are two main theories. The first attributes it to the peculiarities of metabolism under conditions of high pressure. The second suggests that the combination of

Figure 4-17 Large female angler fish, *Linophryne argyresca*, with attached small parasitic male. (From C. T. Regan and E. Trewavas, Deep sea angler-fishes *(Ceratoidea)*, Dana report no. 2, 1932, The Carlsberg Foundation and Oxford University Press.)

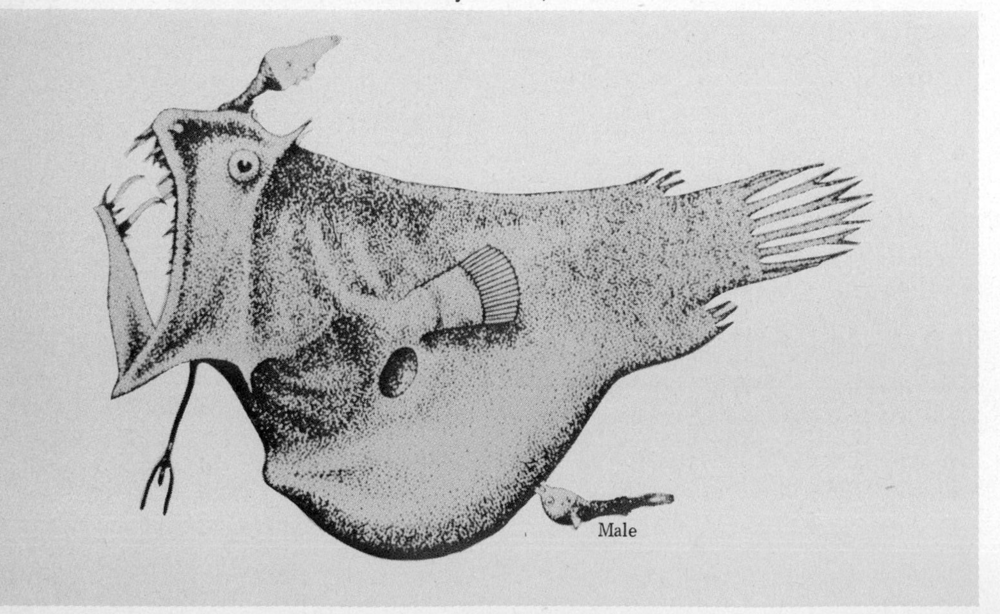

low temperature and scarce food reduces the growth rates in these crustaceans and increases the longevity and the time it takes them to reach sexual maturity, such that larger size is obtained. The extreme expression of this combination would be the "abyssal giants."

The large size may also be the result of natural selection operating. Large size, long life, and delayed sexual maturity confer certain advantages to organisms in the deep sea. These include production of larger eggs and subsequently larger young, which can then feed on a wider range of food sizes, thus precluding special food in this food-short area. Large animals are also more mobile and can cover more area in search of both food and potential mates. Increased longevity means a longer period of sexual maturity and hence greater time to find mates.

While these abyssal giants have captured much attention, they are rare. Most of the deep-sea benthic infauna (polychaete worms, crustaceans, and mollusks) are much smaller than their shallow water counterparts. In fact, the small size of benthic invertebrates is actually the major characteristic of deep-sea fauna, not gigantism. Given the vast area of the deep sea, some giants should be expected and similar examples of giant taxa can also be found in the less extensive shallow water areas, for example, the giant clams found on tropical reefs.

The bottom substrate of most deep ocean areas is a soft ooze. Benthic organisms inhabiting this soft material tend to have delicate bodies, to possess long legs or, in the case of sessile animals, to have long stalks to raise them above the ooze (Fig. 4–18). Some fishes have long, narrow fins that serve the same purpose.

Biochemical adaptations to the deep sea have been difficult to ascertain in the past because of the aforementioned lack of live animals. However, recent data on chemical composition suggest profound changes may occur. In fishes, Childress and Nygaard (1973) report that the water content of the body tissues increases with increasing depth, while the lipid and protein concentrations decrease. In other words, they become more like jellyfish. In crustaceans, Childress and Nygaard (1974) found that the protein content also decreased with depth. Presumably, this decrease in both groups is related to the scarcity of food from which to produce the protein. Caloric content also decreases.

A final set of adaptations in deep-sea organisms has to do with bioluminescent organs and will be taken up separately.

BIOLUMINESCENCE IN THE DEEP SEA

Marine bioluminescence is not confined to deep-sea organisms. Indeed, it is a phenomenon of widespread occurrence in the sea, and many people have observed it in the form of "phosphorescent seas," which occur as a result of light production in the surface waters by untold millions of dinoflagellates. Whereas bioluminescence is of wide occurrence, it is in the deep-sea pelagic organisms that it reaches its highest and most complex development; there we find the greatest number of organisms that have the ability to produce light. Hence it is appropriate to discuss the phenomenon at this time.

(*A*)

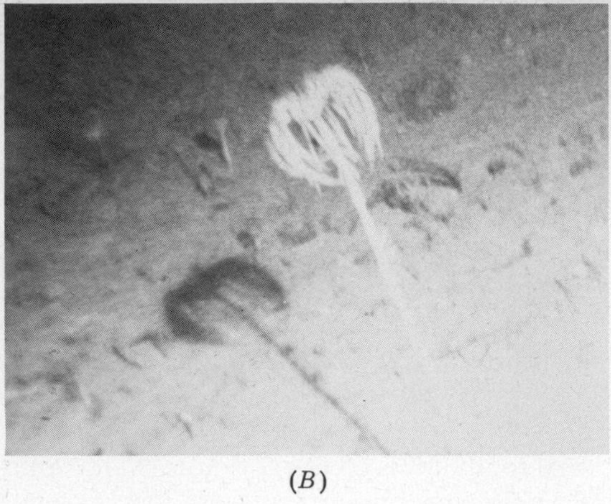

(*B*)

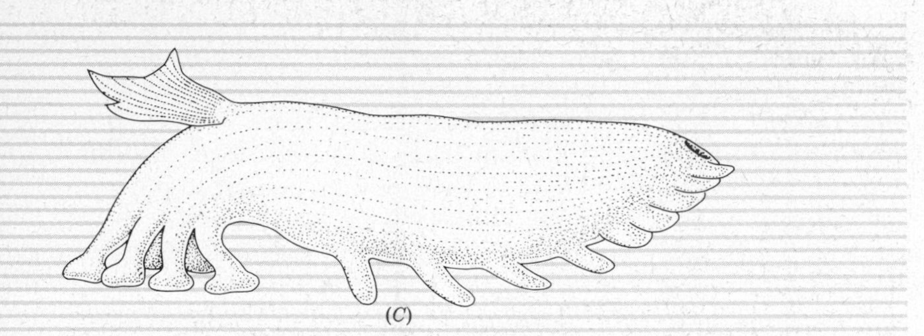

(*C*)

Figure 4-18 Adaptations to raise animals above the ooze. *(A)* Cirrate octopod. *(B)* A stalked crinoid. *(C)* A sea cucumber of the genus *Peniagone*. (*A*, courtesy of Dr. Clyde Roper. *B*, courtesy Dr. H. Mullins. *C*, Redrawn from Hansen, Galathea Report, vol. 2.)

Bioluminescence is simply the production of light by living organisms. The mechanism of light production is well known from studies of terrestrial organisms like fireflies, and the same mechanism is used by aquatic organisms. The spectrum of color produced varies from species to species, but in total, encompasses the visible range from violet to red.

Not only are large numbers of deep-sea organisms capable of producing light, they also have developed the most elaborate organs for producing this light. These light-producing organs are called *photophores* and are particularly abundant in fishes and squids, but are present in other invertebrates as well (see Plates 3 and 4).

The largest number of animals with photophores is found in the upper areas of the deep sea, the mesopelagic and upper bathypelagic. The incidence of organisms with bioluminescence decreases in the deepest parts of the sea.

Photophores in deep-sea organisms are of several types and range in structure from simple to complex. The simplest photophores consist of a series of glandularlike cells that produce the light or else a simple glandular cup holding a bacteria culture that produces the light (see also pp. 381–382). In either case, the cells or cup are surrounded with a screen of black pigment cells. The more elaborate photophores have in addition one or more of the following: lenses to focus the light, a color filter, and an adjustable diaphragm of pigment cells (Fig. 4-19). In some fishes and crustaceans, the photophores may be covered by a flap of flesh, which can be adjusted to turn the light either on or off. Still others are able to move the photophores by muscular action.

That the production of light is of considerable adaptive significance to its possessor is demonstrated by (1) its great prevalence in the deep sea, and (2) the complex anatomical organs evolved to accommodate its production. It is also

Figure 4-19 Structure of the ventral photophore of the squid *Abralia trigonura*. The size of the entire photophore is 180 microns. Abbreviations: *ac*, apical chromatophore; *ax*, axial stack: *bv*, blood vessel; *c*, photogenic cone; *cac*, distal cap chromatophore; *cc*, proximal cup chromatophore; *co*, core; *cr*, crystalloid; *crc*, crystalloid cell; *dc*, distal cap; *dr*, distal reflector; *gr*, girdle; *icc*, immature crystalloid cell; *lm*, lamella; *m*, muscle; *mc*, mitochondrial cell; *n*, nucleus; *ne*, nerve; *os*, orbital space; *pr*, proximal reflector; *ri*, ribbon; *s*, collagen sheath; *to*, torus; *v*, vesicles; *sur*, surface. (Drawing courtesy of Richard Young, University of Hawaii.)

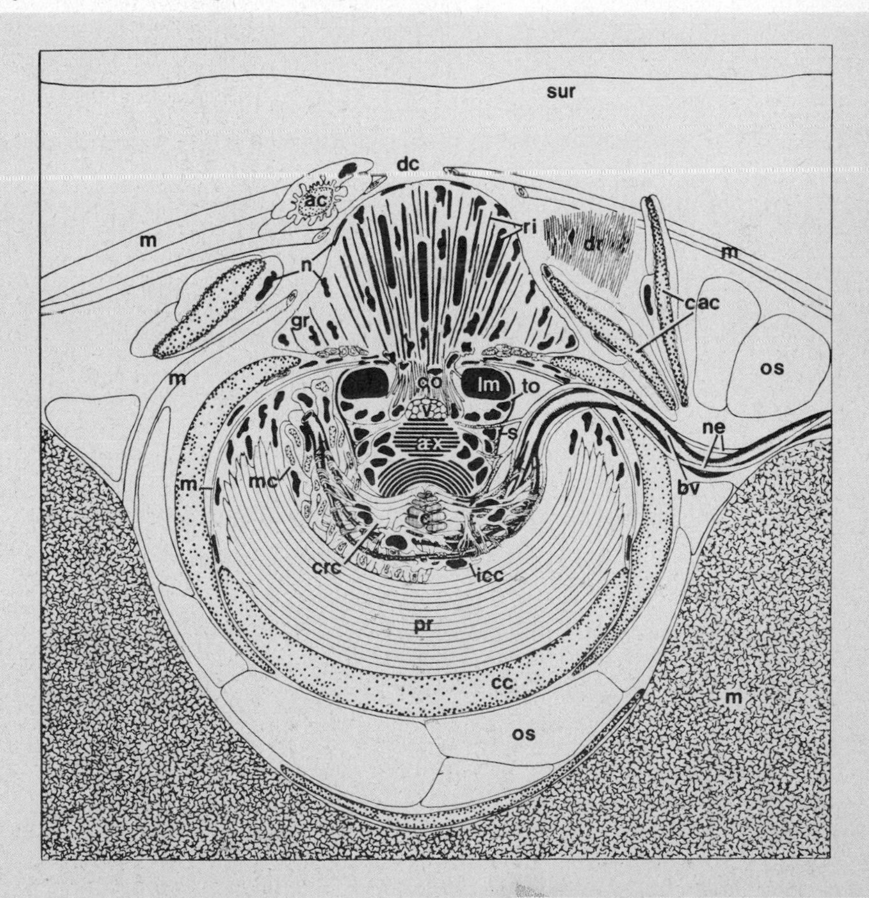

apparent that bioluminescence probably serves different functions in different animals. Several different hypotheses have been suggested for the use of bioluminescence in the deep sea and these will be reviewed here.

Many organisms, particularly fishes and squids, have their photophores placed ventrally. For those animals living in the upper reaches of the mesopelagic zone, where some light downwells from the surface, a predator looking up at them could probably see a silhouette. The photophores, when turned on to produce a similar intensity as the downwelled light, would cause the silhouette to disappear and make the animal less vulnerable to predation. This is, in effect, camouflage. Since some organisms have been shown to be able to adjust their downward photophores to the incident radiation level from above, this is a viable function.

Photophores might also be used to produce a "blinding" flash of light, which would momentarily stun a potential predator and thus allow disengagement and escape. Some deep-sea squids (*Histioteuthis dispar*) distract potential predators by shooting out a luminescent cloud rather than a black "ink" cloud as is done by squids in lighted waters. This cloud masks the escape of the squid.

As noted before, photophores might also be employed as lures to attract potential prey items within range of the predator. This is undoubtedly the function of the elaborate light organ (esca) on the modified dorsal fin (illicium) of the angler fish.

Photophores may also be used for simple illumination of an area such that a predator may pick out prey items.

A final set of uses has to do with recognition. In a perpetually dark area such as the deep sea, the particular pattern of distribution of photophores over the body of the animal would be visible at a distance. Elaborate patterns of photophores occur on fishes in particular. Furthermore, the pattern is different for different species. The existence of these patterns would thus allow individuals to recognize their own species. This may be of importance in maintaining schools and also in finding potential mates for reproduction (Fig. 4-20). In the latter case, certain fishes also have photophore patterns which differ between males and females to further ensure mate selection. It is interesting to note here that one of the first scientists to actually observe living mesopelagic fishes *in situ*, William Beebe, who made his observations from a bathyscape, noted that he could tell the different myctophid (lantern fish) species in the dark from their photophore pattern. If he could, it seems likely the fish can also.

Other uses of photophores may occur and more may come to be recognized with increased knowledge of this area, but the above represent the current most plausible hypotheses.

COMMUNITY ECOLOGY OF THE BENTHOS

Although we have samples of organisms from the pelagic environment of the deep sea, the problems of capture and net avoidance make the available

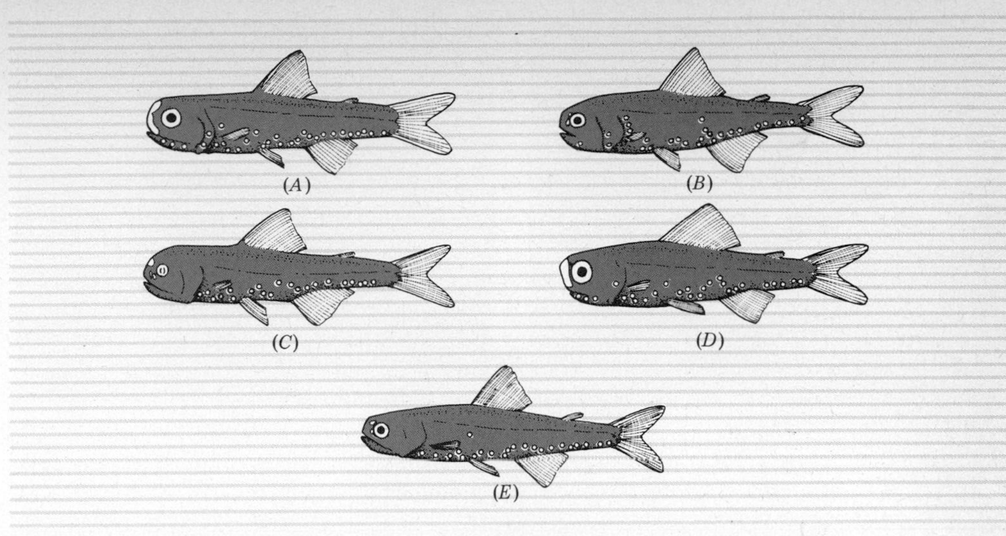

Figure 4-20 Species-specific light organ patterns (open circles) in the lantern fish genus *Diaphus*. *(A) D. macrophus. (B) D. lucidus. (C) D. splendidus. (D) D. garmani. (E) D. effnlgeus.* (From N. B. Marshall, Aspects of deep sea biology, 1954, Philosophical Library.)

information on the community composition of the pelagic fauna incomplete at best and suspect at worst. The exception to this is the mesopelagic, where many samples and observations are available. Considerably more is known about the deep-sea benthic communities because the sedentary and sessile nature of the animals makes net avoidance less important. Thus, the samples taken and the observations of this area by deep submersibles and cameras have given us a reasonable idea of its composition.

FAUNAL COMPOSITION

Virtually all the major animal groups have representatives in the deep sea, but the relative abundance of groups varies. Crustaceans, particularly isopods, amphipods, tanaids, and cumaceans, are common in the deep sea. In the Atlantic abyssal area, they constitute 30–50 percent of the fauna. Polychaete worms are also abundant, making up from 40–80 percent of the fauna in the Atlantic. Particularly common in abyssal areas are sea cucumbers (Holothuroidea), usually of large size, and brittlestars (Ophiuroids). Holothurians are often the most common organisms in deep-sea photographs. This, plus the fact that they form 30–80 percent of the biomass of living organisms taken in some trawls, suggests they are a major component of the abyssal community. Since they are deposit feeders, the abyssal oozes are likely an excellent food source, hence their abundance.

Starfish (Asteroidea), sea lilies (Crinoidea), and sea urchins (Echinoidea) are also present, but not in abundance (Fig. 4-21). Among the sponges, the deep sea is populated by the glass sponges (Hexactinellida), a group rarely found in shallow water. Sea anemones (Anthozoa), sea pens (Pennatulacea), and sea fans (Gorgonacea) are the only common representatives of the phylum Cnidaria. A number of fishes are also present, but their general mobility makes it difficult to

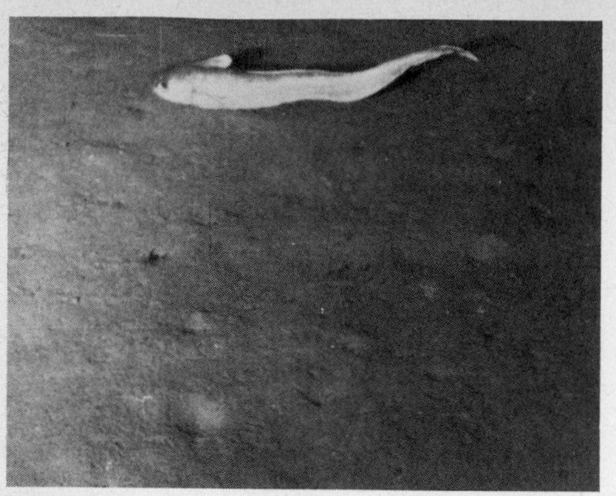

(*A*) (*B*)

Figure 4-21 Two views of the bottom at 1,300 m on the Gay Head-Bermuda transect showing variation in abundance of organisms. *(A)* Area barren of large benthic animals. *(B)* Concentration of the echinoid *Phormosoma placenta*. (Photos courtesy of Dr. Fred Grassle; reprinted with permission from J. F. Grassle et al., Deep sea research, vol. 22, copyright 1975, Pergamon Press Ltd.)

estimate relative abundance. Table 4-2 summarizes the types of organisms found on the deep-sea floor and Figure 4-22 diagrammatically summarizes zonation changes and typical abyssal animals.

In the trenches (hadal zone) the available data suggest that there is a higher percentage of pericaridean crustaceans, polychaetes, and holothurians and a lower percentage of sea stars, sea urchins, and brittlestars in comparison with the abyssal.

The major fishes of the bottom waters of the deep sea are rat tails (Macrouridae), brotuilds, liparids, and certain eels (*Synaphobranchus*, *Cyema*).

DIVERSITY

Because of the low density of benthic populations, the small size of most invertebrates, and the paucity of adequate samples, the deep sea was considered a biological desert for many years. In the early 1960s, this notion was abandoned as sampling efforts intensified and methods of dredging and sample processing improved. As a result, the desert idea was replaced by the opposite belief, that of a highly diverse community. This concept of species diversity is founded on the intuition that large groups of co-occurring species are characterized by a greater complexity of biological interactions. As noted in Chapter One, species diversity has two important components: the number of species and their patterns of relative abundance. For example, if two areas have the same number of species, one area has a greater diversity if it has a more even distribution of individuals among the resident species. The idea of a highly diverse deep-sea fauna is based entirely on their relative abundance patterns. While the total number of individuals is sparse, each deep-sea species is commonly represented by only a few individuals. In other words, most deep-sea species are rare. This pattern had led to the popular contention that deep-sea diversity is as high as that of coral reefs, or among the highest in the world. In contrast, shallow water

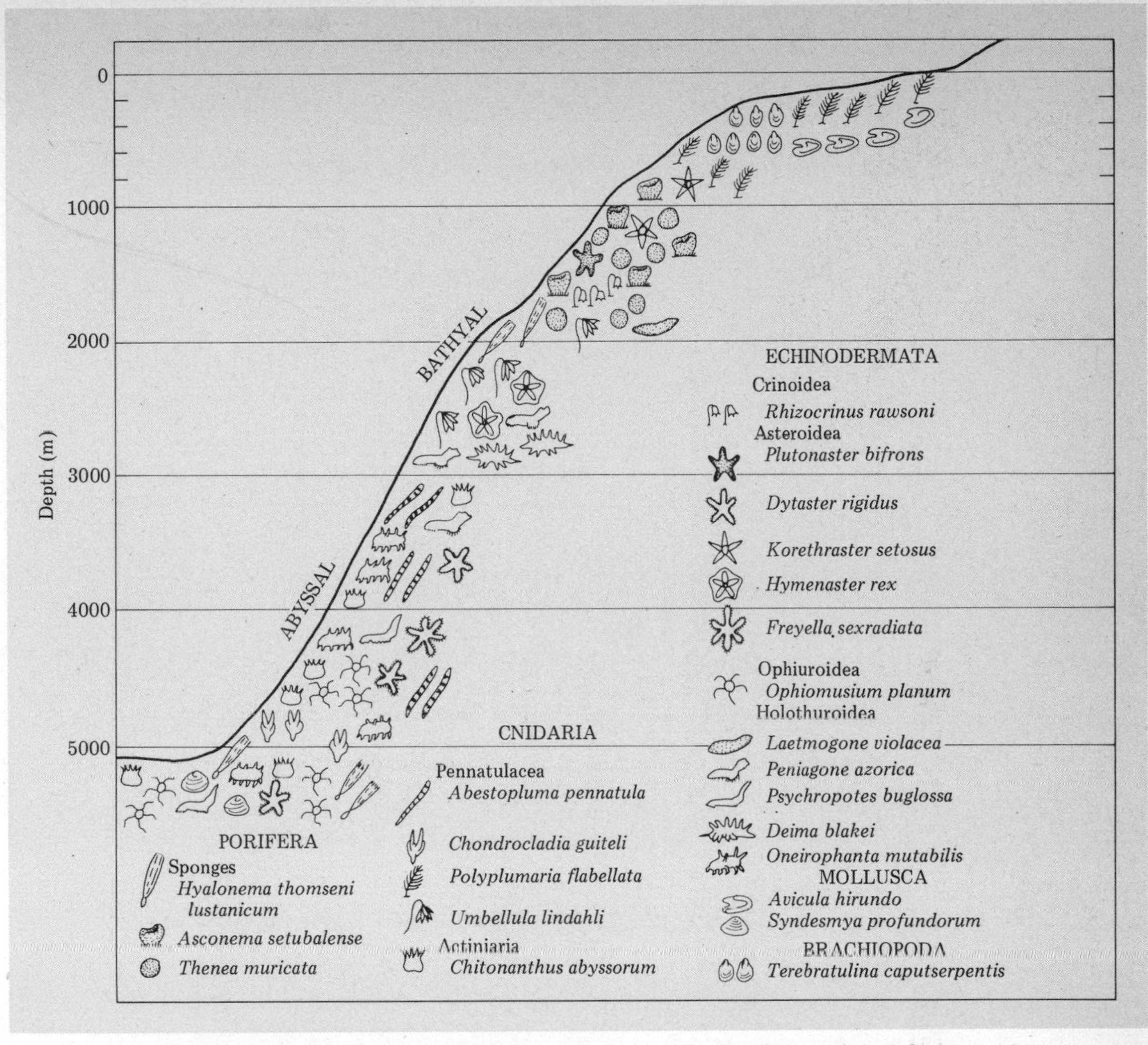

Figure 4-22 Zonation of benthic animals with depth on the Atlantic continental margin of Europe. (From B. Heezen and C. Hollister, The face of the deep, Oxford University Press.)

assemblages are often dominated by one or several very abundant animals, but there are also many rare species.

Despite the popular belief in deep-sea diversity, there is a major unresolved problem. The most fundamental component of diversity is simply the number of species found in a particular area. Unfortunately, there are no comparisons of the actual density of species found in deep and shallow waters. Thus, there is no answer to the extremely important question: Are the numbers of species from ecologically equivalent bottom areas the same or different in deep and shallow habitats? Until this problem can be resolved, the idea of a highly diverse deep-sea fauna must remain speculative. But clearly, the deep sea is not a biological desert.

If one accepts the idea of a highly diverse deep sea, there are several conflicting hypotheses attempting to explain this diversity. The first is termed

TABLE 4-2 Types of Deep Sea Organisms

PHYLUM / A. SUBPHYLUM / SUPERCLASS / 1. CLASS / I. Subclass / (a) Order		Captured: Non-parasitic	>200 meters	>5000 meters	> 1 centimeter	>10 centimeters	>20 centimeters	Common	Abundant	Photographed: Photographed	5–10 photos	>10 photos	>50 photos	Major Groups		
	PROTOZOA	■	■												Sponges	MOST ARE SESSILE
	PORIFERA, Sponges	■	■	■	■	■	■	■		■	■	■				
CNIDARIA	1. HYDROZOA, Hydroids	■	■	■	■	■	■			■					CNIDARIANS	
CNIDARIA	2. SCYPHOZOA, Jellyfishes	■	■	■	■	■										
CNIDARIA, 3. ANTHOZOA, I. ALCYONARIA	(a) Stolonifera	■														
CNIDARIA, 3. ANTHOZOA, I. ALCYONARIA	(b) Telestacea	■														
CNIDARIA, 3. ANTHOZOA, I. ALCYONARIA	(c) Alcyonacea	■														
CNIDARIA, 3. ANTHOZOA, I. ALCYONARIA	(d) Coenothecalia	■														
CNIDARIA, 3. ANTHOZOA, I. ALCYONARIA	(e) Gorgonacea	■	■	■	■	■	■	■		■				Gorgonians		
CNIDARIA, 3. ANTHOZOA, I. ALCYONARIA	(f) Pennatulacea	■	■	■	■	■	■	■		■	■	■		Pennatulids		
CNIDARIA, 3. ANTHOZOA, II. ZOANTHARIA	(a) Actinaria, Sea Anemones	■	■	■	■	■	■	■		■	■	■		Actinarians		
CNIDARIA, 3. ANTHOZOA, II. ZOANTHARIA	(b) Madreporaria, Stony Coral	■														
CNIDARIA, 3. ANTHOZOA, II. ZOANTHARIA	(c) Zoanthidea	■														
CNIDARIA, 3. ANTHOZOA, II. ZOANTHARIA	(d) Antipatharia, Black Coral	■	■	■	■	■	■	■		■				Antipatharians		
CNIDARIA, 3. ANTHOZOA, II. ZOANTHARIA	(e) Ceriantharia	■	■	■	■	■										
	CTENOPHORA, Comb Jellies	■	■													
	NEMERTINEA and ACANTHOCEPHALA	■	■	■	■	■	■	■								
	BRYOZOA	■	■	■	■	■	■			■						
	PHORONIDA	■														
	BRACHIOPODA	■	■	■	■											
ECHINODERMATA	1. CRINOIDEA, Sea Lilies	■	■	■	■	■	■	■		■	■	■		Crinoids	ECHINODERMS	
ECHINODERMATA	2. ASTEROIDEA, Starfishes	■	■	■	■	■	■	■		■	■	■		Asteroids		
ECHINODERMATA	3. OPHIUROIDEA, Brittle Stars	■	■	■	■	■	■	■	■	■	■	■	■	Ophiuroids		
ECHINODERMATA	4. ECHINOIDEA, Sea Urchins	■	■	■	■	■	■	■	■	■	■	■	■	Echinoids		
ECHINODERMATA	5. HOLOTHUROIDEA, Sea Cucumbers	■	■	■	■	■	■	■	■	■	■	■	■	Holothurians		

ABYSSAL ANIMALS

Group	Organism														
	CHAETOGNATHA, Arrow Worms	■													
MOLLUSCA	1. MONOPLACOPHORA	■	■	■	■	■									
	2. POLYPLACOPHORA	■	■	■	■	■									
	3. SCAPHOPODA, Tooth Shells	■	■	■	■	■									MOLLUSCS
	4. GASTROPODA, Snails, Slugs	■	■	■	■	■									
	5. PELECYPODA, Bivalves	■	■	■	■	■									
	6. CEPHALOPODA, Squids, Octopuses	■	■	■	■	■	■			■				Cephalopods	
ANNELIDA	1. ARCHIANELLIDA	■													ANNELIDS
	2. POLYCHAETA, Sand Worms	■	■	■	■	■	■	■	■	■	■	■	■	Polychaetes	
	SIPUNCULA, Peanut Worms	■	■	■	■	■	■								
	PRIAPULIDA	■	■	■	■	■									
	ECHIURA	■	■	■	■	■	■								
ARTHROPODA 1. CRUSTACEA	I. Branchiopoda	■	■												
	II. Ostracoda	■	■	■											
	III. Copepoda	■	■	■											
	IV. Cirripedia, Barnacles	■	■	■	■	■									
V. Malacostraca	(a) Mysidacea	■	■	■	■	■									
	(b) Cumacea	■	■	■											
	(c) Tanaidacea	■	■	■											
	(d) Isopoda	■	■	■	■			■		■				Isopods	ARTHROPODS
	(e) Amphipoda	■	■	■											
	(f) Stomatopoda	■													
	(g) Euphausiacea	■	■	■	■										
	(h) Decapoda	■	■	■	■	■	■			■				Decapods	
	2. PYCNOGONIDA (Sea-Spiders)	■	■	■	■	■	■	■		■				Pycnogonids	
	POGONOPHORA	■	■	■	■	■	■			■					
	HEMICHORDATA	■	■	■	■	■	■			■				Hemichordates	
CHORDATA	C. CEPHALOCHORDATA	■													
	D. AGNATHA	■	■	■	■	■									CHORDATES
	E. TUNICATA, 1. ASCIDIACEA	■	■	■	■	■				■				Tunicates	
	F. PISCES (Fish)	■	■	■	■	■	■	■		■	■	■	■	Fish	
	G. TETRAPODA, 1. MAMMALIA (Whales)	■	■	■	■	■									

MOST ARE MOBILE

NOTE: Read this key from left to right. Thus if an organism, regardless of size, is not found in deep water it is eliminated from subsequent columns.

After Bruce Heezen and Charles Hollister, 1971.

the *stability-time* hypothesis, first suggested by Sanders (1968). This says that high diversity occurs because the highly stable environmental conditions have persisted over long periods of time and have allowed species to evolve that are highly specialized for a particular microhabitat or food source. Since most of the benthic animals are detritus feeders in the deep sea, this means that they have become very specialized for a particular narrow range of particle sizes. Such specialization is virtually unknown among deposit feeders elsewhere in the oceans. Moreover, the stability-time hypothesis has been falsified as an explanation of diversity patterns among the insects found in sugarcane fields and among the marine crustaceans inhabiting coral heads.

A second hypothesis is the "cropper" or *disturbance theory* suggested by Dayton and Hessler (1972). The basic premise here is that organisms generally increase in numbers until they reach the limit of some resource that is in least abundance. At that point, competition occurs, and in the ensuing competition, species are eliminated (principle of competitive exclusion). Since, in the deep sea, animals are very low in numbers compared to the total amount of space available to inhabit, competition for space does not occur. The only other resource likely to be competed for is food. As noted above, it is in very short supply. The cropper theory suggests that none of the deep-sea animals are food specialists, and that they are generalists feeding on anything they can engulf, not distinguishing whether it is living or dead; hence, the term "cropper," as one that feeds indiscriminately on anything smaller than itself. High diversity is maintained because the intense "cropping" by all levels and sizes of animals prevents any species from building up its population to the point where it would be competing with another for the food. High diversity is due, then, to intense predation at all levels, which allows a large number of species to persist, eating the same food, but never becoming abundant enough to enter into competition. A central argument against this theory is that heavy predation should result in the evolution of definite life history features. These include production of large numbers of young, a short life span, and fast growth. As far as is known (see below), these are the exact opposite of the life history characteristics found in the deep-sea organisms, which grow slowly, live long, and produce few offspring. However, if cropping results in differential mortality of age groups, either set of life history traits can, in theory, evolve.

A final hypothesis, which does adequately explain the original reports that diversity increases with increasing depth, is the *area hypothesis*. This is a very simple idea. Since species number is positively correlated with area (i.e., more area equals more species), diversity is highest in the deep sea because it covers the greatest area. Both diversity and area of sea bottom decrease with decreasing depth. This correlation is highly significant. So, although the actual numbers of species per unit bottom area have not been established, trends in relative abundance patterns can be related to the size of large biogeographical areas. But this correlation applies only to samples taken in the Atlantic Ocean. Workers in the Pacific Ocean have not found a general increase in diversity with increasing depth. On the contrary, both components of diversity (species density and relative abundance) are highest at intermediate bathyl depths, rather than in the abyss.

The present situation can be adequately summarized by admitting confusion. The idea of a highly diverse deep-sea fauna has intrigued many ecologists and is central to theories of community and environmental stability. Additional sampling and speculation have blurred the original pattern and undercut popular explanatory schemes.

LIFE HISTORY PATTERNS

In shallow water areas and terrestrial areas, there are usually cyclic changes in some environmental parameter that serve to initiate and control reproduction. As a result, reproduction is usually confined to a particular time of year. In the deep sea, where light is absent and all environmental parameters are monotonously constant, it would appear that a cyclic pattern of reproductive activity should be absent. What little evidence we have thus far suggests that this indeed is the case. Reproductive data from benthic organisms in the deep sea studied by Rokop (1974) show that reproduction is constant and continuous throughout the year, with no seasonal peaks.

Since food is scarce, it might be suspected that most deep-sea organisms would produce few offspring that would not spend time as larvae in the open water (see Chapter Five, 165–167, on larval strategies). However, our current state of knowledge has little information about eggs and larvae of deep-sea organisms except for fishes, and we can neither completely confirm nor deny that assumption. What we do know suggests that many deep-sea benthic animals do have few large eggs, suggesting they are hatched in an advanced state and have a short or no larval life. Most fishes, on the other hand, have pelagic eggs and larvae.

Two general reproductive patterns exist in the deep sea. In the first case, the early stages of the life history are spent in the lighted surface waters, and the juveniles migrate down to adult depths as they mature or metamorphose. Such is the pattern exhibited by certain deep-sea angler fish (Fig. 4-23), myctophids, stomiatids, and certain squids. In the other, no migration occurs and the young stages are spent in the same area as the adult. Such is the case with the unique cephalopod *Vampyroteuthis infernalis* and certain prawns.

With respect to feeding types among deep-sea organisms, knowledge is, as yet, fragmentary, but Sanders and Hessler (1969) suggest that deposit feeders greatly dominate the benthic animals. They constitute over half the fauna in some areas. In midwater or pelagic areas, carnivores dominate the fauna.

THE GALAPAGOS RIFT ZONE

Perhaps one of the most significant and exciting discoveries in deep-sea biology, indeed in marine biology in general, occurred in 1977. At that time, a series of dives with the deep submersible "Alvin" was made in the Galapagos Rift Zone

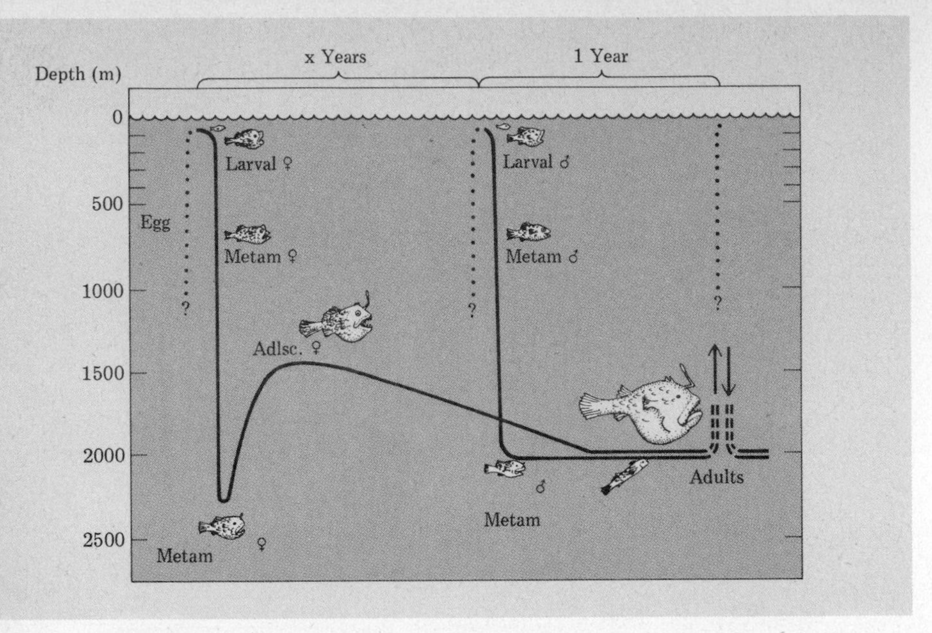

Figure 4-23 Life cycle of an angler fish. The fish spends its larval life at the surface and descends on metamorphosis. (From N. B. Marshall, Aspects of deep sea biology, 1954, Philosophical Library.)

some 200 miles northeast of the islands. As reported by Ballard (1977), these dives, to 2,700 m, revealed an abundance of hitherto unknown marine animals living in and around four hot water geysers in an otherwise barren ocean floor. The water temperature in these vent areas was 8–16°C, or considerably above the normal temperature prevalent at the bottom at these depths. This warm water was found to be very high in reduced sulfur compounds, namely, H_2S. The hydrogen sulfide was used as an energy source by bacteria (see Chapter Six on sulfide bacteria in mud shores), which were then able to produce large amounts of organic matter, which, in turn, served as the basis of a food chain in the otherwise dark, cold, and energy-poor area. The result was a spectacular abundance of animals living in these "oases" (Fig. 4-24). The vent areas were alive with huge clams, crabs, tube worms, various limpets, giant pogonophoran worms, and other as yet unidentified forms. What has made this discovery so exciting is that most of the animals found are new to science and several are giants compared to their relatives. Even more fascinating is that all of this occurred within only 50 m of the vents, an unbelievably small area for an evolutionary oasis. Subsequently, similar vent systems and associated fauna have been found in several other locations between 20°N and 20°S latitude in the Pacific Ocean suggesting a more widespread phenomenon.

MIDWATER COMMUNITY ECOLOGY

Because of the increased difficulty in sampling pelagic deep-sea organisms due to net avoidance and migration, the lack of observations on these animals from deep submersibles, their avoidance of light, and their small size, we know less

Figure 4-24 A cluster of giant sea worms found in the hot spring vents of the Galapagos Rift. The worms live in 1-inch diameter tubes which may be more than 8 ft long. A worm 5 ft long was recovered during a 1979 expedition. The closest relatives of these worms normally reach a maximum length of 12 inches. Clams and mussels are in the foreground; some measure nearly 1 ft. (Photo courtesy of Kathleen Crane, Woods Hole Oceanographic Institution.)

about the ecology of midwater animals than we do about the deeper but sedentary and sessile benthos.

DEEP SCATTERING LAYERS

Early in World War II, when the use of echo-sounders was being experimented with in the oceans as a means to detect submarines and the bottom, the devices (now known as sonar) regularly began to pick up echoes. They came, not from the bottom but from sources in the water column itself, hundreds or even thousands of feet above the known bottom. These areas of sound reflection occurring in midwater were termed *deep scattering layers* (DSL). The depths of the DSLs varied. During the day, there were often two or three layers varying in depth from 200 to 700 m (Fig. 4-25). At night, these different layers migrated toward the surface where they often merged into a single broad band. At dawn, the layers returned to depth. Although first interpreted as the result of some physical discontinuity in the water, the DSLs are now known to be concentrations of midwater animals, and the movement of the layers toward the surface at night is the result of a vertical migration similar to that undertaken by surface zooplankton (see pages 75–78). Presumably, these midwater animals are migrating into the surface waters at night in order to feed on the abundant plankton.

The major animals forming the DSLs include the lantern fish (Myctophidae) and other similar mesopelagic fishes (Fig. 4-9), shrimplike crustaceans such as

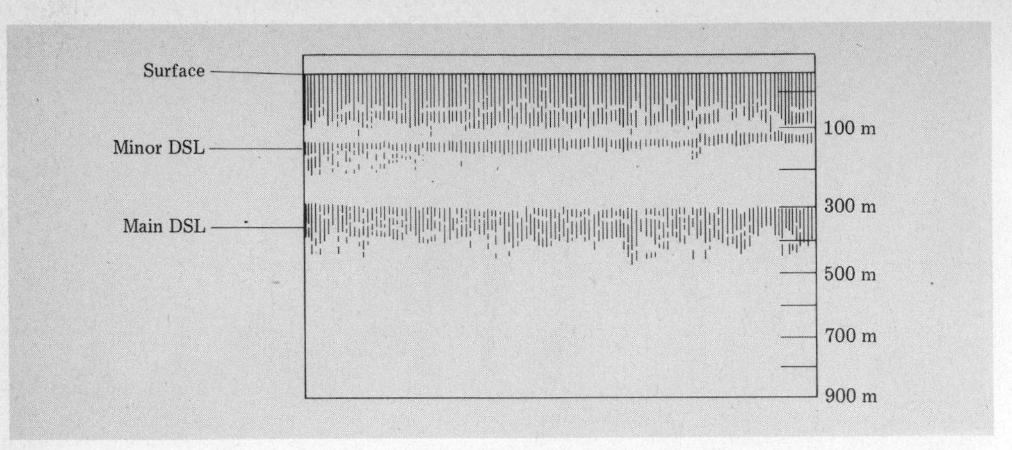

Figure 4-25 A sonogram recording showing the presence of the deep scattering layers in the Indian Ocean. (After G. B. Farquhar editor, 1970, Proceedings of an international symposium on biological sound scattering in the Ocean. U.S. Government MC Report 005.)

euphausiids ("krill"), sergestids and pasaphaeids, various squids, and certain deep-floating hydrozoan cnidarians called siphonophores. The composition of the scattering layer varies with geographical area.

Deep scattering layers have been found in all oceans except the Arctic. They are best developed in areas with high surface productivity and are faint in areas of low productivity.

SPECIES COMPOSITION AND DISTRIBUTION

The pelagic zones of the deep sea are dominated primarily by small fishes of the families Myctophidae, Gonostomatidae, and Sternoptychidae, and euphausiid and decapod crustaceans (Fig. 4-26). Blackburn (1977) has noted that, together, these groups constitute more than 80 percent of the biomass, at least in the mesopelagic zone (Fig. 4-27). One estimate for the Pacific Ocean has fishes making up 20–45 percent of the biomass, shrimps 15–25 percent, and euphausiids 35–50 percent in the upper 1000 m of water. Minor components include other crustaceans such as mysids, amphipods, and copepods, chaetognaths, scyphozoan jellyfish, and cephalopods. This dominance of the pelagic deep-sea fauna by small fishes, euphausiids, and shrimps seems to be widespread in all oceans, but as Omori (1974) has noted, the relative abundance of each group changes with latitude and with depth (Fig. 4-27). Seasonal changes in the composition of the assemblage appear to be minor and have received little study.

There are two components of the pelagic species assemblages in the mesopelagic zone. One includes those animals that do not migrate vertically and includes certain fishes (*Cyclothone*, *Sternoptyx*), mysids, some shrimps (sergestids, penaeids, carideans), and a few euphausiids, amphipods, cephalopods, and cnidarians. The other component is called the *interzonal fauna* and undergoes a diurnal vertical migration. These animals live below 450 m by day and migrate into surface waters by night. This component includes most of the mesopelagic fishes (Myctophidae, Gonostomatidae) and most euphausiids and

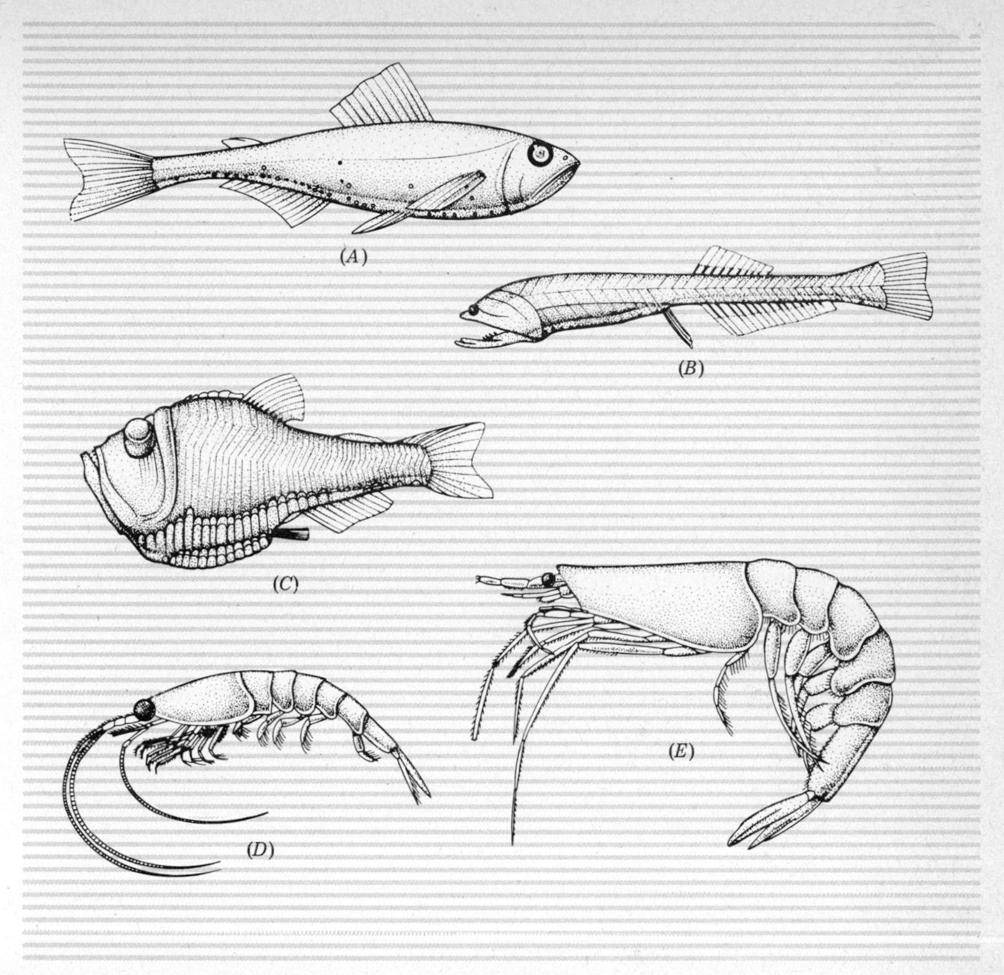

Figure 4-26 Representatives of the major groups of organisms of the pelagic zones of the deep sea. *(A)* Lantern fish, *Tarletonbeania crenularis*. *(B)* Gonostomatid fish, *Cyclothone pallida*. *(C)* Hatchet fish, *Argyropelecus affinis*. *(D)* Euphausid, *Euphausia pacifica*. *(E)* Sergestid shrimp, *Sergestes similis*.

decapods. Hence, the composition of the community changes depending on whether it is sampled at night or during the day (Fig. 4-27).

FOOD AND FEEDING

Information on the food of midwater organisms is sparse, but is most adequate for two dominant groups, the fishes and decapod crustaceans. Studies of the diet of midwater fishes have been concentrated on the numerically abundant families Myctophidae, Gonostomatidae, and Sternoptychidae. Whereas there is often considerable variation in species composition in the diets of these fishes, the principal components in all cases, as noted by Hopkins and Baird (1977), are smaller crustaceans such as copepods, euphausiids, ostracods, amphipods, and small decapod shrimps.

Changes in the diet of these fishes occur with age. In general, the larger and older individuals take a greater proportion of the larger crustaceans than do the

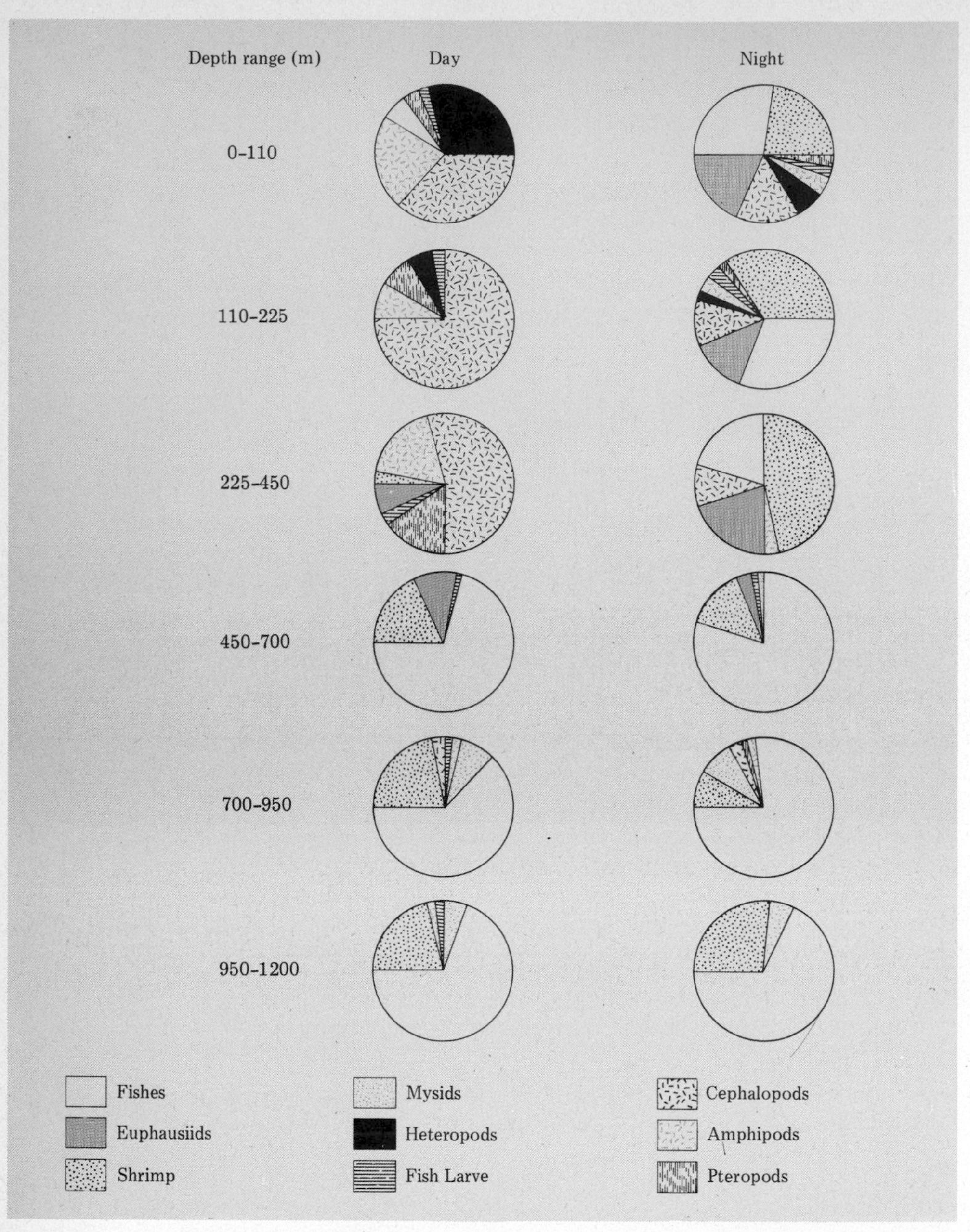

Figure 4-27 Differences in biomass composition of the midwater community in the tropical Pacific with depth and between day and night. (Adapted from N. R. Andersen and B. J. Zahuranec, editors, Oceanic sound production, Plenum.)

smaller fishes (Table 4-3). Seasonal changes in diet have also been observed for some fishes and they generally reflect availability of prey. Geographical variation in diet composition also occurs in widely distributed species. Midwater fishes are thus best characterized as opportunistic predators, in which the diet composition strongly reflects the prey availability.

TABLE 4-3 Changes in Diet with Size and Age in Six Midwater Fishes

	Size Range (mm)	Percent of Food <5mm	Percent of Food >5mm	Trend in Diet Composition (increasing percent)
Argyropelecus aculeatus	10–20	99	1	Ostracods, copepods
(WN Central Atlantic)	50–60	12	88	→ fish
Argyropelecus affinis	21–34	94	6	Copepods, ostracods
(off Southern California)	52–68	68	33	→ euphausiids, salps, chaetognaths
Argyropelecus sladeni	10–20	66	34	Ostracods, copepods,
(off Southern California)	30–50	24	76	chaetognaths → euphausiids
Sternoptyx diaphana	10–24	100	0	Copepods, ostracods
(Pacific subantarctic)	40–80	60	40	→ euphausiids
Sternoptyx obscura	10–20	86	14	Copepods → ostracods,
(off Southern California)	30–50	63	37	amphipods, chaetognaths
Valenciennellus tripunctulatus	15–20	85	15	Small copepods →
(E. Gulf of Mexico)	30–35	58	42	larger copepods;

Modified from Hopkins and Baird, 1977.

Omori (1974) reports most decapod crustaceans are also predators or omnivores utilizing various available food items. Analyses of stomachs of midwater shrimps suggest that copepods, euphausiids, and other small crustaceans constitute the majority of the food. There is also evidence for a certain amount of cannibalism. As with fishes, the shrimps are not specialists, but opportunists, taking what is available within a certain size range.

In spite of the paucity of food in the pelagic areas of the deep sea, a condition that, as we have seen, causes the organisms to have the varied diets observed among fishes and crustaceans, there is a certain amount of selectivity and food resource partitioning among animals. The most common form of food resource partitioning occurs when species feed in different depth zones in the water column. Different prey items or different prey abundances occur in different depth zones; thus, predators feeding in these different zones encounter and consume prey populations of differing composition. Even within a given depth zone, the prey composition of the resident predators varies, indicating a certain degree of resource partitioning. This is probably due to anatomical differences in the prey capture mechanisms of the predator. Such structural modifications allow them to prey only on certain prey sizes or shapes. Since many organisms migrate through depth zones on a diurnal basis, the available prey composition varies over a 24-hour period at any given depth. Therefore, a predator that feeds during the day at a given depth may encounter different prey than one feeding at the same depth at night.

LIFE HISTORY PATTERNS OF MIDWATER ORGANISMS

Most fishes and decapod crustaceans of the mesopelagic zone are short-lived creatures. According to Mauchline (1977), myctopid and gonostomatid fishes tend to become sexually mature at from one to three years and probably live no more than two to four years. Decapod shrimps of the dominant mesopelagic genera *Sergestes*, *Pasaphaea*, and *Gennadas* mature in their first or second year and have life spans of one to three years. Euphausiids of midwater zones generally mature in their first year and live perhaps two years. On the other hand, the bathypelagic and abyssal pelagic crustaceans are estimated to live considerably longer, as much as two to seven times longer.

The lower temperatures and the greater paucity of food in the bathypelagic and abyssal pelagic zones probably decrease the rate of growth, delay the onset of sexual maturity, and therefore, increase longevity of these animals. The extreme expression of this trend is the aforementioned "abyssal gigantism" (pp. 140–141).

The reproductive cycle among midwater fishes is not as yet well known. Data from Clarke (1973, 1974) on species of myctopids and stomiatoid fishes from the central North Pacific Ocean suggest that the fishes have a seasonal reproductive cycle and spawn in the spring and summer. Juveniles do not appear to undergo vertical migration as do adults, and they are distributed in shallower water than the adults.

Mesopelagic decapod shrimps also show seasonal reproductive activity, usually spawning in late winter or spring in temperate zone seas. In bathypelagic shrimps, it appears that spawning is prolonged over more of the year. Little is known about the movements and distribution of larval shrimps.

References

Anderson, N. R., and B. J. Zahuranec (eds.). 1977. Oceanic sound scattering prediction. Plenum, N.Y. 859 pp.

Brauer, R. (ed.). 1972. Barobiology and the experimental biology of the deep sea. North Carolina Sea Grant Program, Univ. of North Carolina, Chapel Hill. 428 pp.

Brunn, A. 1957. Deep-sea and abyssal depths, chpt. 22, pp. 641–672. *In*:

Hedgpeth, J. E. (ed.). The treatise on marine ecology and paleoecology. Vol. I, Ecology. Memoir 67, Geol. Soc. of Amer.

Dietz, R. S. 1962. The sea's deep scattering layers. Sci. Amer., 207(2):44–50.

Hardy, A. 1965. The open sea. Vol. I, The world of plankton; chpt. 12, Life in the depths. Houghton Mifflin, N.Y. 385 pp.

Heezen, B. C., and C. D. Hollister. 1971. The face of the deep. Oxford Univ. Press, N.Y., 659 pp.

Idyll, C. P. 1976. Abyss. Crowell, N.Y. 428 pp.

MacDonald, A. G. 1975. Physiological aspects of deep sea biology. Cambridge Univ. Press. 450 pp.

Marshall, N. B. 1954. Aspects of deep-sea biology. Philosophical Library, N.Y. 380 pp.

Marshall, N. B. 1980. Deep sea biology: Developments and perspectives. Garland STPM Press, N.Y. 566 pp.

Mauchline, J. 1972. The biology of bathypelagic organisms, especially crustacea. Deep-Sea Res. 19:753–780.

Menzies, R. J., R. Y. George, and G. T. Rowe. 1973. The abyssal environment and ecology of the world oceans. Wiley, N.Y. 488 pp.

Omori, M. 1974. The biology of pelagic shrimp in the ocean. Adv. Mar. Biol. 12:233–324.

Young, R. E. 1975. Function of the dimorphic eyes in the midwater squid, *Histioteuthis dofleini*. Pac. Sci. 29(2):211–218.

Chapter Five
SHALLOW WATER SUBTIDAL BENTHIC ASSOCIATIONS AND LARVAL ECOLOGY

In this chapter, we return from the deep sea to consider the bottom communities and conditions for life in the shallow waters fringing the land masses of the world, namely, the continental shelf region of the world's oceans. This is a region about which we have considerably more knowledge, because it is more accessible and because the continental shelf areas are the major fishing grounds of the world and have thus been the object of much study. This chapter covers major subtidal associations except for coral reefs, which, because of their many unique properties, are covered in Chapter Nine. The shallow subtidal area is contiguous at its upper end with the intertidal regions of the world's oceans, and at least some of the principles which we discuss in this chapter will be applicable to the intertidal zone also.

COVERAGE AND DEFINITIONS

This chapter covers that area of the oceans that lies between the area of lowest low water on the shore to the edge of the continental shelf at a depth of about 200 m. In our scheme of classification, this area is known as the *sublittoral*. Overlying it are the waters of the *neritic* zone (see Chapter One, Fig. 1-17, on the scheme of zonation). Most of this zone is composed of soft sediments, sand and mud, and a much lesser area of hard substrate. Ecologically, two quite different groups of organisms inhabit this area. The *epifauna* are all those benthic organisms that live on, or are otherwise associated with, the surface. The *infauna* is the term used to describe those organisms that live in the soft substrate. These two terms are applicable to all benthic habitats. A third group comprises the large, mobile predators such as fishes and crabs. This chapter will deal primarily with infauna, since it occupies perhaps 90 percent of the sublittoral area.

Infauna organisms are usually divided into categories based upon size. *Macrofauna* are those organisms that are greater than 1 mm in size. The term *meiofauna* is applied to those organisms that lie within the size range of 1.0 mm to 0.1 mm. A final size class is called *microfauna*, which is organisms below 0.1 mm in size. This group is primarily composed of protozoa and bacteria. This chapter will deal only with macrofauna organisms. Meiofauna and microfauna are dealt with in Chapter Seven.

ENVIRONMENTAL CONDITIONS

The continental shelf waters are less constant and show more variability in environmental conditions than either the epipelagic of the open ocean or the deep sea. Perhaps the most important physical factor that acts on the bottom communities is turbulence or wave action. In these shallower waters, the interaction of waves, currents, and upwelling all act to create turbulence. This turbulence generally keeps inshore waters from becoming thermally stratified except for brief periods, at least in the temperate zone. Thus, nutrients are rarely limiting or locked up in a bottom reservoir. Productivity is much higher than in similar waters offshore because of nutrient abundance, both from runoff from land and recycling (see Table 2-1). This high productivity sustains high populations of both zooplankton and benthic organisms.

Wave action is an important factor in this area. Long period ocean swell and storm waves have an effect that can extend to the bottom in these shallow waters. In soft bottoms, the passage of such waves may cause large surging motions in the bottom, which greatly affects the stability of the substrate. The substrate particles may be moved around and resuspended. This has a profound effect on the infauna animals in the substrate. Wave action also determines the type of particles that are present. Heavy wave action removes fine particles by keeping them in suspension, leaving mainly sand; thus, fine silt sediments can occur only in areas that have low wave action or are deeper than wave action can affect.

Salinity in this region is more variable than in the open ocean or deep ocean, but except for areas where large rivers discharge massive amounts of fresh water, the salinity does not change enough to be of ecological significance.

Temperature is also more variable in inshore waters and shows a definite seasonal change in the temperate zone. These temperature changes may be used by organisms as cues to begin or end various activities such as reproduction.

Light penetration in these turbulent waters is reduced when contrasted with open ocean areas. The combination of large amounts of debris, both from land and from breakup of kelp and sea grasses, plus the high plankton densities due to abundant nutrients, act to reduce light penetrations to a few meters.

The food supply is abundant in this area, due partly to the increased productivity by the plankton, but also due to the production of attached plants such as kelp and sea grasses. This is one of the few areas in the sea where macroscopic plants have had any significance in production. A final food source is runoff from land. Although there are large plants in the sublittoral shelf, there are relatively few large grazing animals and the major use of kelps and sea grasses as food occurs only after they have been broken up into detrital particles.

The soft bottoms in the sublittoral are essentially without topographic diversity and the vast expanses extend monotonously for long distances (Fig. 5-1). Lacking any topographic relief, the only apparent difference from one place to another is that of substrate grain size. Subtidal hard substrates, on the other hand, may have considerable relief with many potential habitats. The lack of relief in the infaunal areas generally means fewer different habitats for animals to occupy and fewer potential ways for making a living. The number of infauna species is thus generally less than the number of epifauna species. There are, if you wish, fewer niches available. Most infauna animals are deposit feeders, ingesting the abundant detritus that rains down, or else are suspension feeders, filtering out of the water column the abundant plankton or floating detritus. Bottom fishes, on the other hand, are predominantly carnivores.

BENTHIC ECOLOGY

INFAUNAL ASSOCIATIONS

The dominant groups of organisms that constitute the macrofauna present in sublittoral soft bottoms belong to four main taxonomic groups: class Polychaeta, class Crustacea, Phylum Echinodermata, and Phylum Mollusca. Polychaete worms are represented by numerous tube-building and burrowing species. The dominant crustaceans are the larger ostracods, amphipods, isopods, tanaids, mysids, and a few of the smaller decapods. They mainly inhabit the surface of

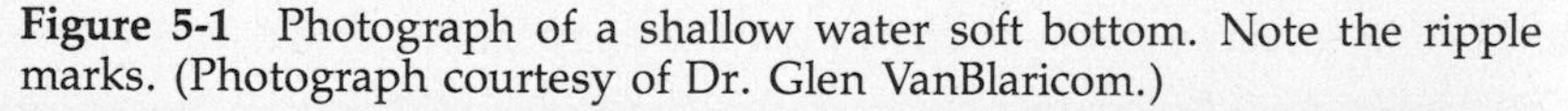

Figure 5-1 Photograph of a shallow water soft bottom. Note the ripple marks. (Photograph courtesy of Dr. Glen VanBlaricom.)

the sand and mud. Mollusks are mainly represented by various burrowing bivalve species with a few gastropods at the surface. Echinoderms common in the subtidal benthos include particularly brittlestars and echinoids (heart urchins and sand dollars).

It is of interest that the first quantitative work done in marine ecology was done on the sublittoral benthic infauna. Back in the first years of this century, C. G. Joh. Petersen, a Danish biologist, began his investigations on the benthos of the Danish seas. He was interested in evaluating the role of bottom organisms in supporting the fish populations on which the Danish fishing industry depended. In order to do this on a quantitative basis, he invented a grab, the now famous "Petersen grab," which picked up a definite area of the bottom, usually 0.5 or 0.1 m^2. Since he had a sample that covered a known area of bottom, he could then count the numbers of organisms in those samples and from them, extrapolate the numbers in the whole bottom area. Once this was done, he would have some estimate of the amount of food available to the fishes and could guess at the biomass of fish that could be supported.

Analyzing many of these grab samples over the years, Petersen (1918, 1924) observed that extensive bottom areas were occupied by recurrent groups of species and that other areas were inhabited by different associations of species. In all cases, he observed that a relatively few species made up most of the individuals and biomass. This was in contrast to nonquantitative dredge samples taken in the same waters. In the latter case, the faunal lists might be nearly identical in the two areas. The value of the quantitative study was that for the first time it gave a way to evaluate differences between communities based on relative abundance, not just presence or absence of species.

After several years of evaluating these quantitative samples, Petersen observed that the different areas and their dominant organisms remained relatively constant and uniform over time. He then proposed these associations as communities and named them on the basis of the dominant animals. For example, he had a *Macoma balthica* community from inner Danish waters in 8–10 m of water, which was composed of *M. balthica* as the dominant but was also characterized by *Arenicola marina*, *Mya arenaria*, and *Cardium edule* (Fig. 5-2). In this same manner, he named and characterized a whole series of communities from the sublittoral in Scandinavian waters.

Following Petersen's work, various benthic biologists began to investigate other shallow water benthic areas around the world, of which the leader was G. Thorson. What Thorson (1955) discovered, particularly for the temperate zone waters, was that communities of organisms similar to those found by Petersen were found in similar habitats around the world. Thus arose the concept of *parallel bottom communities*. In this ecological concept, similar sediment types at the same depths around the world contain similar communities. Granted, the species are not the same, but they are closely similar ecologically and taxonomically (Fig. 5-2). They occupy similar niches. Additional work in

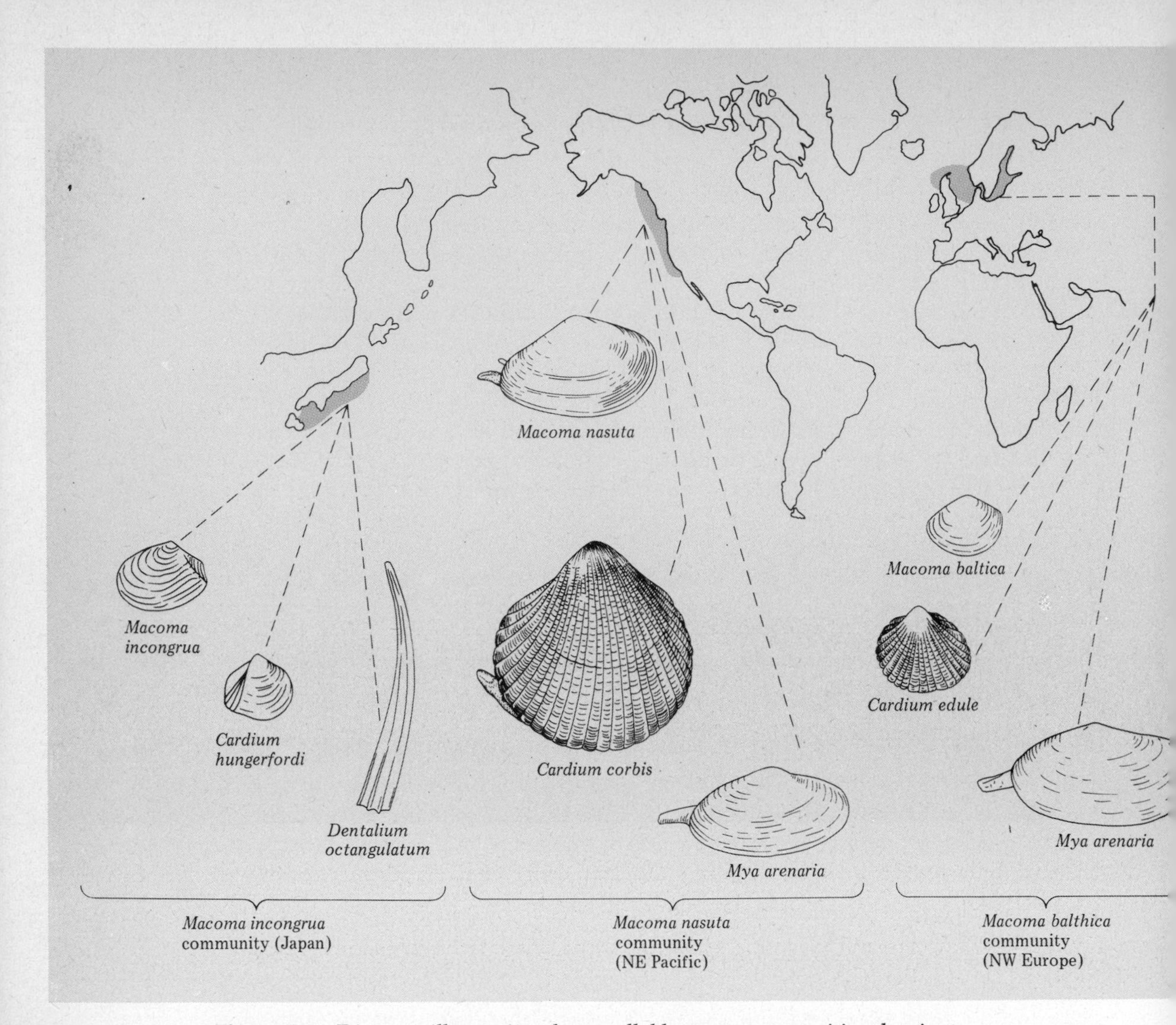

Figure 5-2 Diagram illustrating the parallel bottom communities dominated by the bivalve *Macoma* in three different areas in the North Temperate Zone. (Modified from G. Thorson, Bottom communities, *in:* The Treatise on marine ecology and paleoecology, vol. 1, Ecology, edited by J. Hedgpeth, The Geological Society of America.)

tropical waters by Thorson (1966) revealed that this concept did not extend to such areas, at least not in the same way.

That certain recurrent groups occupy the same substrate types over large areas of the world's ocean seems established. This suggests that such associations are not random, but represent real interacting systems in which some combination of factors acts to ensure continued persistence of the community.

Petersen described the communities, and others later validated their continued existence over time and in various parts of the world. What was lacking in all this descriptive material was an explanation of how such communities persisted and what kinds of interactions ensured continued success. Since Petersen's time, answers to these questions have come to light and will be discussed in the following sections.

LARVAL ECOLOGY

Some of the major factors involved in the persistence of marine benthic communities are those involved with various aspects of larval ecology. Most benthic infaunal and epifaunal communities, throughout the world's oceans, are composed of species that reproduce by producing various larval types that undergo a free-swimming stage in the plankton before metamorphosing to produce the benthic adults. Therefore, an understanding of larval ecology is central to understanding how such communities persist. We discuss the basic features of larval ecology here, but it must be remembered that thses principles will apply to all subsequent benthos as well (Chapter Six through Nine).

Larval Types and Strategies. As noted by Vance (1973), there are three possible paths that development may take in a benthic invertebrate with a given amount of energy available with which to produce young. One way would be to produce a great many very small eggs. Such eggs would hatch quickly into larvae that would be free swimming in the plankton. Because so little yolk was put into each egg, the larvae would be dependent for nutrition on the plankton. Such larvae are called *planktotrophic*. A second strategy would be to produce fewer eggs and endow each with somewhat more energy in the form of yolk. Such eggs would hatch into larvae that, because of their yolk reserve, would not feed in the plankton and spend less time in the plankton before settling. Such larvae would use the plankton phase mainly for dispersal. Larvae of this type are called *lecithotrophic* larvae (Fig. 5-3).

A final option would be to dispense with the larval phase altogether. In this strategy, the adult produces a very few eggs, each with a very large amount of yolk. Such eggs would undergo a long development without additional energy sources. They would pass through the larval stages in the egg and hatch as juveniles. There would be no free-swimming larval stage. Such hatchlings are called *nonpelagic larvae* or *juveniles*.

Each of the above developmental pathways has both advantages and disadvantages. In the case of the planktotrophic larvae, the advantages are that a very large number of young can be produced for a given amount of available energy, and wide dispersal is assured through the long time spent in the plankton. The disadvantages are that the larvae depend for nutrition on the plankton, a notoriously unpredictable situation, and that long residence in the plankton increases the chances that a predator will consume the larvae. The chances of being consumed by a predator in the plankton are extremely high and the pelagic zone is thus a very dangerous habitat.

Lecithotrophic larvae have the advantage of spending less time in the

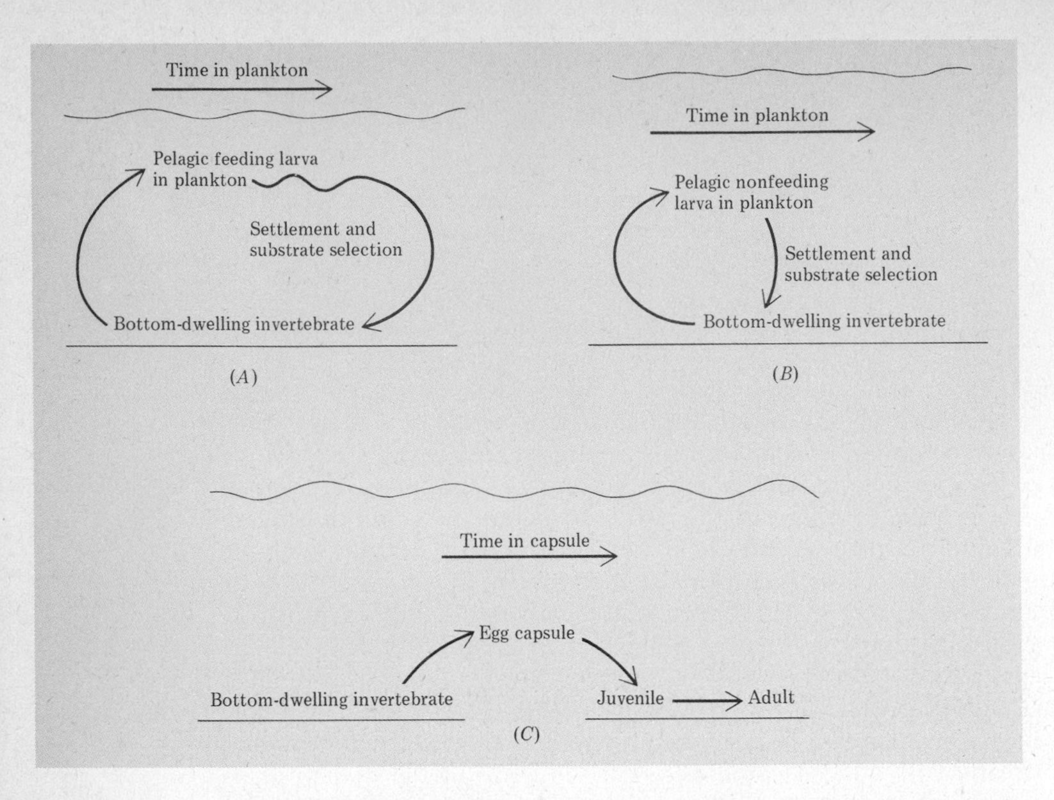

Figure 5-3 Representations of the three patterns of development found among benthic invertebrates. *(A)* Planktotrophic. *(B)* Lecithotrophic. *(C)* Nonpelagic.

plankton and thus have less chance of being consumed. They also are not dependent on the vagaries of the plankton for food. The disadvantage of this strategy is that, because of the greater amount of energy placed into each egg, fewer eggs and larvae can be produced per unit of available energy. A shorter time in the plankton also means less dispersal ability.

The nonpelagic development has the advantage of reducing the plankton mortality to zero. In this case, however, the disadvantages are that only a few eggs can be produced because of the great amount of energy each takes, and that dispersal is nil.

Since all three types of development exist, it can be assumed that under certain conditions, one or the other type is favored. Even though we are currently ignorant of all the reasons why some areas and conditions favor one larval type over another, it is possible to suggest some which are valid.

In polar waters, where the productivity is reduced and confined to a narrow summer peak (pp. 68–69) and where it takes a long time for benthic animals to obtain sufficient energy to reproduce, nonpelagic development is common. Such development avoids the problem of precise timing to hit the very short polar bloom period, where failure would mean the death of most, if not all, larvae, and it ensures the highest survival rate of the few eggs which can be produced.

Lecithotrophic larvae are common under somewhat similar conditions as above. Lecithotrophy is preferred over nonpelagic development if dispersal is of prime concern to the survival of the organism in question. Lecithotrophy is also more energy efficient than planktotrophy. Both lecithotrophy and nonpelagic development appear to be common in deep-sea organisms.

The advantages of planktotrophy are that eggs are cheap to produce, can be produced in large numbers, and have high dispersal rates. The disadvantages are primarily the great attrition rate suffered while in the plankton and the unpredictable plankton food source. Planktotrophic larvae are common then, where the plankton food is predictable over long periods, where dispersal is of importance, and where development time is short. The area which fulfills most of these criteria is the tropics, and it is there we find most forms with planktotrophic larvae.

This is not to imply that each area has only a single developmental type. Most temperate zone areas, for example, have a mixture. Why should this be? We can only suggest that for each species, there is a necessity to balance all the factors, primarily the need for dispersal, against the need for larger numbers of young. The type of reproduction that has evolved is that which balances dispersal on one hand with survival on the other. This is reinforced by the finding of certain benthic infauna species which are able to alter their reproductive strategy among the three developmental patterns depending on prevailing conditions.

Larval Ecology and Community Establishment. Many, if not most, marine benthic communities are composed primarily of species that have a free-swimming larval stage. The exception to this rule appears to be the aforementioned polar waters. Since over larger areas the benthic communities seem to persist over time (Petersen's communities can still be found 50 years later), but are dependent for this continued existence on the settlement of larvae from the plankton, and since the adults live only a short time, how is this stability maintained? In other words, what mechanisms are there that will act to ensure that the "right" larvae will settle and mature to keep the community going? A partial answer to this is to be found in the characteristics of larvae.

Since the work of D. P. Wilson (1952), many studies on larvae have disclosed certain aspects of their physiology and behavior that result in maintaining communities. Larvae have been shown to have preferences as to the areas in which they will settle. In other words, larvae do not just settle out of the water column on whatever substrate is available when the time comes for them to metamorphose into adults. Larvae have the ability to "test" the substrate, probably responding to certain physicochemical factors; if the substrate is unsuitable, they do not settle out or metamorphose. This means that certain substrates will always be attractive to certain larvae and repellent to others. Larvae also respond to the presence or absence of adults of their own species. Many larvae preferentially settle where the adults are living. They are, as Crisp and Meadows (1962) have shown, attracted to the area by a chemical or *pheromone* secreted by the adult. Since the presence of the adult means that the area is suitable for habitation, this mechanism is one that acts to ensure survival of the young. It also ensures persistence of the community.

The larvae of many invertebrates also have the ability to delay their own metamorphosis for a certain period of time if they cannot find suitable substrates at the time they are supposed to settle. This delay of metamorphosis is a finite period, however, and after a certain time, metamorphosis will take place even over unsuitable substrates. The ability to delay settlement, however, is yet another factor that can be employed to ensure that the larvae will settle in the proper place.

Larvae, as Thorson (1966) has noted, also respond to other physicochemical factors such as light, pressure, and salinity. Many free-floating larvae are positively phototactic in the early stages of their larval life. This ensures that they will reside in the upper, faster-moving waters where dispersal is greatest. Later, as time for settlement approaches, they become negatively phototactic and migrate downward toward the bottom. Some larvae are very sensitive to light and pressure such that they inhabit only certain levels in the water column, areas with a definite light regime and pressure. Segregation of larvae into various layers in the water column also means that wherever a given layer impinges on the bottom, it will have only a certain complement of larvae available for settlement, a further way of ensuring a nonrandom bottom community (Fig. 5-4).

Finally, different species have different times of reproduction and hence, different times of the year during which larvae are in the water column. As we shall see subsequently, this may also have an effect on the associations which develop.

Despite the considerable abilities of larvae to discriminate among substrates and select places to settle, there is often considerable variability from year to year in both species abundances and composition within bottom infaunal communities. Samples taken close to each other often show considerable differences in species composition and abundance. If the larvae have such selective powers, why then is there such variability? The answer to this lies in the life histories of the various invertebrates composing the community, their interaction with each other and with the physical environment, and the effects of predators.

LIFE HISTORIES

According to MacArthur (1960), it is possible to recognize two quite different types of life history patterns among organisms in any habitat. These are introduced here and will be referred to in subsequent chapters. One type can be called *opportunistic*. Opportunistic species are characterized by having short life spans, rapid development to reproductive maturity, many reproductive periods per year, larvae present in the water during much or all of the year, and high death rates. Usually, they are small animals and often are sedentary or sessile.

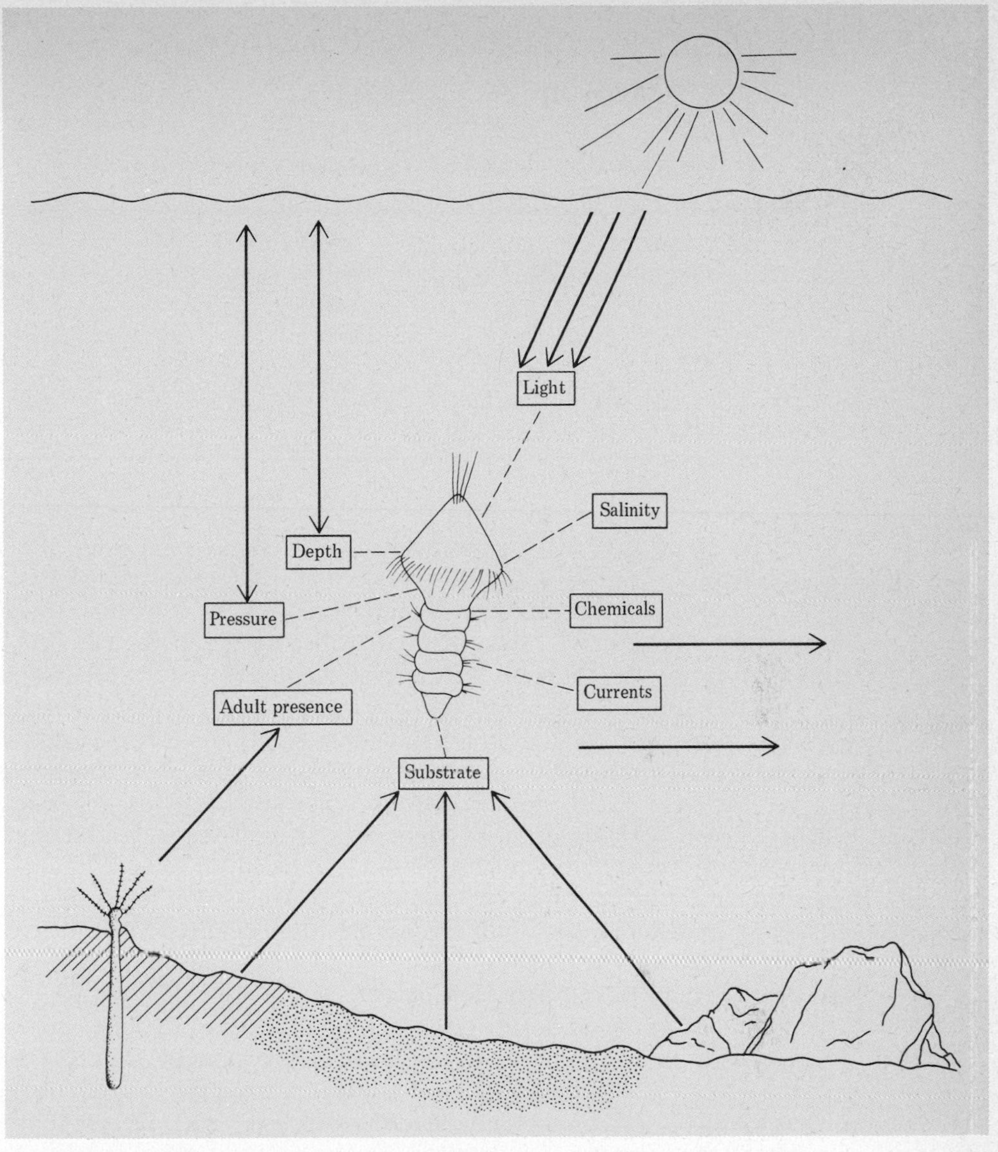

Figure 5-4 Various factors of the environment to which larvae respond in selecting a suitable site for settling.

The second type of life history can be termed *equilibrium*. Equilibrium species are those that have long life spans, relatively long development time to reach reproductive maturity, one or more reproductive periods per year, and low death rates. Usually, these species are somewhat larger in size than opportunists and often mobile (Table 5-1).

The above characteristics define the two ends of a continuum. It must be remembered that, whereas some species will have all characteristics of an opportunistic species and others those of equilibrium species, still others will fall somewhere in the middle and have varying mixtures of these features. For

TABLE 5-1 Summary of the Characteristics of Opportunistic and Equilibrium Species

Features	Equilibrium	Opportunistic
1. Reproduction periods	Few per year	Many per year
2. Development	Slow	Rapid
3. Death rate	Low	High
4. Recruitment	Low	High
5. Colonizing time	Late	Early
6. Adult size	Generally large	Generally small
7. Mobility	High	Low

purposes of discussion, however, it is easier to consider the ends of the spectrum.

Opportunistic species tend to be favored where the substrate is subjected to frequent disturbance from some particular agent. In the case of shallow water soft bottoms, the common disturbance agent is waves or other water motion that stirs up the bottom or carries away the upper layer of sediment. Another disturbance may be the rapid deposition of sediment on the bottom. In certain cases, biological activity may also disturb the bottom, for instance when large fish such as rays dig into the bottom. When such a disturbance occurs, it usually causes the mortality of the resident organisms. The result is an open area, which can be resettled by organisms. Because of their attribute of having larvae in the water most of the time, opportunists quickly settle these open areas. Furthermore, if disturbances occur frequently, the opportunists will have the advantage, because they can quickly mature and reproduce again before being destroyed by the next disturbance.

Equilibrium species, on the other hand, tend to inhabit areas that are not subject to frequent disturbance, thus allowing them to complete their life cycles. Too frequent disturbance would mean destruction before they could reach reproductive maturity, and hence, no larvae would be present in the water column to settle the area again.

Whenever the bottom is subject to frequent disturbance, the benthic communities are usually composed primarily of opportunists, whereas in areas where disturbance is less frequent, more equilibrium species are found.

Because of their different characteristics, there are differences in abundances between these two groups. Communities in which the number of opportunists is high show great fluctuations in numbers of individuals over the year, whereas communities composed of equilibrium species are more constant in abundances.

If the above situation prevails, it would appear that communities would be composed of either opportunists or equilibrium species, depending on frequency of disturbance. In reality, samples of the bottom usually contain representatives of both groups. How does this happen? Apparently, the equilibrium species are slower to colonize disturbed areas than are opportunists, but they are better competitors once they are there. That is to say, once there, they tend to push out (outcompete) the opportunists. Over time, then, there is a sequence in which immediately following a disturbance, the area is settled by opportunists. As time goes on, the equilibrium species settle in, and finally, they outcompete the opportunists. When sampling, then, differing numbers of both groups would result, depending on the time since the last disturbance. Thus, varying numbers of both groups can be seen in bottom samples (Fig. 5-5).

The reason for different numbers of individuals and species from different areas in the same bottom is at least partially due to the fact that disturbances do not affect the whole bottom equally, but rather act selectively. The result is to produce a patchy array of communities, each in varying degrees of recovery from disturbance and with differing numbers of opportunists and equilibrium species.

Since wave-induced disturbance decreases with depth, it is not surprising that deeper communities tend to be inhabited by more equilibrium species than shallow water unless another type of disturbance creates unstable conditions. Such a disturbance could be biological; for example, the reworking of sediment by fishes or burrowing deposit feeders.

BIOLOGICAL INTERACTIONS

Within any community, the species do not exist in isolation. They interact with other species in the same area. This interaction is also important in determining the composition of the community.

Figure 5-5 Changes in a soft bottom community with time after a physical disturbance. (From D. C. Rhoads, P. L. McCall, and J. Y. Yingst, Disturbance and production on the estuarine seafloor, *In:* American Scientist, September-October, 1978. Reprinted by permission of *American Scientist*, journal of Sigma Xi, The Scientific Research Society.)

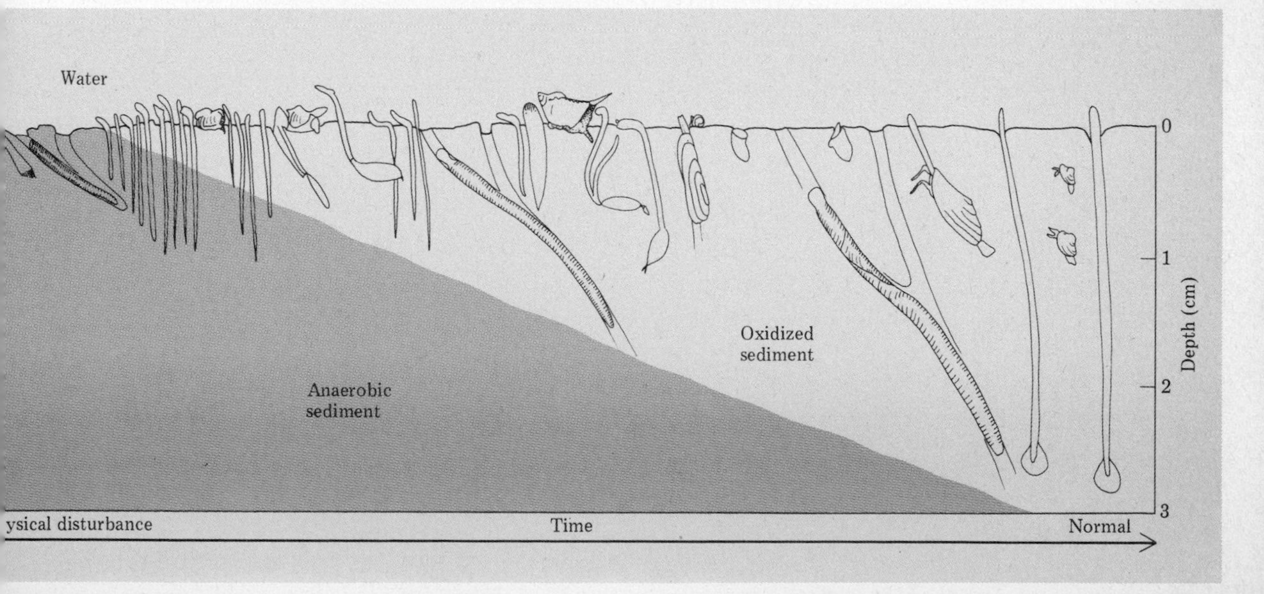

In addition to the opportunist-equilibrium classification established above, Woodin (1976) has suggested that it is also possible to classify infauna organisms into burrowing deposit feeders, suspension feeders, and tube builders of various types. This classification is entirely independent of the opportunist-equilibrium classification (Fig. 5-6).

Burrowing deposit feeders tend to be most abundant in soft, muddy sediments, areas with a high concentration of organic matter. It has been observed in several areas that where burrowing deposit feeders are prevalent, suspension feeders are rare or absent. What happens is that the deposit feeding organisms burrow through the top few centimeters of bottom and in so doing, create a very loose, unstable layer of fine particles including fecal pellets. This layer is then easily resuspended by the slightest water motion. The resuspended sediment has the effect of clogging up the fine filtering structures of the suspension feeders, rendering them unusable. Furthermore, the constant reworking and settlement of resuspended particles tend to lead to burying of the newly settled larvae of suspension feeders, thus killing them. It does not kill the larvae of the deposit feeders because they burrow into the more compact substrate underneath. The combination of clogged filtering surfaces making feeding impossible plus the suffocation of settling larvae excludes the suspension feeders. This exclusion of one group by modification of the environment by another is called

Figure 5-6 Representation of the three major infauna groups: burrowing deposit feeders, suspension feeders, and tube builders.

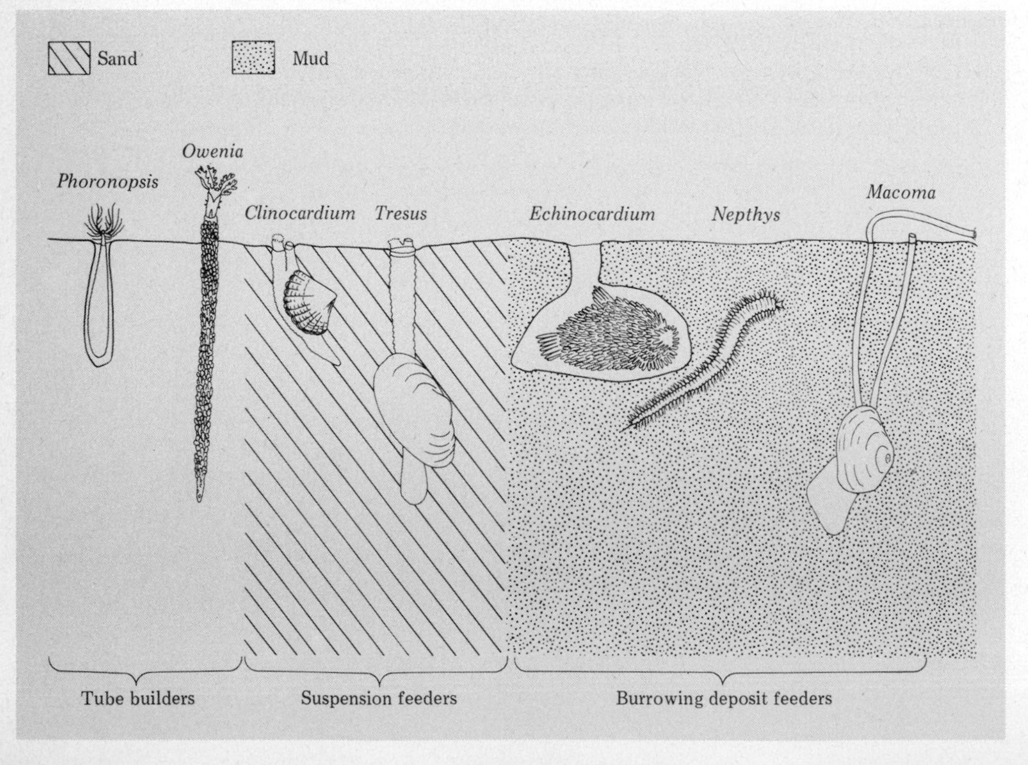

by Rhoads and Young (1970) *trophic group amensalism*. In this way, then, deposit feeders maintain their own communities and exclude suspension feeders.

Suspension feeding organisms seem to be more abundant where substrate is more sandy, where the organic matter is less, and thus where deposit feeders would find less food and more difficult burrowing. Since the substrate is often more stable, the suspension feeders can establish themselves. Once established, they may also exclude any potential deposit feeders by filtering larvae out of the water.

Tube-building organisms may be either suspension feeders or deposit feeders. These animals build tubes in the substrate in which they live. The tubes have the ability to stabilize the substrate. They also extend down into the substrate. By stabilizing the substrate, they prevent resuspension of fine particles and allow the presence of suspension feeding organisms. At the same time, the presence of the tubes in the substrate serves to restrict the available space for burrowing deposit feeders. The deposit feeders cannot burrow through because they collide with the rigid tubes. Thus, the tube builders act indirectly to restrict burrowing deposit feeders. Tube dwellers may be found on either mud or sand, but where found in mud, the presence of their tubes act to exclude burrowing deposit feeders and enhances the area for suspension feeders. Such exclusion by interference with normal activities is termed *competitive interference*.

As a result of these interactions among different types of organisms and the substrate, different associations of organisms can occur. These are summarized in Table 5-2.

Another biological factor that is important in determining the species structure of infaunal communities is predation. Both invertebrate and vertebrate predators take a toll of the infauna organisms. Predator activity may be responsible for clearing certain small areas of macrofauna and thus creating a disturbance that will be followed by a recolonization sequence. This is one way that patchy distributions occur on the bottom. Such clearing of areas is known to result from feeding activities of certain flatfish and rays. In the case of rays, their habit of burrowing into the bottom destroys rather extensive areas and thus allows a recolonization to occur.

It is difficult to determine the relative roles of predator-induced and wave-induced disturbances on the structure of soft-bottom communities. This is mainly because it has been difficult in the past to devise good experiments that will exclude predators from areas of the bottom while allowing wave action to proceed. Recently, Virnstein (1977) and other ecologists have devised various "cages" that, when placed over areas of the bottom, exclude predaceous fishes and invertebrates. When properly done, such studies have produced increased numbers of infauna organisms within the "cage" as compared to outside (Fig. 5-7). This suggests that predation is important in determining at least the numbers of infauna organisms present and perhaps also the species composition. Further studies are needed, however, before we can say more about the exact effects of predation on communities. In addition to various fishes that prey

TABLE 5-2 Characteristics of the Three Types of Infauna

Functional Groups	Ingests or Disturbs by Its Feeding Activities Surface or Near-Surface Larvae	Filters Larvae Out of Water Prior to Settlement	Changes Sediment
Deposit feeding bivalve	Yes	No	Makes more easily resuspended
Suspension feeding bivalve	No	Yes	No, or much more slowly
Tube-building forms	Yes	Depends on trophic type	Stabilizes; reduces space below surface; increases epifaunal space on surface due to its tubes

After S. A. Woodin, 1976.

on infauna organisms, the dominant invertebrate predators appear to be various crabs and starfish and, if the community has many bivalves, carnivorous gastropod mollusks.

COMPETITION AND VERTICAL DISTRIBUTION

Competition among benthic marine invertebrates is generally restricted to food and space. In the subtidal soft-bottom areas of the oceans, where the dominant

Figure 5-7 Effects of a cage excluding predaceous starfish and fish from the infauna, thus increasing the infauna density inside the cage as compared to outside.

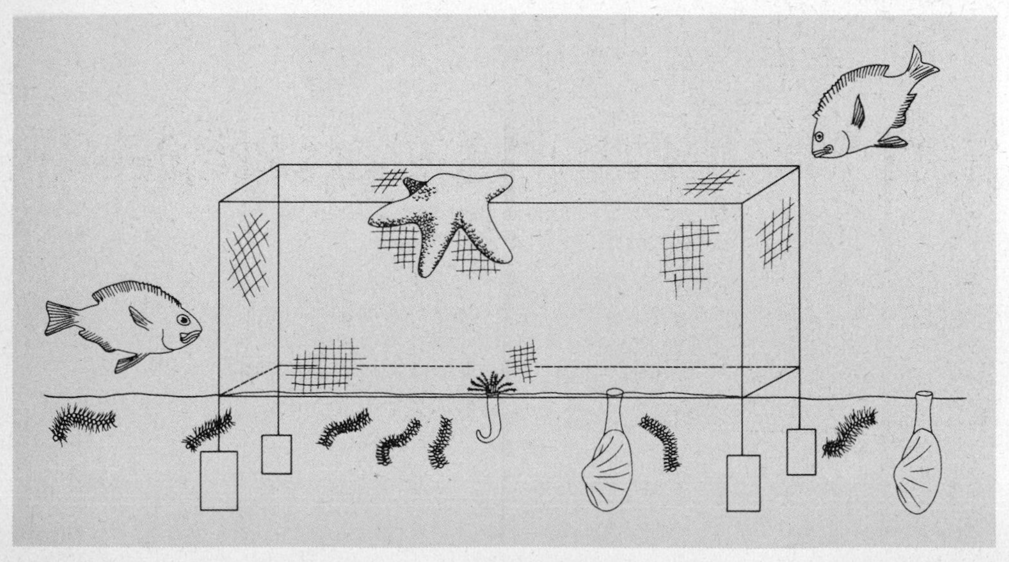

Larval Type at Settlement	Predicted Dense Co-occurring Forms
May be surface or may burrow below surface	Burrowing polychaetes
Surface	None
Surface	Epifaunal bivalves and tube epifauna

species are suspension and deposit feeders, competition is usually for space. Food, in the form of either detritus or plankton, is abundant. Organisms that are suspension feeders must, of course, have their feeding structures in the water column. However, it is possible to get more suspension feeders in a given area of bottom if instead of placing them all on the surface, they are distributed vertically in the substrate with only the feeding organ at the surface (Fig. 5-8). The feeding structure takes up less room than the whole animal. Deposit feeders may feed not only at the surface, but at any level in the substrate that contains organic material. They do, however, require a connection to the surface to obtain oxygen from the overlying water. In a similar manner then, more deposit feeders could be accommodated and direct space competition reduced if the different species were distributed vertically in the substrate.

Such vertical distribution of species is a well-known feature of soft bottoms. It occurs in both mud and sand (Fig. 5-9) and in both suspension and deposit feeding organisms. Experiments by Peterson (1977), in which certain species were removed from given depth strata, resulted in an increase in abundance of other species occupying the same depth level. Still other experiments, in which the density of given species at given levels has been increased above the natural levels, have resulted in rapid emigration out of the high density areas by the species. These experiments together suggest that there is definite competition for space among species, and it results not only in vertical spacing of species, but also in maintenance of certain densities.

Competition for space may also be the reason for the occurrence of *commensalism* among certain species (see Chapter Ten for discussion). That is to say that

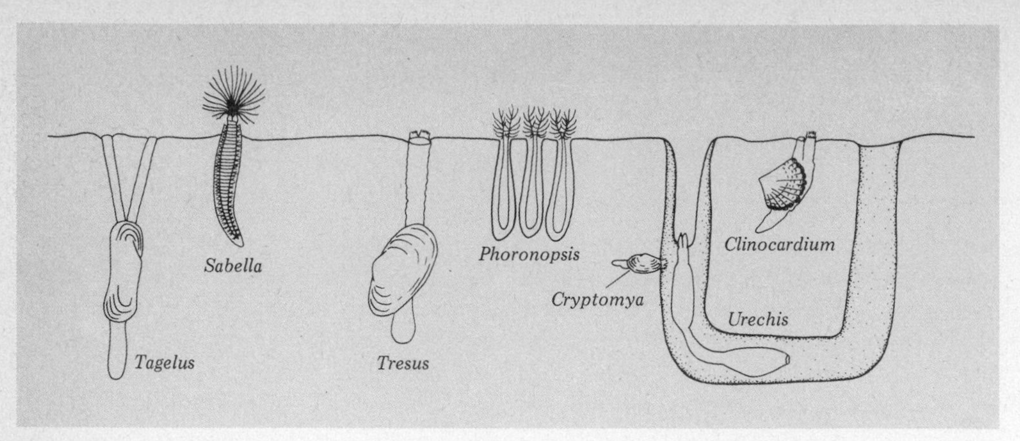

Figure 5-8 Various infauna suspension feeding types distributed at different depths in the substrate.

competition is so severe that it selects for organisms that share space. An example is the clam *Cryptomya californica*, which lives with its siphons in burrows of the shrimp *Callianassa californiensis* or the worm *Urechis caupo*.

In summary, marine subtidal soft-bottom infaunal communities are characterized by patchy distributions of organisms and certain variability in species abundances and composition over time. The communities are maintained by settlement of larvae from the plankton. The ability of the larvae to choose areas for settlement coupled with ability to delay metamorphosis ensures that settlement is not random. The patchiness and variability are the result of continual random disturbances created by water motion or biological activity such as predation, which eliminates the population in a small area (= patch). This is followed by recolonization and change with time due to a complex series of biological interactions involving different life history strategies, opportunist versus equilibrium; competitive interference; and exclusion and predation. The result is that different areas have different species abundance and composition depending on the time from the last disturbance and the relative stability of the substrate. Competition for space is ameliorated through vertical stratification of the organisms.

KELP BEDS

Whereas soft substrates are the dominant type on the continental shelf areas of the world, there are other areas which have a hard substrate. Throughout a large part of the cold temperate regions of the world, such hard substrates are inhabited by very large brown algae known collectively as *kelps* (Fig. 5-10).

STRUCTURE AND DISTRIBUTION

Kelps are attached to the substrate, not by roots, but by a structure called a *holdfast* (Fig. 5-11). From the holdfast arises a stemlike or trunklike structure

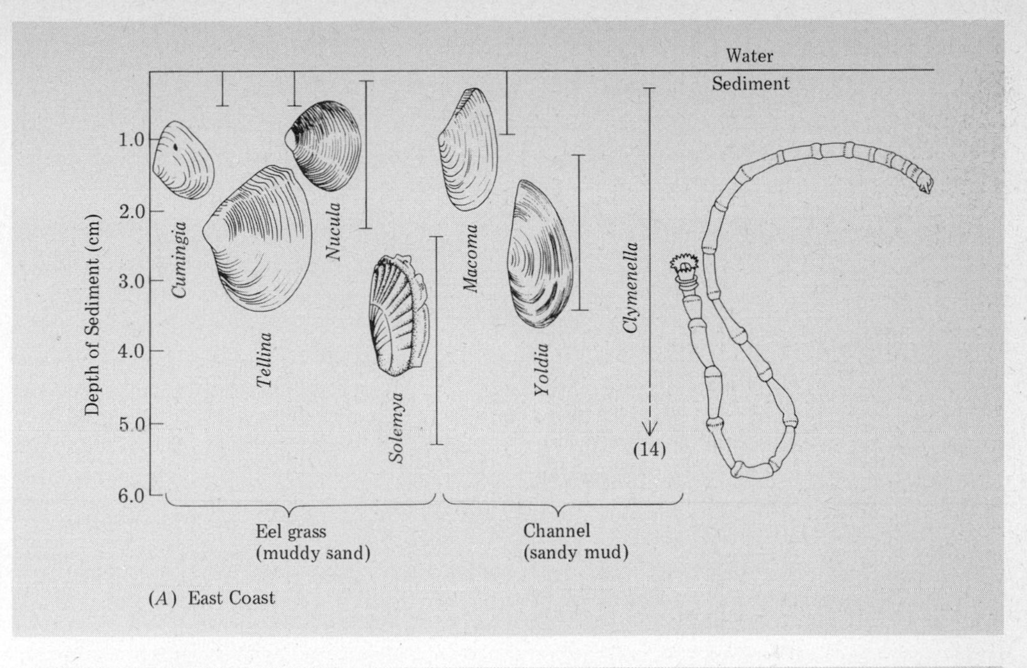

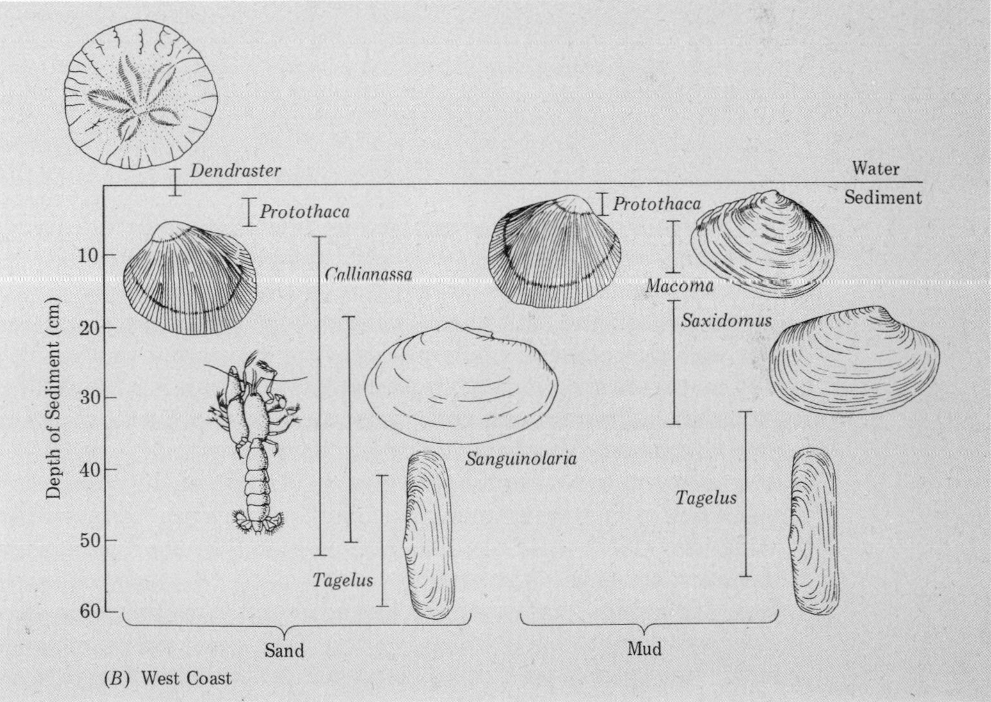

Figure 5-9 Vertical distribution of benthic infauna in (A) an Atlantic Coast bay, Quisset Harbor, Massachusetts, and (B) a Pacific Coast lagoon, Mugu Lagoon, California. (Modified from J. Levinton, Ecology of shallow water deposit feeding communities, Quisset Harbor, Mass., *in:* Ecology of marine benthos, edited by B. Coull, University of South Carolina Press; and from C. H. Peterson, Competitive organization of the soft bottom macrobenthic communities of southern California lagoons, Marine biology, vol. 43, 1977.)

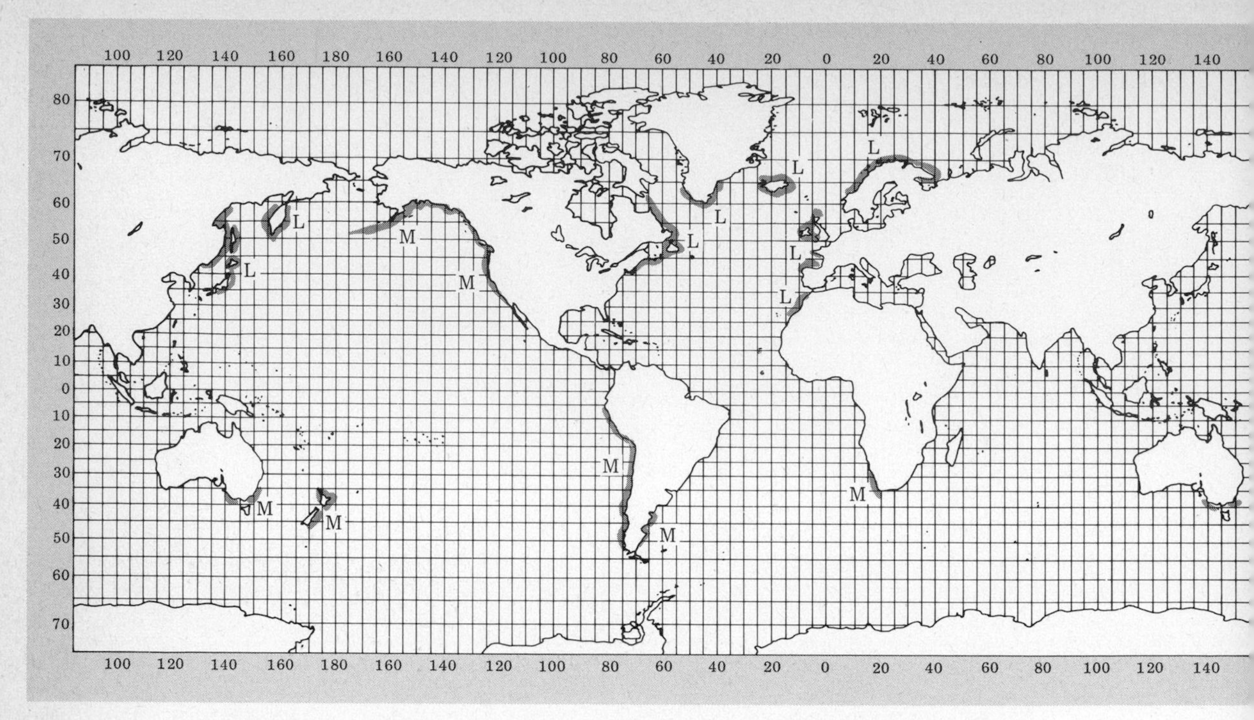

Figure 5-10 Distribution of kelp beds in the world indicating the dominant genera. L. *Laminaria;* M, *Macrocystis.* (Modified from K. H. Mann, 1973, Seaweeds: Their productivity and strategy for growth, *Science.*)

called a *stipe*. The stipe is terminated by one or more broad, flat *blades*. At the base of the blade is a *pneumatocyst* or *float*, which keeps the blade at the surface. Kelps obtain their nutrients directly from the sea water as do the phytoplankton. They depend on the constant movement of water past them to avoid nutrient exhaustion. Since these shallow waters are constantly moved by wave action and currents and the nutrients replenished by turbulence, upwelling, and runoff from land, nutrient depletion rarely occurs and luxuriant growth occurs, forming what may justifiably be called "kelp forests."

In contrast to the monotonous, level landscape of the soft bottom, the kelps form an extensive three-dimensional habitat composed of several vertical layers or strata (Fig. 5-12). As a result, a large number of potential habitats is available and the variety of life is greater.

The major kelps that dominate the beds and form the basic structure of the "kelp forest" are the genera *Macrocystis*, *Nereocystis*, and *Laminaria*. The Pacific coast of both North and South America is dominated by *Macrocystis*, while *Laminaria* is dominant in Atlantic waters (Fig. 5-10). In contrast to most algae, which are small, generally never exceeding a few centimeters or a decimeter in length, the major kelps are giants, with lengths that are the equivalent of trees on land. *Macrocystis* and *Nereocystis* on the Pacific coast of North America may reach 20–30 m in length. Such massive plants grow upward from the bottom and

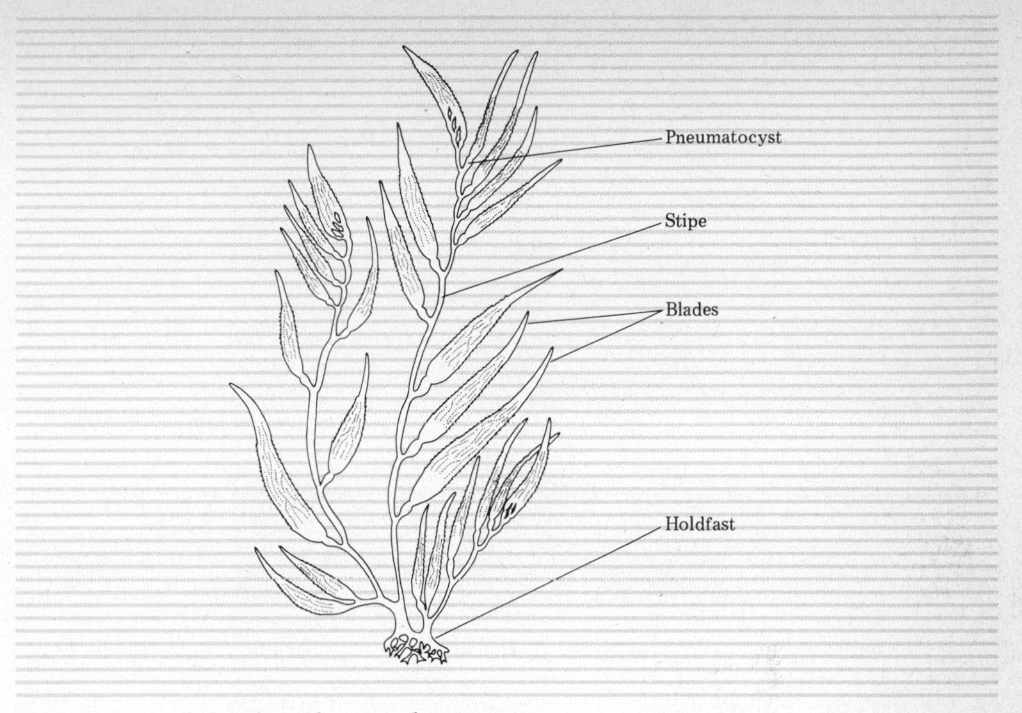

Figure 5-11 A kelp plant showing the structure.

spread their blades at the surface of the water, where they obtain the maximum light. These blades form a "canopy" similar to that of terrestrial forests, cutting off light to the substrate below (Fig. 5-13). Below the canopy, extending up from the bottom, is a set of "understory" algae forming another layer (Fig. 5-14).

Figure 5-12 Cross section of a kelp bed showing four different vertical strata. (From M. Foster, Marine biology, vol. 32, no. 4, 1975-)

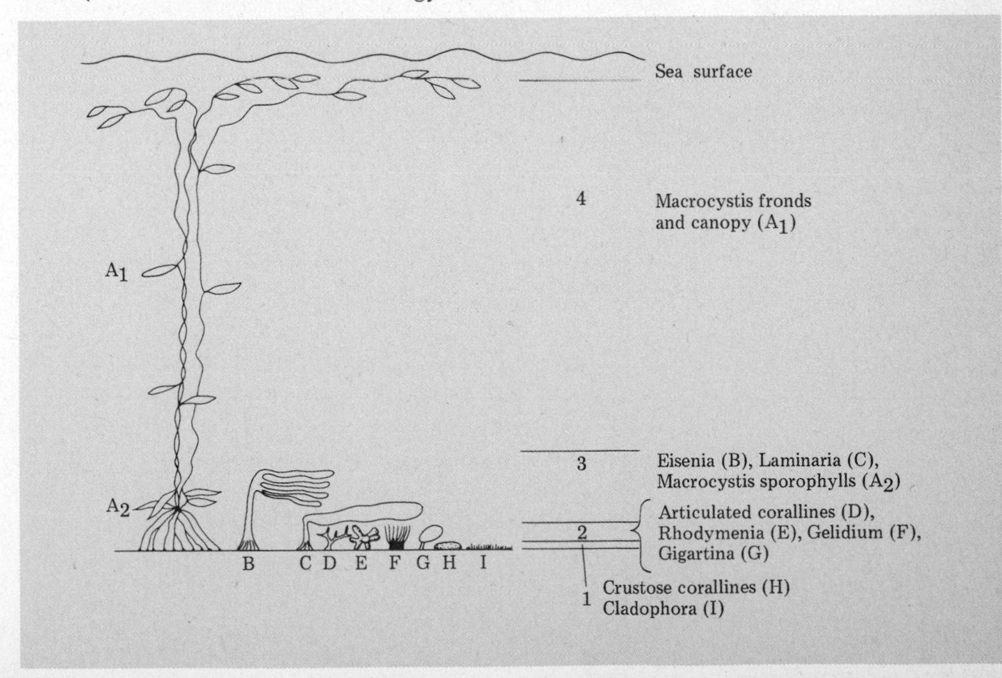

(A)

(B)

Figure 5-13 *(A)* Photo of the surface of a kelp bed showing blades. *(B)* Photo of the interior of a California kelp bed. (Photos courtesy of Michael Foster.)

The extent of the kelp beds on various coasts depends on several factors. In the first place, a hard substrate must be available for attachment. Secondly, the kelps must have light and can establish themselves only in water depths where the young, small plants receive enough light to grow. The amount of light received at any depth is a function of water clarity, and where clear water prevails, kelp beds extend out from shore to depths of 20–30 m. Where such shallow water is extensive, the beds may extend out as much as several kilometers from shore. In more turbulent, murky water, the depth of the beds may be more restricted. Finally, kelps seem limited by temperature. Kelp beds occur throughout the world wherever the water is cool, but are absent from warm temperate and tropical areas. Their considerable extent along the Pacific coast of North and South America is due to the cold upwelling on the North American coast and the cold water of the Humboldt Current along the South American coast.

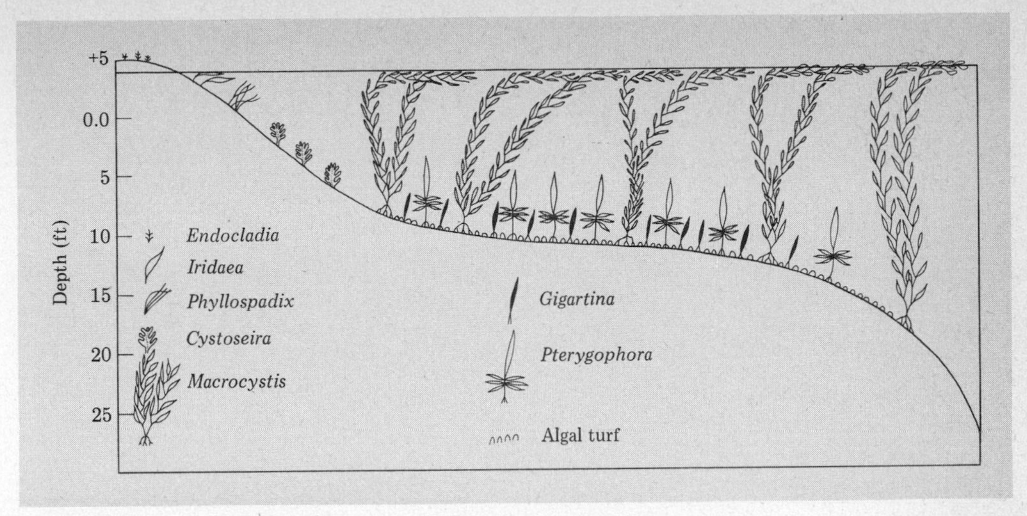

Figure 5-14 Generalized cross section of a kelp bed in California.

ECOLOGY AND LIFE CYCLE

Kelps have a phenomenal growth rate. *Nereocystis luetkeana* has been recorded by Scagel (1947) as growing in length at 6 cm/day, while, according to North (1971), *Macrocystis pyrifera* commonly grows 50 cm/day on the California coast. Such growth rates mean that plants grow from the bottom to the surface in a very short period of time. Kelp beds are also extremely productive. Mann (1973) reports that the net annual productivity ranges from 800 g C/m^2 in California kelp beds to perhaps 2,000 g C/m^2 in Indian Ocean kelp beds. These figures are several times the production of the phytoplankton on the same per-area basis.

Kelps tend to be perennial plants, often losing stipes and blades but regrowing new ones from the holdfast. The life span for *Macrocystis pyrifera* plants in California ranges from three to seven years maximum. The life cycle of kelp alternates between an asexual macroscopic stage termed the *sporophyte* and a sexual microscopic stage termed the *gametophyte* (Fig. 5-15). Only the sporophyte stage is considered here because little is known about the gametophyte in the natural environment.

The kelps provide the framework for the kelp bed community. Associated with these dominants are many other species of algae, invertebrates, and fishes.

Despite the enormous productivity of these kelps, relatively few herbivores graze directly on the plants. It has been estimated that only 10 percent of the net production enters the food webs of the kelp bed through grazing. The remaining 90 percent enters the food chains in the form of detritus or dissolved organic matter.

Since the general economy of a kelp bed depends on the continued existence of the giant kelps, it is of interest to know the causes of kelp mortality. Adult plants are apparently too large to be destroyed by grazing herbivores, but they are vulnerable to destruction by mechanical forces, mainly wave action. Since they occur in quite shallow water, often on open coasts, storm waves can have a devastating effect. The waves rip the plants up from the substrate and cast them up on the beach. Furthermore, once a plant has been uprooted, its movement through a bed, as Rosenthal and associates (1974) have demonstrated, often

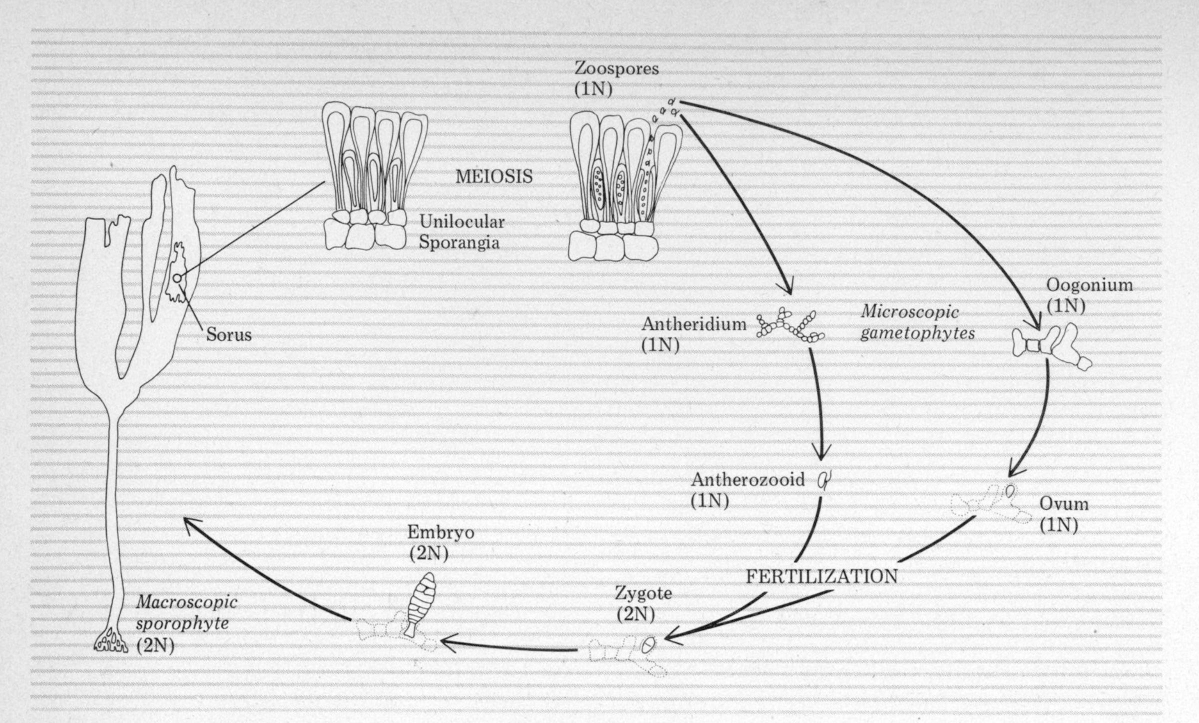

Figure 5-15 Life cycle of a typical kelp plant.

results in entangling among the stipes and blades of other plants. The result is that even more plants are pulled away from the substrate and destroyed. Very severe storms can virtually denude kelp beds of the stipes and blades, but if the holdfast and some meristem tissue remain, new stipes and blades can be produced.

A second major cause of mortality, at least in some areas such as the Atlantic and Pacific coasts of North America, is the grazing of the juvenile plants by sea urchins. If the densities of sea urchins get very high, even large plants will be attacked. It has been shown that certain kelps such as *Laminaria* are the preferred food of sea urchins. In recent years, population explosions of sea urchins have occurred in southern California with the result that, as food became scarce, the starving urchins moved into kelp beds in the area and severely reduced or destroyed them. Similarly, kelp-free patches on the Atlantic coast of Canada have been found where there are high densities of sea urchins. There seems little doubt that large numbers of urchins are capable of destroying kelp beds, but under most natural conditions, do not. Why?

The population explosions of sea urchins are triggered by a reduction in the density of their major predators, or possibly by an increase in organic pollution, such as domestic sewage. Predators normally keep the sea urchin population at a level where the kelp forest maintains itself. On the Pacific coast of North America, one dominant predator of urchins is the sea otter (*Enhydra lutris*). Estes and Palmisano (1974) have demonstrated that high population numbers of sea otters ensure that urchin densities are kept low. The low urchin densities in turn mean that the kelp beds and associated communities remain in good health. The almost complete destruction of the sea otter during the nineteenth century along most of the Pacific coast allowed the increase in sea urchin numbers and the gradual decrease in the kelp beds. The recent recovery of the sea otter in California and Alaska has been followed by decreased sea urchin numbers and increased extent of kelp forests in a few areas studied.

On the Atlantic coasts, a similar situation is thought to exist, except that in that case, the major predator of sea urchins is the lobster (*Homarus americanus*). The fishery activities of humans have reduced the lobster populations and allowed the increase in sea urchin populations.

Other predators of sea urchins exist in kelp beds, notably several starfish species and, in southern California, the fish the sheephead (*Semicossyphus pulcher*); in some areas, they may be as important as otters in controlling urchin populations (Fig. 5-16).

The role of other organisms in structuring the kelp bed community is not as well known. A generalized food web for invertebrates is given in Figure 5-16 for a California kelp bed.

In this qualitative food web, it can be noted that the primary producers have only the two species of urchins as major grazers. The other grazer, *Astraea undosa,* is a small herbivorous gastropod that is unlikely to exert any considerable effect on the kelp. Sea otters are not present in this area, and urchin control is through the activities of the four predaceous starfish, octopus, and one fish.

Kelp beds also show a pattern of zonation (Fig. 5-17). Zonation is determined

Figure 5-16 The food web of a southern California kelp bed. (From R. J. Rosenthal, W. D. Clarke, and P. K. Dayton, *Ecological Monographs*, vol. 44, no. 1, 1974. Copyright 1974, the Ecological Society of America.)

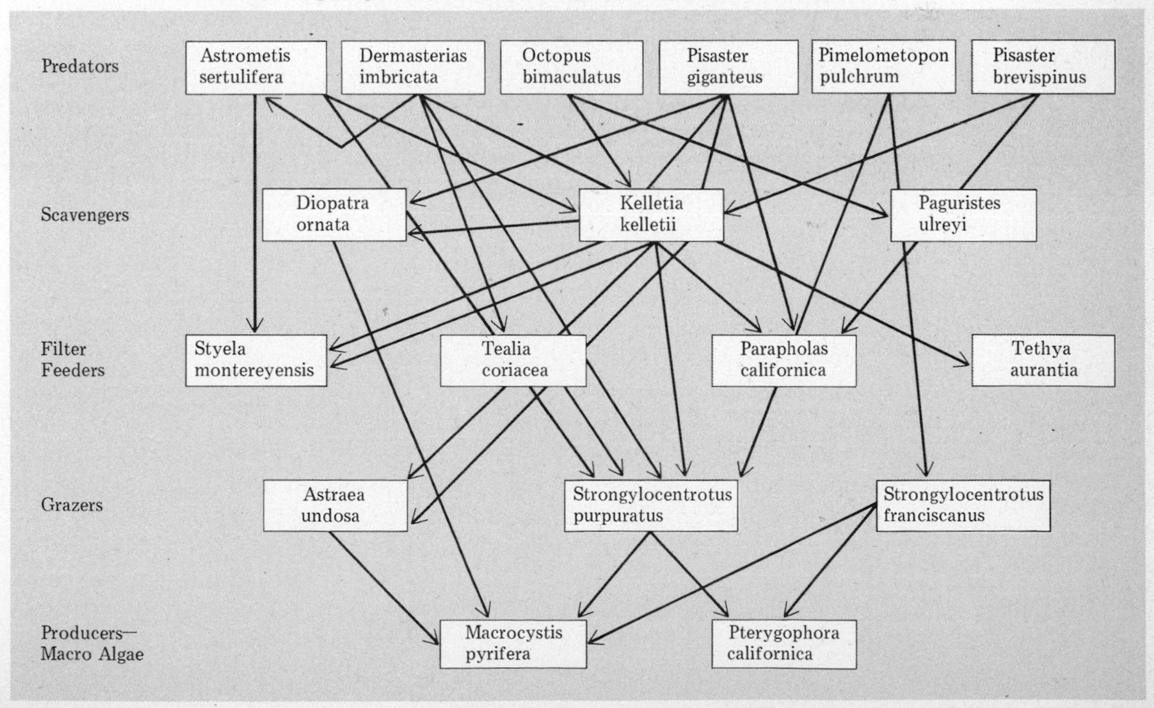

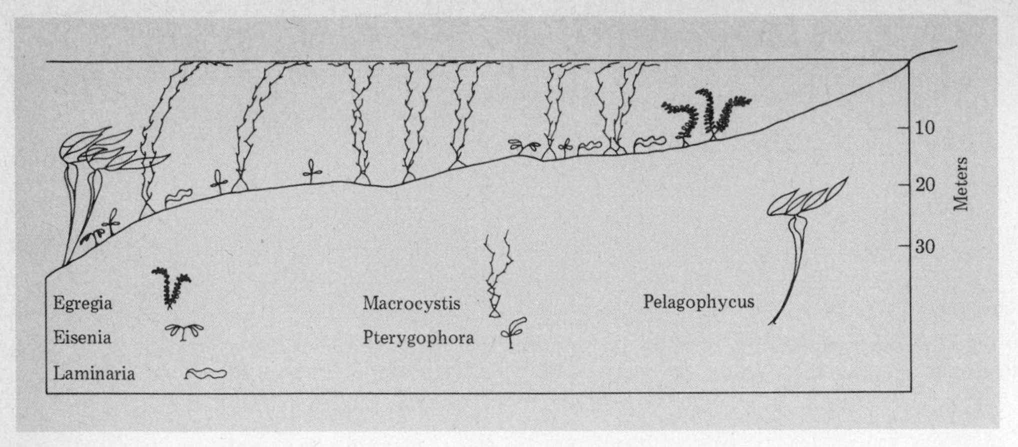

Figure 5-17 Zonation pattern of the dominant large algae in a southern California kelp bed. (From E. Y. Dawson, M. Neushul, and R. D. Wildman, Seaweeds associated with kelp beds along southern California and northwestern Mexico, Pacific naturalist, vol. 1, no. 14, 1960.)

by changes in physical factors with depth, such as light and wave action, and partly by competition among species. On the California coast, the brown alga *Egregia menziesii* is abundant in the most turbulent shallow waters, whereas *Macrocystis pyrifera* is restricted to deeper waters where wave action is not as great. As greater depths are approached, the light intensity becomes limiting and the edge of the kelp bed seems to be set at that point where there is insufficient light for kelp growth.

SEA GRASS COMMUNITIES

Many areas of the shallow sea bottom are covered with a lush growth of aquatic "grasses," which are collectively called *sea grasses*. Sea grasses are flowering plants that are adapted to live submerged in sea water.

COMPOSITION AND DISTRIBUTION

Sea grasses worldwide encompass only about 50 species, according to den Hartog (1977), a small number in comparison to their ecological importance. Sea grass beds form dense carpets of as many as 4,000 plants per square meter over extensive areas of the bottom, making them one of the most conspicuous communities of the shallow waters of temperate and tropical seas. They may have a standing biomass of 2 kg/m^2.

Most sea grass species are very similar in gross external morphology. They have long, thin, straplike leaves that have air channels and a monopodial growth form. The plants arise from a creeping rhizome (Fig. 5-18). When compared to fresh water aquatic plants, the sea grasses are considerably fewer in number of species and also very much less diverse in morphology.

Sea grasses occur from the mid-intertidal region down to depths of 50 or 60 m. They seem to be most abundant, however, in the immediate sublittoral area. The

Figure 5-18 Eelgrass, *Zostera marina,* at low tide. (Photo by the author.)

number of species is greater in the tropics than in the temperate zones. All types of substrates are inhabited by these grasses, from soupy mud to granitic rock, but the most extensive beds occur on soft substrates.

PRODUCTIVITY

Because sea grass beds are densely covered with the plants themselves and because they cover such extensive areas in continental shelf waters, they have a very high productivity rate and contribute significantly to the total production of inshore waters. McRoy and McMillan (1977) have estimated the production of these beds to be from 500 to 1,000 g C/m^2/year, or among the most productive areas of the oceans (compare with plankton productivity, Table 2-1 and kelp, p. 181.). Most of these data on productivity, however, are for the two most studied species. *Zostera marina* (eelgrass) of the North Temperate Zone and *Thalassia testudinum* (turtle grass) of the tropics. Little is known about the other species.

In contrast to other productivity in the ocean, which is confined to various species of algae dependent upon nutrient concentrations in the water column, sea grasses are rooted plants that absorb nutrients from the sediment or substrate. They are, therefore, capable of recycling into the ecosytem nutrients that would otherwise be unavailable because of being trapped in the bottom.

Despite the obvious position of sea grass beds as key primary production units in inshore waters, relatively little is known concerning the importance of this energy in the economy of coastal ecosystems. In contrast to the situation in the terrestrial environment, where grasses are heavily grazed by a variety of vertebrate and invertebrate herbivores, sea grasses are consumed directly by surprisingly few animals. Perhaps the major ones are sea urchins, a few fishes (Scaridae, Acanthuridae), turtles, and sea cows. Of the large energy reserve built up by photosynthesis, very little is transferred directly to the various inshore food chains. How is this large food resource channeled into the coastal ecosystem? Fenchel (1977) suggests that this energy enters the system via detritus; that is, the sea grasses die and decompose, breaking into smaller particles, and it is this decomposition material that is then picked up by various detritus feeding organisms and enters into the food chains in this way (Fig.

5-19). This was the mechanism proposed by C. H. Joh. Petersen 60 years ago. It still appears to be true. This detrital material may be transported and enrich areas at considerable distance from the sea grass beds themselves. It has been suggested as a source of energy for organisms as far removed as the abyssal benthos.

ECOLOGY

Sea grass communities generally have only one or, at most, a few dominant species of sea grasses present and are, in appearance, rather simple in structure and homogeneous. In reality, these beds are complex communities consisting of large numbers of epiphytic and epizoic organisms, burrowers, and other organisms associated with the sea grasses, where they find shelter and/or food.

Ecologically, sea grass beds serve a number of important functions in inshore areas. They are a major source of primary productivity in shallow waters around the world and thus an important source of food for many organisms (in the form of detritus). Secondly, they function to stabilize the soft bottoms, on which most species grow, primarily through the dense, matted root system. This stabilization of the bottom by roots is extremely durable, being able to withstand storms as severe as hurricanes! This system, in turn, serves to shelter many organisms.

Figure 5-19 The pathway for the channeling of eelgrass into the food web. (Modified from McRoy and Helfferich, 1977, p. 136.)

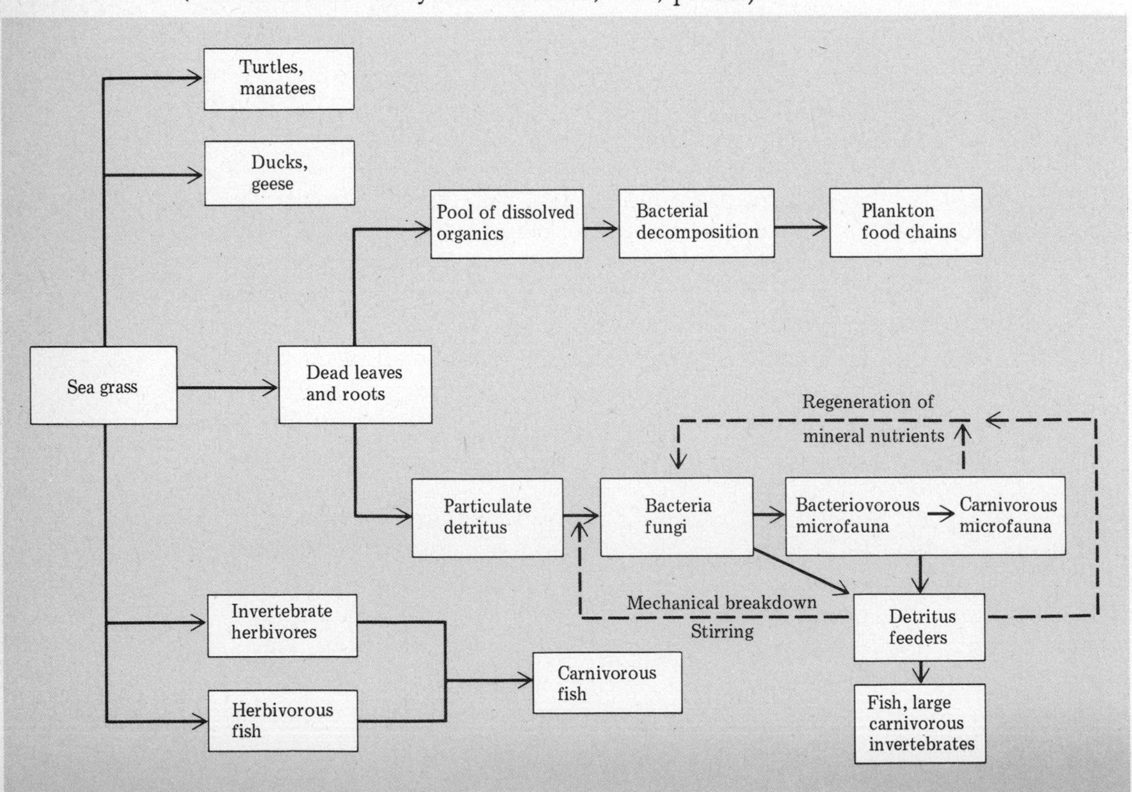

Thus, there are many animals that are commonly found in sea grass beds but that do not have a direct trophic relationship to the grasses themselves. The beds serve as nursery grounds for many species that spend their adults lives in other areas. Included among these latter species are several of commercial importance, such as the shrimp *Penaeus duorarum* of southern Florida. Destruction of sea grass beds, therefore, may have a serious effect on certain commercial fisheries.

Sea grass beds may also act as traps for sediment and, therefore, act to build up the bottom. Such a situation has been recorded for the *Posidonia* beds in the Mediterranean Sea, which build extensive terraces up from bottom. When such a situation prevails and the buildup approaches the surface, the floating leaves act to break the force of the waves, forming a calm water habitat on the bed.

The leaves of the sea grasses also serve as a protective canopy, shielding the inhabitants of the bed from the effects of strong sunlight. Where the beds become intertidal, the leaves may cover the bottom substrate at low tide, protecting the inhabitants from dessication.

The importance of sea grass beds and what might happen when they are destroyed was given dramatic emphasis by a natural destructive event that occurred to the dominant sea grass *Zostera marina* in the North Atlantic Ocean. In the early 1930s, the *Zostera* beds on both sides of the Atlantic Ocean were massively destroyed by a mysterious "wasting disease" of epidemic character. There is still no universal agreement as to the causative agent for this destruction, but most scientists have felt it was a combination of a fungus and a slime mold. Other theories included a bacteria attack and changes in the hydrographic regime, mainly temperature. The temperature change has perhaps the most plausibility, since *Z. marina* exists only between narrow temperature bounds. The *Zostera* beds in the Pacific Ocean were not affected.

The effects of the *Zostera* disappearance were different in different areas. In Europe, the most immediate results were that the level of the beach fell; with the loss of sediment, many bared rocks appeared that spouted a new growth of algae in place of *Zostera*; and there was the formation of many new sandbars (Fig. 5-20). The organisms also changed, but less than was anticipated. Whereas the eelgrass beds had been dominated by detritus feeding animals, after the loss, the same bottom area had more suspension feeders and in general more species. In Europe very few annual species associated with *Zostera* actually disappeared completely. In the United States, on the other hand, destruction of the *Zostera* beds meant the almost complete disappearance of the animals associated with the beds. The reason for this difference probably has to do with the fact that in European waters, the alga *Fucus* quickly replaced *Zostera*, thus offering the same protected environment, whereas this was not the case in the northeastern United States. Since the 1930s, the *Zostera* beds have slowly returned to reoccupy their former habitat.

For some sea grass communities, den Hartog (1977) reports a successional

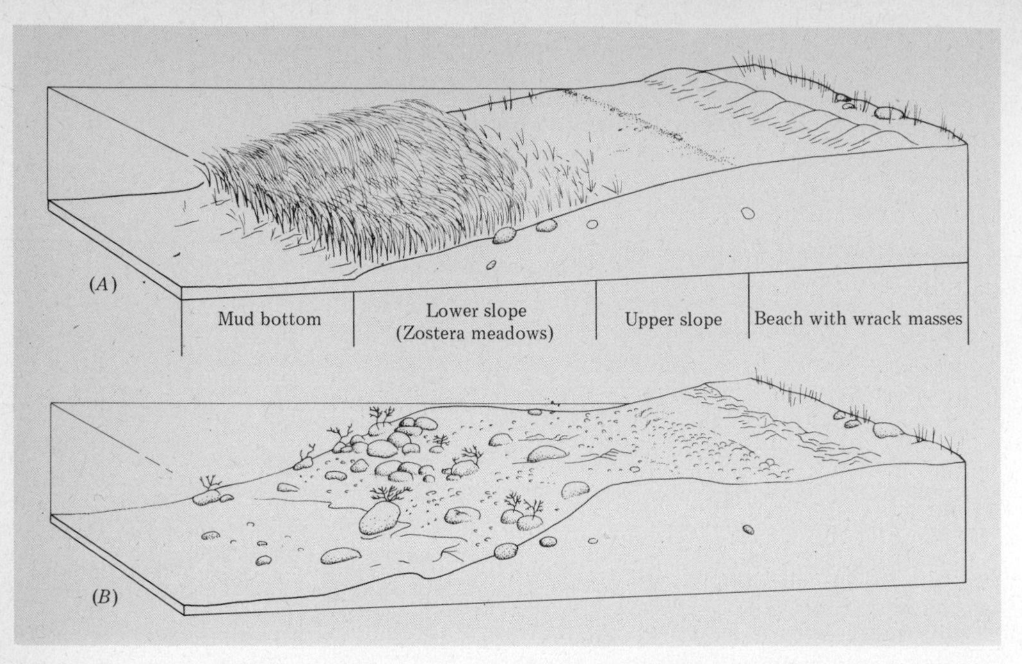

Figure 5-20 Representation of: *(A)* the conditions of the eelgrass beds of Scandanavia before the onset of the wasting disease, and *(B)* the extreme conditions following the complete destruction of the beds. (From McRoy and Helfferich, 1977, Seagrass ecosystems.)

pattern such as has been found for terrestrial communities (Fig. 5-21). An example is the sea grass beds of the tropical Caribbean. Here, the climax community is dominated by *Thalassia testudinum*, but there are a series of stages leading up to *Thalassia* that are different, depending primarily on the substrate. On the other hand, in the temperate Atlantic where the dominant sea grass is *Zostera marina*, no series of successional stages has been uncovered and *Z. marina* colonizes directly; that is, it is both pioneer and terminal stage. In the temperate Pacific Ocean, the sea grass *Phyllospadix torreyi* occurs on hard substrates on wave-swept shores (*Zostera* occurs on soft bottoms in protected areas.) In this case, *P. torreyi* may occasionally be a successional stage in a sequence that terminates in a community dominated by the giant kelps (*Macrocystis*, see pp. 176–184). *Phyllospadix* does, however, maintain itself wherever the conditions are unsuitable for kelp development, usually the boundary area between the lowest part of the intertidal and the sublittoral.

SOME SPECIAL COMMUNITIES

In addition to the widespread communities described above, there are a few of more restricted occurrence that have received considerable study; hence, they deserve mention here as examples of how communities are organized, how they may vary temporally, and how they are maintained.

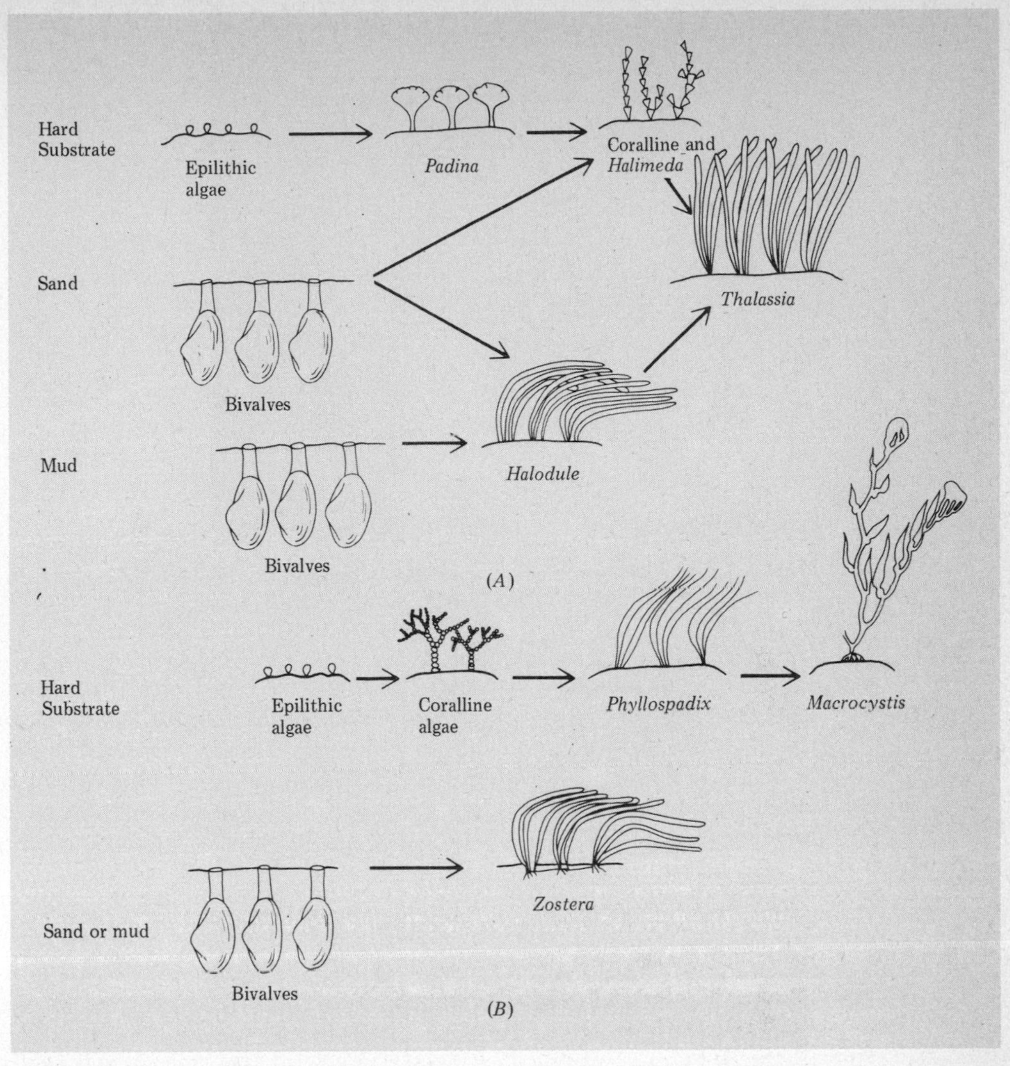

Figure 5-21 Successional sequences in sea grass beds. *(A)* Tropical areas. *(B)* Temperate areas.

A SEA PEN COMMUNITY

In Puget Sound, Washington, at depths of 10–50 m on soft bottoms, there exists a community dominated by dense stands of the long-lived cnidarian sea pen *Ptilosarcus gurneyi* (Fig. 5-22). In an extensive study, Birkeland (1974) found that *Ptilosarcus gurneyi* provides the major food source for seven predators: four starfish (*Hippasterias spinosa, Dermasterias imbricata, Crossaster papposus,* and *Mediaster aequalis*) and three opisthobranch mollusks (*Armina californica, Tritonia festiva,* and *Hermissenda crassicornis*). *Ptilosarcus* itself has a life span of about 15 years, takes several years to reach sexual maturity, has irregular, unpredictable recruitment resulting in stands of differing age structure, and never reaches sizes at which it is immune to predation.

Figure 5-22 Photo of a North Pacific sea pen *(Ptilosarcus)* bed. (Photo courtesy of Dr. Charles Birkeland.)

Adult *Ptilosarcus* are readily consumed by three starfish, *Hippasterias*, *Dermasterias*, and *Mediaster*, plus the opisthobranch *Armina*, which seems to prefer the largest sea pens. Of these four predators, *Hippasterias* has the most restricted diet, feeding only upon *Ptilosarcus*. The others are capable of consuming other prey, although in Puget Sound at least, *Armina* also seems to feed exclusively on *Ptilosarcus*. Juvenile *Ptilosarcus* are preyed upon by *Hermissenda*, *Tritonia*, and *Crossaster*, but this is uncommon.

In this community, the predator specialists are *Armina* and *Hippasterias*, preying only on *Ptilosarcus*. *Dermasterias*, *Mediaster*, and *Crossaster* are more generalists, capable of feeding upon other invertebrates. One might expect that the generalists, because they are not dependent completely on *Ptilosarcus*, would maintain high population numbers and thus reduce the *Ptilosarcus* population so low that the specialists would have nothing left. In other words, they could outcompete them. What is to prevent this? The answer lies with yet another starfish, the top predator, *Solaster dawsoni*, which preys upon *Mediaster*, *Dermasterias*, and *Crossaster*, reducing their numbers so competitive exclusion does not occur (Fig. 5-23), allowing *Hippasterias* to persist.

This community is thus maintained by a complex food web relationship that has as its key industry species the sea pen *Ptilosarcus*, which is preyed upon by several specialist and generalist predators. The generalists (*Mediaster*, *Dermasterias*, and *Crossaster*) are unable to obtain numerical dominance and thus outcompete the specialists (*Hippasterias* and *Armina*) because a top predator, *Solaster*, keeps their numbers in check.

A SEA PANSY COMMUNITY

Off San Diego in southern California in 5–10 m on a sand bottom exists quite a different community. This community was first studied by Fager (1968) during

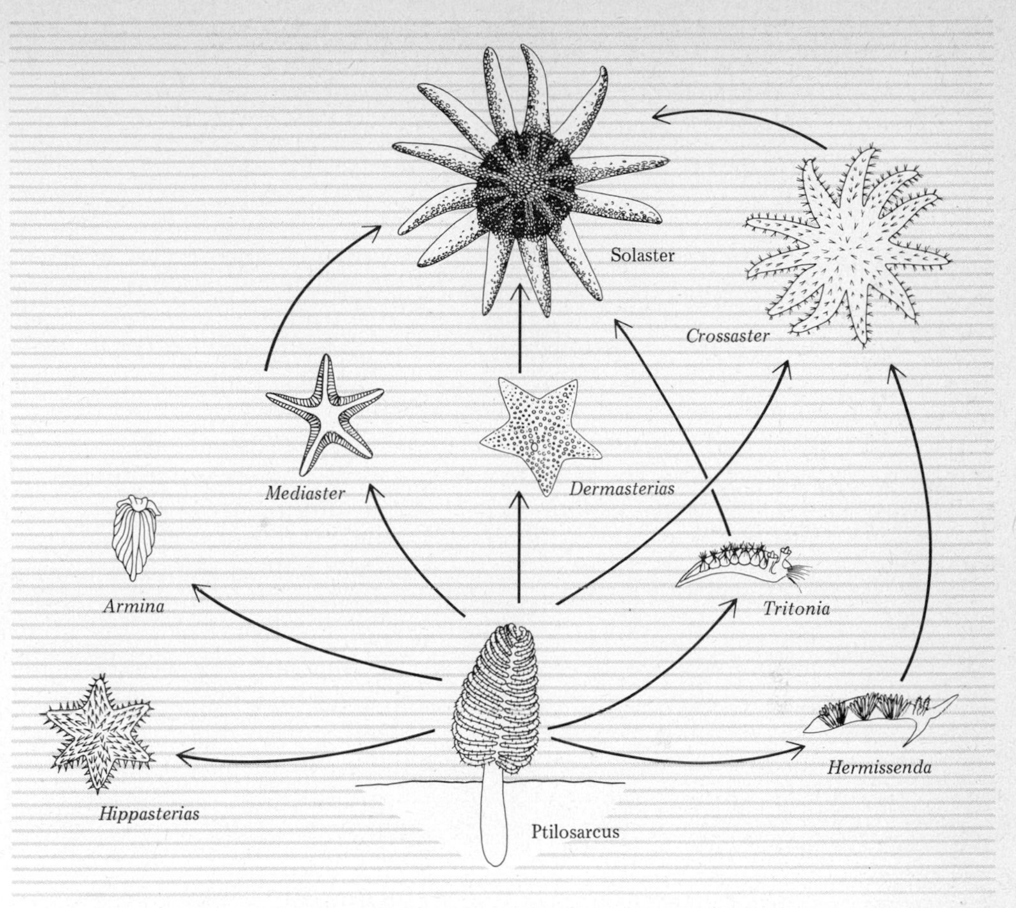

Figure 5-23 The food web of a North Pacific sea pen community.

the years 1957–1963, and was then dominated by cnidarians, primarily the sea pansy, *Renilla kollikeri*, and the anemones *Harenactis attenuata* and *Zaolutus actius*. Associated with these were three gastropods (*Nassarius fossatus, Nassarius perpinguis*, and *Polinices recluzianus*) and two echinoderms, *Amphiodia occidentalis* and *Astropecten armatus*. In this community, Fager found the pattern of distribution of the major organisms to be primarily aggregated and the populations remained constant over the six-year period. Furthermore, there was little evidence of any interaction among the species and no evidence that these large macroinvertebrates were either preyed upon by mobile fishes or themselves preyed upon the smaller infauna organisms.

Eleven years later, in 1974, Davis and Van Blaricom (1978) repeated the study at the same site. Whereas Fager had observed essentially nonvariant densities of the major species over six years, Davis and Van Blaricom found both long- and short-term changes in the abundances and densities of seven of the dominant species. These studies suggest that shallow water benthic communities are not completely stable over long periods of time, even though they may appear so for several years. Furthermore, it appears that a much longer time interval is necessary to evaluate what is happening in such areas.

A SAND DOLLAR COMMUNITY

As evidence that similar sand bottoms at the same depth in the same geographical area may have different communities and different forces acting upon them, it is useful to contrast the Fager study above with a study by Kastendiek (1976) on a community at the same depth on sand at Zuma Beach, California, some miles north. The dominant species are the sea pansy *Renilla kollikeri*; the sand dollar, *Dendraster excentricus*; the nudibranch *Armina californica*; and the starfish *Astropecten armatus*. Off Zuma Beach, *D. excentricus* forms a dense band (up to 1,500 individuals/m^2), which lies parallel to the shore and runs the entire length of the beach. The band moves into more shallow water during calm periods of the year and offshore during stormy times. *Renilla kollikeri* occupies an area inshore of the sand dollar bed. *Armina californica* is a specialist predator of *R. kollikeri*, and *A. armatus* feeds on both *R. kollikeri* and *D. excentricus* (Fig. 5-24).

In this sytem, *Renilla kollikeri* is prevented from moving further shoreward because the increased wave motion in shallower water prevents it from anchoring itself. It cannot extend further seaward because of the sand dollar bed. When *R. kollikeri* enter the sand dollar bed, the *Dendraster excentricus* uproot and eliminate them. Hence, during that time of year when the sand dollars move shoreward, that portion of the *R. kollikeri* bed it occupies is destroyed.

Armina californica and *Astropecten armatus* have a synergistic effect. *A. armatus* is repelled by the expanded autozooids of the *Renilla kollikeri*. However, *R. kollikeri* has a characteristic behavior when attacked by *A. californica*. This is to close up all polyps and raise itself up off the substrate. The colony thus acts like a sail. The turbulent water then uproots the *R. kollikeri* and tumbles it away from the *A. californica*. Thus, *A. californica* gets only one or two bites before the escape response. However, a closed-up *R. kollikeri* is readily eaten by *A. armatus*; so escape from one predator is not escape from the other. *Astropecten armatus*,

Figure 5-24 The food web of a southern California subtidal sand flat.

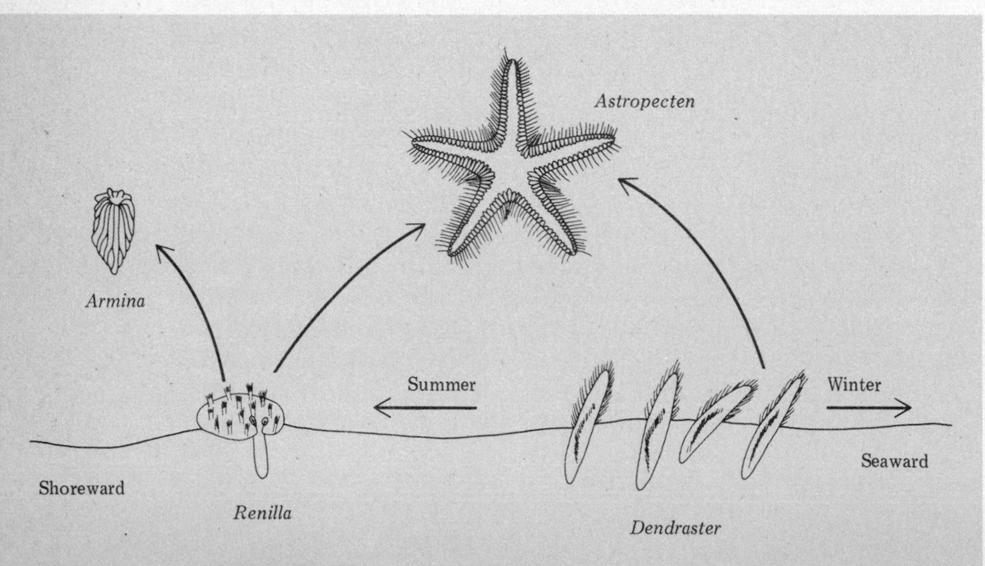

however, is capable of ingesting only *R. kollikeri* up to 40 mm in rachis diameter. *Armina californica*, in turn, is able to consume *R. kollikeri* colonies if they become lodged against objects such as sand dollars.

The sand dollar bed forms a barrier to *Astropecten armatus*, which normally is distributed seaward of the bed and is not able to penetrate shoreward of it. *Astropecten armatus* also consumes sand dollars. What controls the edges of the sand dollar beds is not known.

Hence, in this sytem, there is a physical and biological control on the community. *Renilla kollikeri* is controlled seaward by *Dendraster excentricus* and shoreward by waves, and the population is further regulated by predators, of which one is unable to attack it due to the sand dollar bed. Should the sand dollar bed disappear, it is felt by Kastendiek that *Astropecten armatus* and *Armina californica* would destroy the *R. kollikeri* population.

BENTHIC COMMUNITIES IN ANTARCTIC SEAS

Relatively few thorough ecological studies exist for polar bottom communities other than those that are simply lists of species found. Perhaps because of the profound interest and support of the U.S. Government through its Antarctic research programs, we have a better understanding of Antarctic communities than those of the Arctic.

Polar shallow water areas are subject to a stress factor unknown in other areas. That factor is ice. During the winter months, ice forms a thick layer over the continental shelf areas extending down several meters. Wherever this ice layer makes contact with the bottom, the fauna is destroyed, either by mechanical action of the ice grinding against the bottom or by freezing the animals. This means that in polar areas, the shallowest subtidal areas are devoid of permanent communities down to the level to which the ice extends. This area may, however, be inhabited during ice-free times of the year by mobile or transient organisms.

In Antarctic seas, in addition to the action of the thick layer of annual sea ice affecting the bottom, there is an additional area below that subject to a different kind of ice action. This area, which extends down to 30 m or so, is one in which ice platelets begin to form on the bottom around any convenient nucleus. As these platelets grow and increase in size, they may surround various sessile and sedentary invertebrates living on the bottom. Since ice is generally less dense than the surrounding water, as these grow they also tend to lift off the bottom and rise toward the thick sea ice above, carrying with them the entrapped invertebrates. Such organisms become trapped in the sea ice and are permanently lost from the community. Such ice formation is termed *anchor ice* (Fig. 5-25). The result of this anchor ice formation and its tendency to remove organisms is that there is a definite zonation of the benthic fauna with depth, which is directly related to the action of this ice formation.

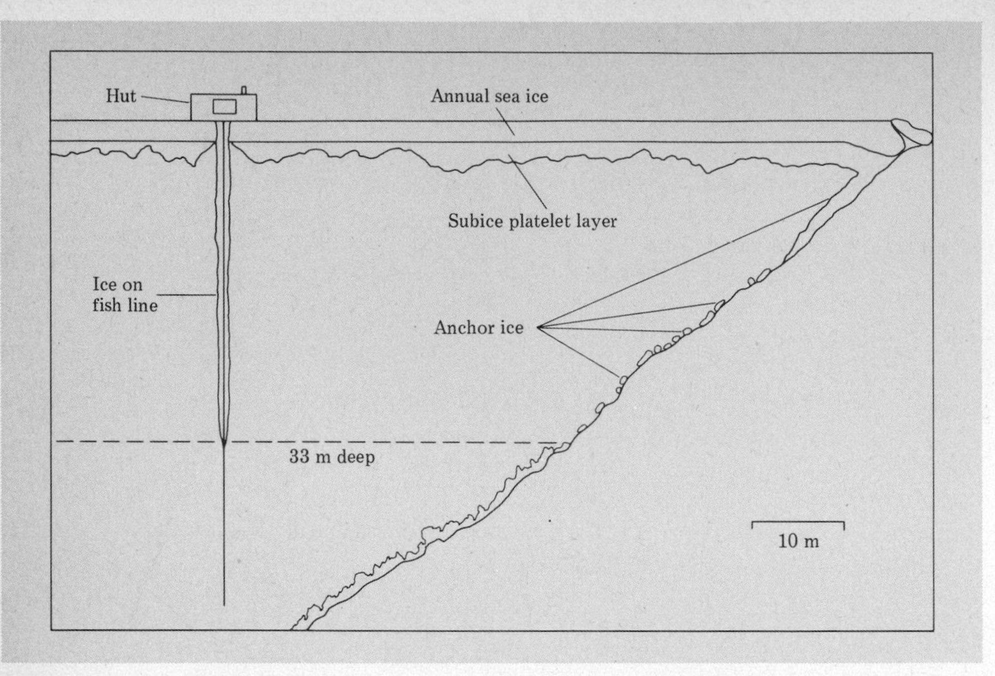

Figure 5-25 Anchor ice area in the Antarctic. (From P. K. Dayton, G. Robilliard, and A. L. DeVries, Anchor ice formation in McMurdo Sound, Antarctica, and its biological effects, *Science,* vol. 163, 1969. Copyright 1969 by the American Association for the Advancement of Science.)

Dayton, Robilliard, and Paine (1970) recognize three zones (Fig. 5-26). Zone I extends from 0 to 15 m and is essentially devoid of life, due to scouring by sea ice and the almost universal coverage by anchor ice. Only in the ice-free periods are mobile organisms able to enter.

Zone II extends from about 15 m down to the limit of anchor ice formation at 30 m or so. This zone experiences less anchor ice and has a fauna of numerous sessile animals, mainly anemones and other cnidarians. The anemones here often feed upon rafted-in jellyfish or sea urchins, while the other cnidarians are suspension feeders. The anemones appear to be territorial.

Zone III begins abruptly where anchor ice ceases and continues down to an undetermined depth. The substrate here changes from a cobblyrocky bottom to one that is a thick mat of sponge spicules. Sponges dominate this area, but the area has a rich and diverse fauna of other invertebrates as well, of which starfish are the most conspicuous motile animals. This zone is noteworthy because of its extreme physical stability, which has resulted in a complex, highly diverse community that, according to Dayton and associates (1974), seems to be regulated by biological factors (predation) rather than by physical factors as in Zones I and II.

The dominant organisms in this area are very slow growing, so slow, in fact, that in many cases, the growth rates are too slow to measure in one year. One exception is the sponge *Mycale acerata*, which has a very rapid growth rate and seems capable of outcompeting all other sessile invertebrates and taking over the substrate space. *Mycale* is prevented from dominating the substrate because it is

Figure 5-26 Vertical zonation of invertebrates in Antarctic shelf areas. A few motile animals forage into Zone I, which is otherwise barren of sessile animals. Cnidarians dominate the sessile animals of Zone II and sponges Zone III. (With permission from M. W. Holdgate, *Antarctic Biology*, vol. I, 1970, Academic Press. Copyright by Academic Press Inc. [London] Ltd.)

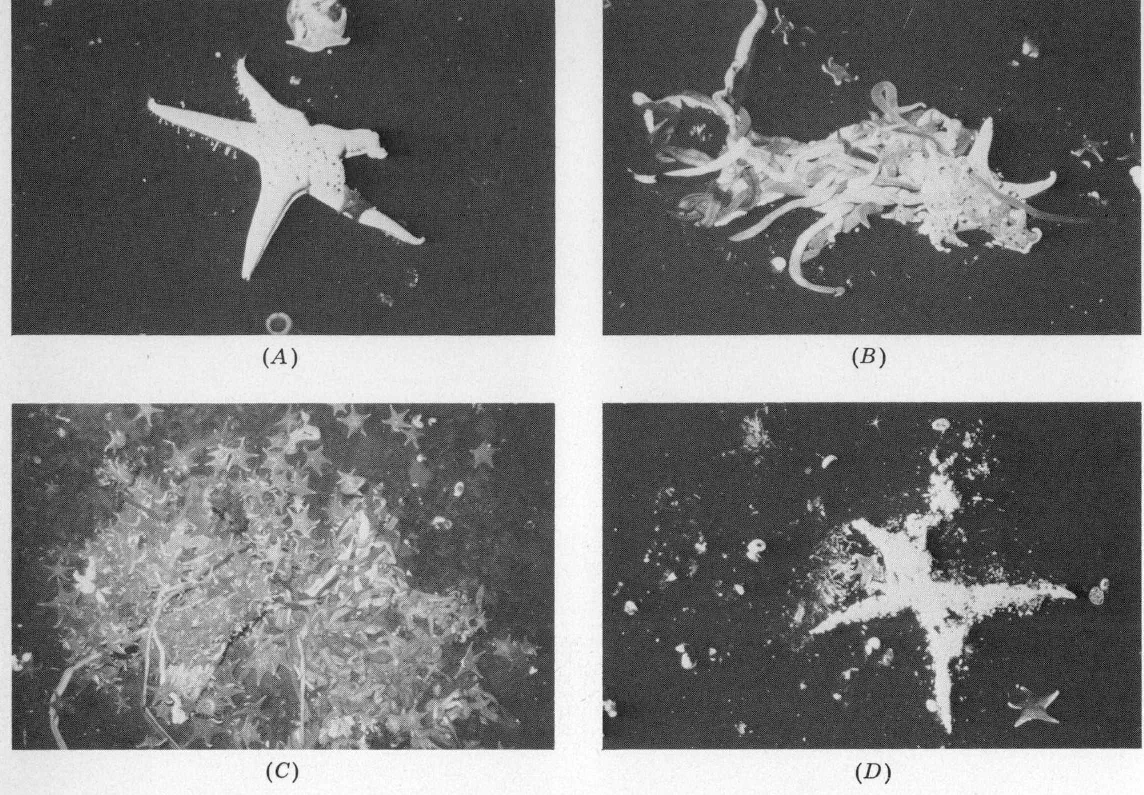

Figure 5-27 Sequence showing the consumption of the large Antarctic starfish *Acodontaster conspicuus* by the small starfish *Odontaster validus* and the nemertean *Lineus corrugatus*. (From P. K. Dayton, et al., 1974.)

preyed upon by two sea stars, *Perknaster fuscus* and *Acodonaster conspicuous*, which keep it in check. *Acodonaster conspicuous* also preys on several other dominant sponges, but is prevented from reaching population densities that would allow it to destroy the sponge community by yet another asteroid, *Odontaster validus*, which consumes the larvae, young, and even adults of *A. conspicuous* (Fig. 5-27). Thus, we have a community that is regulated through finely-tuned biological interactions, primarily predation.

References

Coull, B. C. (ed.). 1977. Ecology of marine benthos. Univ. of South Carolina Press, Columbia. 467 pp.

Dayton, P., G. Robilliard, and R. T. Paine. 1970. Benthic faunal zonation as a result of anchor ice at McMurdo Sound, Antarctica, pp. 244–258. *In*: Antarctic ecology, Vol. I. Academic Press, N.Y.

Estes, J., and J. Palmisano. 1974. Sea otters, their role in structuring near-shore communities. Science 185(4156):1058–1060.

Fager, E. W. 1968. A sand bottom epifaunal community of invertebrates in shallow water. Limnol. Oceanogr. 13(3):448–464.

Jones, N. S. 1950. Marine bottom communities. Biol. Rev. 25:283–313.

Keegan, B. F., P. O. Ceidigh, and P. J. S. Boaden (eds.). 1977. Biology of benthic organisms. Pergamon Press, N.Y. 630 pp.

Mann, K. H. 1973. Seaweeds: Their productivity and strategy for growth. Science 182(4116):975–983.

McRoy, C. P., and C. Helfferich (eds.). 1977. Seagrass ecosystems. Dekker, N.Y. 314 pp.

Phillips, R. C., and C. P. McRoy (eds.). 1980. Handbook of seagrass biology. Garland STPM Press, N.Y. 353 pp.

Thorson, G. 1950. Reproductive and larval ecology of marine bottom invertebrates. Biol. Rev. 25:1–45.

Thorson, G. 1957. Bottom communities (sublittoral or shallow shelf), pp. 461–534. *In:* The treatise on marine ecology and paleoecology, Vol. I. Memoir 67, Geol. Soc. of Amer.

Woodin, S. A. 1976. Adult-Larval interactions in dense infaunal assemblages: Patterns of abundance, Jour. Mar. Res. 34(1):25–41.

Chapter Six
INTERTIDAL ECOLOGY

Although it constitutes by far the smallest area of all in the world's oceans, existing as an extremely narrow fringe a few meters in extent between high and low water, the intertidal zone is perhaps the best known and studied of all areas of the sea because it is the most accessible to humans. Here, and only here, can direct observations be made of aquatic organisms during the period of low tide, without special equipment. The intertidal zone has been of interest and use to humans since prehistoric time.

Despite its extremely restricted area, the intertidal has within it the greatest variations in environmental factors of any marine area, and these can occur within centimeters of each other. Coupled with this is a tremendous diversity of life, which is greater than that found in the more extensive subtidal areas.

It should be emphasized that this area is truly an extension of the marine environment and is inhabited almost exclusively by marine organisms. Although this area is terrestrial as much as half the time, the terrestrial fauna and flora have not invaded it to any extent, except at the uppermost fringes.

The richness, the diversity of environmental factors, and the ease of access act together to attract to this area an inordinate amount of scientific attention. More knowledge is thus available for this small area and its organisms, and their interactions are better known than in any other area. Thanks to such interest and knowledge, this area has produced more unifying concepts with respect to the organization of marine communities than any other. As a result, a disproportionate amount of space in this book is devoted to a discussion of it.

ENVIRONMENTAL CONDITIONS

In part, the tremendous array of environmental factors and ranges found in the intertidal is due to the fact that this zone is exposed to the air for a certain amount of time during the year and most physical factors show greater range in air than in water (see Chapter One, pp. 2–5).

TIDES

The periodic, predictable rise and fall of the level of the sea over a given time interval is called a *tide*. Certainly, this is the most important environmental factor influencing life in the intertidal zone. Without the presence of the tide or some other means of inducing a periodic rise and fall in the water level, this zone would not exist as it is, and many of the other factors would cease to be of influence. This is because the great range observed in many of the physical factors in the intertidal is a result of alternating exposure to air and to water. If there were no tide, these great fluctuations would not occur.

With very few exceptions, most shore areas of the world experience tides. The only major seas that virtually lack tidal action are the Mediterranean and the Baltic. In these areas, the water level on the shore line fluctuates primarily due to the effects of wind action pushing water. This does not mean, however, that all seashores experience the same tidal range, or even the same type of tide. The reasons for the occurrence of different tides and different ranges are complex, having to do with the interaction of the tide-generating forces, the sun and moon, the rotation of the earth, the geomorphology of the ocean basins, and the natural oscillations of the various ocean basins. Most of the explanation is beyond the scope of this text, but a simplified explanation will be given here.

Tides occur due to the interaction between the gravitational attraction of the sun and the moon on the earth and the centrifugal force generated by the rotating earth and moon system. As a result of these forces, the water in the ocean basins is pulled into bulges. Gravitational attraction of one body for another is a function of the mass of each body and their distance apart. In the case of the sun and the moon, the gravitational generating force of the moon on the earth is about twice that of sun, even though the sun is many times more massive than the moon. This is due to the great distance of the sun from the earth relative to the moon.

The earth and moon form an orbiting system that revolves around their common center of mass. Because of the large size of the earth relative to the moon, this point is located inside the earth. The revolution of the earth-moon system creates a centrifugal force (acting outward) which is balanced overall by the gravitational force acting between the two bodies. However, gravitational force is much stronger than centrifugal force on the side of the earth facing the moon than on the side opposite. As a result, the side facing the moon has the water pulled into a bulge (= high tide). On the opposite side of the earth, the gravitational force of the moon will be the least, and the stronger centrifugal force will pull water into a bulge away from earth (another high tide) (Fig. 6-1). Thus, we have the two high tides. These then circle the earth following the position of the moon as the earth revolves on its axis once every 24 hours. Low tides are about halfway between the high tides.

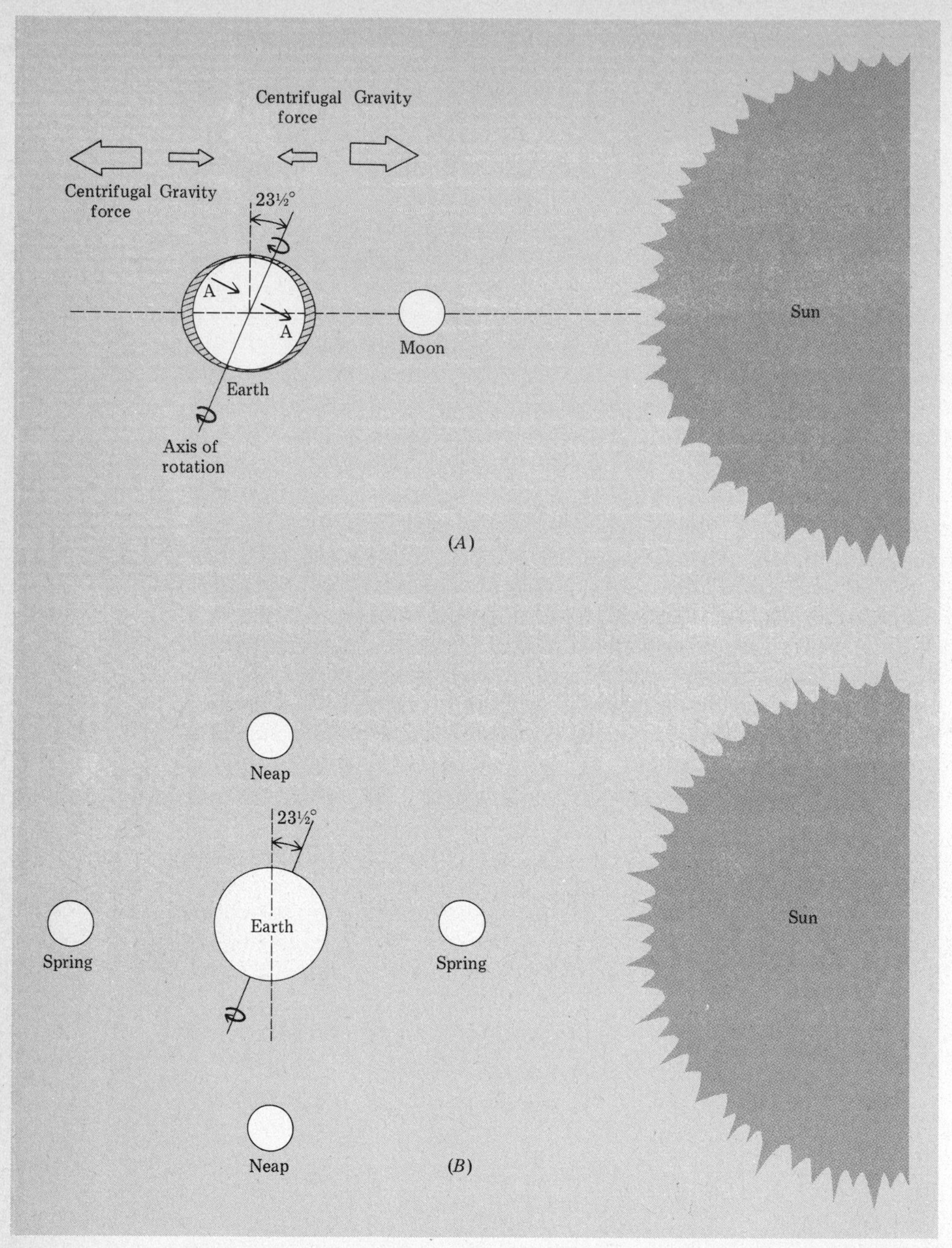

Figure 6-1 Origin of tides. *(A)* Moon acts to raise a bulge on the side of earth nearest it due to greater gravitational force than counteracting centrifugal force. On the opposite side the centrifugal force is stronger and throws another bulge outward. Because of the inclination of the earth on its rotational axis, point A, as it rotates, will experience two high tides of different height. *(B)* Position of the moon and sun at neap and spring tides.

The system as described above would give two high tides and two low tides of equal magnitude each day. As we know, however, tides are not equal, nor are there always four tides per day. How do these discrepancies arise? In part, this is due to the fact that the earth is not vertical with respect to its plane of orbit around the sun. It actually is inclined 23½° from the vertical. As a result, during the earth's rotation on its axis, a given point on the earth's surface is subjected to different tidal heights, as Figure 6-1 shows. It should also be noted from Figure 6-1 that, at the highest latitudes (polar seas), only a single high tide would occur. Further changes in the height of the tides result from changes in the moon relative to the earth as the moon moves in its orbit around the earth. Since the orbit of the moon is not circular but elliptical, there are times when the moon is closer to the earth (*perigee*) and others when it is further away (*apogee*). Tides are greater at perigee and diminished at apogee.

The effect of the sun is seen in spring and neap tides. Spring tides are tides that show the greatest range (both high and low) and result when the moon and sun are directly aligned so these two forces are combined. Neap tides, on the other hand, are tides showing minimum range and result when sun and moon are at right angles to each other and thus counteract each other.

Differences both in number of high and low tides per day and in height in various parts of the world are due to peculiarities of the various ocean basins in which the tides occur. The well-known extremely high tides of the Bay of Fundy or Cook Inlet in Alaska are the result of the basic tide forces acting in the geometry of the basin.

Tides that are expressed as a single low and high per day are called *diurnal tides* (Fig. 6-2). Tides that have two highs and lows per lunar day are termed *semidiurnal tides*. Where there is a mixture of diurnal and semidiurnal we have what are called *mixed tides*. The heights of the highs and lows vary from day to day as the positions of the sun and moon relative to each other change, each being alternately fully aligned and 90° out of alignment every 14 days.

Perhaps the most significant effect of the tide with respect to intertidal organisms and communities is that it subjects them to periodic exposure to the air, with its considerably wider range of physical parameters. It is this exposure to aerial conditions that necessitates most of the adaptations required by intertidal organisms in order to inhabit this zone. Physical factors, the extremes of which are reduced for organisms inhabiting water, become limiting or lethal when the insulating water is removed (see pp. 2–5 on physics and chemistry of water).

Tides in combination with time can, however, have two significant direct effects on the presence and organization of intertidal communities. The first effect results from the time that a given area of the intertidal is exposed to the air relative to the time that it is submerged in water. It is the *duration of exposure* to air that is most important, because that is when the marine organisms will be subjected to the greatest temperature ranges and the possibility of desiccation (= water loss). The longer the duration of exposure, the greater the chance for

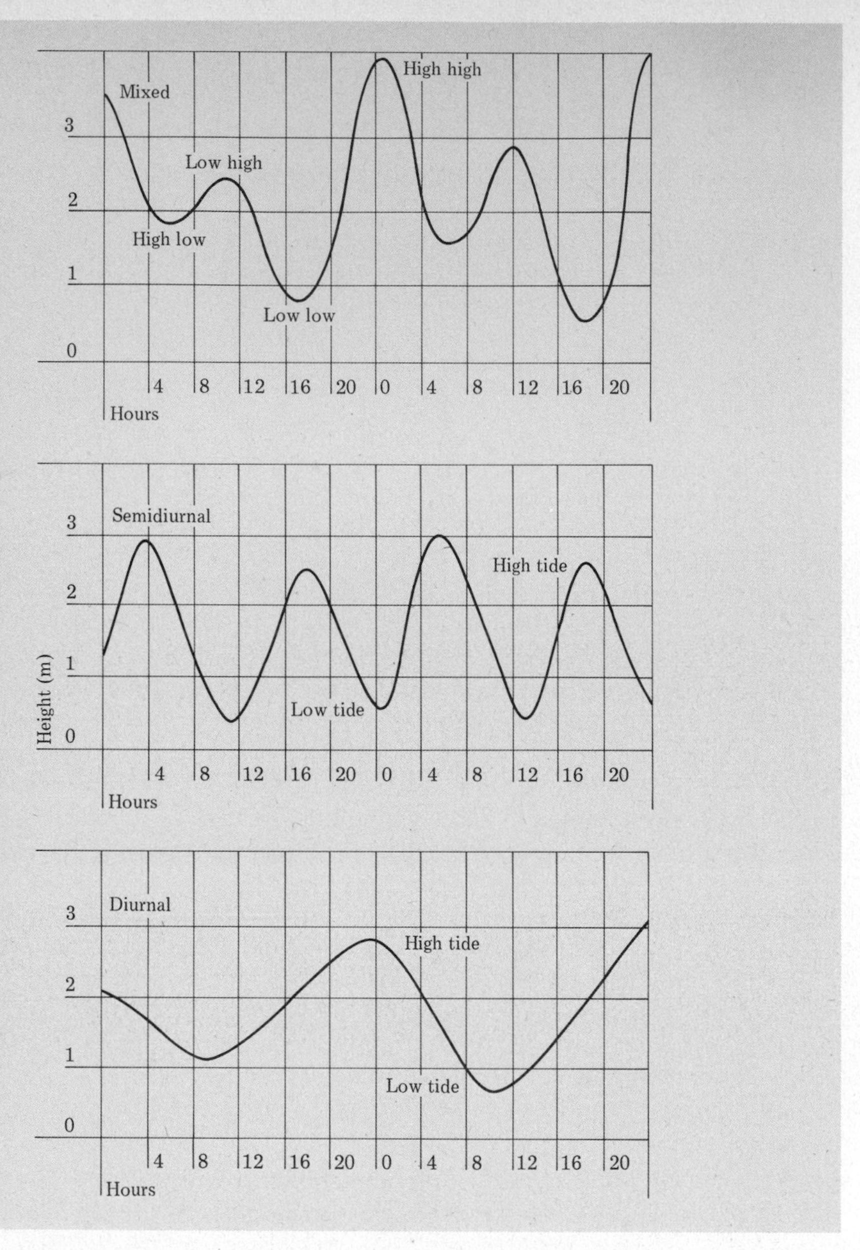

Figure 6-2 Semidiurnal, diurnal, and mixed tide curves. (Partly after D. A. Ross, Introduction to oceanography, Prentice-Hall, 1970.)

encountering a lethal temperature or desiccating beyond tolerable limits. Since most of these animals also must wait until covered with water to feed, the longer the duration, the less opportunity to feed and thus obtain sufficient energy. Animals and plants of the tidal zone vary in their ability to tolerate immersion in air and this differential is part of the reason for the different pattern of distribution of organisms observed on many rocky shores (see pp. 211–214).

In certain areas of the world there are two high and two low tides per day, and where the two highs or two lows are not equal (areas with *mixed tides*), it can be shown that the duration of immersion is not smooth going from low to high tide levels and that there are break points where the duration changes greatly in a short distance. What this means is that an organism living just a few inches above one of these points would be subjected to a greatly increased immersion relative to one just a few inches below. The existence of these *critical tide levels* was early suggested as the reason for the sharp changes in fauna and flora distribution observed on rocky shores (see Fig. 6-10), since they were thought to delineate an important break point.

The second effect is the result of the time of day during which the intertidal is exposed to the air. As before, the potentially lethal ranges for physical parameters are encountered by marine organisms only when exposed to air and in any daily cycle, these vary greatly at different times of the day. If, for example, the low tides in the tropics occurred only during the hours of darkness, the intertidal fauna and flora would be exposed to much lower temperatures and less desiccation than if exposed during midday. Hence, we might expect a greater diversity of organisms in the intertidal of a tropical area where the tides regularly occurred during night or early morning or evening than in an area where they occurred during midday. In the cold temperate zones, the reverse would be true, because in winter, the lowest temperatures would occur in early morning or at night and may act to freeze the animals, whereas at midday, temperatures would be high enough not to do so.

A final effect of the tide is that, because of its great predictability, it tends to induce certain rhythms in the activities of shore organisms. These rhythms may be with respect to time of spawning such as that seen in the grunion, a fish of the Pacific coast of the United States that spawns on the beach only on certain nights of the highest spring tides (Fig. 6-3), or with respect to feeding or other

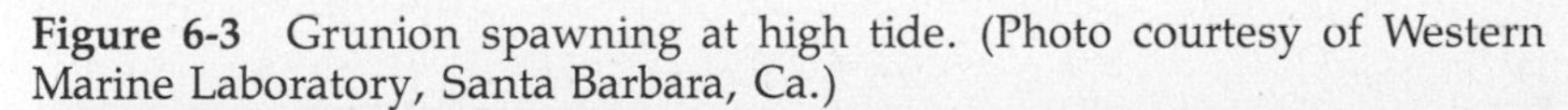

Figure 6-3 Grunion spawning at high tide. (Photo courtesy of Western Marine Laboratory, Santa Barbara, Ca.)

activity. Most intertidal organisms are quiescent when the tide is out and resume normal activitity, such as feeding, only when the tide is in.

TEMPERATURE

Because of its inherent physical characteristics (see Chapter One), water, particularly large bodies of water such as the oceans, shows a minimum range of change in temperature. Furthermore, this range rarely exceeds the lethal limits for the organisms. Intertidal areas are, however, regularly subjected to aerial temperatures for varying periods and these temperatures have a much wider range, on both a daily and a seasonal basis. The range may often exceed the lethal limits of the marine organisms. Should low tide occur when the air temperature is at a minimum (cold temperate, polar) or a maximum (tropics), the lethal limits may be exceeded and the organisms will die. Even if death does not occur immediately, the organisms may be so weakened by the extreme temperatures that they cannot resume normal activities and will suffer mortality from secondary causes.

Temperature also has an indirect effect. Marine organisms are subject to death by desiccation. Desiccation may be hastened by increased temperatures.

WAVE ACTION

It is in the intertidal zone that wave action exerts the most influence on organisms and communities of any area in the sea. This influence is manifested directly and indirectly. Wave action affects shore life directly in two major ways. In the first place, it has a mechanical effect, which acts to smash and tear away objects with which it makes contact. One can appreciate this action by noting the many pictures and stories of destruction wrought upon man-made structures due to storm waves of varying types. The action is similar in the intertidal zone, and hence, any creatures that inhabit this zone must be adapted to this persistent force in some way (see pp. 209–210). On shores that are made up of loose sand or gravel, the wave action is even more profound in that it moves the entire substrate around (see pp. 234–238), influencing the actual shape of the zone. Wave exposure may thus prove to be limiting for certain organisms that are unable to sustain the force, and necessary to others that cannot exist other than in heavy wave areas.

Secondly, wave action acts to extend the limits of the intertidal zone. It does this by throwing water higher on the shore than would normally occur as a result of the tide alone. This continual splashing allows the marine organisms to live higher in wave-swept areas than in calm areas within the same tidal range (Fig. 6-4).

Wave action has some other, perhaps minor, effects also. It acts to mix atmospheric gases into the water, thus increasing the oxygen content so

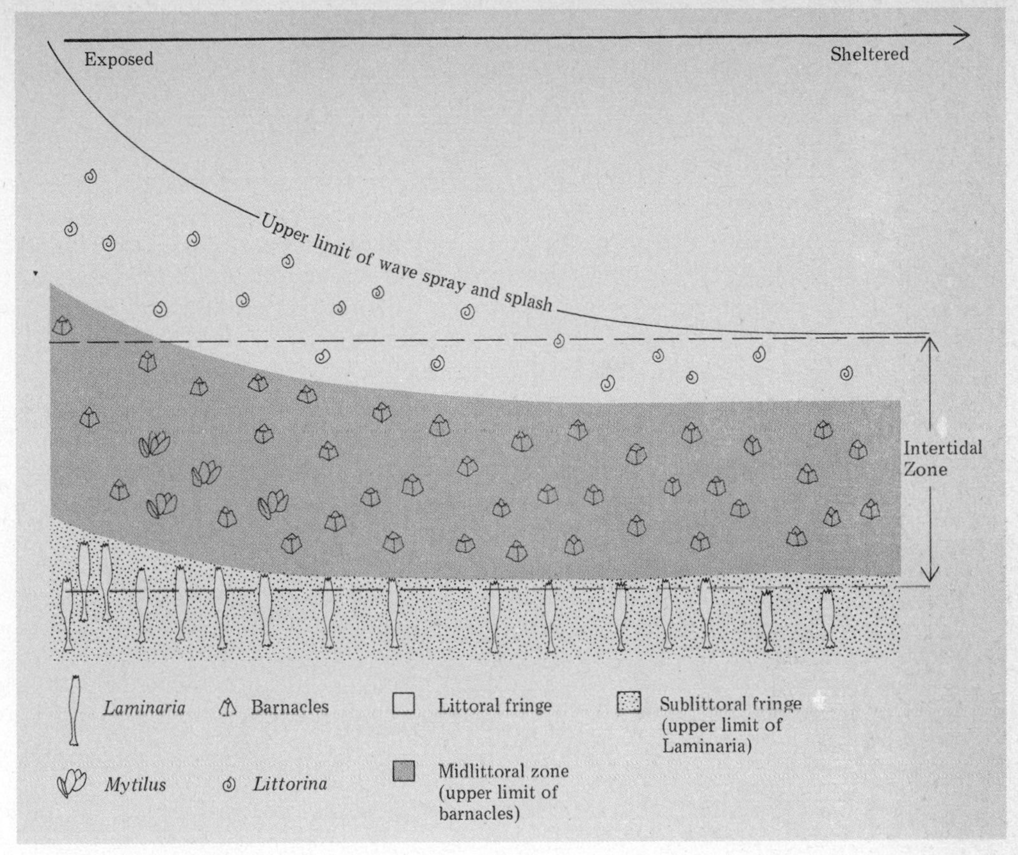

Figure 6-4 Changes in the extent of vertical zonation with change in exposure to wave action.

wave-washed areas never lack for oxygen. Because of the constant interaction with the atmosphere and the formation of bubbles plus the stirring up of the substrate, wave-swept areas may have decreased penetration of light. This is, however, of little ecological significance.

SALINITY

Salinity changes may occur in the intertidal in two situations that may affect the organisms. In the first place, the intertidal may be exposed at low tide and subsequently flooded by heavy rains or runoff from heavy rains. This means they will be subjected to greatly reduced salinities. In certain cases, this reduction in salinity may exceed the limits and the organisms may die, as most intertidal organisms show a limited tolerance to decreased salinity. The second situation has to do with tidepools, those areas that retain sea water at low tide. These may either be flooded with fresh water runoff from heavy rains, thus reducing salinity, or they may show increased salinity when exposed to heavy evaporation during the day (see pp. 232–234).

OTHER FACTORS

Different substrates—rock, sand, mud—have very different faunas and community structure in the intertidal, but these are taken up separately in this chapter. Oxygen seems not to be a factor exerting a limiting effect except in certain special cases (see pp. 232–234 on tidepools), and pH and nutrients are also unimportant to organisms and community structure.

ADAPTATIONS OF INTERTIDAL ORGANISMS

Since the intertidal organisms are primarily marine in origin, the adaptations that we observe have to do primarily with avoiding or minimizing the stresses imposed by the daily exposure to the terrestrial environment. The major stress from the marine environment is wave action.

RESISTANCE TO WATER LOSS

As soon as marine ogranisms are removed from the water into air, they begin to lose water. If they are to survive in the intertidal, this loss must somehow be minimized and/or the organisms must have body systems that will tolerate considerable losses during the hours of exposure to the air.

The simplest mechanism for avoiding water loss is seen in mobile animals such as crabs. These animals simply move from the exposed surface areas of the intertidal into very moist cracks, crevices, or burrows, where water loss is reduced or absent. Alternatively, they may seek refuge under the moist covering of algae. In other words, these animals avoid the adverse environmental condition of the shore by actively selecting suitable microhabitats (Fig. 6-5). A similar situation occurs in some species of anemones such as *Anthopleura xanthogrammica* of the Pacific coast of North America. It is soft bodied, with no way of preventing water loss. However, it normally is found only among barnacles or in crevices where water loss is reduced. Physiological adaptation is not needed.

Another simple mechanism is that seen in several of the high intertidal algae genera such as *Porphyra*, *Fucus*, and *Enteromorpha*. These plants cannot move and have no mechanism to avoid water loss. They simply are adapted to withstand a severe loss of water from their tissues. Specimens of these genera can be found that are dry and brittle after a long exposure at low tide, yet will quickly take up water and resume normal body processes when the tide returns. Kanwisher (1957) found that they can tolerate as much as a 60–90 percent loss of water and still recover. A similar tolerance to great water loss is seen in some intertidal sedentary animals. Boyle (1969), for example, reports chitons can tolerate a 75 percent water loss and limpets, according to Davis (1969), from 30–70 percent, depending on the species (Fig. 6-5).

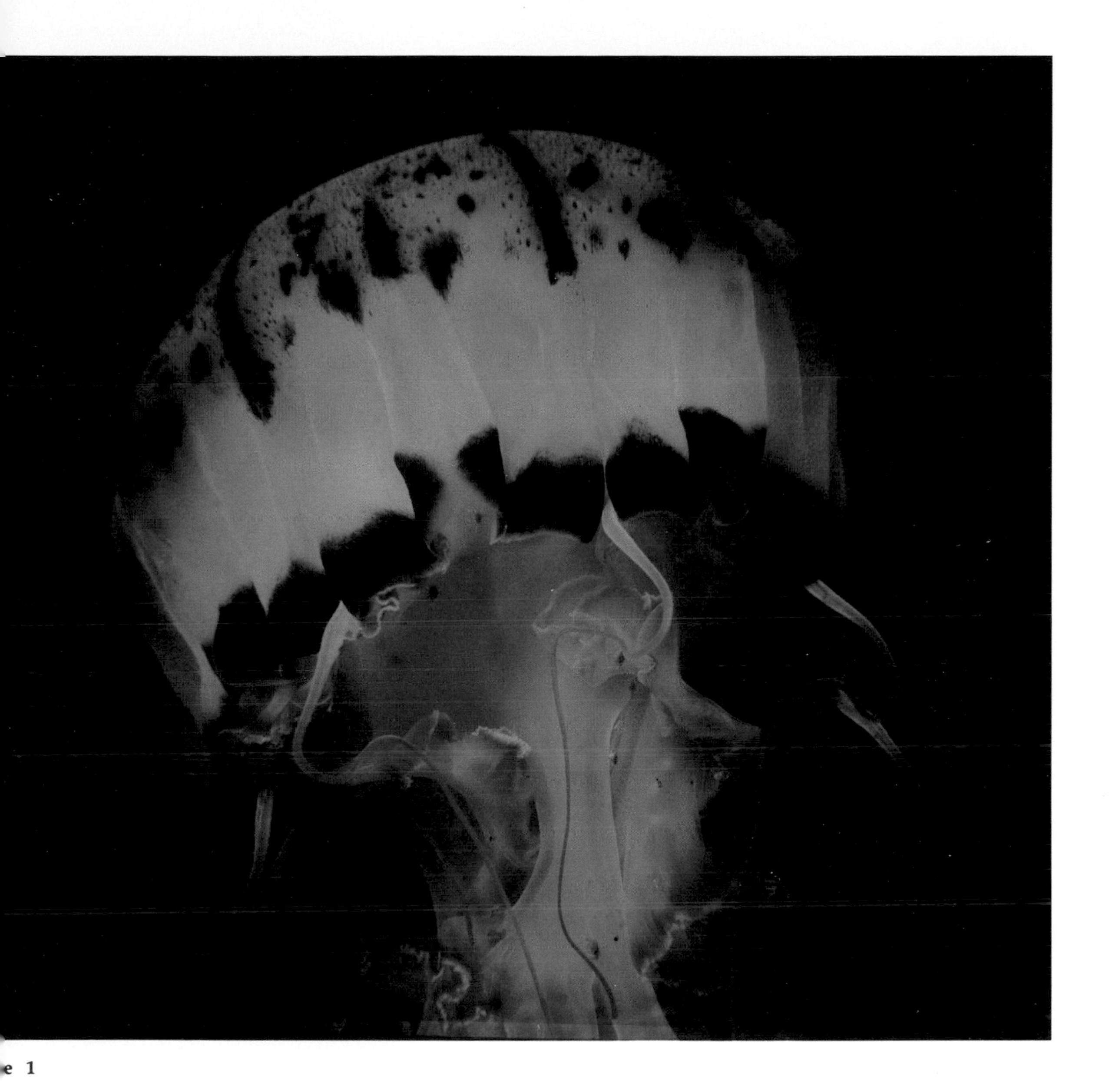

e 1

Plate 2

Plate 1 Planktonic scyphozoan jellyfish *Pelagia colorata* with commensal juvenile cancer crabs. (Photo courtesy of Dr. Lovell Langstroth.)

Plate 2 *Anoplogaster cornuta*, a "daggertooth" fish from the mesopelagic zone. Notice the very large mouth and large, sharp teeth for seizing prey. (Photo courtesy of Michael Kelly.)

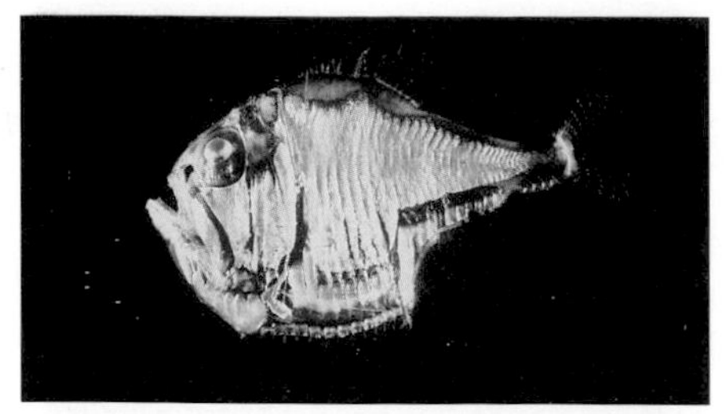

Plate 3

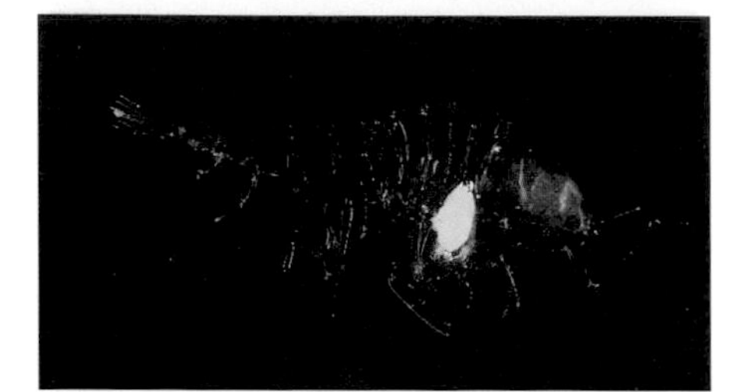

Plate 4

Plate 5

Plate 6

Plate 7

Plate 3 *Argyropelecus lychnus,* a mesopelagic hatchetfish. Note the lines of black photophores along the ventral portions of the body, the tubular eyes, and the silvery reflecting scales. (Photo courtesy of Michael Kelly.)

Plate 4 A mesopelagic amphipod, *Cystisoma fabricii,* showing the transparent body and huge white light organ. (Photo courtesy of Michael Kelly.)

Plate 5 A mysid crustacean of the genus *Boreomysis* showing the characteristic red-orange color of mesopelagic crustaceans. (Photo courtesy of Gary McDonald.)

Plate 6 A killer whale, *Orcinus orca,* surfacing among the pack ice in Antarctica. (Photo courtesy of Dr. John Oliver.)

Plate 7 The large plankton-eating manta ray, *Manta hamiltoni,* in the Gulf of California. (Photo courtesy of Alex Kerstitch.)

e 8

e 8 Commensal barnacles, *Cryptolepas* *hianecti,* and "whale lice," *Cyamus* *mmoni,* on the skin of a California gray le, *Eschrichtius robustus.* (Photo by the or.)

e 9 Female northern fur seals, *orhinus ursinus,* and single black pup on breeding grounds of St. Paul Island in Bering Sea. (Photo by the author.)

e 10 Harbor seal, *Phoca vitulina,* resting rock underwater. (Photo courtesy of Dr. ell Langstroth.)

e 11 Anemone of the genus *Tealia* and hydrocoral *Stylantheca porphyra* with ounding tunicates, sponges, and coralline e on subtidal rocks of central California. oto courtesy of Dr. Lovell Langstroth.)

Plate 9

Plate 10

Plate 11

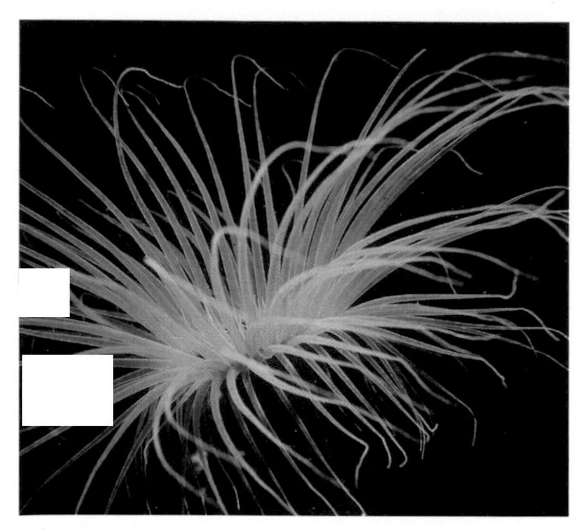
Plate 12

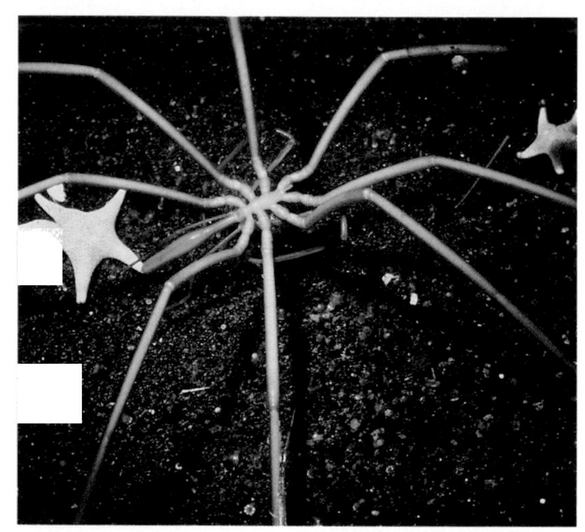
Plate 13

Plate 14

Plate 15

Plate 16

Plate 17

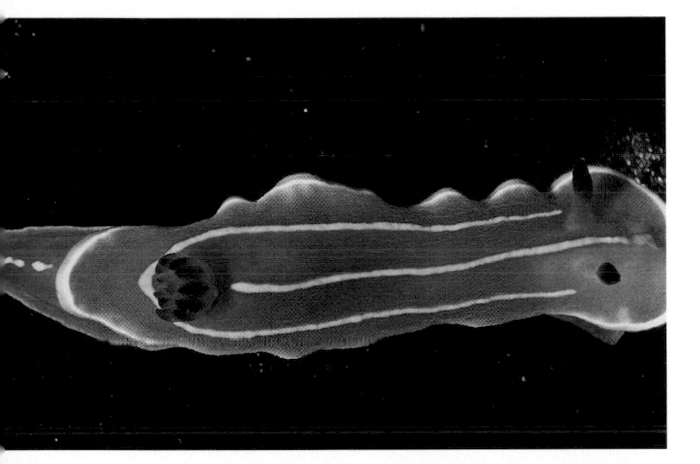

Plate 18

Plate 19

Plate 12 Subtidal tube dwelling anemone *Pachycerianthus fimbriatus* from the Pacific Coast of North America. (Photo courtesy of Dr. Lovell Langstroth.)

Plate 13 A giant sea spider of the genus *Colossendeis* on the bottom in McMurdo Sound, Antarctica. (Photo courtesy of Dr. John Oliver.)

Plate 14 A large mass of the black mussel *Mytilus californianus* with a large aggregation of its major predator, the sea star *Pisaster ochraceus* at Neptune Beach, Oregon. Notice the large areas of clean rock at the lower right where the sea stars have already consumed the mussels. (Photo by Dr. Richard Mariscal.)

Plate 15 The sea star *Pisaster giganteus* with the small aggregating anemones *Corynactis californica* on the rocks of the Monterey Peninsula in California. (Photo courtesy of Dr. Lovell Langstroth.)

Plate 16 The large green Pacific Coast anemone *Anthopleura xanthigrammica.* This anemone contains symbiotic zooxanthellae and zoochlorellae in its tissues which provide part of the nutritional requirements. (Photo courtesy of Dr. Lovell Langstroth.)

Plate 17 The sea palm, *Postelsia palmaeformis,* characteristic only of areas of direct wave exposure and high wave shock on the Pacific Coast of North America. (Photo by Dr. Richard Mariscal.)

Plate 18 A brilliantly colored intertidal nudibranch mollusk, *Chromodoris macfarlandi.* (Photo by the author.)

Plate 19 The eolid nudibranch mollusk *Phidiana crassicornis.* (Photo courtesy of Dr. Lovell Langstroth.)

Plate 20

Plate 21

Plate 22

Plate 23

Plate 20 The neogastropod mollusk *Mitra idae* with the commensal mesogastropods *Crepidula adunca* adhering to the shell. (Photo courtesy of Dr. Lovell Langstroth.)

Plate 21 Underwater shot of the crowding of corals and gorgonians on a coral reef in Palau. (Photo courtesy of Dr. Lovell Langstroth.)

Plate 22 Coral colony showing the characteristic appearance when the polyps are expanded (Palau). (Photo courtesy of Dr. Lovell Langstroth.)

Plate 23 Expanded polyps of the small nonreef building coral *Tubastrea* on a reef in Palau. (Photo courtesy of Dr. Lovell Langstroth.)

e 24 Close-up of the bright red polyps and te spicules of the skeleton of a shallow er gorgonian in Palau. (Photo courtesy of Lovell Langstroth.)

e 25 Expanded polyps of a gorgonian coral coral reef in Palau. (Photo courtesy of Dr. ell Langstroth.)

e 26 Moorish Idols, *Zanclus canescens,* on a l reef in Palau. (Photo courtesy of Dr. ell Langstroth.)

e 27 The giant clam *Tridacna gigas* on a u coral reef. The patterned mantle tissue panded. This mantle tissue contains symbiotic zooxanthellae which furnish clam with a certain portion of its nutrition. oto courtesy of Dr. Lovell Langstroth.)

e 28 A crinoid or feather star crawling over gonians on a coral reef in Palau. (Photo rtesy of Dr. Lovell Langstroth.)

Plate 26

Plate 27

e 24

Plate 28

e 25

Plate 29

Plate 32

Plate 30

Plate 31

Plate 29 A nocturnal coral reef fish, *Myripristis murdjan*. Note the large eyes (Photo courtesy of Dr. Lovell Langstroth

Plate 30 A hawkfish swimming among coral branches on a reef in Palau. (Photo courtesy of Dr. Lovell Langstroth.)

Plate 31 An anemone fish of the genus *Amphiprion* nestled in the tentacles of an anemone. (Photo courtesy of Dr. Lovell Langstroth.)

Plate 32 Mangrove seedlings coming up among the stilt roots of the adult trees. Lizard Island, Australia. (Photo by the author.)

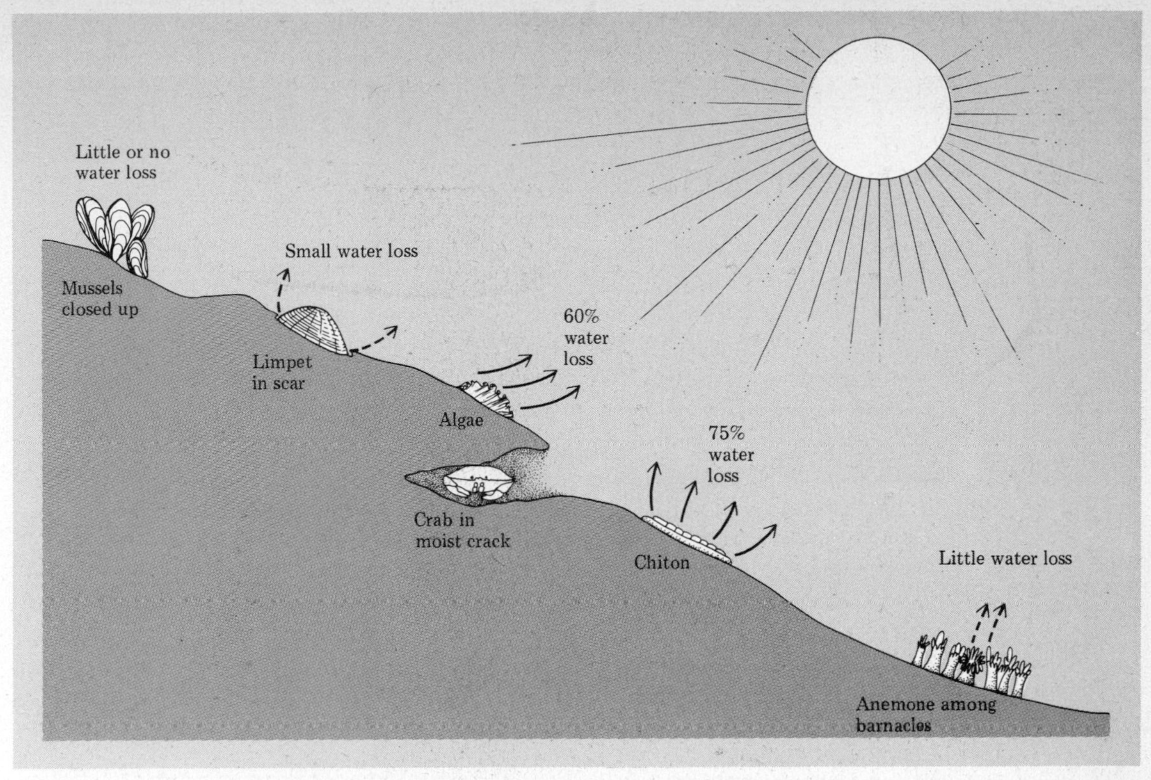

Figure 6-5 Representation of the mechanisms of adapting to water loss in intertidal organisms.

Many species of intertidal animals, in contrast to the above, have mechanisms for prevention of water loss. These mechanisms may be structural, behavioral, or both. Barnacles of many species are predominant members of the intertidal zone all over the world. These animals are sessile and avoid water loss by closing up within their shells at low tide. Here, the presence of an impermeable shell cuts down the evaporative water loss. Limpets of the genera *Patella*, *Acmaea*, and *Collisella* are also dominant animals of the rocky intertidal. Certain species of these limpets have a "home scar" into which their shells exactly fit. At low tide, they return to these "homes" and by fitting into these grooves, greatly reduce water loss. Other limpets that have no scar simply clamp down tightly against the rock so no tissue is exposed, only shell. Other gastropods such as the periwinkles (*Littorina*) have opercula which completely seal off the aperture to their shells. At low tide, they may pull into the shell, closing the aperture with the operculum and thus reduce water loss. Certain bivalves, such as *Mytilus edulis*, are able to survive in the intertidal because of their ability to close their valves tightly to prevent water loss. Still other ogranisms, such as the anemone *Actinia* and the hydroid *Clava squamata*, produce mucus that prevents water loss. Many inhabitants of sand or mud simply burrow into the substrate to prevent desiccation.

MAINTENANCE OF HEAT BALANCE

Intertidal organisms are also exposed to extremes of heat and cold and show behavioral and structural adaptations to maintain their internal heat balance. Though deaths by freezing have been recorded for intertidal organisms, extreme low temperatures seem to be less of a problem to shore organisms than high temperatures. This may be due to the fact that organisms are often living much closer to their upper lethal temperature than to their lower lethal one. Hence, most mechanisms for heat balance concern the avoidance of too high temperatures. This may be accomplished by (1) reducing heat gain from the environment and (2) increasing the heat loss from the body of the animal. Heat gain from the environment is reduced in several ways. One way is to have a relatively large body size when compared to similar species, either lower in the intertidal or in the subtidal. Large body size means less surface area relative to volume and hence, less area to gain heat. At the same time, a larger body takes longer to heat up than a smaller one. It has been recorded for gastropod mollusks such as *Littorina littorea* and *Olivella biplicata* that the larger individuals occur higher in the tidal zone and the smaller ones lower. Yet another mechanism to reduce heat gain is to reduce the area of body tissue in contact with the substrate. This is, however, difficult to achieve for most intertidal animals, since they need the extra purchase provided by a large area of tissue in contact with the substrate to avoid being swept off by waves. Yet, there are some shore animals that do have such a small area of attachment or none at all. These occur where wave action is not severe, and, in those animals that lose attachment, the ability to reattach as the tide comes in is well developed.

Heat loss can be attributed to a number of causes. One mechanism, found in hard-shelled organisms such as mollusks, is greater elaboration of ridges and other sculpturing on the shell. These act as radiator fins and facilitate heat loss. Examples of such sculptured mollusks are best found in the tropics and include *Tectarius muricata* and *Nodolittorina tuberculata*. Enhancement of heat loss occurs when the organism is light in color. (Dark organisms gain heat by absorption.) Many tropical and subtropical snails of the high intertidal such as *Nerita peleronta* (Caribbean) and *Littorina unifasciata* (New Zealand) are much lighter in color than their lower level relatives, presumably then, facilitating heat loss (Fig. 6-6).

Finally and most importantly, heat loss can occur through evaporation of water. Unfortunately, again, intertidal organisms must face the problem of desiccation and can ill afford to cool themselves through evaporation or else chance a lethal desiccation. Many intertidal organisms show a strategy of adaptation that allows them some cooling by evaporating water and, at the same time, avoiding excessive desiccation (see Segal and Dehnel, 1962). To facilitate this balance, many intertidal animals have an extra water supply from which cooling may take place. This extra water is held in the mantle cavity of barnacles and limpets, and it is above the amount the animal needs to survive desiccation. Secondly, most high intertidal animals can withstand a remarkable amount of

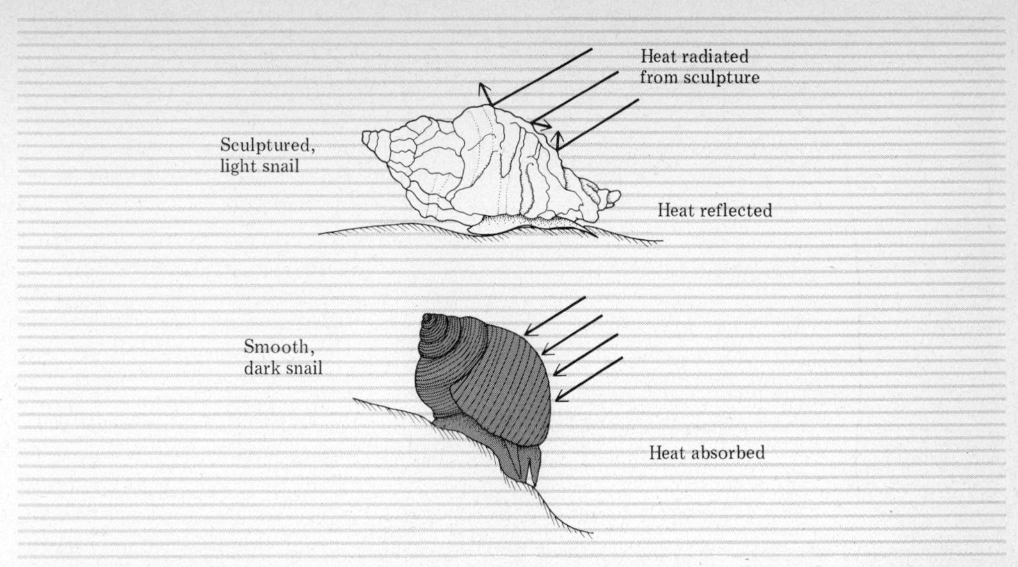

Figure 6-6 Differences in heat absorption between smooth, dark shells and sculptured, light shells.

desiccation (see previous section); and finally, many animals have the ability to reduce the rate of water loss from the tissues when necessary.

MECHANICAL STRESS

As noted previously, wave action reaches a zenith in the intertidal zone. As a result, it is necessary for any organism that lives intertidally to adapt in order to resist the smashing and tearing effects of waves. Wave action has different effects on rocky shores as opposed to sandy shores and requires different adaptations, which are discussed subsequently. However, some generalities may be noted here.

In order to maintain position in the face of wave action, several adaptations have been developed by intertidal organisms. One such adaptation, observed in barnacles, oysters, and serpulid polychaete worms, is simply to become fixed to the substrate. Similarly, the majority of algae in the intertidal are fused via a holdfast to the bottom.

Still other organisms develop attachments that are strong but not permanent, thus giving limited movement. Such attachments are the byssal threads of *Mytilus*, which anchor the animals securely but still can be broken and remade to allow limited slow movement.

The dominant intertidal mollusks, such as the various limpets and chitons, resist wave action through a strong, enlarged foot which clamps them to the substrate.

Motile organisms such as crabs have no structural mechanism to resist being swept away and survive only because they seek shelter from the waves in crevices or under rocks.

Most intertidal mollusks are adapted to wave shock by possessing thicker shells than their subtidal relatives and lack delicate sculpturing, which is easily broken in the waves.

RESPIRATION

Since the animals inhabiting the intertidal are marine, they have respiratory surfaces adapted to extract O_2 from water. Usually, these are thin-walled extensions from the body surface. These respiratory organs are thus very susceptible to desiccation in air and do not function unless immersed in water. Such organs would be a distinct disadvantage in the intertidal. Among intertidal animals, there is a tendency to enclose the respiratory surfaces in a protective cavity to prevent them from drying. This is seen particularly in various mollusks where the gills are in a mantle cavity, which itself is protected by the shell. A similar situation prevails in barnacles where the mantle tissue acts as a respiratory organ.

Because animals with protected respiratory organs must also conserve water at low tide, they often close up (operculum) or clamp down (chitons, limpets) such that gaseous exchange is reduced. Thus, to conserve O_2 and water, most animals are quiescent during low tide.

FEEDING

All intertidal animals must expose the fleshy parts of their bodies to feed. This means exposing those parts most susceptible to desiccation. As a result virtually all intertidal animals are active only during the time the tide is in and they are covered with water. This is true whether the animals are grazers, filter feeders, detritus feeders, or predators.

SALINITY STRESS

As noted above, the intertidal zone may be subjected to flooding by fresh water, which would create an osmotic stress on the intertidal organisms adapted to be submerged only in sea water. This is particularly true since most intertidal organisms do not show adaptations to tolerate salinity changes, certainly not like estuarine organisms do (see Chapter Eight). Most have no mechanism to control the salt content of their body fluids and hence, are *osmoconformers*. The only adaptations observed are the same as those for preventing desiccation, such as closing up valves or shells in barnacles and mollusks. Perhaps this is why there are records of catastrophic mortality of intertidal organisms following heavy rain or runoff. Apparently, however, such events are so rare or unpredictable that special mechanisms have not evolved.

REPRODUCTION

Because so many of the intertidal organisms are sedentary or sessile, they must rely upon eggs and/or larvae that are free floating in the plankton for dispersal.

Hence, we see a very high incidence of planktonic larvae among intertidal organisms. These larvae are of many types (see Fig. 2-9 on meroplankton).

A second reproductive adaptation attributable to the intertidal position is that most organisms have a breeding cycle that is closely synchronized with the occurrence of certain tides, such as spring tides. An example is *Mytilus edulis*, in which gonads mature during periods of spring tides and spawning occurs on subsequent neap tides. In *Littorina neritoides*, the eggs are spawned at time of spring tides.

ROCKY SHORES

Of all the intertidal shores, those composed of hard material, the rocky shores, are the most densely inhabited by macroorganisms and have the greatest diversity of animal and plant species. They stand in striking contrast to the almost barren appearance of sand and mud shores. It is these densely populated, topographically diverse, and species-rich rocky areas that have fascinated marine biologists and ecologists for many years. In recent years, these areas have been the subject of several classic studies that have enhanced our understanding of how these associations of species interact with each other to maintain or change the community.

ZONATION

One of the most striking features of any rocky shore anywhere in the world, when seen at low tide, is the prominent horizontal banding or *zonation* of the organisms. Each zone or band is set off from those adjacent by differences in color, morphology of the major organism, or some combination of color and morphology. These horizontal bands or zones succeed each other vertically as one progresses up from the level of the lowest low tides to true terrestrial conditions (Figs. 6-7, 6-8). This zonation observed on intertidal rocky shores is similar to the zonation pattern that one observes with increasing elevation on a mountain, where the different horizontal zones of trees and shrubs succeed each other vertically until, if one progresses far enough, permanent snow cover is reached. The major difference between these two areas is the scale, mountain zones being perhaps kilometers in width as opposed to intertidal zones extending a few meters vertically.

Rocky intertidal zones vary in vertical extent, depending upon the slope of the rocky surface, the tidal range, and the exposure to wave action. Where there is a gradual slope to the rock, individual zones may be broad, whereas under similar tidal and exposure conditions on a vertical face, the same zone would be narrow. In the same manner, exposed areas tend to have broader zones than protected shores, and shores with greater tidal ranges have broader vertical zones (Fig. 6-4).

Of course, these striking bands may be interrupted or altered in various places

Figure 6-7 Photograph of a rocky shore at low tide illustrating the conspicuous banding or zonation of organisms. (Photo courtesy Dr. Michael Foster.)

along a rocky shore wherever the rock substrate itself shows changes in slope, composition, or irregularities that change its exposure or position relative to the prevailing water movement.

The fact that these prominent zones can be observed on nearly all rocky shores throughout the world under many different tidal regimes led the team of Stephenson and Stephenson (1949) to propose, after some 30 years of study, a "universal" scheme of zonation for rocky shores (Fig. 6-9). This "universal" scheme was really a framework using common terms that would allow comparison of diverse areas. It is characterized by establishing the zones on the basis of the distributional limits of certain common groups of organisms and not on the basis of tides. It thus reflects the knowledge of the Stephensons, and other intertidal ecologists, that distribution patterns of the organisms and zones vary not only with tides, but with slopes and exposure. Thus, under similar tidal conditions, there could be different band widths due to different exposures or slopes of rocks. It was this universal scheme that established a standard format for describing shore zonation, replacing a bewildering host of schemes and names established by earlier biologists.

The Stephenson scheme has three main divisions of the intertidal area. The uppermost is termed the *supralittoral fringe*. Its lower limit is the upper limit of barnacles, and it extends to the upper limit of snails of the genus *Littorina* (periwinkles). The dominant organisms are the littorine snails and black encrusting lichens (*Verrucaria* type). Part of this zone is reached by extreme high water of spring tides, but must of its water comes from wave splash. Above this zone is the terrestrial supralittoral zone.

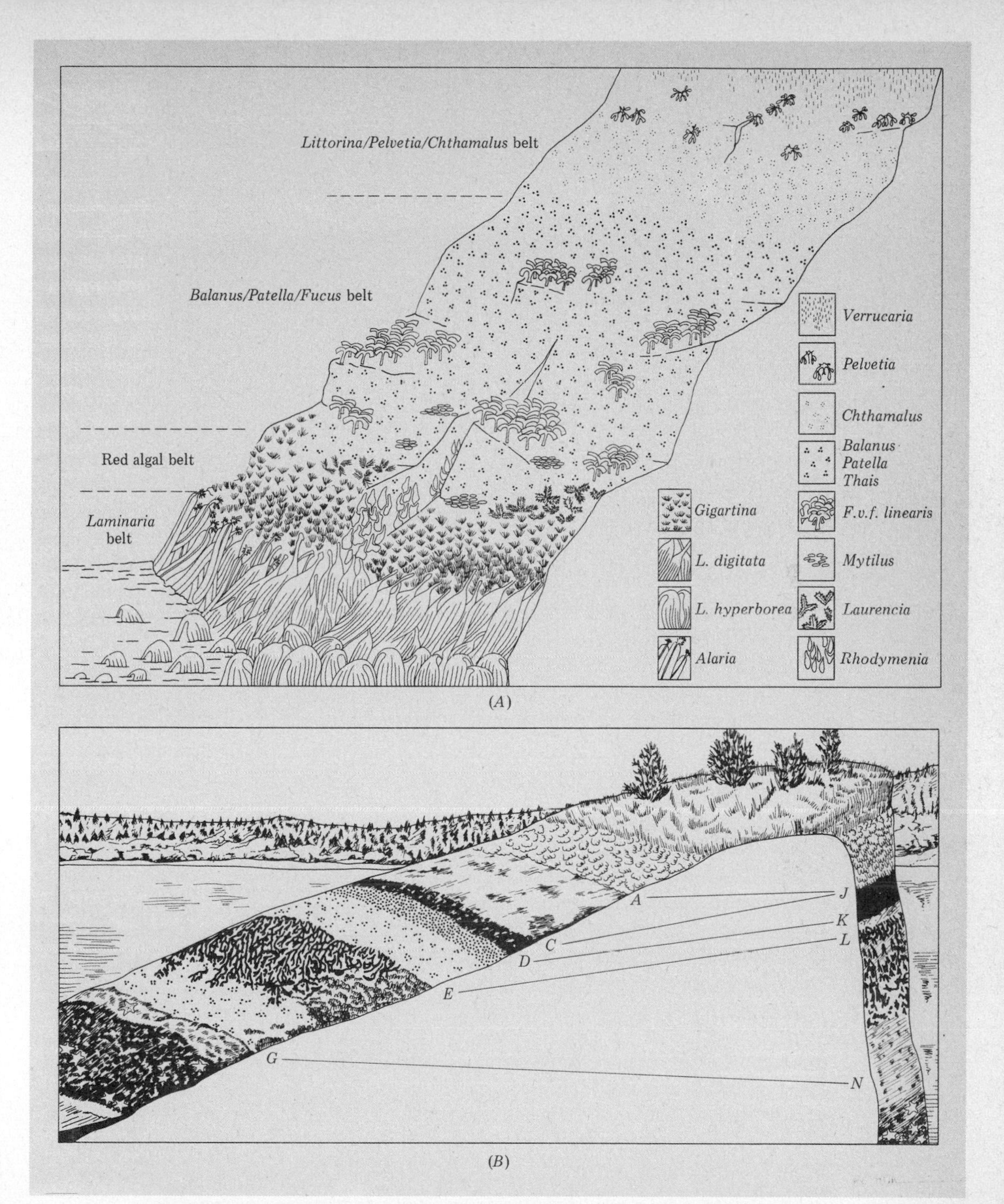

Figure 6-8 Zonation in the rocky intertidal. *(A)* An intertidal area in Scotland. *(B)* An intertidal area in British Columbia. Note in *(B)* the difference in the extent of zones. In *(B)*, line *A–J* indicates the lower limit of lichens; line *C–J*, the upper limit of the "black" zone; line *D–K*, the upper limit of barnacles; line *E–L*, the upper limit of *Fucus*; line *G–N*, the lower limit of barnacles. (*A*, from J. R. Lewis, The ecology of rocky shores, 1964, Hodder & Stoughton Educational. *B*, from T. A. Stephenson and A. Stephenson, Life between the tidemarks on rocky shores, 1972, Freeman.)

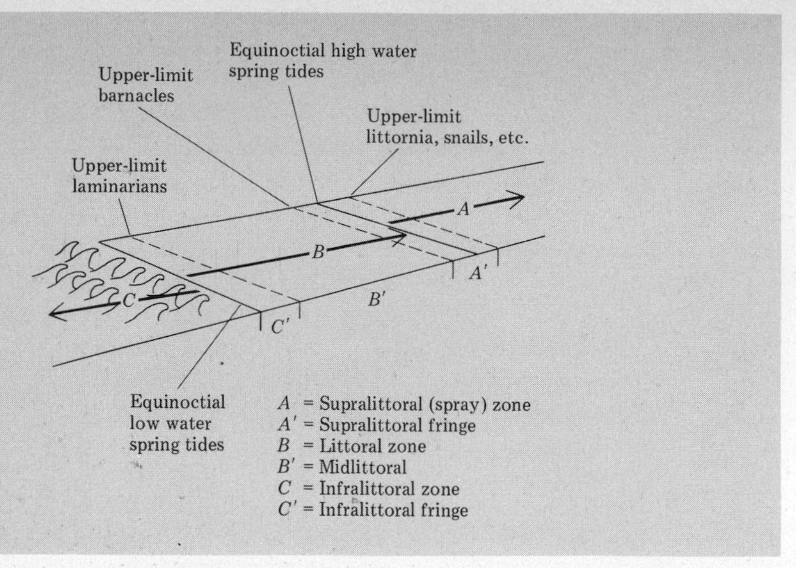

Figure 6-9 The Stephenson "Universal" scheme of zonation for rocky shores. (From R. L. Smith, Ecology and field biology, Harper & Row.)

The middle part of the intertidal is termed the *midlittoral* zone and is the broadest in extent. Its upper limit is coincident with the upper limit of barnacles, while its lower limit is that point where large kelps (*Laminaria*, etc.) reach their uppermost distribution. This zone is often subdivided and contains a host of different organisms. Perhaps the only universally present dominant group are the barnacles.

The lowermost zone of this scheme is the *infralittoral fringe*, which extends from the lowest low tide up to the upper limit of the large kelps. This is an extremely rich zone composed of organisms that can tolerate only limited exposure to air. It is really an intertidal extension of the *infralittoral zone* (Stephenson term) or what we know as the sublittoral area.

Although the above scheme does set forth a means for describing zonation on rocky shores, it does not offer an explanation as to why the zonation occurs. It is this explanation of zonal patterns that intrigues current marine biologists.

CAUSES OF ZONATION

Whereas it is fairly easy to recognize and measure the extent of the zones on a rocky shore, it is more difficult to come up with suitable explanations as to why organisms are distributed into these zones. Two sets of factors can be considered to explain the phenomenon: physical and biological. We shall take each up in turn.

Physical Factors. The most obvious explanation for the occurrence of the zones is that they are a result of the tidal action on the shore and therefore reflect the different tolerances of the organisms to increasing exposure to the air and the resultant desiccation and temperature extremes. One difficulty with this explanation is that the rise and fall of a tide tends to follow a smooth curve with no obvious sharp breaks, which would correspond to the often sharp boundaries observed in the intertidal zones. If, however, one observes a whole series of such tidal curves, such a series of breaks does become apparent. For example, Fig. 6-10A gives a typical tidal curve for a mixed tide area on the California coast and Fig. 6-10B gives the maximum time of continuous submergence for various tide levels. As can be seen from the graph, there are certain points on this curve that reflect sharp increases in exposure to air. This can also be deduced from Fig. 6-10A, which represents a single typical lunar tide cycle. Consider an organism at point x on the graph. At this point, the tide will cover the organism at least once every six hours. If, however, the organism moves up only a few inches, the tide will not cover it for 12 hours. Thus, there is a great change in exposure time with a very short vertical movement. Points exhibiting sharp increases in exposure time over short distances have been termed *critical tide levels* by Doty (1946) and were offered as one of the early explanations for the zonation patterns described above. In this explanation, it is important to remember that it is not the tide per se that causes the limit, but rather the fact that, at these critical points, the organisms are subjected to greatly increased time in the air and hence to greater temperature fluctuations and desiccation.

The critical tide hypothesis has been tested in various places by several scientists since its original promulgation by Doty in 1946. In general, it is difficult to find good correlations, particularly at the lower tide levels. This lack of good correlation can be partly attributed to the varying topography of the various shores and to the variation in exposure. Thus, a species may well be able to exist above the critical tide level if the rocky shore is exposed to persistent violent wave action, which would throw water up higher and thus decrease desiccation.

Figure 6-10 *(A)* Typical tidal curve for a mixed tide area on the Pacific coast of North America. *(B)* Maximum time of continuous submergence for various tide levels on the Pacific coast of North America. *(B, from J. Nybakken, Readings in marine ecology, Harper & Row.)*

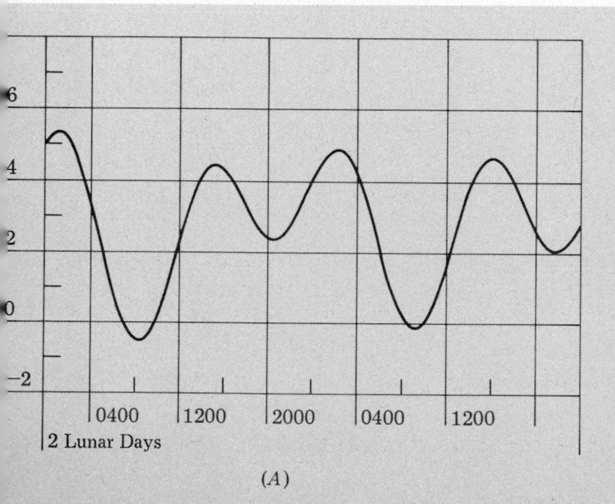

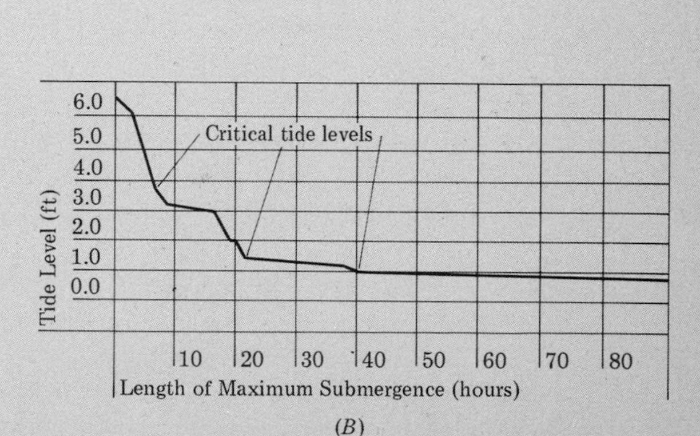

Similarly, caves, overhangs, and crevices remain moist when exposed areas are dried out, thus allowing persistence of organisms above the level predicted by the critical tides. The effects of these topography and exposure differences suggest, however, that the real upper limits are set by certain physical factors, namely desiccation and temperature.

Since the intertidal organisms are marine, desiccation is a serious problem. Several natural observations as well as field experiments have suggested that desiccation can set the upper limits to organisms and zones. In the North Temperate Zone, for example, rocks which have a north-facing slope often have the same organism occurring higher than do adjacent south-facing slopes. Since such rocks experience no difference in wave exposure or tides, the only reason for the height difference is that north faces dry out more slowly than south faces. Similarly, field experiments by Frank (1965) and others in which water was slowly dripped down slopes have resulted in the organisms extending higher up (Fig. 6-11). In other experiments, organisms were transplanted above their normal positions. Notable in this respect are barnacles which, of course, cannot move. When transplanted by Foster (1971) above their normal tidal height, they died, with younger ones dying more quickly than older ones. Again, desiccation can be suspected.

Along with desiccation and often acting in concert with it, temperature is the other major physical limiting factor. As noted previously, aerial temperatures have ranges that may exceed the lethal limits and hence, intertidal organisms may die from either freezing or "cooking." Upper limits to zones may therefore

Figure 6-11 Zonation change in the presence of a water seep.

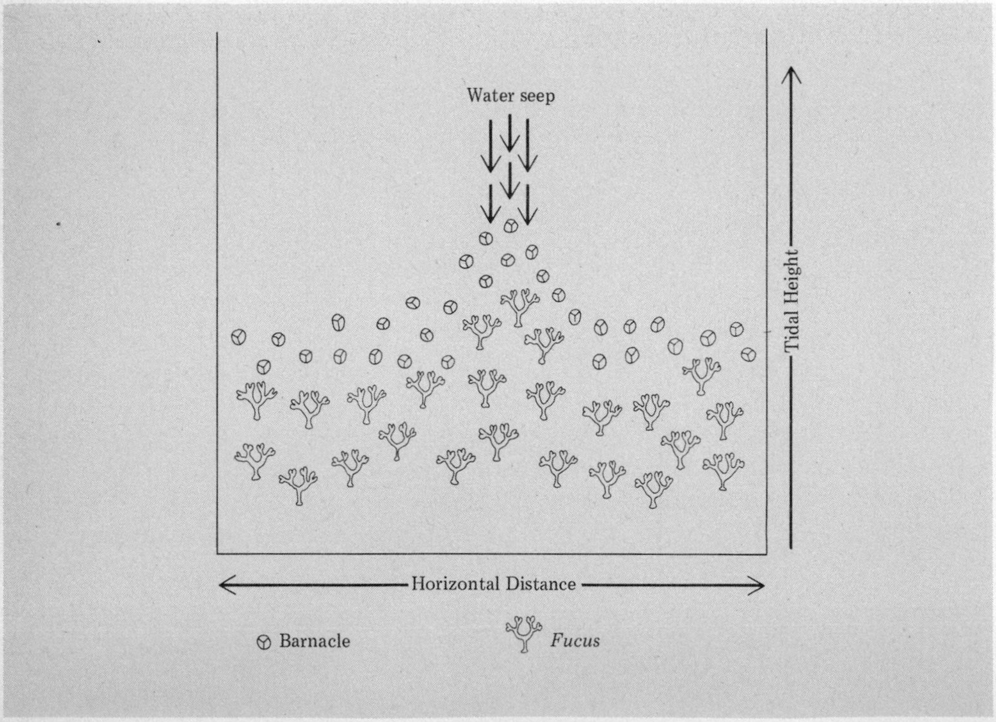

be partly attributable to the tolerable temperature limits of the intertidal organisms. In addition, high temperature promotes desiccation, and the synergistic effect of these factors may be even more devastating than each acting alone.

Finally, sunlight itself may act adversely to limit organisms on the shore. Sunlight includes wavelengths in the ultraviolet region that are deleterious to living tissue. Water quickly absorbs these wavelengths and so serves to protect most marine animals. However, intertidal animals have direct exposure to such rays at low tide. The higher an organism is in the intertidal, the greater the exposure to these rays. At the present time, there is little information concerning whether such UV radiation is active in controlling distribution of organisms, but the possibility remains. Light also has been suggested as a regulator of the distribution of intertidal algae. This has to do primarily with the spectral quality of the light. As noted in Chapter One, different wavelengths of light are absorbed differentially by water. In the case of intertidal algae, those that need light of longer wavelength (reds) and that are absorbed most quickly by water would tend to be found higher in the intertidal so that when submerged (when they actually photosynthesize), they would not be too deep for the penetration of red light (about 2 m). Since the main intertidal algae belong to three different groups, reds, browns, and greens, each of which has a slightly different absorption spectrum, it might be thought that they would be arranged along a depth gradient. In such a gradient, the green algae would be expected to be the highest, since they absorb mainly red light; the brown next; and the red algae, which absorb mainly the deep, penetrating green light, deepest. This, however, has not been found to be the case, and the intertidal algae are a mixture of all types at most levels. This is likely due to the interaction of other factors and the physiology of the algae, and it points out again the fallacy of attempting to explain patterns of species distribution with only single factors when a multitude are acting at all times.

We can summarize the above by stating that it appears that physical factors are strong determinants of the upper limits of the distribution of intertidal organisms.

Biological Factors. Although the early work in the intertidal focused on the importance of physical factors in setting the zonal patterns, more recent work has begun to clearly establish the great importance of various biological factors in setting various observed distribution patterns. In general, these biological factors are more complex, often subtle, and closely linked to other factors, which is probably why we have only recently begun to understand how they act. The major biological factors are *competition*, *predation*, and *grazing* (= herbivory). We shall take them up in order.

Competition for a certain resource does not occur if the resource is so plentiful that adequate supplies of it are available for all species or individuals. In the rocky intertidal zone, only one resource is commonly in limited supply and that is space. This is perhaps the most restricted area in the marine environment in terms of physical dimensions; at the same time, it is densely populated, at least in the temperate zone. As a result, there is an intense competition for space. This has resulted in observed zonal patterns. On the intertidal shores of Scotland, there is a distinct zonation of barnacles with the small *Chthamalus stellatus* living in the highest zone and the larger *Balanus balanoides* occupying the major portion of the mid-intertidal. Studies done by Connell (1961) showed that *Chthamalus* larvae settled out throughout the zones occupied by both barnacles, but survived to adulthood only in upper zones. The reason for the disappearance in the midlittoral region was due to competition from *Balanus balanoides,* which either overgrew, uplifted, or crushed the young *Chthamalus. Balanus* was prevented from completely eliminating *Chthamalus* because it did not have the tolerance to drying and high temperatures that prevailed at the higher tidal levels and hence did not survive there, whereas *Chthamalus* could survive there. Here, then, is a case where the zonation is at least partially a function of biological competition (Fig. 6-12).

A more complex example concerning competition is that among the mussel *Mytilus californianus* and several species of barnacles on the Pacific coast of North America. In this case, studies by Dayton (1971) have shown that *M. californianus* is the dominant space competitor on open coast shores. Given enough time and freedom from predators, the *M. californianus* eventually overgrow and outcompete all other macroorganisms and take over the substrates throughout most of the mid-intertidal. *Mytilus californianus* takeover is, however, slow. Wherever open space occurs, it may be rapidly colonized by other organisms, including three species of barnacles, *Balanus glandula, B. cariosus,* and *Pollicipes polymerus.* These, in turn, displace any rapidly growing algal species. The barnacles persist only until the mussels enter. The mussels outcompete and destroy the barnacles by settling on top of them and smothering them. Since nothing appears to be large enough to settle and smother the *M. californianus,* they remain in control of the intertidal space. Given this competitive edge, it would appear that eventually, the rocky Pacific coast of North America would be a monotonous band of *M. californianus.* It is also curious that *M. californianus* forms dense clumps or bands only in the intertidal, though it is perfectly capable of living subtidally. Since it is a premier space competitor, why this abrupt lower limit? The reasons for this have to do with other biological factors which act to prevent such resource monopolization and may be grouped under the term *predation.*

Intertidal algae, particularly those on temperate shores, also often show abrupt limits to their upper and lower distribution. Earlier, these limits were ascribed to critical tide levels, but it has also been suggested that this could be due to competition for space and/or access to light. Unfortunately, very little information is available concerning actual or potential competition among algal species to test these hypotheses. In two experimental studies done on the Pacific coast of North America, evidence was found for competition. The first study by

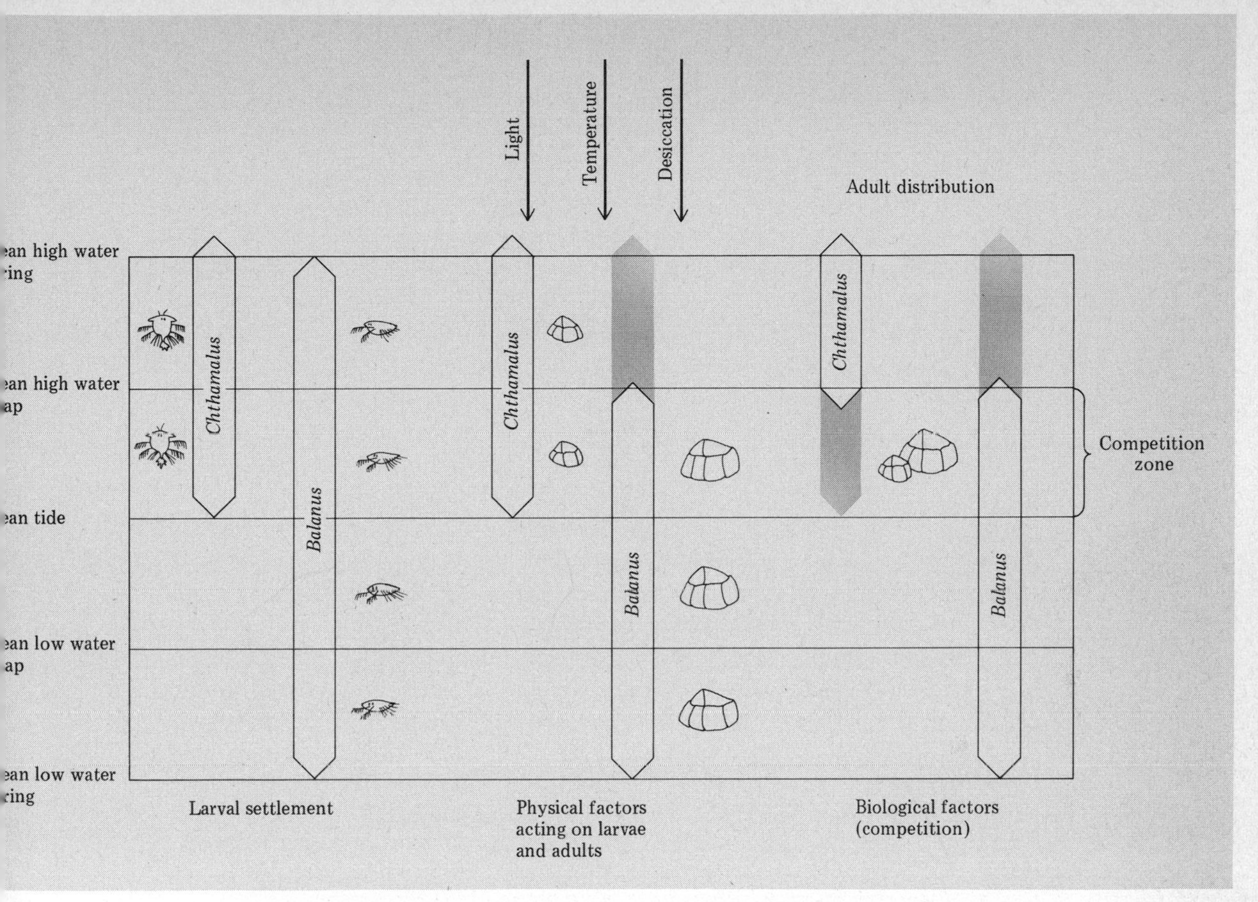

Figure 6-12 Intertidal zonation as a result of the interaction of physical and biological factors. The larvae of two barnacles, *Chthamalus stellatus* and *Balanus balanoides*, settle out over a broad area. Physical factors, mainly dessication, then act to limit survival of *B. balanoides* above mean high water of neap tides. Competition between *B. balanoides* and *C. stellatus* in the zone between mean tide and mean high water of neap tides then eliminates *C. stellatus*.

Dayton (1975) was directed at the larger kelps, and he found that the dominant kelps, *Hedophyllum sessile*, *Laminaria setchelli*, and *Lessionopsis littoralis* all outgrew and outcompeted certain smaller species in the lower intertidal. These smaller species were generally fast-growing species, which quickly colonized open areas. As noted in Chapter Five, they are *opportunistic* or *fugitive* species. In turn, among the three dominants, *Hedophyllum* was outcompeted by the other two such that they came to dominate the areas. The second study was actually done in the subtidal, where Vadas (1968) found that the giant kelp *Nereocystis* outcompeted and overgrew the brown alga *Agarum*. On the New England coast, similar studies by Lubchenko (1978) directed at tidepool algae indicated that *Enteromorpha intestinalis* was a dominant space competitor as opposed to *Chondrus crispus*; in the absence of grazers, *E. intestinalis* would quickly outcompete *Chondrus* and take over the space. These few studies suggest that competition among algal species may be more widespread than originally assumed and may be a fertile ground for future ecological work.

The role of predators in determining the distribution of organisms in the intertidal, and hence, the zonal patterns, is best documented for the Pacific coast of North America and is discussed here as an example of how complex biological interactions serve to give the prevailing observed distributions.

The dominant abundant intertidal animals on the Pacific coast other than the space-dominating *Mytilus californianus* are the barnacles *Balanus cariosus* and *Balanus glandula*. These latter two species occur abundantly in the intertidal region, despite the fact that they are competitively inferior to *M. californianus*, because a predatory starfish, *Pisaster ochraceus*, preys upon *M. californianus*, thus preventing it from completely overgrowing the barnacles (see Plate 14) *Pisaster ochraceus* is a voracious predator of mussels, consuming them at a rate that effectively prevents them from occupying all the space.

At the same time, *Balanus glandula* is found primarly as a band of adults in the high intertidal, while *Balanus cariosus* occurs as scattered, large individuals or clumps in the mid-intertidal. This pattern, as Connell (1970) has shown, is also due to predation. *Balanus glandula*, like *Chthamalus* in England, is capable of living throughout the intertidal zone and indeed settles throughout. The same is true for *B. cariosus*. That both show a restricted distribution is due to predation by three species of predatory gastropods of the genus *Thais: T. lamellosa, T. emarginata*, and *T. canaliculata*. The abundance and motility of these predators is such that they are capable of completely consuming all the young *B. glandula* settling out in the mid-intertidal in 12–15 months. *Balanus glandula* survives only in a narrow band at the top of the intertidal, where the *Thais* species are prevented from entering because of excessive desiccation. In the case of *B. cariosus*, however, the situation is somewhat different. There is no high level refuge for this barnacle, but, as with *B. glandula*, the young are consumed by the *Thais*. The refuge for *B. cariosus* is that once it reaches 2 years of age, it is too large for *Thais* to attack (Fig. 6-13). The only problem, as yet unresolved, though it may have to do with periodic events like big freezes at low tide, is how it manages to survive the *Thais* for two years. As a result, the pattern of distribution for *B. cariosus* is random clumps of large, older barnacles.

It is not known for certain what regulates *Thais* populations, but they are preyed upon by large *Pisaster* and are also vulnerable to periodic events such as freezing during winter low tides. Such periodic events may reduce the population by such a significant amount that the *Balanus cariosus* could successfully establish themselves in an area. This may explain the differences among various shores with respect to the abundance of this barnacle.

Pisaster ochraceus can consume *Thais* and barnacles of any size and is the primary predator of small and medium-sized *Mytilus*. It is the top predator in the system. Because of its ability to influence the structure of the entire community in the intertidal—through its ability to consume *Mytilus* and

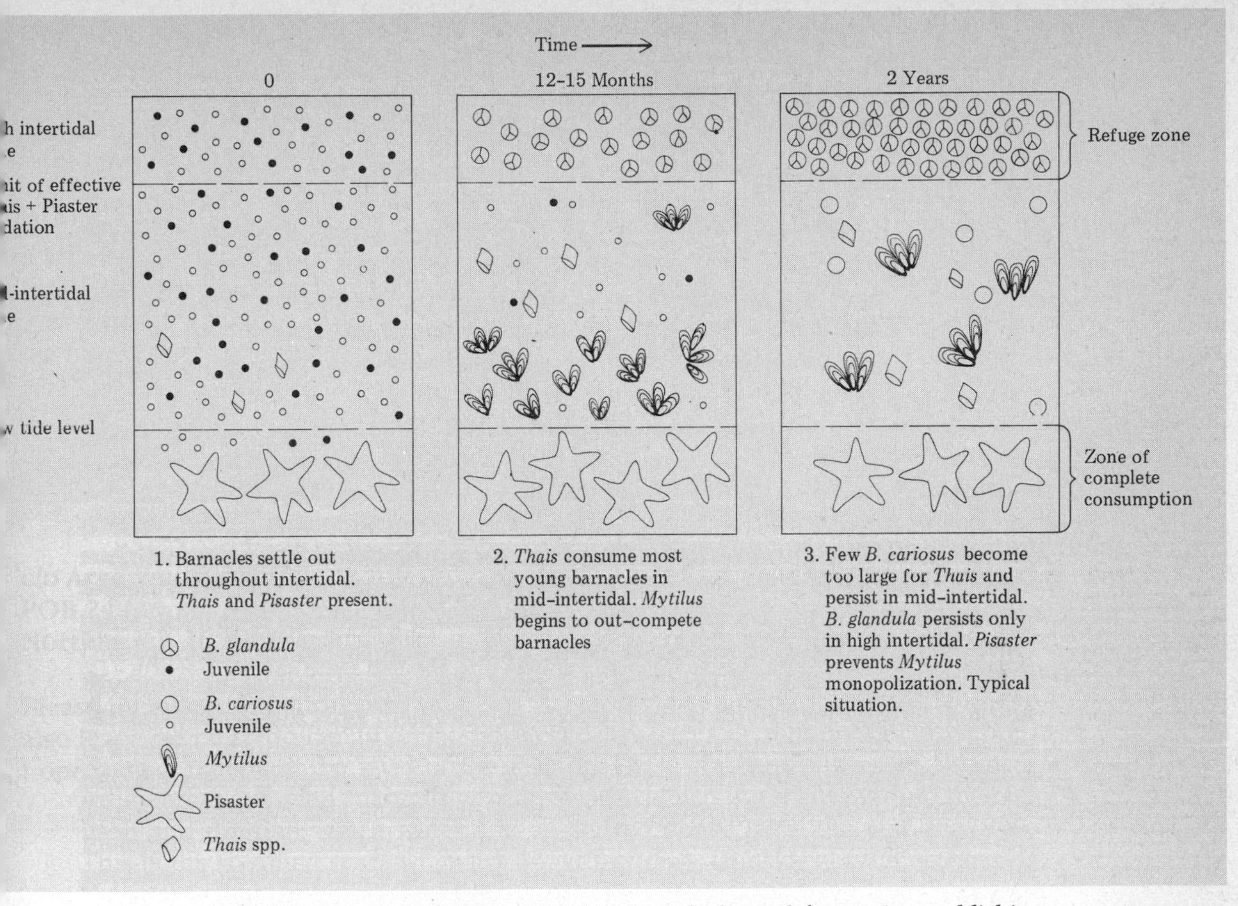

Figure 6-13 Interaction of predation and physical factors in establishing the zonation of the dominant intertidal organisms on the rocky shores of the Pacific coast of North America.

therefore prevent monopolization of the space—it has been called by Paine (1966) a *keystone species* as defined in Chapter One.

The reason that refuges from predators occur in the upper intertidal is that both *Pisaster* and *Thais* can feed only when the tide is in, and they require a long period to successfully attack their prey. The short period of immersion of the upper intertidal does not allow sufficient time for them to make successful forays into that area. In subtidal areas, unlimited time is available; hence, the starfish have sufficient time to attack and consume any prey item. It is probably for that reason that *Mytilus*, for example, do not extend into the subtidal areas, even though they can certainly live there; hence, the reason for the sharp lower boundaries to the zonation of this species (Fig. 6-13).

Lubchenco and Menge (1978) have also demonstrated that predation is important in setting zonal patterns in the intertidal of the North Atlantic Ocean. In the low intertidal of New England, the competitive dominant species is *Mytilus edulis*. *Mytilus edulis* is able to outcompete and eliminate the barnacle *Balanus balanoides* and the alga *Chondrus crispus*. It is prevented from doing this by the starfish *Asterias forbesi* and *Asterias vulgaris* and the snail *Thais lapillus*, all of which prey upon *M. edulis*. In the areas where these predators are absent, namely the most exposed, wave-beaten areas, *M. edulis* eliminates *Balanus* and *Chondrus*. Thus, the exposed wave-beaten areas exhibit a zonal pattern where the low intertidal is a *M. edulis* band, while in protected areas, a diverse lower zone exists with *Mytilus*, *Balanus*, and *Chondrus* present (Fig. 6-14).

The role of *grazers* or herbivores in regulating upper and lower limits of algal species is less well studied and documented, but evidence suggests this process may also be of considerable importance. The dominant groups of intertidal grazers are the various limpets, sea urchins, and littorine snails. In the tidepools and on the shores of New England studied by Lubchenco (1978), one of the dominant algae is *Chondrus crispus* (Irish moss). In the tidepools, but not on the exposed intertidal, it is outcompeted by the green alga *Enteromorpha intestinalis*, but it is able to persist in those pools that have a high population of the grazer *Littorina littorea*, because the latter preferentially consume *E. intestinalis*. Similarly, in England, when Jones (1948) removed all the limpets (*Patella vulgata*) from an intertidal strip, the strip was densely settled by algae and appeared in marked contrast to the surrounding barer areas. Since the only alteration was the

Figure 6-14 Composition of the low intertidal in the North Atlantic Ocean with and without mussel predators.

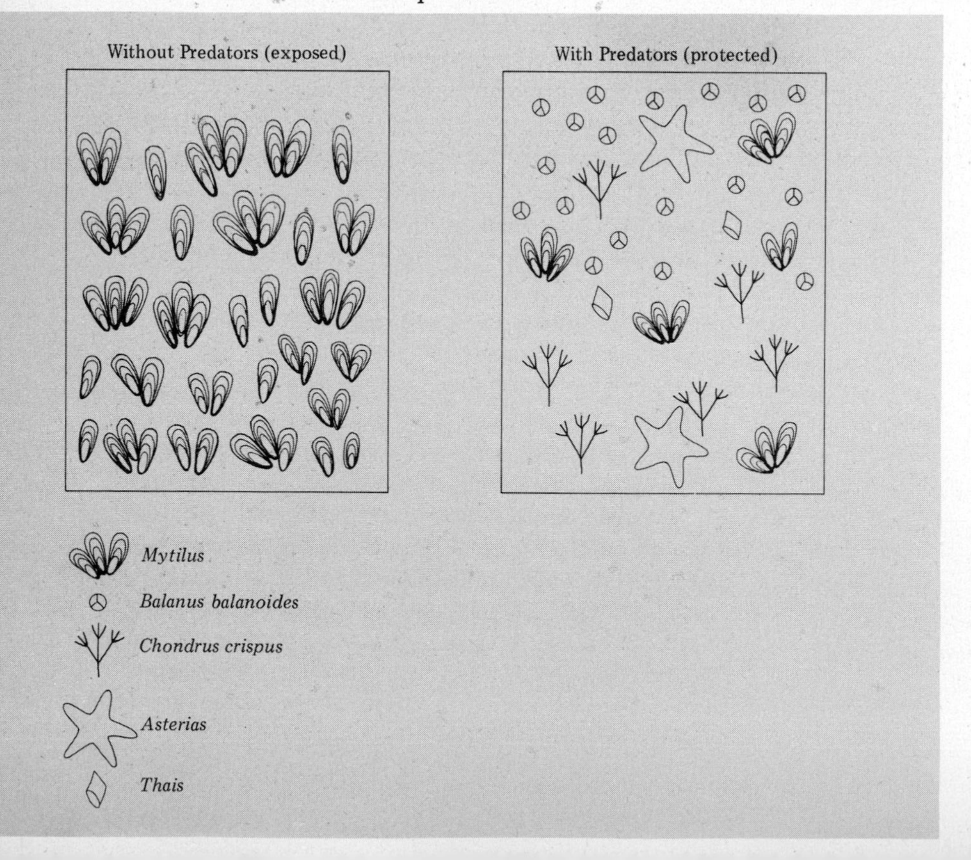

removal of the main grazers, this suggests their great importance in determining at least the abundance of algal species. In this same study, after a few years, the limpets again returned (via settlement) and the algae were consumed until the area appeared like the rest of the shore. In somewhat similar studies on the Pacific coast of North America, elimination of limpets (*Acmaea*, *Collisella*) from intertidal areas has resulted in heavy algal growth where none, or little, had been before.

Sea urchins are large grazers and often occur on shores in dense aggregations. Removal experiments with sea urchins in both the Atlantic Ocean (England) by Jones and Kain (1967) and the Pacific Ocean (Washington coast) by Dayton (1975) have had similar results. In both cases, removal resulted in dense growth of algae where there had been sparse or no growth before (Fig. 6-15). In all of these cases, the small, rapidly growing opportunistic algal species were the first to settle, and later, they were replaced by the dominant, larger algae. What is important here is that the algae that settle in these areas often differ from those in the immediately surrounding areas. This suggests that the urchins prefer to

Figure 6-15 Effect of sea urchin removal on kelp growth on the Isle of Man, Great Britain. (Modified from T. Carefoot, Pacific seashores, University of Washington Press.)

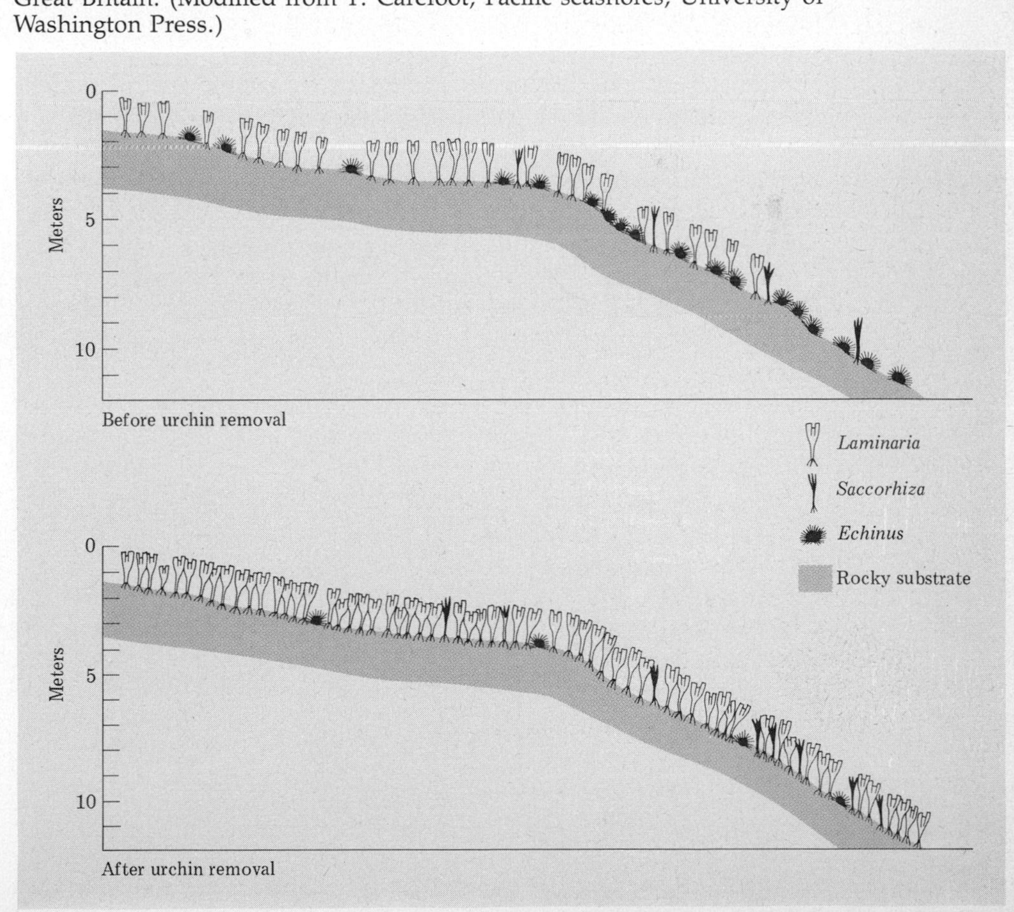

eat certain species, leaving others, and thus effectively alter the appearance of those areas of the intertidal that they occupy. In spite of this evidence as to their ability to remove certain algae and determine the composition of the zones, there is, as yet, no direct evidence that they set the limits to the distribution of any species.

SMALL-SCALE DISTRIBUTION PATTERNS

Although the major large-scale patterns of organism distribution in the rocky intertidal, the zonation, can be explained through some combination of exposure, desiccation, temperature, competition, predation, or grazing, acting separately or together, these vertical bands are never perfect. Often small-scale local variations occur that act to alter the bands and affect the immediate local distribution of certain organisms. Such small-scale or *local distribution* patterns can result from either biological or physical factors.

Perhaps the easiest of these local distribution patterns to explain are those that result from physical factors, primarily the variation in the nature, slope, and exposure of the rock surface. Existence of a different rock type in a local area as compared to the surrounding shore almost always means the presence of different organisms, or at least different abundances. An outcrop of sandstone among an otherwise granite rock shore usually means a different complement of organisms.

Similarly, slopes facing south and those facing north often have different organisms, or the densities and/or levels to which the organisms are found differ (see Fig. 6–8). This reflects the different amount of desiccation and harmful effects of direct exposure to the sun's rays (= insolation). On the Pacific coast of North America, for example, two limpet species, *Collisella* (*Acmaea*) *digitalis* and *Collisella* (*Acmaea*) *scabra* inhabit the high intertidal zone, but, as Haven (1971) has shown, their local distribution within the zone reflects their reaction to the small-scale changes in the substrate. *Collisella scabra* is found primarily on horizontal surfaces where it has a home "scar" or depression in which it remains during periods of low tide. *Collisella digitalis* has no scar and is found on the cooler, moist, shaded vertical slopes. *Collisella scabra* is able to survive on the drier horizontal rocks because of its depression or scar, in which the fit is tight enough to prevent desiccation at low tide. *Collisella digitalis*, on the other hand, without the increased desiccation protection of the scar, must live where desiccation is not as intense (i.e., north-facing slopes and crevices). Similarly, certain species do not prefer to settle on vertical faces or overhangs. These areas are then settled by other species unable to compete on the flat surfaces.

Much small-scale patchiness probably results from the destructive effects of wave action and wave-borne objects, such as logs, that may be hurled against the shore, smashing into small areas and clearing them of organisms. Such battering was studied by Dayton (1971) in the state of Washington, and it was found that any point on an open shore had a 5–30 percent chance of being hit by a log within a three-year period. An additional effect of wave action is that it produces aggregations of mobile species, such as snails and crabs, in crevices or other protected areas. Heavy wave action from storms, even if unaccompanied by battering of objects, also is responsible for removal of organisms from the intertidal. Thus, Menge (1976) found in the exposed intertidal in New England that winter storms could clear as much as 90 percent of the intertidal, whereas during summer and fall, less than 10 percent of the same space was free of organisms.

The premier space competitors, *Mytilus edulis* in New England, or *M. californianus* on the Pacific coast of North America, both tend to form large clumps that grow out from the substrate. As a result, as the clumps get larger, fewer mussels are attached directly to the substrate through their byssal threads and more are attached to each other. This creates an unstable situation in which the battering waves are more likely to tear off a clump, and hence, older, larger clumps become more vulnerable (Fig. 6-16).

Creation of such open spaces in any zone usually results in quick colonization by a series of *opportunistic* or *fugitive* species. These species quickly settle an open area, mature, and reproduce before the slower growing dominant species retake the open space and force them out through competition. The most common of these opportunists are the smaller filamentous green and red algae and smaller animals such as hydroids. These initial species are then replaced successively by competitively better species until the competitive dominant in the area takes over (Fig. 6-17). In any one zone, then, there may be many of these patches, which, depending on the time since their initial clearance, will be in differing stages of *succession* toward the dominant community characterizing the zone. This accounts for the differences one observes in different patches in the same zone in the same area.

Small-scale distribution anomalies may also be the result of biological interactions of various types. Some of these are *passive*, such as the fact that many species of animals must inhabit moist areas in order to survive the exposure to air and hence shelter under the fronds of various intertidal algae. The patchiness of these forms follows the distribution of the algae, but it is not due to any direct

Figure 6-16 Buildup and removal of mussel clumps by wave action.

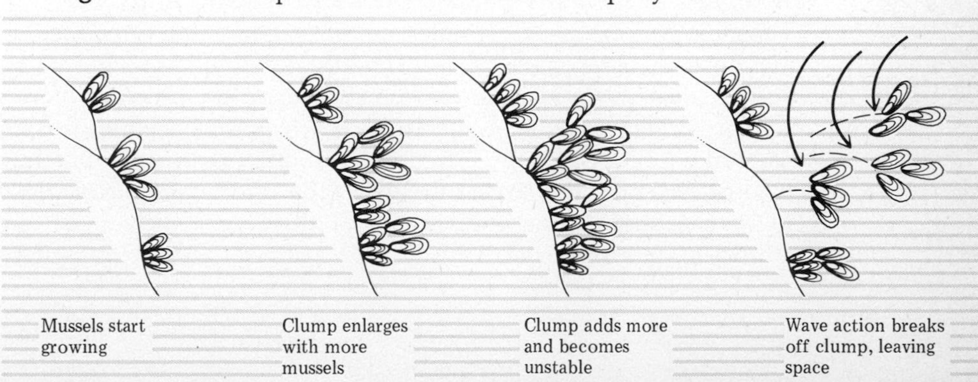

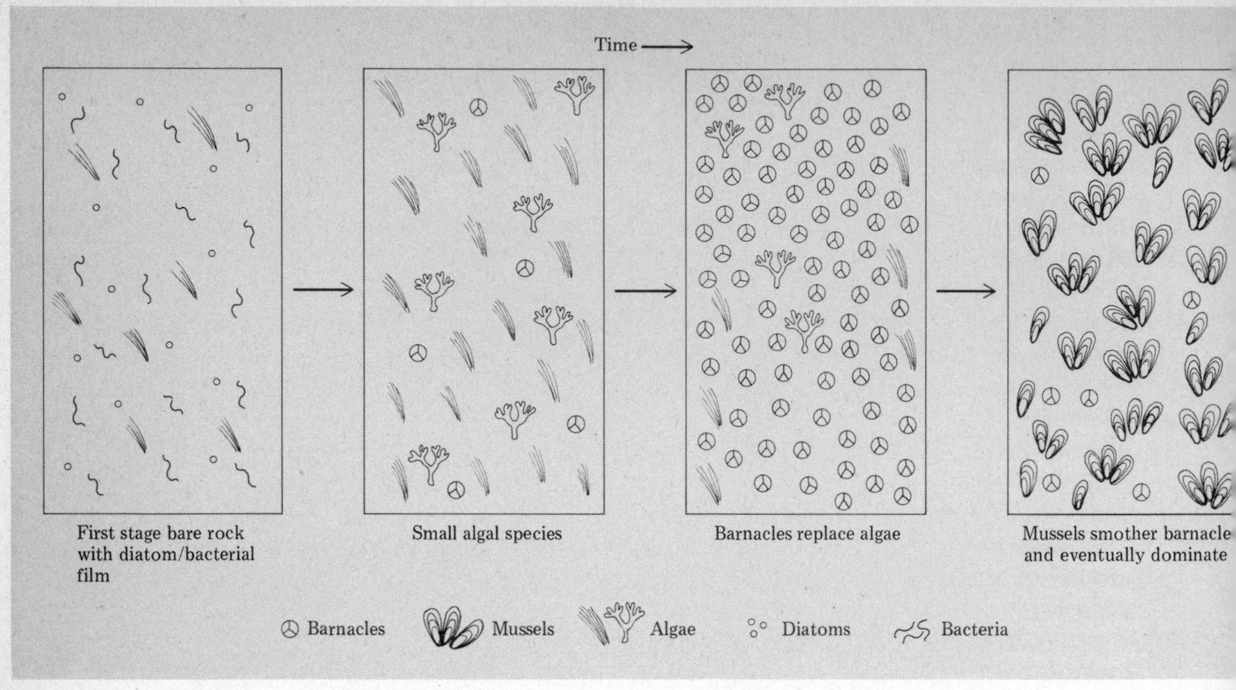

Figure 6-17 Succession in a Pacific coast intertidal mussel bed in the absence of *Pisaster*.

action of the algae. Removal of the algal canopy also results in the disappearance of the sheltering organisms.

Other biological effects are more direct and lead to definite distribution patterns, which can be observed. Thus, for example, the intertidal anemone *Anthopleura elegantissima* of the Pacific coast of North America tends to occur in large clumps, uniformly separated from each other. Each clump has resulted from the continued asexual division of the original anemone, such that each clump consists of genetically identical individuals, a *clone*. Francis (1973) discovered that the sharp boundaries that exist between adjacent clones are the result of active fighting between the outer individuals of each clones (Fig. 6-18). In these fights, special tentacles called acrorhagi are employed and are able to inflict massive tissue damage to other anemones. The result of such clonal "border wars" is the separation of clones by anemone-free strips, which are just the width that prevent two anemones from reaching each other. Thus, clumps of anemones are always separated from other such clumps.

Yet another example of biological interaction that produces spacing is the phenomenon of *territoriality*. Territoriality is defined as an individual or group of individuals actively keeping other members of the same or different species out of an area. This phenomenon is well known among various birds and mammals, but it also occurs among certain invertebrate organisms. On the north Pacific coast of America, the largest limpet in the intertidal is *Lottia gigantea*. Stimson (1970) has shown that individuals of this species create a territory in the intertidal, which they actively defend against other *Lottia* and other limpets.

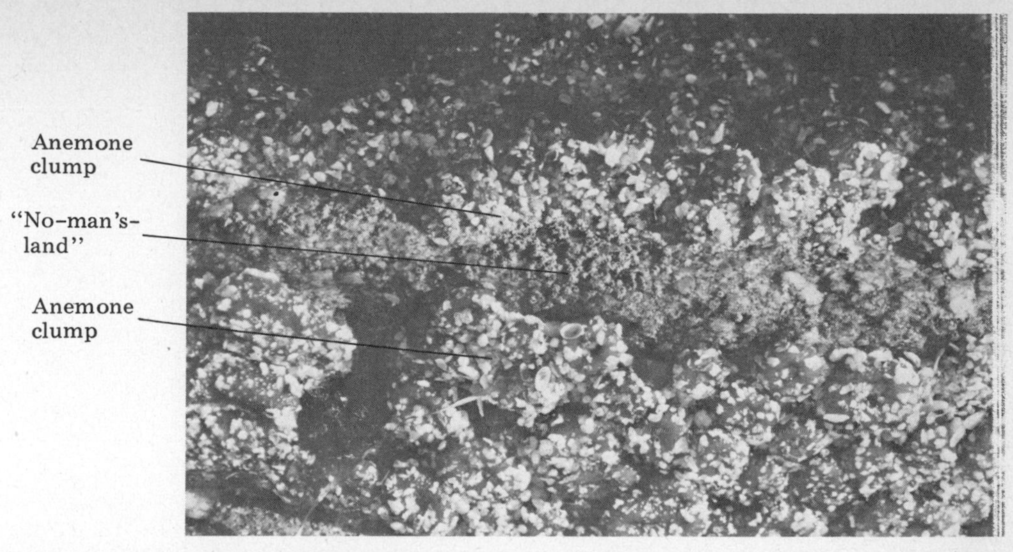

Figure 6-18 Clones of the anemone *Anthopleura elegantissima* showing the separation caused by fighting between adjacent clones. (Photo by the author.)

Should other limpets invade this territory, the resident *Lottia* pushes the interloper out. In this way, each *Lottia* keeps a small patch of rock free of potential competition and ensures that all algae growing on the site will be consumed by itself.

Another type of interaction producing small patches is that observed by Dayton (1973) in the alga *Postelsia palmaeformis*, which occurs in isolated clumps in the most wave-swept portions of the Pacific coast (see Plate 17). This alga creates its own patch by sacrificing a portion of its offspring. Young *Postelsia* plants always settle close to the parent plants. Usually, this means settling on various barnacles, algae, or *Mytilus*. As they grow in the wave-swept area, they eventually produce so much surface of resistance to the wave battering that the young are pulled free at the same time, removing the barnacle, alga, or *Mytilus* to which they were attached. This creates a new open area, which is then invaded by a second wave of *Postelsia* sporelings, which this time attach directly to the rock surface and grow up into adults. Thus, the space around a *Postelsia* patch is maintained, and competitive dominants do not force *Postelsia* out (Fig. 6–19).

Predation and grazing can also create small patches or patchy distribution. Predation is the reason for the occurrence of *Balanus cariosus* (see p. 220) in small clumps in the intertidal; grazing by sea urchins may result in tidepools free of algae, while others without urchins have an abundant growth. On the New England coast, as noted above, tidepools without *Littorina littorea* are dominated by *E intestinalis* whereas those with the snail have *Chondrus crispus*.

A final biological factor contributing to the patchiness, both large- and small-scale, is the preference for settlement shown by various intertidal invertebrate larvae. Most intertidal invertebrates have a larval stage that is free swimming or floating, which serves to allow dispersal of the otherwise sessile or

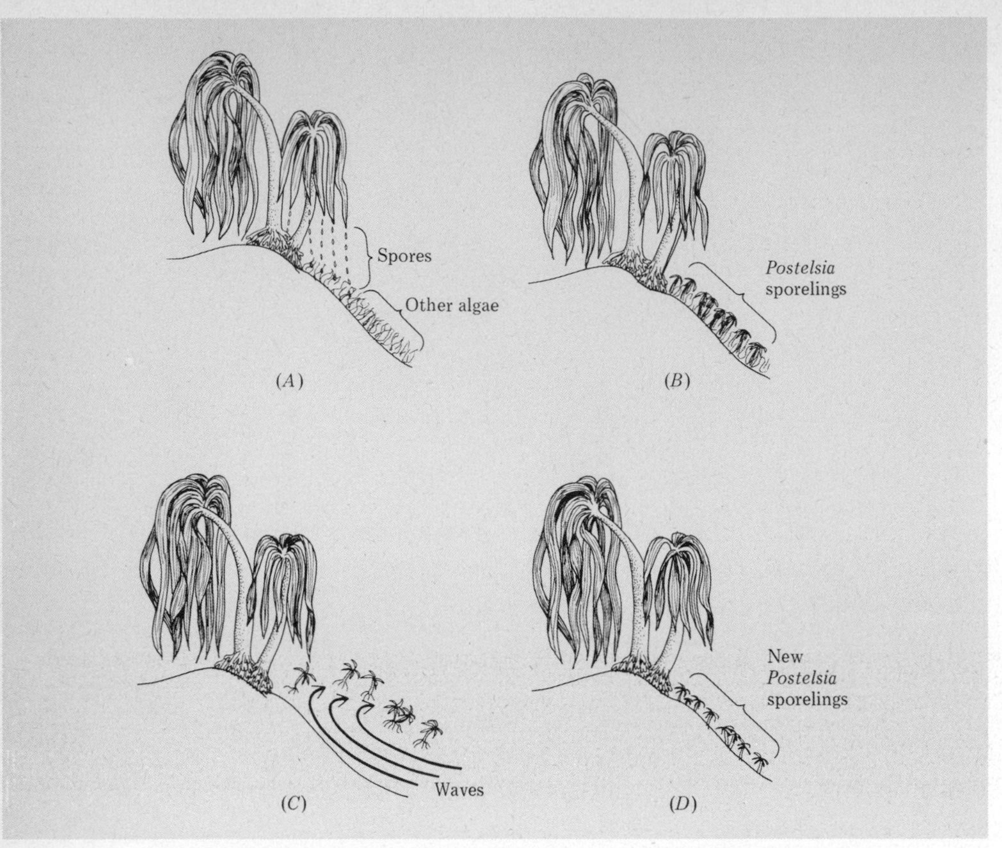

Figure 6-19 Mechanism of perpetuation of a *Postelsia* patch on a rocky shore. (A) Spores released. (B) First crop of young *Postelsia* grow up on top of barnacles and mussels. (C) Wave action carries away first crop of *Postelsia* along with barnacles and mussels. (D) New crop of *Postelsia* grow up on newly freed space. (Modified from T. Carefoot, Pacific seashores, University of Washington Press.)

sedentary adult (Fig. 6-20). Over the years, a great number of studies have been done on various invertebrate larvae (see Chapter Five for a discussion).

INTERACTIONS AMONG FACTORS: A SUMMARY

We have seen how various physical and biological factors each act to cause the large-scale vertical distribution patterns of algae and invertebrates that we call vertical zonation on rocky shores. We have also seen how some of these same factors, plus others, act to introduce small-scale distribution anomalies into this pattern. In general, different factors tend to operate in different areas and at different levels on the shore. The high intertidal tends to have the distribution pattern of its characteristic species set by physical factors, whereas biological factors become more important in setting limits in the low intertidal. It is important to realize that these factors do not act in isolation and the final distribution pattern observed, either on a small or large scale, is often the result of the interaction among two or more factors.

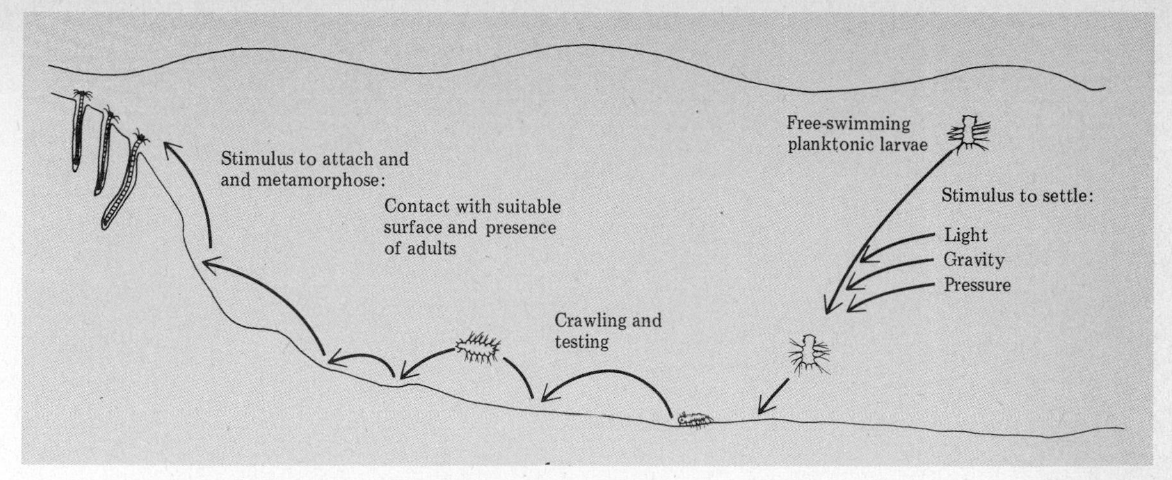

Figure 6-20 Pattern of selection and settlement of a planktonic larva of a benthic intertidal invertebrate. (Modified from T. Carefoot, Pacific seashores, University of Washington Press.)

In the case of *Chthamalus stellatus* and *Balanus glandula*, without a refuge in the highest intertidal (physical factor), both would be eliminated from the intertidal, the former by competition and the latter by predation (biological factor). Similarly, persistence of *B. cariosus* in the mid-intertidal of the Pacific coast depends upon the predation of *Pisaster ochraceus* for, without this predator to remove *Mytilus* and *Thais*, *B. cariosus* would be outcompeted by *Mytilus* or completely consumed by *Thais* (Fig. 6-21).

On the coast of Scotland, where *Balanus balanoides* and *Chthamalus stellatus* occur, the competition between the two barnacle species is ameliorated by the activity of the predator *Thais lapillus*, which consumes *Balanus*.

When several factors begin to act simultaneously, the picture becomes even more complicated and we can begin to appreciate the real complexities that act to give the observed patterns. One of the best examples is the aforementioned patterns worked out for the New England intertidal by Lubchenco and Menge (1978) and summarized here. In the low intertidal of New England, the dominant space competitor is *Mytilus edulis*, which can outcompete other space occupiers such as *Balanus balanoides* and the alga *Chondrus crispus*. On open, exposed rocky headlands, no predators exist for *M. edulis*, and hence, it outcompetes the barnacles and alga to dominate the low intertidal (Fig. 6-14). In protected areas, three predators occur: *Thais lapillus*, *Asterias forbesi*, and *A. vulgaris*, all of which prey upon *M. edulis*, with the result that *Chondrus* is able to persist because *M. edulis* never becomes abundant enough to monopolize the space. The only major herbivore in this system, *Littorina littorea*, prefers to graze on ephemeral algae that compete with *Chondrus*. Thus, the presence of the herbivore reduces competition by other algae on *Chondrus*, allowing it to persist and dominate protected areas. In this same system, the high intertidal is dominated by *B. balanoides*. Its lower limits are apparently set in exposed areas by competition with the lower-living *M. edulis* and in protected areas by predation by *T. lapillus*. The mid-intertidal in this system has a distribution pattern in exposed areas that is determined primarily by competition between *B.*

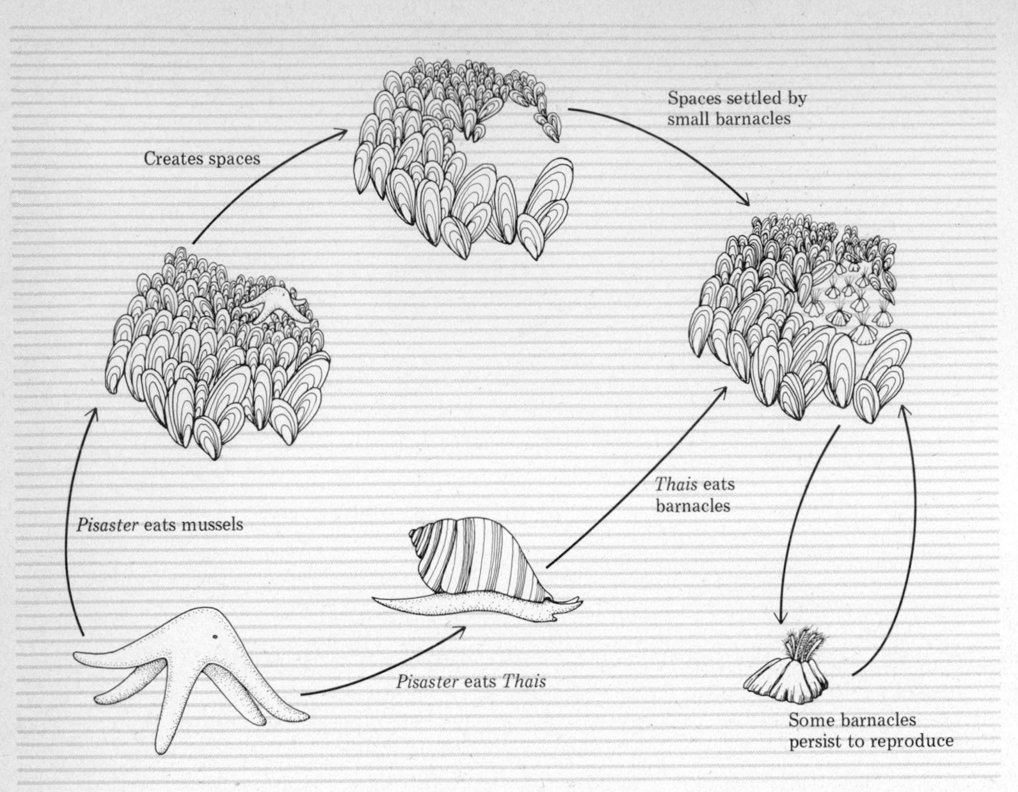

Figure 6-21 Interactions among mussels *(Mytilus)*, barnacles, and their predators on the Pacific coast of North America, which allow barnacles to persist in the intertidal zone.

balanoides and *M. edulis*, but modified by wave action. Thus, winter storms clear much open space, which is first settled by *B. balanoides* in the spring. The barnacles are, in turn, outcompeted by the *M. edulis*, which settle in the summer so that by the onset of winter again, the area is dominated by *M. edulis*. In protected areas, predators become abundant and heavy predation by *Thais* prevents either the barnacle or the mussel from monopolizing the space. Thus, in this relatively simple system, the final structure of the intertidal and its changes depend upon the interaction of wave action, predation, grazing, and several levels of competition. As we learn more about the intertidal in other parts of the world, it can be anticipated that more such examples will come to light.

HORIZONTAL DISTRIBUTION PATTERNS

As we noted in the preceding section, the appearance of the various vertical bands or zones changes as one moves from areas of high wave exposure to protected and quiet water. This is true not only for the New England example, but for other areas as well. Partly, this is due to the action of water restricting certain species in or out of wave-swept areas. In other words, some species of algae or invertebrates simply cannot tolerate the wave action and are absent from wave-swept areas. Usually, these are species with delicate structure or with poor means of holding on to the substrate. In a similar fashion, certain species cannot tolerate quiet water areas. This may be due to their inability to compete, to lack of enough or proper food, or to the presence of predators that are not present in wave-swept areas. Thus, we have seen that *Mytilus edulis* outcompetes *Balanus balanoides* and *Chondrus crispus* in exposed areas of New England because no predators can tolerate the wave action; in protected areas, predators reduce competition allowing the barnacle and algae to exist and thus altering the appearance of the zones.

SUCCESSION

Succession, as defined in Chapter One, is an ecological concept referring to the orderly progression of communities or *seres,* in which one community or association of species occupies a site and somehow modifies its conditions before the next one can take over. This replacement of one association by another continues until reaching the last or *climax* community, which persists indefinitely under a given climatic regime. Thus, in any one area, when open space appears, there is an orderly, predictable sequence of communities that develops on that site, which will be culminated by a known climax community. A fundamental requirement of this concept is that certain species must be present before others in order to modify the environment for subsequent ones. Does such a predictable sequence occur in the rocky intertidal?

Current thinking has it that succession, as defined above, does not occur in the rocky intertidal. When areas are cleared in the rocky intertidal, there is often a sequence of species that occurs, but in other instances, there is not (Fig. 6-17). The discrepancy appears to be with the size and location of the cleared area. If the area is small and near the dominant association for that part of the intertidal, it is quickly overgrown by the dominant. If, however, the area is large, there may well be a sequence of species occupying it before the dominant species again reoccupy the area. Usually, these initial species are those that we have termed opportunistic or fugitive, and they are gradually replaced by the better competitors. In this sequence, usually the first settlers on cleared areas are colonial diatoms or fast-growing small algal species. These are followed by barnacles and tube worms, which are followed by the dominant large algae and/or mussels. The reason for the sequence is not, however, that the earlier species must be present or modify the substrate, but simply that their larvae are more quickly produced and dispersed and hence, arrive first. *Mytilus,* for example, is perfectly capable of settling on newly cleared areas.

AGE STRUCTURE

As we noted in Chapter One, terrestrial communities or associations consist of a matrix of long-lived plant species, which allow us to characterize terrestrial communities on the basis of the dominant plant species. In the marine environment, especially the intertidal, the long-lived species tend to be various invertebrate animals, whereas the plants and the various algal species tend to be more short lived. We, in turn, tend to name zones in the intertidal on the basis of animals—*Mytilus* zone, barnacle zone, and so forth. Even though the potential longevity is great for various marine invertebrates, it is well to consider under what conditions such longevity is achieved and what the age structure of the animals actually is.

As noted previously, on the Pacific coast of North America, the high intertidal has a band of *Balanus glandula*; below that, any of these species found are usually less than 2 years old. If the age structure of the large barnacles in the band is analyzed, it is generally found to consist primarily of a single age class. Why? Because the physical conditions are so harsh in this area that most years, the juvenile barnacles simply do not survive. As a consequence, the barnacles there result from a single good year in which conditions allowed survival through the vulnerable juvenile stage. Once adult, the barnacles can survive succeeding harsh years. The reason that all of the lower level *B. glandula* are less than 2 years old is that in two years, *Thais* completely consumes all mid-tide barnacles. Similarly, *B. cariosus* populations in the mid-intertidal consist primarily of animals less than 2 years old. However, there are a few that may be large and many years old. Again, the majority are less than 2 years old, because that is the time period during which *Thais* consume them. The few large, old individuals are those that somehow avoided predation until they were 2 years old and thus invulnerable to *Thais*. Unless consumed by *Pisaster*, these individuals may live for many years. Thus, under very stringent environmental conditions or under optimal conditions low in the intertidal, the age structure of the dominant organisms is mainly of single-year classes.

Predators also seem to consist mainly of older individuals. Examples are *Pisaster ochraceus* and *Thais lamellosa*. Herbivores such as *Lottia gigantea* and the turban snail *Tegula funebralis* also have populations dominated by older individuals. Why this should be the case is not known at the present time.

TIDEPOOLS

A characteristic feature of many rocky shores is the presence of tidepools of various sizes, depths, and locations. Certain conditions affecting life of tidepools differ markedly from the surrounding intertidal and hence, necessitate a separate discussion here. Tidepools can, of course, occur at any level in the intertidal; however, low-level pools often are little different from the surrounding sea and are essentially extensions of the sublittoral. They will not be considered here. Only in midlittoral and infralittoral pools do we observe the conditions that characterize these areas. Our remarks here will be restricted to those pools that undergo a complete interchange with the ocean water during the tidal cycle (Fig. 6-22).

At first glance, tidepools would appear to be more or less ideal places for aquatic organisms seeking to escape the harshness of the intertidal during its exposure to air. In reality, however, escape from certain physical factors such as

Figure 6-22 A high intertidal rock pool at low tide. (Photo by the author.)

desiccation may mean exposure to others that operate more severely in tidepools.

Tidepools, of course, vary a great deal in size and, hence, volume of water which they contain. Since water is a great moderator of harsh physical conditions, the larger the pool and the greater the water volume, the less the fluctuations in physical factors.

Three major physical factors are subject to variation in tidepools. The first is temperature. Whereas the ocean itself is a vast reservoir that heats and cools very slowly and usually within very narrow limits, the same is not true of tidepools. These relatively small bodies of water are subject to considerably more rapid changes. Shallow tidepools exposed to the sun on warm days may quickly reach lethal or near lethal temperatures. Similarly, tidepools in cold temperate or subpolar regions may have temperatures in the freezing range in winter. An additional problem is this: The pool may either heat up or cool down over a several-hour period while exposed to the air, but when the tide returns, it will be flooded at some point with ocean water at quite a different temperature; this will suddenly change the temperature of the whole pool. Thus, any organism inhabiting such pools must be adapted to considerable sudden fluctuation.

The second factor to vary in pools is salinity. During exposure at low tide, tidepools may heat up. When that happens, evaporation occurs and the salinity increases. Under hot tropical conditions, the salinity increase can be dramatic enough in some instances to reach the point of precipitating out salt. The opposite situation is the case when heavy rains occur at low tide and flood pools with fresh water, dramatically lowering the salinity. Again, tidepool animals and plants may have to be adapted to wider ranges in salinity than typical marine or intertidal organisms. As before, when the tide returns, the pool will be flooded with sea water at some point and again, there will be a sudden, abrupt return to normal conditions.

The final physical factor undergoing change in pools is oxygen concentration. Since the amount of oxygen that can be held in sea water is a function of temperature, it follows that those pools that heat up during exposure to the air will lose oxygen. Under normal conditions, this may not be serious enough to produce oxygen stress, but if the pool is crowded with organisms, it may well produce a stress situation. For example, a pool filled with algae that was exposed at night would produce a situation in which the lack of photosynthesis coupled with high respiration could reduce the oxygen level significantly. Oxygen has been recorded as falling to but 18 percent of saturation in tropical tidepools.

Tidepools are thus areas of refuge from desiccation for intertidal organisms, but, in turn, these organisms suffer from rapid changes in temperature, salinity, and occasionally oxygen, thus restricting the fauna and flora to those organisms that are able to tolerate such ranges.

SANDY SHORES

Intertidal sand beaches are common throughout the world and are certainly better known than rocky shores to most humans, because they are the substrate of choice for various recreational activities. At the same time, they present a very marked change from the previously described rocky shores because, in contrast to the crowded life on the latter, sand beaches appear devoid of macroscopic life. They are not, of course, but the environmental factors acting on these shores create conditions where virtually all organisms bury themselves in the substrate.

In this section, we shall explore the action of physical factors on sand beaches and see how these factors act to impose certain adaptations upon the larger organisms found in this area. The very specialized conditions and organisms existing in the tiny spaces between the sand grains are the subject of the next chapter.

ENVIRONMENTAL CONDITIONS

As with rocky intertidal shores, sand beaches are subject to a similar array of physical factors, but the relative importance of these factors in structuring the community and their effect on the substrate differ.

Perhaps the most important physical factor governing life on sand beaches is *wave action* and its attendant effect on *particle size*. Anyone who has visited marine beaches is aware of the fact that the particle size of the sand can differ among beaches and also among seasons on the same beach. Particularly in the temperate zone, the profile of the beach also changes between winter and summer. The particle size of sand on a beach is a function of wave action on that beach. Where wave action is light, the particles are fine, but where wave action is heavy and strong, the particles are coarse, forming deposits called gravel or shingle rather than sand. The importance of the particle size to organism distribution and abundance rests with its effect on *water retention* and its suitability for burrowing. Fine sand, through its capillary action, tends to hold much water above the tide level in its interstices after the tide has retreated. Coarse sand and gravel, on the other hand, allow water to drain away quickly as the tide retreats. Since the organisms inhabiting the intertidal are aquatic, they are well protected against desiccation in a fine sand beach but subject to desiccation in a coarse gravel beach. This makes the latter less hospitable. Fine sand is also more amenable to burrowing than coarse gravel (Fig. 6-23).

A second important physical factor in sand beaches, which was not of concern in rocky shores, is also the product of wave action. That is wave-induced *substrate movement*. The particles on sand or gravel beaches are not large enough to be stable when waves strike the shore. As a result, with each passing wave, the substrate particles are picked up, churned in the water, and redeposited (Fig. 6-24). Particles are thus being continually moved and sorted. The reason that fine sand beaches occur only where wave action is light and coarse ones where it is heavy is that, in heavy wave action, the smaller particles remain in

Figure 6-23 Comparison of the physical conditions between fine-grained and coarse-grained beaches.

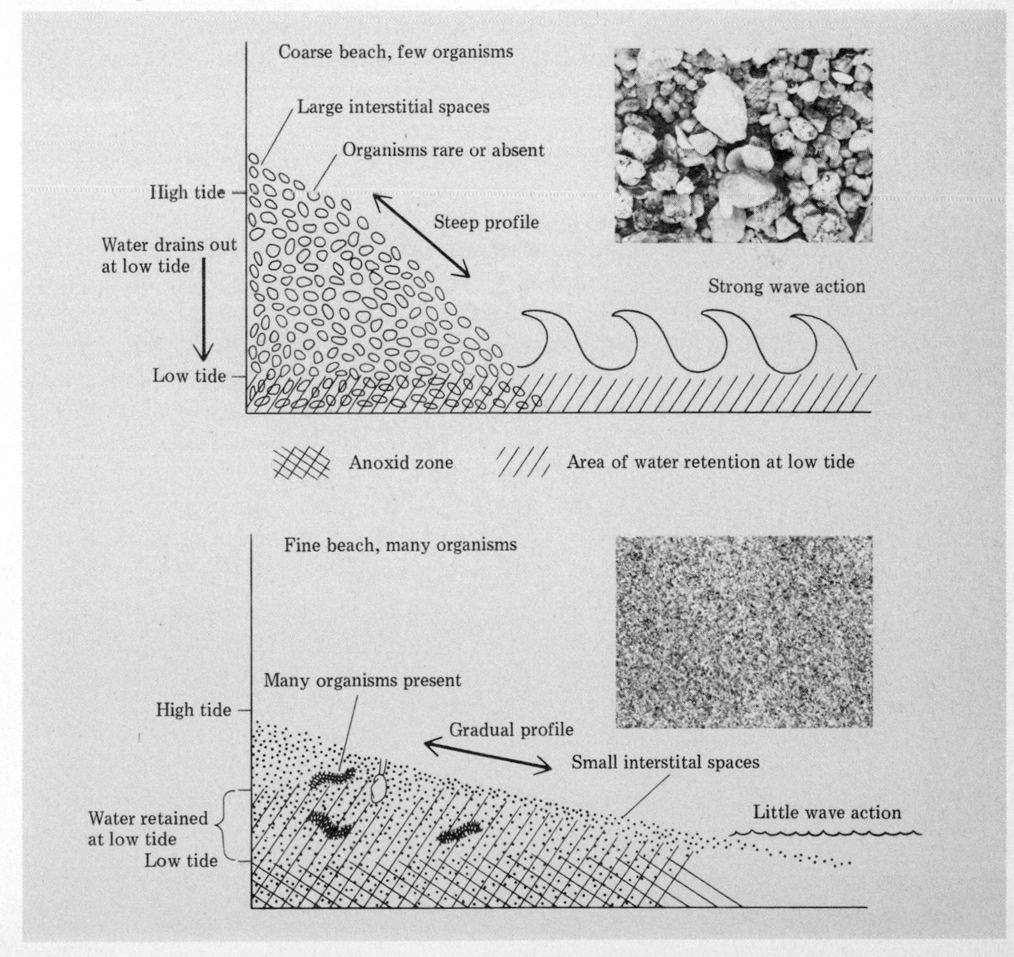

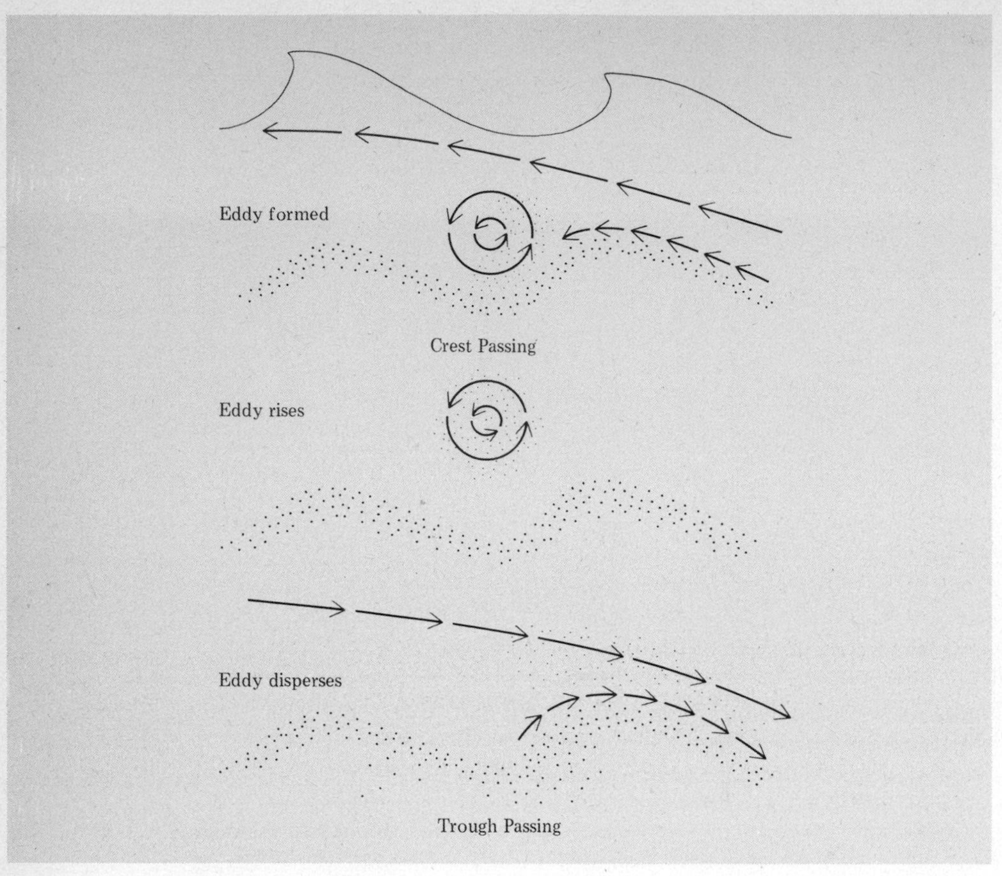

Figure 6-24 Transport of sand particles by wave action. (After Hedgpeth, 1957.)

suspension so long that they are carried away from the beach, leaving only those heavy enough (= coarse) to settle immediately. Light wave action also means that a smaller depth of particles is affected by the passage of a wave, whereas heavy wave action disturbs the substrate to a greater depth. Many beaches show a gradation of particle sizes from fine near low water to coarser at the high tide mark. This reflects the distribution of wave energy on the beach, being greater near the high tide mark, where waves break more often, and lesser with increasing depth in the tidal zone.

Since the substrate itself, at least the surface layers, is in constant motion and may be picked up and held for greater or lesser amounts of time in suspension, depending upon the strength of the waves; it follows that any change in wave intensity on a given beach will mean a change not only in grain size, but also in the profile or shape of the beach. This comes about when gentle wave action is replaced by heavier waves, which dig deeper into the beach, pick up more and heavier grains, and hold them in suspension longer, so they are deposited some distance away, usually offshore. This changed profile is a common seasonal occurrence on many temperate zone beaches, where a gentle slope of fine sand occurs during the summer months and is replaced by a steep, coarse beach during winter storms (Fig. 6-25). Significant for organisms is the fact that the depth of substrate that may be physically moved can be several feet!

Because of this condition of ceaseless movement of the surface layers of the sediment, few large organisms have the capability of permanently occupying the surface of sand or gravel beaches. This is the reason for the barren appearance of such beaches.

In contrast to rocky shores, sand beaches usually have a rather smooth, uniform profile and thus lack the great topographic diversity of rocky shores. There are, for example, no crevices, no slopes facing different directions and offering different moisture conditions, nor are there overhangs or tidepools. As a result, the environmental factors such as temperature, desiccation, wave action, and insolation act uniformly at each tidal level on the beach.

Wave action is the dominant environmental factor acting on the sand beaches, creating the special conditions that make it difficult, or impossible, for many organisms to inhabit this area. At the same time, sand beaches offer some positive advantages to marine organisms with respect to other physical factors. Sand is an excellent buffer against large temperature and salinity changes. Measurements of temperature with depth on sand beaches at low tide have demonstrated that, below the first few centimeters, the temperature is very nearly that of the surrounding sea water. This amelioration of temperature is partly due to the insulating properties of sand itself and partly due to the water held in the interstices of the deeper layers. In a similar manner, salinity changes are also minimal below 10 or 15 cm on beaches, even if the upper layers have fresh water flowing over or falling on them. Again, this is due to the fact that the water held in the interstices is salt water, which has a higher density than fresh water. As a result, the fresh water remains perched on the surface. Sand is further a barrier to any harmful effects of exposure to direct sunlight (insolation)

Figure 6-25 Beach profiles showing the difference between winter and summer.

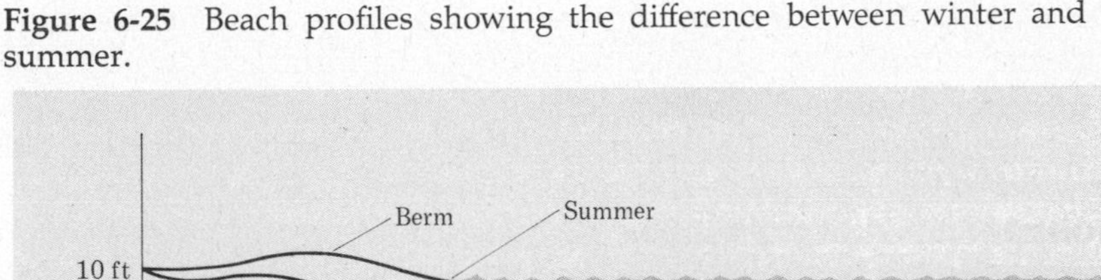

for any organism living in the sand, simply because the sand is opaque to light and reflects or absorbs it in the surface layers. Desiccation is not a problem as long as the beach sand is fine enough to hold water by capillary action during periods of low tide. Thus, any organism burrowing into the sand will be constantly moist.

A final physical factor acting on beaches is *oxygen content*. Oxygen is never limiting in the water bathing the beach because the turbulence of wave action ensures constant saturation. Where it might become limiting is in the substrate itself. The water held in the beach and responsible for ameliorating the changes in temperature and salinity also contains oxygen, which is available for use by the organisms. This supply, however, is used up by respiration of the organisms and must be replenished. This replenishment comes through interchange of the water with the sea above, and is dependent on the fineness of the sediments. Fine sediments have a slow rate of exchange and coarse sediments a rapid exchange. Thus, in fine-grained beaches, water interchange is slow and may result in reduced oxygen supplies.

ADAPTATIONS OF ORGANISMS

The dominant environmental factor acting on sand beaches is wave action, which creates the unstable, constantly moving substrate. In order to inhabit this area, organisms must first be adapted to tolerate these features. There are two routes that may be taken by organisms in adapting. The first is to burrow deeply enough into the substrate such that the organism is deeper than the depth of sediment affected by the passing wave. This strategy is employed by many large clams such as *Tivela stultorum*, the Pismo clam (Fig. 6-26). Such animals are usually also aided by developing a very heavy shell, which helps in keeping the animal down in the substrate. The only difficulty with this type of adaptation is that a very severe storm may generate waves large enough to pull these animals out and throw them on the beach. This is the reason for the catastrophic destruction that occasionally occurs in these forms.

The second route of adaptation is the ability to burrow in very quickly as soon as the passing wave has removed the animal from the substrate. This appears to be the more common mechanism and is employed by many annelid worms, small clams, and crustaceans. A good example of this type of adaptation can be found in the various sand crabs of the family Hippidae found in many beaches around the world. These animals have a short body with limbs highly modified to dig quickly into wet sand. As soon as they are pulled out of the substrate by a passing wave, they reburrow before the water motion can carry them off. Small clams of the genus *Donax* also do the same thing, and razor clams of the genus *Siliqua* are extraordinarily fast at reburrowing.

Other adaptations are correlated with the above. For example, most burrowing mollusks tend to have very smooth shells that reduce the resistance to burrowing into the sand. Similarly, shore echinoderms such as sand dollars have much reduced spines to allow them to burrow into the sand.

Figure 6-26 Some sandy beach animals of the Pacific coast of North America. *(A) Tivela stultorum* (Pismo clam). *(B) Olivella biplicata*. *(C) Blepharipoda occidentalis* (sand crab). (Photos by the author.)

A special adaptation observed by Chia (1973) in small sand dollars (*Dendraster excentricus*) is that they tend to accumulate iron compounds in a special area of their digestive tracts, which serve as a "weight belt" to keep them down in the presence of wave action.

A final set of adaptations has to do with the problem of preventing clogging of respiratory surfaces by the resuspended sand. In order to prevent this from occurring, the intake siphons of sandy beach clams are often fitted with various screens, which prevent entrance of sand but allow passage of water. Similarly, in the sand crabs, the antennae held together form a tube to the surface through which water enters the branchial chamber. These are densely clothed with closely spaced hairs designed to prevent entrance of sand.

TYPES OF ORGANISMS

The most noticeable group that is absent from sand beaches is the large plants. No macroscopic plants occur on open sand beaches, presumably because there is no way for them to attach and maintain themselves in the wave action. Small benthic diatoms, however, may be present on the sand grains. The second major group of organisms conspicuous by its absence is the sessile animals such as barnacles and mussels, so dominant on rocky shores. Again, there is no place for them to attach. The sand beach is dominated by representatives from three invertebrate classes: polychaete worms, bivalve mollusks, and crustaceans. Various combinations of species from these three groups dominate sand beaches throughout the world.

FEEDING BIOLOGY

The absence of large multicellular plants means that there is very little primary productivity. Although diatoms are present, the opaqueness of the sand ensures that even the diatom population is restricted to the very surface layers. As a result, there are no macroscopic herbivores on a sand beach. Since there is virtually no primary productivity, the animals living on the beach must depend for food upon either the phytoplankton carried in sea water above the beach, the organic debris brought in by the waves, or consuming the other beach animals. There are relatively few true carnivores among beach animals, because to be such would require active movement across the substrate in search of prey. This is probably not feasible in the face of the heavy wave action. As a result, most of the organisms in a beach are either suspension or detritus feeders.

Suspension feeding animals are those that filter particles out of the water column. These particles may be plankton organisms or, particularly on the sand beach, they may be various organic particles resuspended from the bottom by the passing waves. The dominant group of suspension feeders on sand beaches are the bivalve mollusks such as razor clams (*Siliqua*, *Ensis*), surf clams (*Tivela*, *Spisula*), and coquinas (*Donax*).

Sand beaches tend to have lesser amounts of organic detritus than muddy shores, but enough debris from various sources finds its way to the beach to be a reliable source of food for certain organisms. Since this detrital material is often carried up and down the beach suspended in the wave wash rather than being deposited on the bottom, the mechanisms employed by detritus feeders on the beach are often different from those employed by detritus feeders on muddy shores.

The sand crabs of the family Hippidae, such as *Emerita analoga* and *Blepharipoda occidentalis* of the Pacific coast of America, employ a unique mechanism to trap debris in the wave wash (Fig. 6-26). MacGinitie and MacGinitie (1949) observed that *Emerita* orient so that their heads point shoreward. They remain completely buried as the incoming wave passes over them shoreward. After passage, as the water begins to run down the beach, they stick out their very large second antennae. These are highly clothed in hairs forming a net that, when spread into the backwash, intercepts all particles within a certain size range. The particles are then wiped off by the mouth and ingested. As the tide rises and falls, the position of the wave wash front also moves. These crabs take advantage of this to maximize their feeding time by being moved up and down the beach with the tide. Movement is effected by surfacing as a wave rolls in and being carried up some distance by the wave and then quickly reburrowing. In a similar manner, as the tide recedes, they emerge and roll down the beach before reburrowing and commencing feeding.

A similar method of feeding is employed by the gastropod mollusk *Olivella columellaris* on the coast of Central America. In this case, a mucus net is produced by the animal and held across the wave backwash. This net acts like a filter and captures small particles. Since the animal feeds by consuming the entire net with particles, it must produce a new net at frequent intervals.

Most sand beaches also include one or more polychaetes, which burrow through the sand ingesting the sand and digesting out the organic particles. Much like earthworms on land, however, they are rarely abundant on open sand beaches because the amount of organic material is less than in mud shores.

Sand dollars, one of the echinoderms common on sandy beaches, are generally deposit or detritus feeding animals. These animals generally burrow through the sand. As they do so, the finer organic particles, but not the sand grains, fall down between the short spines, are trapped in mucus, and carried to the mouth. Some species, such as the Pacific coast *Dendraster excentricus,* according to Timko (1976), have even modified this basic feeding process to remove suspended particles by projecting part of the body up into the water column.

Carnivores are rare but the few found usually are polychaete worms or gastropod mollusks. The polychaetes, such as species of the genera *Nephtys* and *Glycera,* actively burrow through the sand and seize small prey items with their eversible proboscis. The carnivorous gastropods are represented most commonly by burrowing snails of the family Naticidae (moon snails). These animals burrow slowly just below the surface layers of the sand in search of bivalve mollusks, which they consume by boring holes through the shell and eating out the contents. A number of transient predators also occur. At low tide, various shorebirds actively feed on beach organisms, and at high tide, fishes invade the area (Fig. 6-27).

ZONATION AND COMMUNITY ORGANIZATION

Since there are no organisms visible on the surface of the intertidal sand beaches, these areas do not display the surface zonation patterns so obvious on rocky shores. Furthermore, since the various physical factors such as desiccation and temperature that act to limit organisms on rocky shores are of far less importance to organisms living in the sand, it might be suspected that zonation of organisms would be reduced or absent. In fact, there is a zonation of sandy shore organisms, but it is neither as clearly defined nor as well understood as that for rocky shores. The lack of clear zonation patterns is due partly to the habit of some dominant organisms moving up and down the beach for feeding

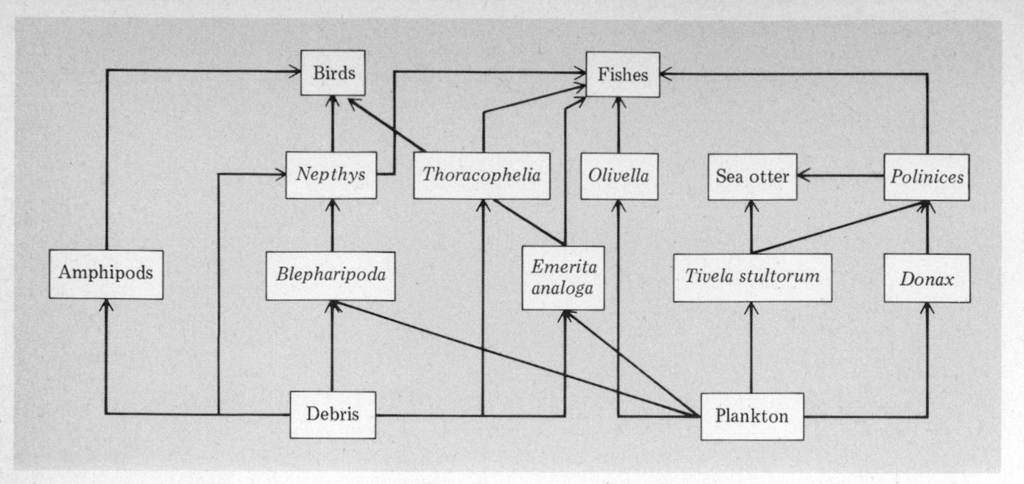

Figure 6-27 Generalized food web for a sand beach in California.

and partly to the dearth of studies on the determinants of distribution patterns in this area. Certainly one of the major contributors to the lack of understanding is the inability of researchers to set up field experiments in this area to determine the results of excluding certain organisms. Wave action precludes such work.

From the few data that are available, it is possible to establish a three-zone division of most sandy beaches that follows quite closely the universal scheme of Stephenson's, mentioned earlier. The highest parts of the sandy beach, corresponding to the supralittoral fringe, are usually inhabited by talitrid amphipod crustaceans (beach hoppers) in the temperate zone and by the fast-moving ghost crabs (*Oxypode*) in the tropics. Both groups excavate burrows and are scavengers. The broad mid littoral area, which corresponds to the area inhabited by barnacles and mussels on rocky shores, is much more variable. One group found here is isopods of the family Cirolanidae. Also coming into this zone when it is the area of wave swash are the various sand crabs and those other animals that feed in a similar way (see previous section). The lowest zone, or infralittoral fringe, is inhabited by the greatest number of species, including the large surf clams (*Tivela, Spisula*), sand dollars (*Dendraster*), various polychaete worms, a host of crustaceans, and the larger carnivorous snails (*Natica, Polinices*) (Figs. 6-28, 6-29).

In direct contrast to rocky shores, where experimental work has demonstrated the importance of physical factors, competition, and predation in determining the observed zonation patterns, the explanations for the establishment of the pattern in sandy beaches have not been subjected to rigorous experimental analysis. As a result, we know little of the factors or their interactions that establish the sandy beach distribution patterns.

The extremely crowded conditions for organisms observed on rocky shores are not found on sand beaches. Indeed, the fauna is very sparse and certainly not occupying all available space. This immediately suggests that competition for space is not a major contributor to the observed distribution pattern. Competition for food also appears negligible, since the sparse populations and

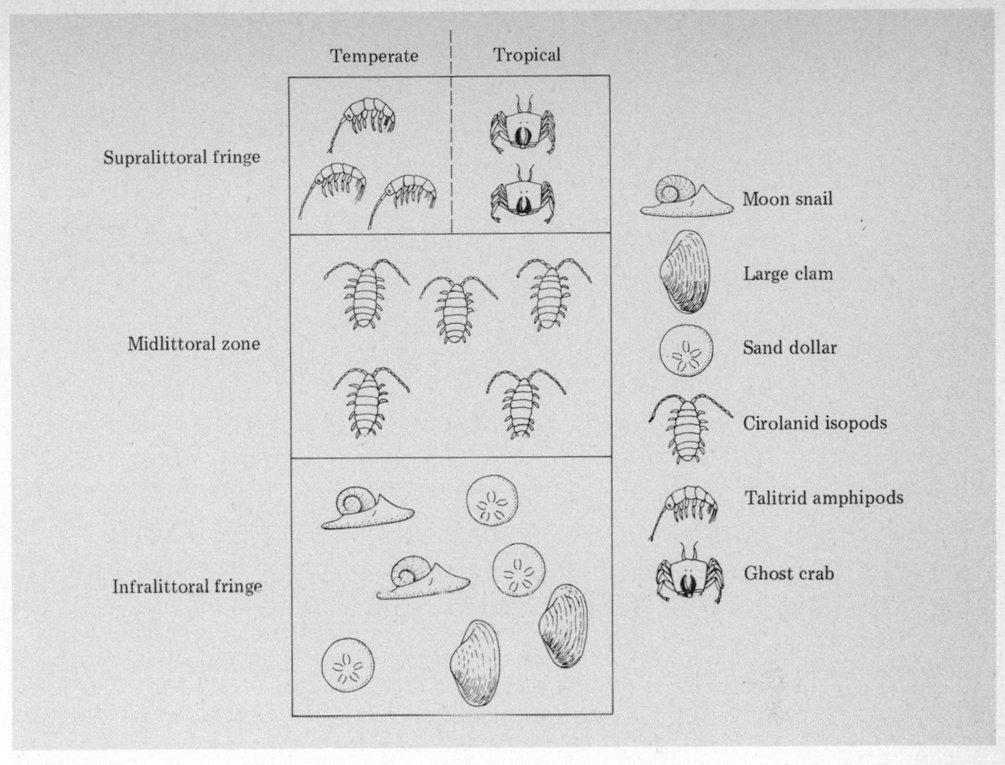

Figure 6-28 Generalized zonation patterns for sandy beaches.

abundant plankton make competition unlikely. Similarly, there are relatively few indigenous invertebrate predators, and it seems unlikely that they are responsible for any major distributional pattern. The effect, however, of large vertebrate predators such as birds and fishes remains to be investigated. This leaves the physical factors as probable major contributors to the patterns observed.

MUDDY SHORES

Whereas sharp boundaries exist between rocky shore and sandy shore, such that they are easily defined and recognizable, a similar situation does not prevail when trying to differentiate between sandy and muddy shores. Indeed, it is not possible to draw sharp boundaries between the two, because as shores become more protected from wave action, they tend to become finer grained and accumulate more organic matter and thus become more "muddy." Sand and mud shores are therefore the opposite ends of a continuum, the sand beaches having larger grain sizes and the muddy areas the finest grain sizes.

Since sharp boundaries do not exist and one grades into the other along a gradient of increasing protection from wave action, the fauna and flora also show a change from organisms typical of open sand beaches to those typical of muddy shores along the same gradient.

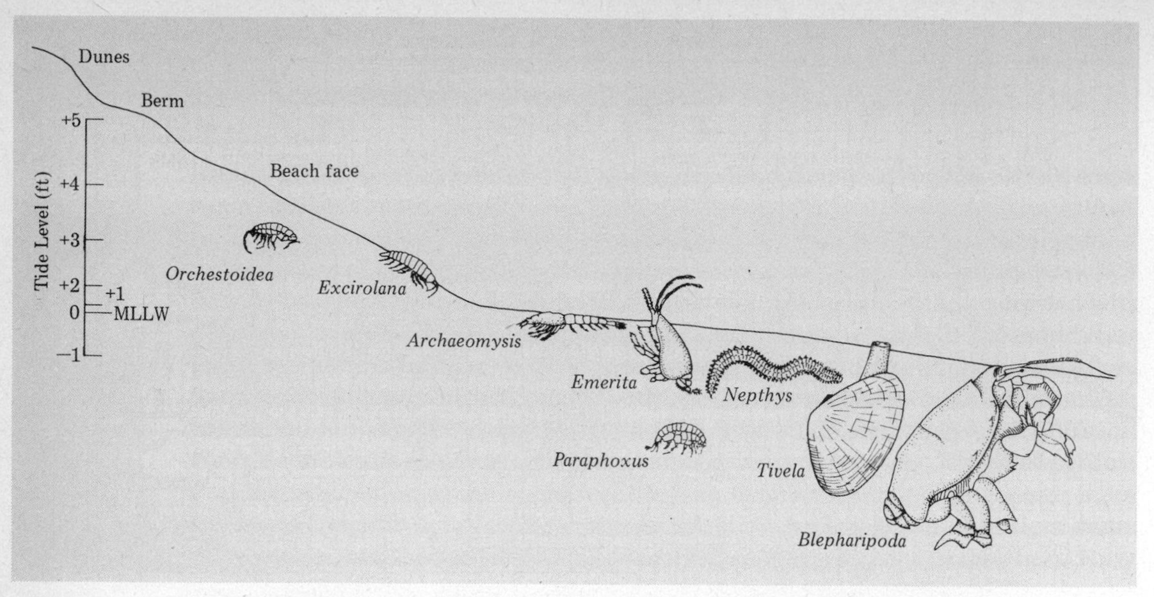

Figure 6-29 Zonation of the numerically dominant invertebrate genera on a central California intertidal sandy beach. *Orchestoidea, Paraphoxus* (Crustacea, Amphipoda); *Excirolana* (Crustacea, Isopoda); *Archaeomysis* (Crustacea, Mysidacea); *Emerita, Blepharipoda* (Crustacea, Decapoda); *Nephtys* (Annelida, Polychaeta): *Tivela* (Mollusca, Bivalvia). (Not to scale.)

This section will discuss the features characteristic of typical muddy shores; that is, those at the opposite end of the continuum from the open sand beaches. It should be borne in mind, however, that various transitional communities exist that will show mixtures of organisms from both extremes. Since muddy shores are also composed of sediments, similar factors will apply here as did with sandy shores and will not be repeated.

Muddy substrates are also characteristic of estuaries and salt marshes (Chapter Eight). Indeed, most of the muddy shores of the world are associated with estuaries and similar embayments. Hence, many of the types of organisms and adaptations are similar between the two areas. To avoid the problem of overlap in discussing these two areas, this section will treat the infaunal organisms of mud flats, their adaptations, feeding biology, and trophic structure. These considerations will, of course, apply to the estuarine infauna of Chapter Eight. The role of other organisms, productivity and ecological relationships are taken up in Chapter Eight.

PHYSICAL FACTORS

The major contrast with open sand beaches is that muddy shores cannot develop in the presence of significant *wave action*. Therefore, muddy shores are restricted to intertidal areas completely protected from open ocean wave activity. Muddy shores are best developed where there is a source of fine-grained sediment particles. Muddy shores are located in various partially enclosed bays, lagoons, harbors, and especially estuaries (see Chapter Eight for further discussion).

Since these areas are developed where water movement is minimal, the slope of mud shores tends to be much flatter than that observed for sand beaches. This is why these areas are often referred to as "mud flats."

The very fine particle size coupled with the very flat angle of repose of these sediments means that water in the sediments does not drain away and is held within the substrate. This long retention time for water, coupled with a very poor interchange of the interstitial water with the sea water above and a high internal bacterial population, usually results in complete depletion of the oxygen in the sediments below the first few centimeters of the surface. *Anaerobic* conditions thus prevail within the sediment and this is one of the most important characteristics of the muddy shore.

Muddy shores tend to accumulate organic material, which means that there is an abundant potential food supply for the resident organisms; but the abundant small organic particles "raining" down on the mud flat also have the potential to clog respiratory surfaces.

ADAPTATIONS OF ORGANISMS

As we noted for sandy shores, the surface of muddy shores is also often rather barren, as few animals inhabit the surface of the mud flat. Most organisms inhabiting these areas show an adaptation to burrowing into and through the soft substrate or else inhabit permanent tubes in the substrate. In contrast to sand beaches, however, the presence of the organisms in the mud flat is advertised on the surface by the presence of various holes of differing sizes and shapes. Thus, one of the primary adaptations of organisms in mud flats is the ability to burrow into the substrate or to form permanent tubes.

A second major adaptation concerns the anaerobic conditions that prevail in the substrate. If organisms are to survive while burrowed into the substrate, they must either be adapted to live under anaerobic conditions or else must have some way of bringing the overlying surface water with its oxygen supply down to them. Since most multicellular organisms cannot survive in the absence of oxygen, the latter is the more common adaptation. It is to obtain this oxygen-rich surface water and food that the various burrows, holes, and tubes appear on the surface of the mud flat.

Although most mud flat organisms are intolerant of completely anaerobic conditions, many have adaptations that permit them to exist at lower oxygen tensions than similar forms that live on open sand beaches. The most common adaptation to low oxygen supply is the development of carriers (e.g., hemoglobin) that will continue to pick up oxygen at concentrations well below that of similar pigments in other organisms (Fig. 6-30).

Since wave action is essentially absent on these mud flats, there is no necessity

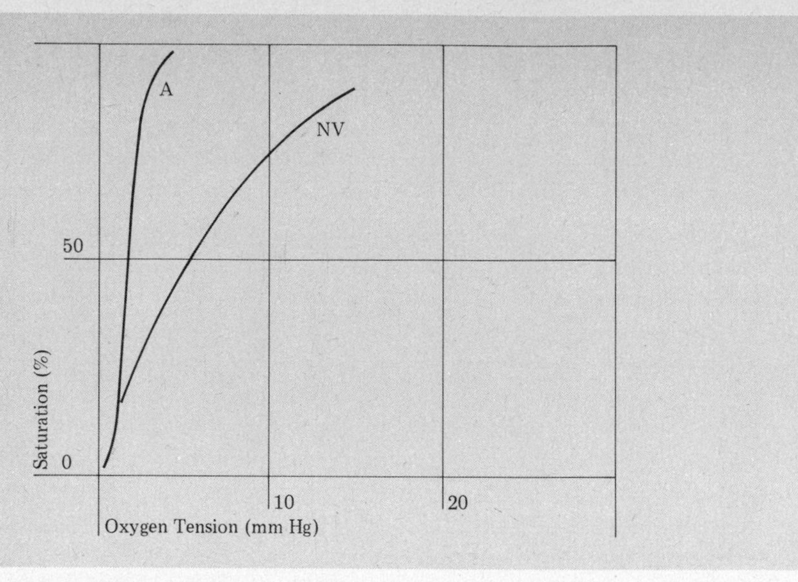

Figure 6-30 Oxygen dissociation curves for the blood of *Arenicola (A)* and *Nephtys (NV)*. (Modified from A. E. Brafield, Life in sandy shores, 1978, Edward Arnold.)

for development of either rapid burrowing or heavy bodies to maintain positions, as is the case for sandy shores.

TYPES OF ORGANISMS

In contrast to sand beaches, muddy shores often develop a substantial growth of various plants. On the bare mud flats, the most abundant plants are diatoms, which live in the surface layers of mud and often give a brownish color to the surface of the mud at low tide. Other plants include large macroalgae such as species of *Gracilaria* (red algae), *Ulva* and *Enteromorpha* (green algae). These large algae often undergo seasonal cycles of abundance, becoming common in the warmer months and virtually disappearing in colder months. Other areas, particularly the lowest tidal levels, may be covered with a growth of various sea grasses such as the genus *Zostera* (see Chapter Five for a discussion of these areas). As a result of the occurrence of these various primary producers, there is substantial primary productivity in the mud flats.

Mud flats contain large numbers of bacteria, which feed on the abundant organic matter. Bacteria are the only abundant organisms found in the anaerobic layers of the mud shore and constitute a significant biomass. Among the dominant bacteria inhabiting this area are a number capable of utilizing the potential energy of the various reduced chemical compounds abundant here. These *chemosynthetic* or *sulfur bacteria* obtain energy through the oxidation of a number of reduced sulfur compounds, such as various sulfides (for example H_2S). These organisms are thus primary producers of organic matter analogous to green plants. They produce organic matter using energy obtained from the oxidation of the reduced sulfur compounds, whereas plants produce organic matter using energy obtained from sunlight.

Since these autotrophic bacteria are located in the anaerobic layer of the mud, mud flats are unique among marine environments in that they have two separate layers in which primary productivity occurs: the surface, where diatoms, algae, and marine grasses carry on photosynthesis, and a deep layer, where bacteria conduct chemosynthesis.

The dominant macrofaunal groups on muddy shores are the same as those encountered on sand beaches; namely, various polychaete worms, bivalve mollusks, and small and large crustaceans, but of quite different sorts.

FEEDING BIOLOGY AND TROPHIC STRUCTURE

Because of the greater amount of organic matter present in and on muddy shores and because of increased productivity due to both bacteria and plants, there is vastly more food available on muddy shores than on sand beaches. This permits more large organisms to live on mud shores and, indeed, mud flats may be very densely populated.

The dominant feeding types on mud flats are deposit feeders and suspension feeders, a situation similar to that on sand beaches. Deposit feeders are particularly abundant due to the large amount of organic material and the large populations of bacteria in the sediments. Deposit feeding polychaetes are represented by genera such as *Arenicola* and *Capitella*. Both feed by burrowing through the substrate, ingesting it and digesting out the organic matter (or bacteria), and passing out the undigested material through the anus. *Arenicola* spp., however, form a U-shaped burrow in which one arm of the U is a permanently open shaft to the surface, while the other arm is filled with sediment that the worm ingests (Fig. 6-31). The worm consumes sediment from

Figure 6-31 Two common polychaete worms of mud flats: *Arenicola*, in its U-shaped burrow, and *Capitella*, burrowing through the substrate.

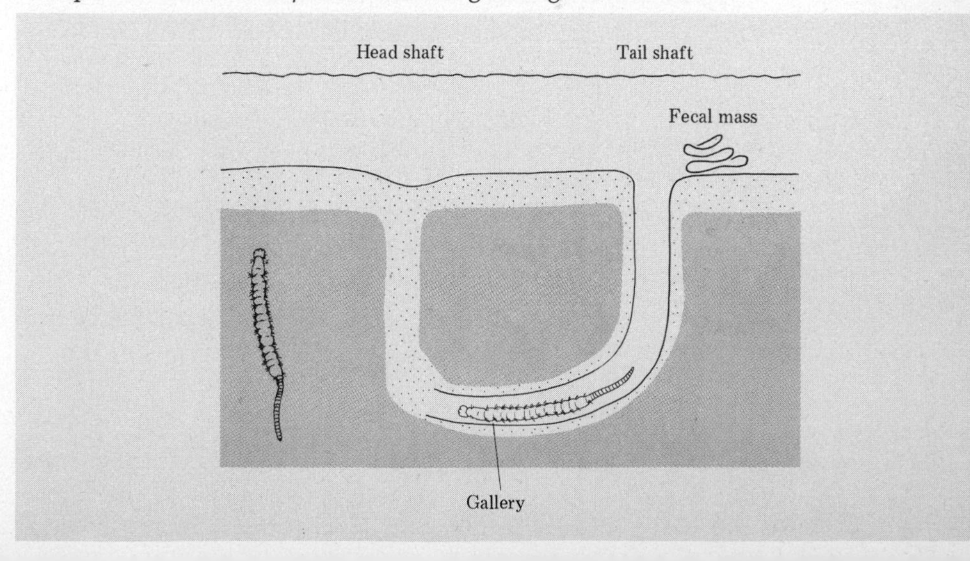

the filled shaft, passes it through its gut, and then moves up in the open burrow to defecate at the surface, leaving characteristic mounds. Capitellids, on the other hand, form no permanent tubes, but move like aquatic earthworms through the surface layers of the substrate, ingesting it.

Deposit feeding bivalves are also common in these areas. Temperate zone mud flats often have large numbers of small tellinid clams of the genus *Macoma* or *Scrobicularia*. These bivalves have siphons that are separated. Each clam lies buried in the substrate with its siphons extended to the surface. There, the intake siphon is moved over the surface like a vacuum cleaner, ingesting organic particles and bringing them down to the clam for digestion (Fig. 6-32).

Suspension feeders include various other species of clams, some crustaceans, and numbers of polychaetes. Suspension feeding mechanisms are not different from those observed in other areas. There is, however, some question as to how exclusively these forms are feeding on suspended plankton. The fine sediment particles are easily stirred up on mud flats and remain resuspended in the water. Hence, suspension feeders may possibly take in amounts of these particles in addition to various plankton organisms. Perhaps most of the suspension feeders on mud flats also partially take in resuspended sediments and hence, are actually feeding on both deposited and suspended material.

In general, deposit feeders are more common in fine-grained shores and suspension feeders become more abundant in coarser sediments, where there is little organic matter.

The major carnivores on mud shores are often the fishes, which feed when the tide is in, and the birds, feeding when the tide is out (Fig. 6-33). Indigenous mud flat predators include a few polychaete worms such as *Glycera* spp., moon snails (*Polinices*, *Natica*), and crabs.

Figure 6-32 *(A) Macoma nasuta* in the substrate. *(B) Macoma nasuta* feeding with its incurrent siphon. (Modified from A. E. Brafield, Life in sandy shores, 1978, Edward Arnold.)

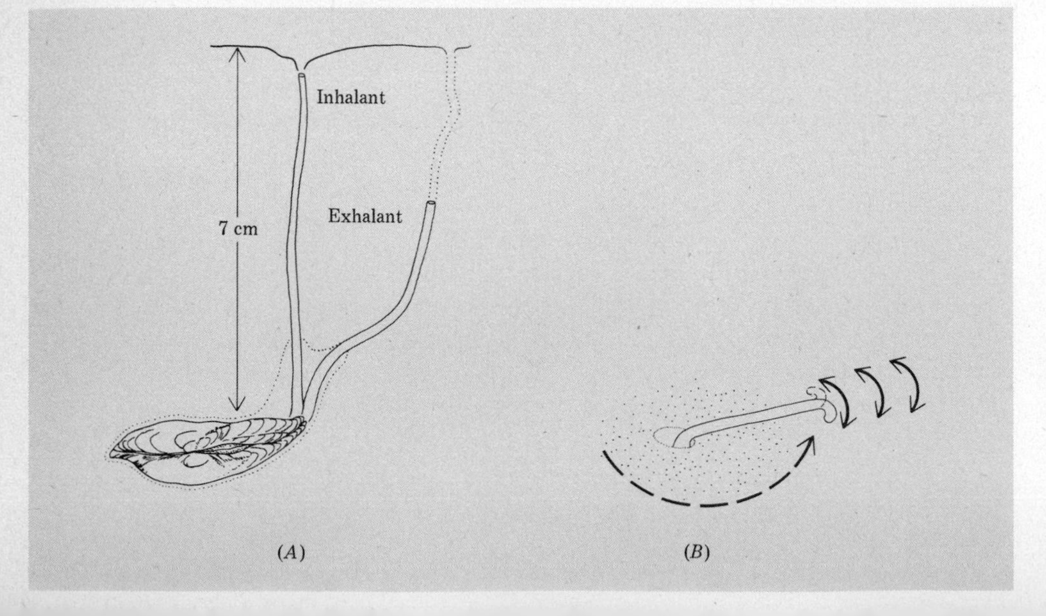

Figure 6-33 Birds feeding on a mud flat. (Photograph courtesy of Bruce Stewart.)

Despite the relative abundance of plant material on muddy shores, there are very few herbivores, and most plant material finds its way into the food chains only after it has been broken down into small pieces to enter the deposit feeding food web.

Thus, the trophic structure of a mud flat is often built up from two bases: a detritus-bacteria base and one based on plants. The detritus base web is derived from plants and other organic sources and includes the bacteria that live upon the detrital particles. Detrital particles can thus either be taken in directly by the large macrofaunal invertebrates or else be consumed by bacteria. An additional bacterial component at the base of this trophic pyramid is the sulfur bacteria.

Bacteria are consumed commonly by various nematode worms, which thus occupy a position similar to that of herbivores in other areas. Bacteria are also consumed by deposit feeders as they ingest the organic particles upon which the bacteria are found. Nematodes and deposit feeders are, in turn, consumed by various carnivores, including predatory invertebrates (moon snails, *Glycera*) as well as birds and fishes (Fig. 6-34).

A second food web is based primarily on the microscopic diatoms as the autotrophic base. The diatoms are consumed by several different polychaetes, mollusks, and crustaceans (Fig. 6-34). These, in turn, are consumed by the large predatory birds and fishes.

ZONATION

Little information exists regarding the zonation of mud flats. The very gentle slope of these areas means that the intertidal is often very extensive, more so than rocky or sandy shores. The upper area, the supralittoral fringe, is often inhabited by various species of crabs, many of which burrow into the substrate. The very extensive midlittoral area is the home of most of the common species of

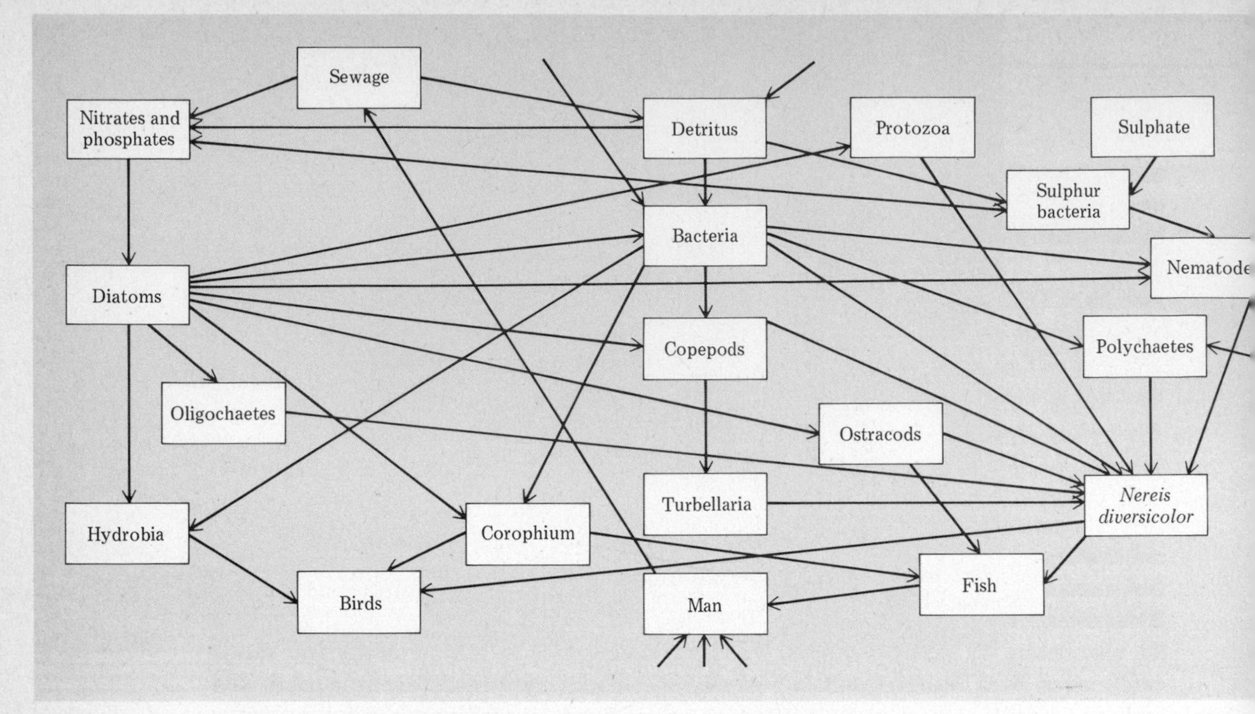

Figure 6-34 Generalized food web of a muddy shore. (From S. K. Eltringham, Life in mud and sand, 1971, Crane, Russak.)

clams and polychaetes. There is no sharp boundary with the infralittoral fringe, and similar organisms are encountered.

The role of physical factors and biological factors in structuring these communities has been little studied. Some studies have demonstrated a separation among the species by depth in the substrate, implying a "niche" separation. Experimental removal of a species or increased densities usually has resulted in corresponding changes in density of other species, which suggests some sort of biological interaction. However, this is not well worked out at this time, and hence, we are not able to discuss how this interaction may produce the pattern we have observed. Since this area is often similar to the substrate found in shallow water subtidal soft bottoms, and estuarine areas, it is likely that a similar set of factors is acting. (See Chapters Five and Eight for a discussion.)

INTERTIDAL FISHES

Whereas a great deal of research has been concentrated on the ecology of invertebrates and plants of the intertidal zone on all three types of shores, there are very few accounts of the fishes of these areas or of the role they may play in community organization as grazers or predators. Since fishes are often present in considerable numbers in the intertidal and since it is known that they have a significant effect on other communities such as coral reefs (see Chapter Nine), it is reasonable to assume that future research may elucidate their role here as well.

Most intertidal fishes, because of the turbulent environment, are of small size. The body shape is usually compressed and elongate (Blenniidae, Pholidae) or depressed (Cottidae, Gobiesocidae), which allows them to inhabit holes, tubes, crevices, or depressions for protection against both desiccation and wave action (Fig. 6-35). Most also lack swim bladders and are closely associated with the substrate. Many of these fishes are adapted to withstand greater ranges of salinity and temperature than their subtidal relatives. A few are even adapted to spend some time out of the water (see Chapter Nine on *Periophthalmus*).

Most intertidal fishes in the temperate zone have been shown to be visual carnivores, again suggesting a potentially significant role in intertidal community organization.

The life history patterns of the few species that have been investigated are all largely similar. Eggs are demersal and laid on stones, rocks, or submerged vegetation. Often, the eggs are guarded by the male. The eggs hatch after a few weeks into planktonic larvae. The planktonic period varies, depending on the species. It may be as long as two months. During this period, the larvae gradually acquire the adult features and, at the end, become benthic. Life spans of the adults are generally short, from two to ten years, and sexual maturity occurs in the first or second year.

Some intertidal fishes are migratory, moving either tidally, diurnally, or seasonally.

BIRDS

At low tide, a considerable variety of birds are often associated with the intertidal zone. Surprisingly, the effect of these birds on the invertebrate

Figure 6-35 Examples of intertidal fishes from the four dominant families: *(A)* Blenniidae (blennies). *(B)* Gobiidae (gobies). *(C)* Gobiesocidae (clingfishes). *(D)* Cottidae (cottids). (Modified from R. N. Gibson, The biology and behavior of littoral fish, *Oceanography and Marine Biology, Annual Review,* vol. 7, 1969, George Allen & Unwin.)

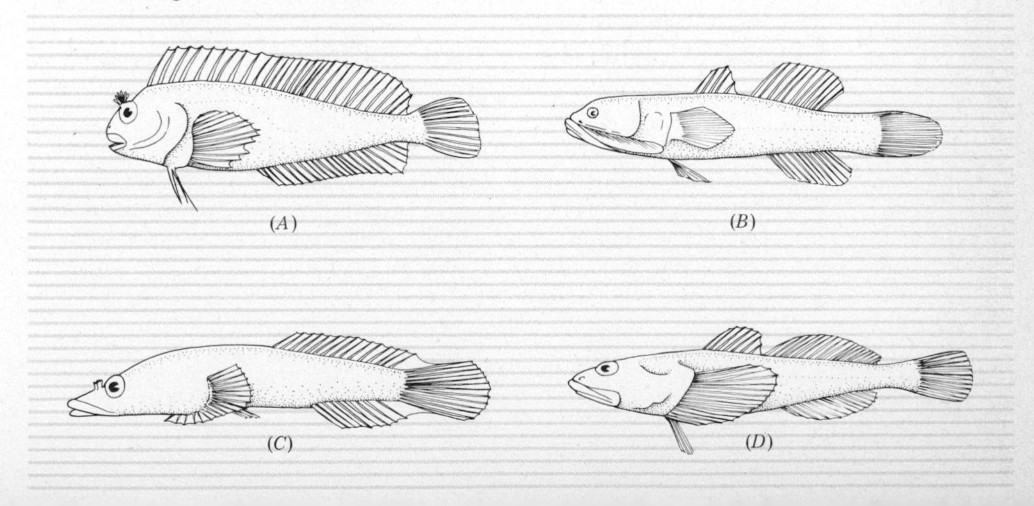

communities is little known at the present time, except for their role on estuarine mud flats (see Chapter Eight). Most birds associated with the intertidal are carnivores or omnivores. Our knowledge of the diet of these birds indicates that they consume various intertidal invertebrates, which, in turn, suggests they may have a potentially important role in the ecology of intertidal areas.

References

Brafield, A. E. 1978. Life in sandy shores. Studies in biology no. 89. Edward Arnold, London. 60 pp.

Carefoot, T. 1977. Pacific seashores, a guide to intertidal ecology. Univ. of Wash. Press, Seattle. 208 pp.

Eltringham, S. K. 1971. Life in mud and sand. Crane, Russak, N.Y. 218 pp.

Gibson, R. N. 1969. The biology and behavior of littoral fish. Oceanography and Marine Biology, Ann. Rev. 7:367–410.

Hedgpeth, J. W. 1964. Introduction to seashore life of the San Francisco Bay region and the coast of northern California. Calif. nat. hist. guides: 9. Univ. of Calif. Press, Berkeley. 136 pp.

Kozloff, E. N. 1973. Seashore life of Puget Sound, the Strait of Oregon and the San Juan Archipelago. Univ. of Wash. Press, Seattle. 282 pp.

Lewis, J. R. 1964. The ecology of rocky shores. The English Universities Press, Ltd., London. 323 pp.

Newell, R. C. 1970. Biology of intertidal animals. Elsevier, N.Y. 555 pp.

Newell, R. C. 1979. Biology of intertidal animals. Marine Ecological Surveys Ltd., Faversham, Kent, U.K. 781 pp.

Ricketts, E., and J. Calvin. 1968. Between Pacific tides, 4th ed.; rev. by J. Hedgpeth. Stanford Univ. Press, Stanford. 614 pp.

Southward, A. J. 1965. Life on the seashore. Harvard Univ. Press, Cambridge. 151 pp.

Stephenson, T. A., and A. Stephenson. 1972. Life between the tidemarks on rocky shores. Freeman, San Francisco. 425 pp.

Voss, G. L. 1976. Seashore life of Florida and the Caribbean. E. A. Seeman Publ., Inc., Miami. 168 pp.

Yonge, C. M. 1963. The sea shore. Atheneum, N.Y. 350 pp.

Chapter Seven
THE ECOLOGY OF INTERSTITIAL ORGANISMS

In addition to the large macrofaunal organisms that inhabit sandy shores, there is another world that exists here, comprising those organisms that occupy the microspaces between adjacent sand grains. These are the *interstitial* organisms. The existence of this unique association of organisms was not realized by biologists until the twentieth century, when European scientists began investigations into these areas. Since then, many scientists throughout the world have made studies of this unique area. The first studies were primarily taxonomic, to determine what kinds of organisms were present and to name them. More recently, ecological studies have been initiated, which have given us a better understanding of the peculiar conditions under which these animals live.

An interstitial fauna and flora assemblage has been found in intertidal and subtidal sandy substrates throughout the world, in both fresh water and salt water. Major interstitial groups have been found by Coull et al. (1977) down to depths as great as 5000 m. Studies of these associations are, however, unequally distributed. The marine interstitial fauna is better known than the fresh water, and the organisms and associations of the European seas are the best known of the various geographical areas. The Pacific coast of North America is practically unknown and unstudied with respect to interstitial fauna, whereas the Atlantic coast of North America has been the site of numerous studies.

This chapter will investigate the principles that govern interstitial life in marine sands.

DEFINITIONS

As with any discipline or subdiscipline, there are a few technical terms that are in general use. The term *interstitial* is a general term that refers to the organisms that occupy the spaces between the sand grains. *Psammon* is a synonym for interstitial organisms, referring to all those that live between the grains. The term *mesopsammon* refers to all interstitial organisms associated with fresh water or brackish water beaches, while *thalassopsammon* refers to those associated with marine beaches and sand areas. *Meiofauna* is a term that is often used synonymously with interstitial or psammon, but it has a more limited application because it refers to a size range (see Chapter Five), whereas interstitial organisms include both the meiofauna and microfauna size classes (see p. 161).

ENVIRONMENTAL CHARACTERISTICS

The conditions that affect the interstitial fauna are somewhat different from those that affect the macrofauna in the same areas. Perhaps the most important factor influencing the presence, absence, and types of interstitial organisms is *grain size*. Grain size is of paramount importance in determining the amount of interstitial space available for habitation. The coarser the grain size, the greater the volume of interstitial space and hence, the larger the interstitial organisms that can inhabit the area. Conversely, the finer the grain size, the lesser the space available and the smaller the organisms must be to inhabit the area. Grain size may thus act as a definite barrier to the dispersal of psammon organisms. Certain organisms will be confined to beaches of certain grain size or areas of a given beach simply because they are too large to fit between the grains of adjacent areas and therefore cannot disperse through the whole area. Psammon organisms show a definite zonation based on grain size.

Grain size is also of great importance because it controls the ability of a beach to retain and circulate water. Interstitial organisms are aquatic and therefore require the presence of water in the interstices of the sand in order to survive (Fig. 7-1). If the grain size becomes too coarse, when the tide ebbs the water will not be held in the sand by capillary action, but will drain away, leaving only a very thin layer coating the grains. On the other hand, fine-grained beaches are able to hold considerable water in the interstices through capillary action.

Circulation of water through the pore spaces in the sand is important because this water movement is responsible for renewing the oxygen supply. Circulation is best in coarse-grained beaches and is reduced in fine-grained beaches. If the grain size becomes really fine, as in mud shores, the circulation practically ceases, and the result is the creation of anaerobic layers in the sediment (see Chapter Six). The maximum number of thalassopsammon have been found where sand grain diameters were between 0.175 and 0.275 mm. Fenchel (1978) reports that the interstitial fauna tends to disappear when the median grain size drops below 0.1 mm.

In addition to grain size, the mineral nature of the grains is also of importance in determining the composition of the psammon. Different associations of organisms may be found in siliceous sand as opposed to carbonate sand.

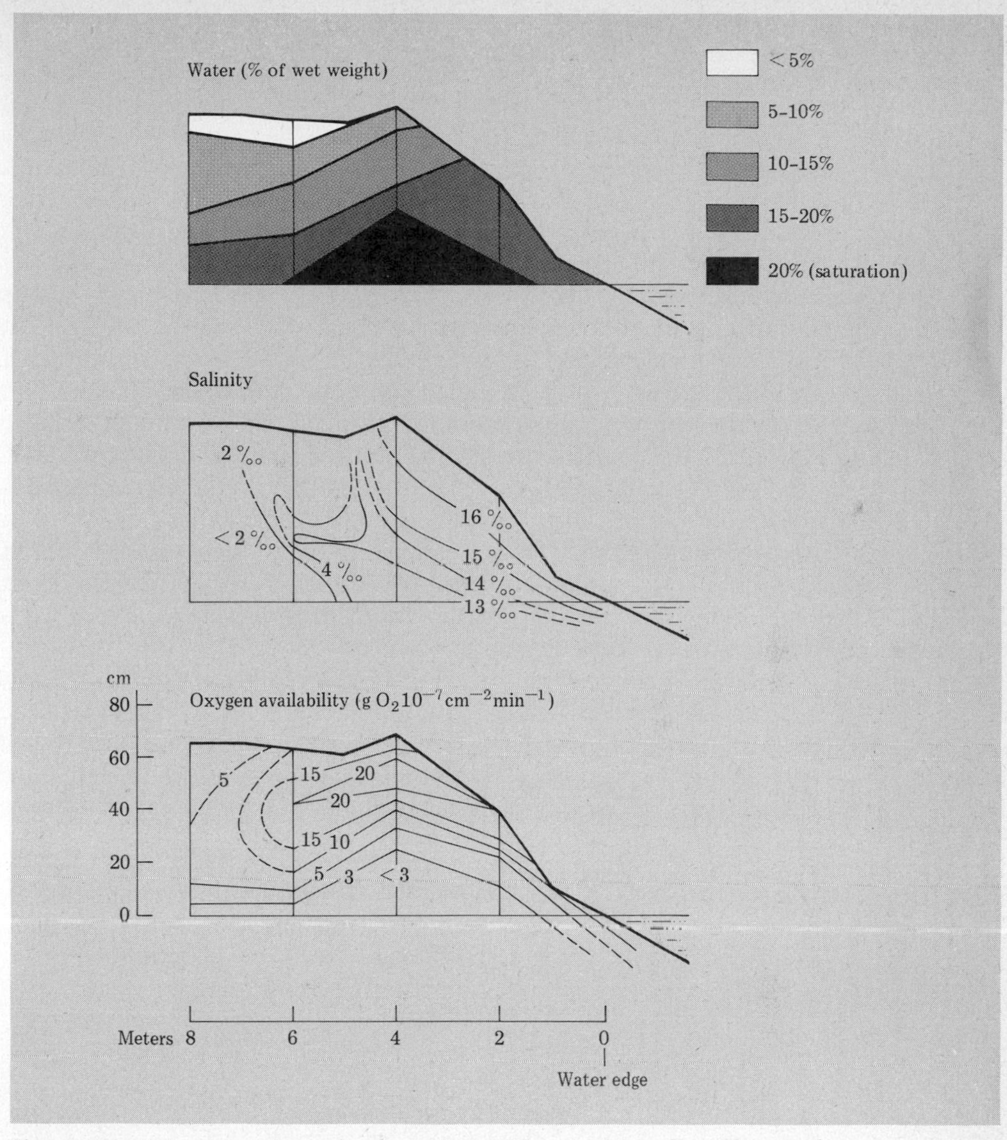

Figure 7-1 A transect through a Danish sand beach showing gradients of water, salinity, and oxygen at low tide. (From T. Fenchel, B. O. Jansson, and W. van Thun, Vertical and horizontal distribution of the metazoan microfauna and of some physical factors in a sandy beach in the northern part of the Oresund, Ophelia, vol. 4, 1967.)

As suggested above, *oxygen* is an important factor in this environment. Practically all marine sediments have an oxidized layer at the surface, beneath which lies a completely anoxic layer. Thus, interstitial organisms living below a certain depth will encounter oxygen-free conditions. The thickness of the oxygenated layer depends on a number of factors such as grain size, amount of organic material, and water turbulence.

Temperature is another environmental factor of importance in determining presence or absence and distribution of interstitial organisms. Temperature ranges are most extreme in intertidal beaches and in the uppermost layers of the beach. Temperature changes are minimal in subtidal sands and in intertidal sands below 10–15 cm in depth (Fig. 7-2). The surface layer of the sand on a beach acts as an insulator to the lower layers, effectively dampening any significant temperature changes. In the surface layers, however, the temperature may change markedly, depending on the air temperature, the effect of wind and rain, the amount of sunlight hitting the surface, and the temperature of the sea water. In cold temperate and polar regions, the temperature may be low enough to freeze the upper layers. Some interstitial organisms have apparently adapted to survive being frozen (e.g., turbellarian *Coronhelmis lutheri*, gastrotrich *Turbanella hyalina*, enchytraeid *Marionina southerni*, and harpacticoid copepod *Parastenocaris phyllura*). Because of the range of temperatures that may occur in the upper layers, the fauna inhabiting these areas may be different from those of lower layers and includes those with a considerable tolerance for temperature change (e.g., = eurythermal).

Another factor acting on interstitial organisms, particularly of intertidal beaches, is *salinity*. Reduced salinities may occur in intertidal beaches due to fresh water runoff over the beach during low tide or else due to heavy rainfall. As noted in the last chapter, such salinity changes are usually confined to the upper layers of the beach because the lower layers retain, through capillary

Figure 7-2 Graph showing the change in temperature at different levels in a sand beach in Denmark during a spring day. (From B. O. Jansson, Diurnal and annual variations of temperature and salinity of interstitial water in sandy beaches, Ophelia, vol. 4, 1967.)

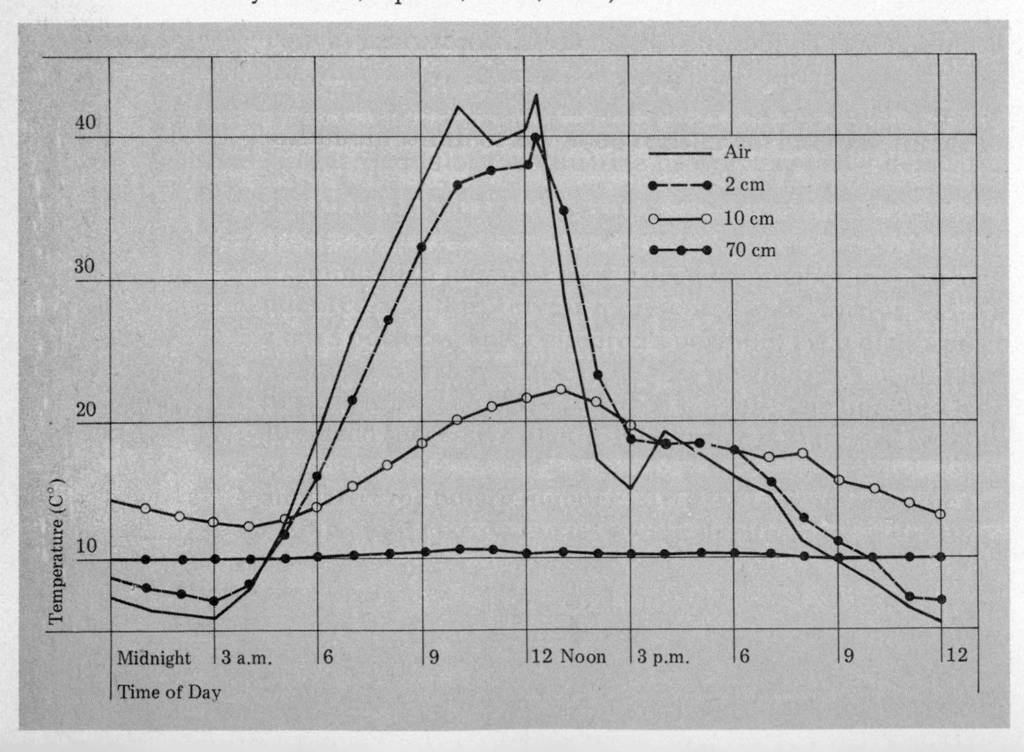

action, elevated levels of salt water. Since fresh water is less dense than salt water, fresh water cannot penetrate below that point where marine water is held by the capillary action. This means that only the uppermost layers are subject to salinity changes. Organisms inhabiting these layers are usually adapted to tolerate considerable salinity changes (e.g., ~ 15 ‰).

A final environmental factor is that of *wave action*. Wave action acts both intertidally and subtidally, but is most dramatic intertidally. Breaking and moving waves have the effect of resuspending the upper layers of sand and, seasonally, of completely removing or depositing large amounts of sand. Whenever the beach is thus churned up, the whole array of internal space is rearranged and the organisms themselves moved about, a process that has profound effects on organisms dependent upon staying in the space between grains. It means that for those organisms, the interstitial space is being constantly rearranged and they are in constant danger of being thrown out into the open water where the risk of predation is very much greater. Special adaptations for coping with this risk are known (see pp. 265–268).

Before leaving the realm of environmental factors, some mention should be made of *light*. Light rarely penetrates below 5–15 mm into sand beaches. This means that microscopic plants should generally be restricted to the very surface layers. However, diatoms have been found at depths considerably below this, raising speculation as to how they are managing to survive. Most of the meioflora and microflora are attached to the surface of the sand grains and form an important food source for meiofaunal grazers.

COMPOSITION OF THE INTERSTITIAL ASSEMBLAGES

The types of organisms constituting the psammon are drawn from a very broad range of invertebrate phyla, but some are represented by one or a few species, whereas others are abundant both in terms of individuals and species.

Those invertebrate phyla that normally have small bodies are, of course, preadapted to live in the small spaces between sand grains and are represented by many species and individuals. The phylum Protozoa is represented abundantly by a large number of species of ciliates, which are very diverse in form and, interestingly enough, often larger than the metazoan animals found with them (Fig. 7-3). The class Turbellaria, of the phylum Platyhelminthes, is also abundantly represented by numerous small, flat, and elongate worms (Fig. 7-4). Most abundant of all, perhaps, are the ubiquitous roundworms, the nematodes (Fig. 7-5).

Several inconspicuous invertebrate phyla are well represented in the psammon, probably because they are normally of small size. These include the phyla Gastrotricha, Tardigrada, and Rotifera. The Gastrotricha are particularly numerous in the thalassopsammon and Rotifera in the mesopsammon (Fig. 7–6).

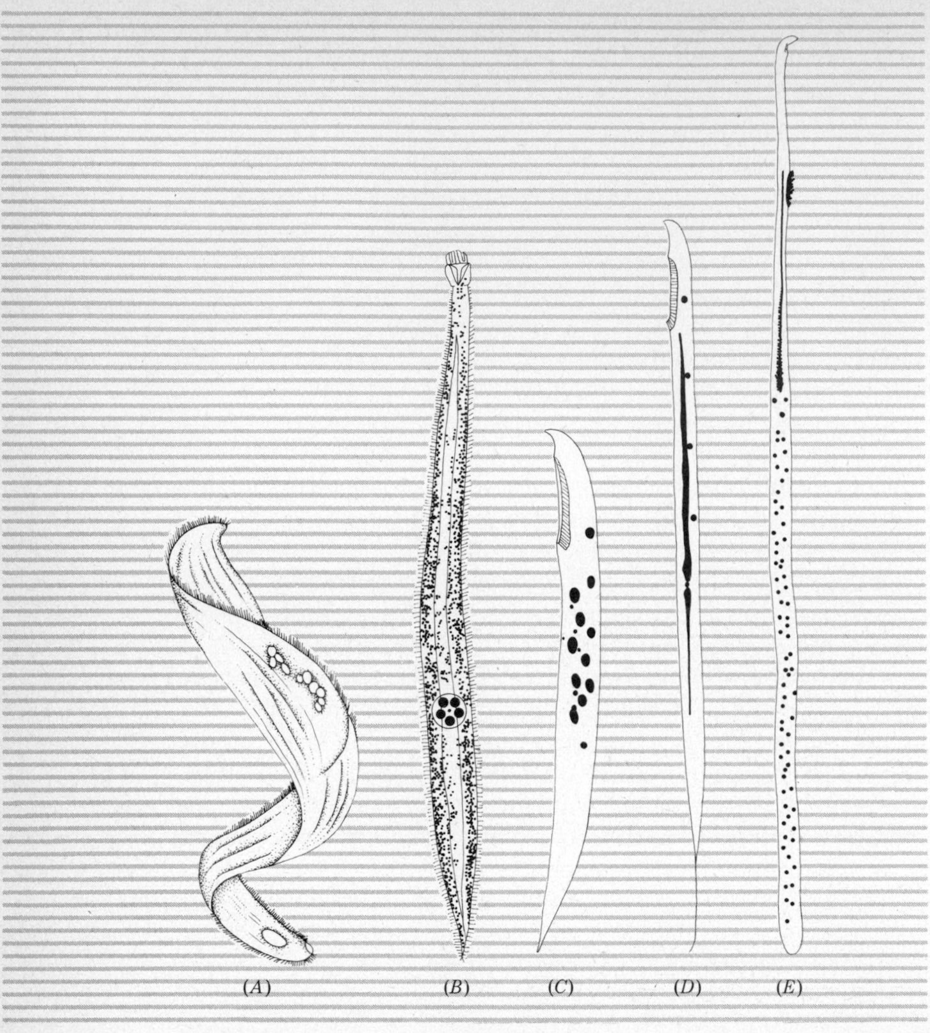

Figure 7-3 Types of interstitial ciliate protozoans. *(A) Loxophyllum verniforme* with a flattened body. *(B) Tracheloraphis remanei* with cylindrical body. *(C) Remanella faurei* with a flattened elongated body. *(D) Remanella caudata* with an elongated body and tail. *(E) Geleia gigas* with a threadlike body. (From B. Swedmark, 1964, The interstitial fauna of marine sand, Biological reviews, vol. 39, Cambridge University Press.)

The worms of the phylum Annelida with their elongated shape are well suited to the interstitial environment and are abundant. Representatives in the meiofauna are from both the classes Oligochaeta and Polychaeta (Fig. 7-7).

Crustaceans are represented primarily by an abundance of harpacticoid copepods and ostracods and a few other small groups such as Mystacocarida (Fig. 7-8).

Phyla that normally consist of large organisms or of sedentary or sessile forms are very poorly represented in the psammon. In the case of large-bodied phyla,

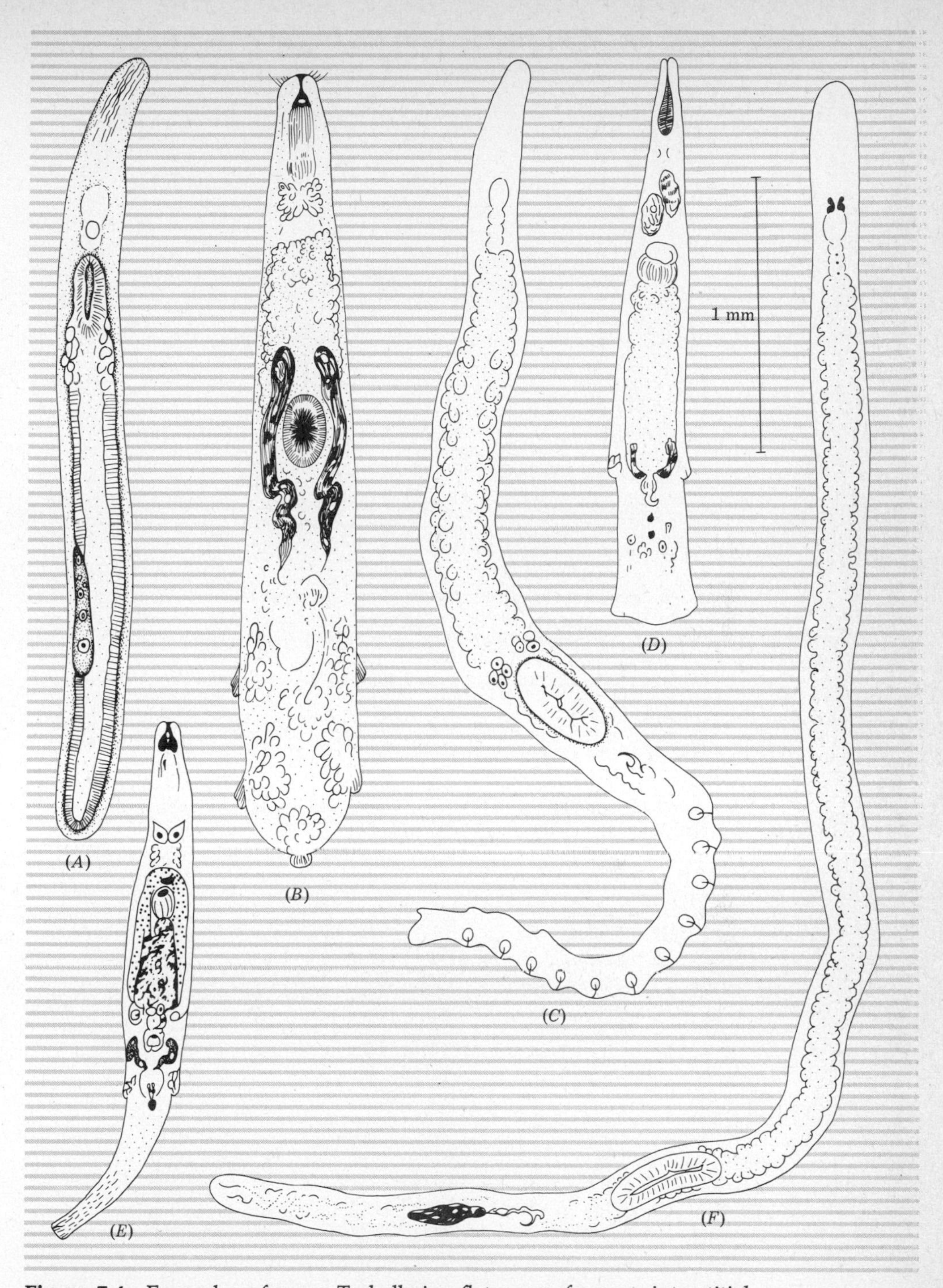

Figure 7-4 Examples of some Turbellarian flatworms from an interstitial marine sand beach in Florida. *(A)* A member of the family Macrostomidae. *(B)* A member of the family Kalyptorhynchidae. *(C) Polystylophora* sp. *(D) Proschizorhynchus* sp. *(E) Cicerina* sp. *(F) Nematoplana* sp. (From L. Bush, 1968, Trans. Amer. Microsc. Soc., vol. 87, no. 2, pp. 244–251.)

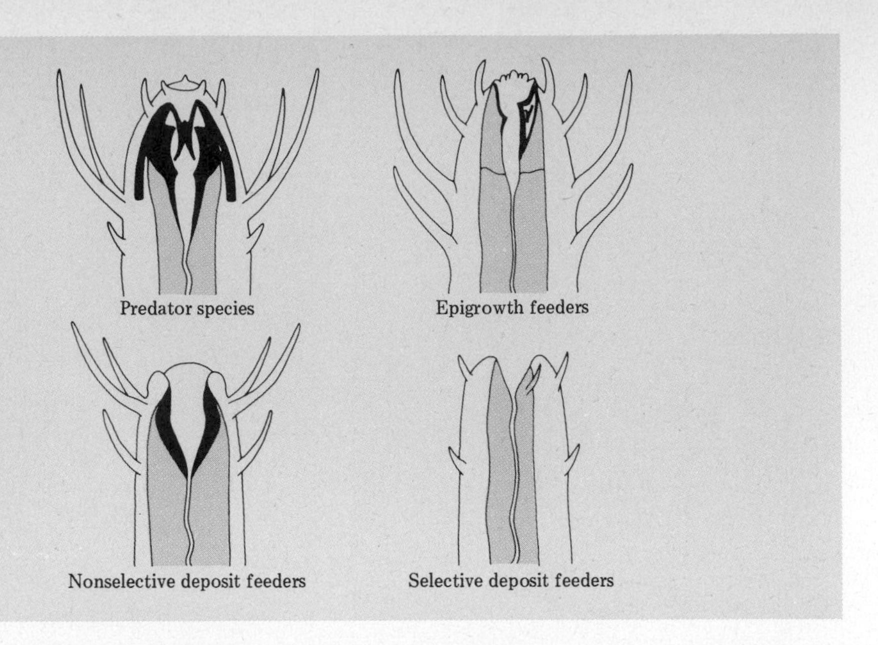

Figure 7-5 Feeding types of interstitial nematodes. (From C. E. King, Some aspects of the ecology of psammolittoral nematodes in the northeastern Gulf of Mexico, Ecology, vol. 43, no. 2, 1962. Copyright 1962, the Ecological Society of America.)

Figure 7-6 Some interstitial gastrotrichs. *(A) Urodasys viviparus. (B) Pseudostomella roscovita. (C) Thaumastoderma heideri. (D) Diplodasys ankeli.* (From B. Swedmark, 1964, Biological reviews, vol. 39, Cambridge University Press.)

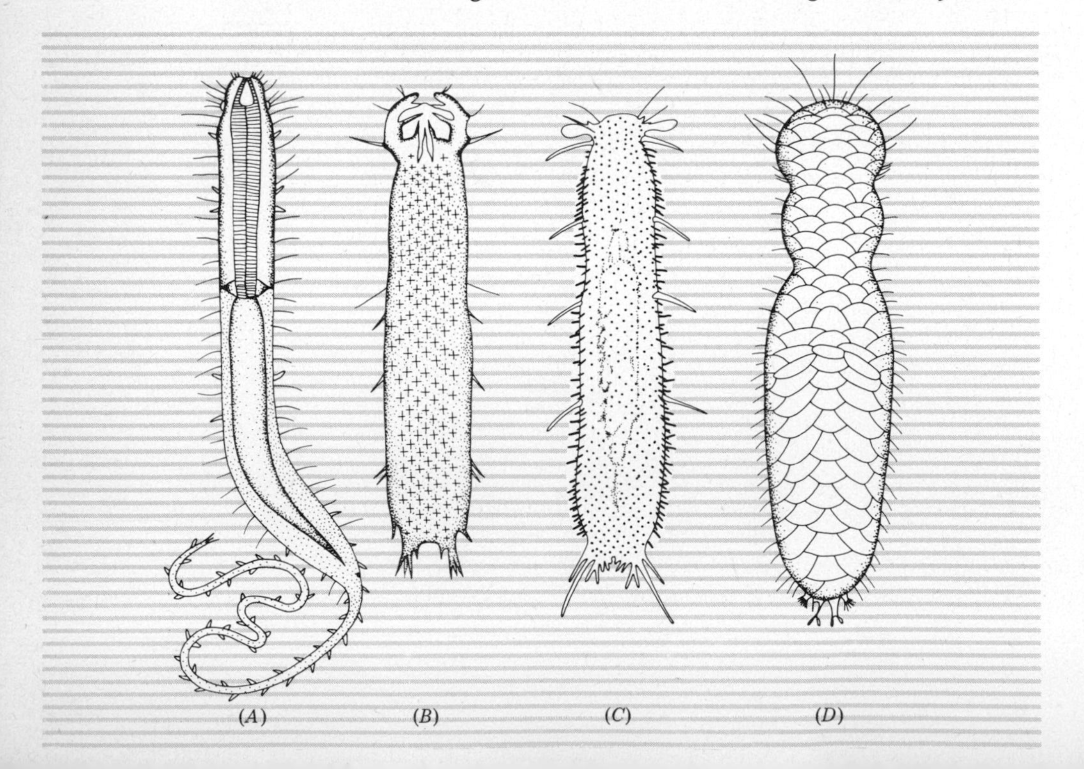

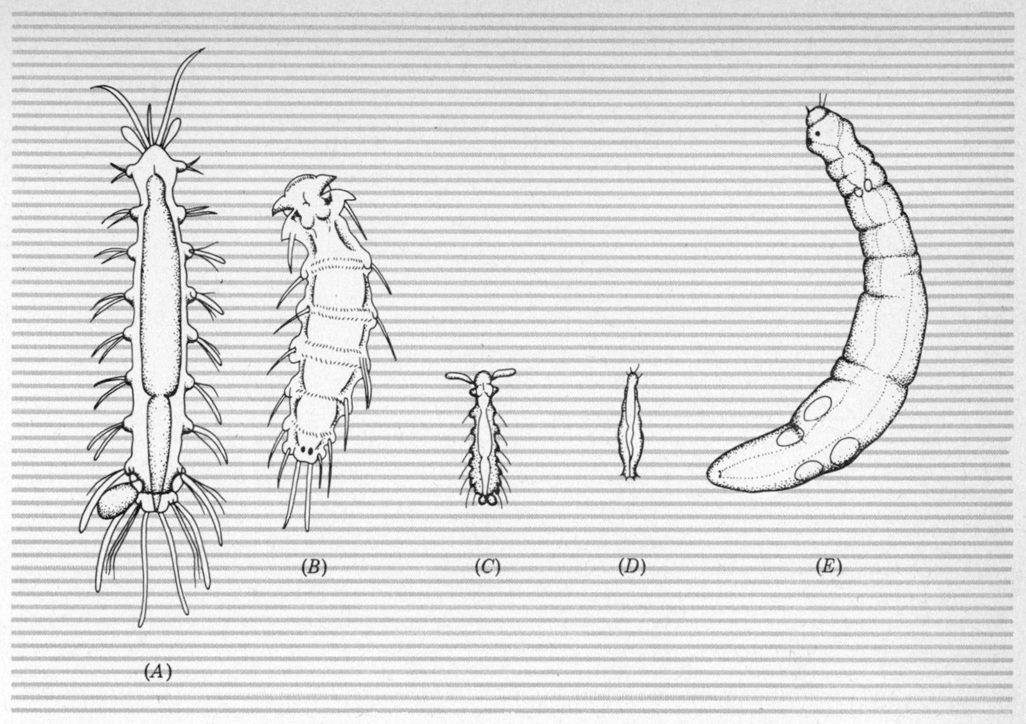

Figure 7-7 Some representative interstitial annelids of the class Polychaeta. *(A) Mesonerilla intermedia. (B) Paranerilla limnicola. (C) Nerillidium simplex. (D) Diurodrilus* sp. *(E) Trilobodrilus* sp. (From N. C. Hulings, editor, Proceedings of the first international conference on meiofauna, Smithsonian Contributions to Zoology, no. 76.)

this is certainly due to the body size, which does not allow them to fit into the small spaces. These groups are, therefore, represented only by a few rather aberrant forms. Examples of such phyla are Echinodermata and Cnidaria (Fig. 7-9). Phyla of sessile or sedentary habit seem to have few representatives, because the dynamic nature of the substrate makes it difficult to exist if one is permanently fixed in one place. Moreover, most of these phyla are suspension feeders, depending upon filtering organisms out of the water. In the interstitial environment, there is relatively little water to filter and very few organisms in the water, as most are attached to or adhere to the sand grains. The sessile groups that are represented by a very few species include the phylum Bryozoa and the class Tunicata (Ascidiacea) (Fig. 7-10).

Absent from the interstitial fauna are the following phyla: Sipuncula, Phoronida, Echiura, Pogonophora, Porifera, Ctenophora, Hemichordata, and Chaetognatha.

SAMPLING AND EXTRACTING PSAMMON

Because interstitial organisms are very small and occupy a unique habitat, special techniques are needed to obtain samples. A great variety of sampling

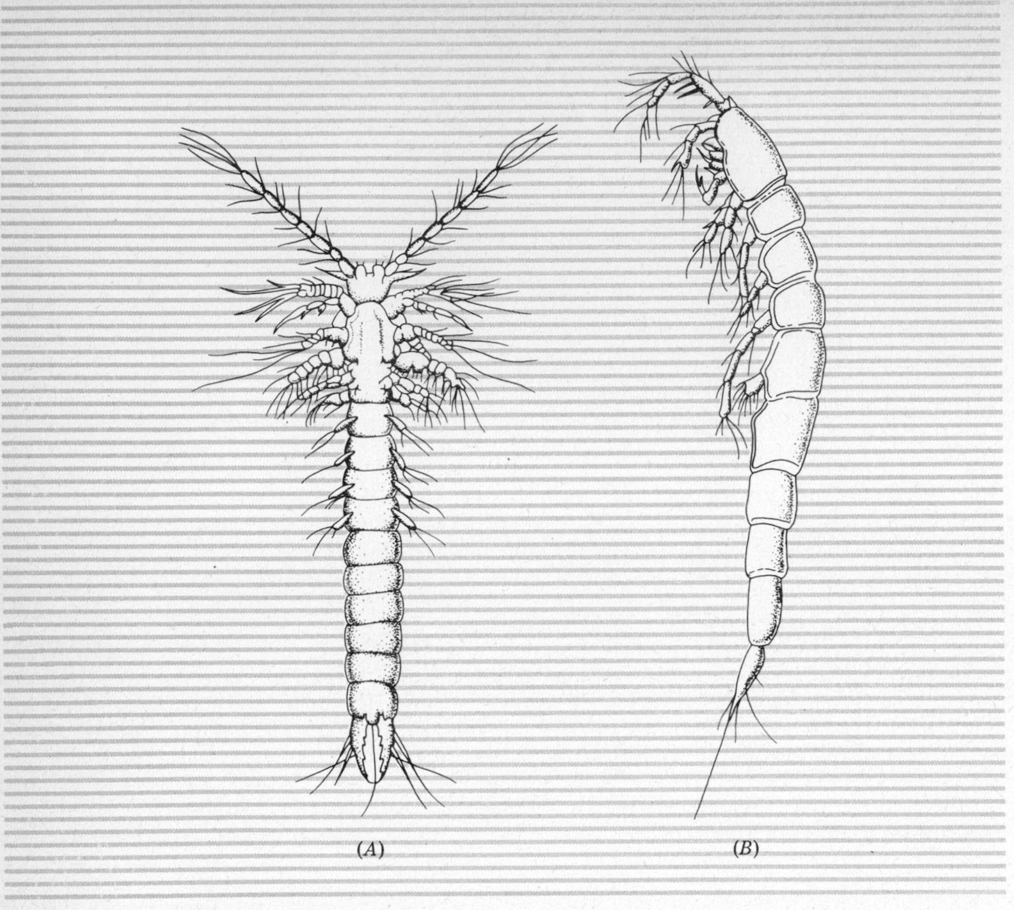

Figure 7-8 Crustacea of the interstitial fauna. *(A) Derocheilocaris remanei* (Mystacocarida). *(B) Cylindropsyllis laevis* (Copepoda). (From B. Swedmark, 1964, Biological reviews, vol. 39, Cambridge University Press.)

and extracting techniques have been employed by meiobenthologists over the years, and none is really standard. Most techniques are, however, simple and may be grouped into general classes.

Because of the small size of the interstitial organisms, only a small sample of the sand substrate is required. This is most conveniently obtained by taking a *core*. Coring devices are usually hollow plastic or metal cylinders that may be pushed into the sand by hand, closed off at one or both ends, and removed from the beach (Fig. 7-11). The enclosed sand column may then be extruded and the organisms extracted. It is also possible to take a small volume or core from a larger grab or dredge sample that may be taken for macrofauna. Since the interstitial fauna exist throughout a number of vertical layers of the sand, it is well to take a core that extends some distance vertically into the substrate. The majority of the meiofauna individuals will, however, be concentrated in the top 10–15 cm, so a depth of 25–30 cm is adequate to sample all but a few forms.

The diameter of the coring device is not critical, but those in common use range in size from 2–4 cm. Larger sizes produce such large samples that extraction is prolonged and difficult.

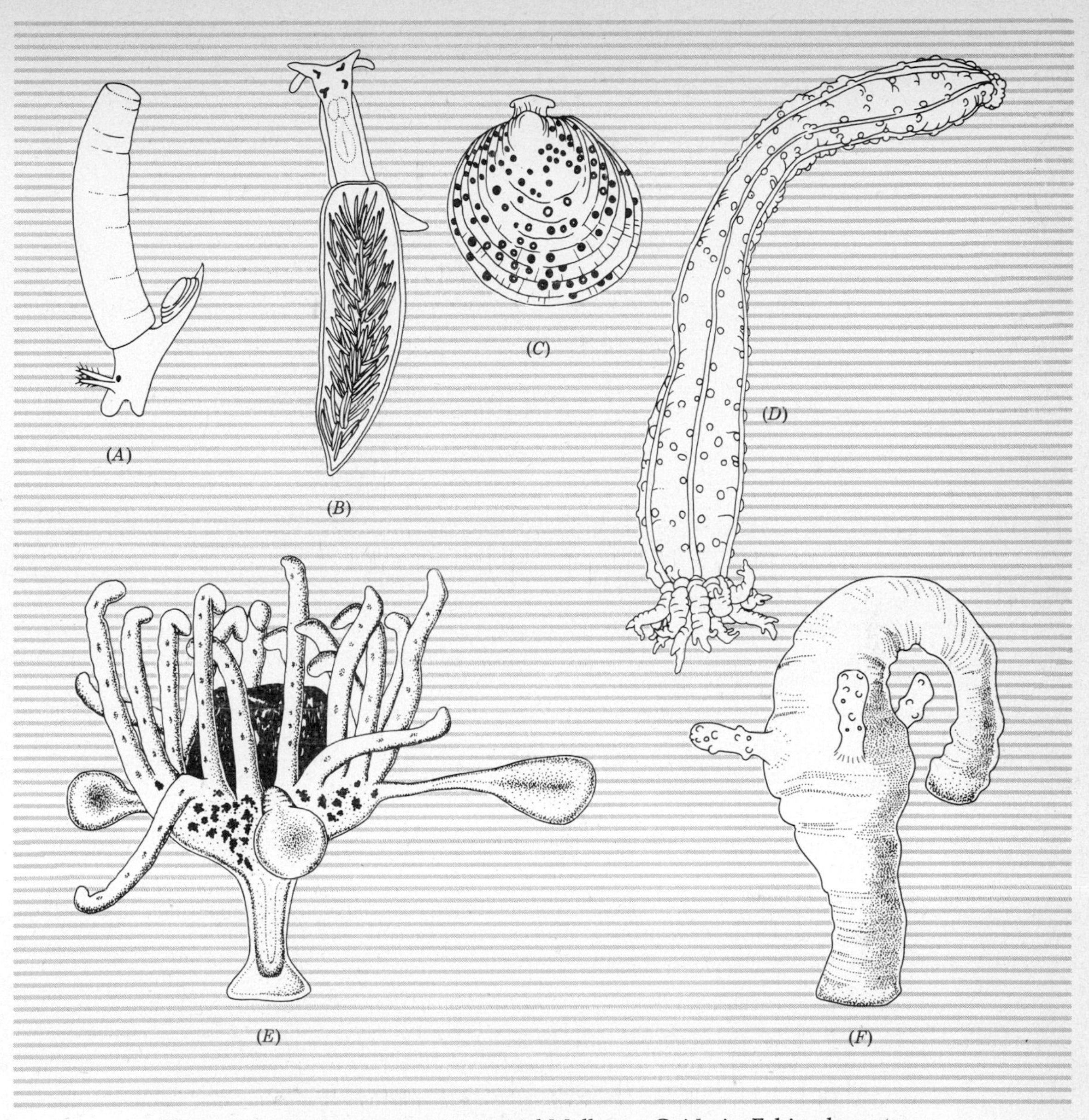

Figure 7-9 Representative interstitial Mollusca, Cnidaria, Echinodermata, and Brachiopoda. *(A) Caecum glabra* (Mollusca, Gastropoda). *(B) Hedylopsis brambeli* (Mollusca, Gastropoda). *(C) Gwynia capsula* (Brachiopoda). *(D) Labidoplax buskii* (Echinodermata). *(E) Stylocoronella riedli* (Cnidaria, Scyphozoa). *(F) Psammohydra nanna* (Cnidaria, Hydrozoa). (From N. C. Hulings, editor, Proceedings of the first international conference on meiofauna, Smithsonian Contributions to Zoology, no. 76.)

Once the sample is in hand, it is necessary to extract the organisms. Interstitial animals are reluctant to leave the interstices and must be removed by special methods in order to be studied. A number of different extraction techniques have been used over the years and each gives somewhat different results. It should be noted that the fauna may be extracted alive or preserved. In most

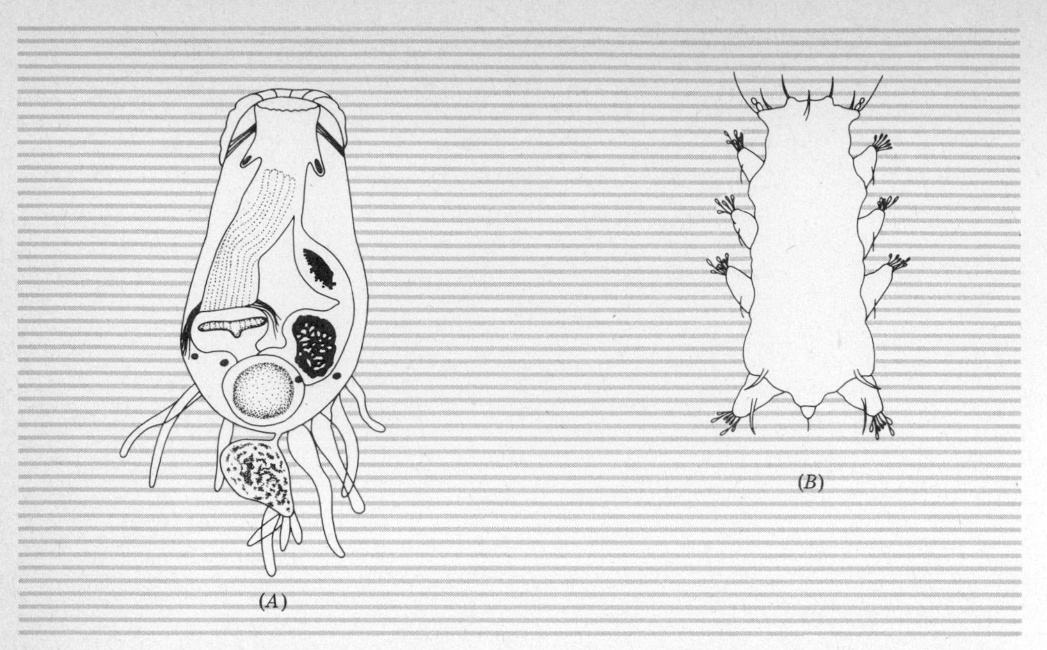

Figure 7-10 Interstitial Bryozoa and Tardigrada. *(A) Monobryozoon ambulans* (Bryozoa). *(B) Batillipes pennaki* (Tardigrada). (*A*, after Hulings, 1971. *B*, after L. W. Pollock, Distribution and dynamics of interstitial Tardigrada at Woods Hole, Massachusetts, U.S.A., Ophelia, vol. 7, no. 2, 1970.)

cases, it is preferable to extract the animals alive, because live animals are much more easily seen and also because groups like flatworms and gastrotrichs are virtually unidentifiable when preserved.

The simplest extraction is to scoop up a bucket of sand and let it stagnate until organisms are driven out. Another simple technique makes use of the response of the organisms to temperature. This is the Uhlig Seawater Ice Technique (Uhlig, 1968). Here, a sand sample is placed into a cylinder that has fine mesh at the bottom. On top of the sand is placed sea water ice. The whole device is then

Figure 7-11 Photograph of a simple coring device for obtaining samples of interstitial organisms.

suspended such that the bottom touches the water surface in a bowl. Over time, the sea water ice melts, runs through the sand, and drives out the animals into the bowl below (Fig. 7-12). Even more efficient is the method of elutriation. In this technique, a sand sample is introduced into a container with openings at the top and bottom. The sand is then constantly stirred up by water entering from below. As the sand stirs, the organisms are dislodged and carried over and out with the water through the upper opening. This water is filtered on a very fine mesh, which then catches the organisms (Fig. 7-12*B*). In this technique, it must be remembered to use only filtered water at the input. Otherwise, the sample will be contaminated by plankton organisms. This technique can be used with either living or preserved material.

ADAPTATIONS

Most of the morphological adaptations observed in the interstitial fauna appear to be correlated most closely with a few aspects of the environment, which can be reviewed here. The environment is a dynamic one in which the individual sand grains of the upper layers are constantly being resuspended and deposited, thus constantly changing the location and amount of interstitial space. In order to survive here, psammon organisms must remain in these spaces in the sand and must avoid both being crushed by moving sand particles and being thrown out into the plankton where survival would be nil.

Perhaps the most obvious adaptation to this environment is that of *size*. All interstitial organisms are very small. Furthermore, in this fauna, one finds the smallest representatives of most large phyla. For example, mollusks, echinoderms, and annelids are usually fairly large animals. All have representatives in the psammon, but they are remarkable for their small size. The holothurian *Leptosynapta minuta* is 2 mm; the gastropod *Caecum glabrum* is 2 mm; the polychaete *Diurodrilus minimus* is but 350 μm; and the hydroid *Psammohydra nanna* but 1 mm. In the case of the ciliate protozoans, however, the size is generally larger than that of ciliates living in other environments. At the same time, however, these larger ciliates are considerably flattened and elongated, which allows them to fit more readily in the interstices. In this case, larger really means longer.

Another adaptation concerns the *shape* of the body. Most of the interstitial organisms have elongated or vermiform bodies, even in those organisms that are normally not vermiform (Fig. 7-3). Another type of body shape observed is represented by organisms which are very flat. Flat organisms fit into the narrow spaces well, and they also give a greater surface area for hanging on to the sand grains. This latter adaptation has survival value in that it permits the organism to stay on the sand grain during resuspension.

Coincident with small body size, there is a reduction in the complexity and number of the body organ systems in metazoan interstitial animals. For

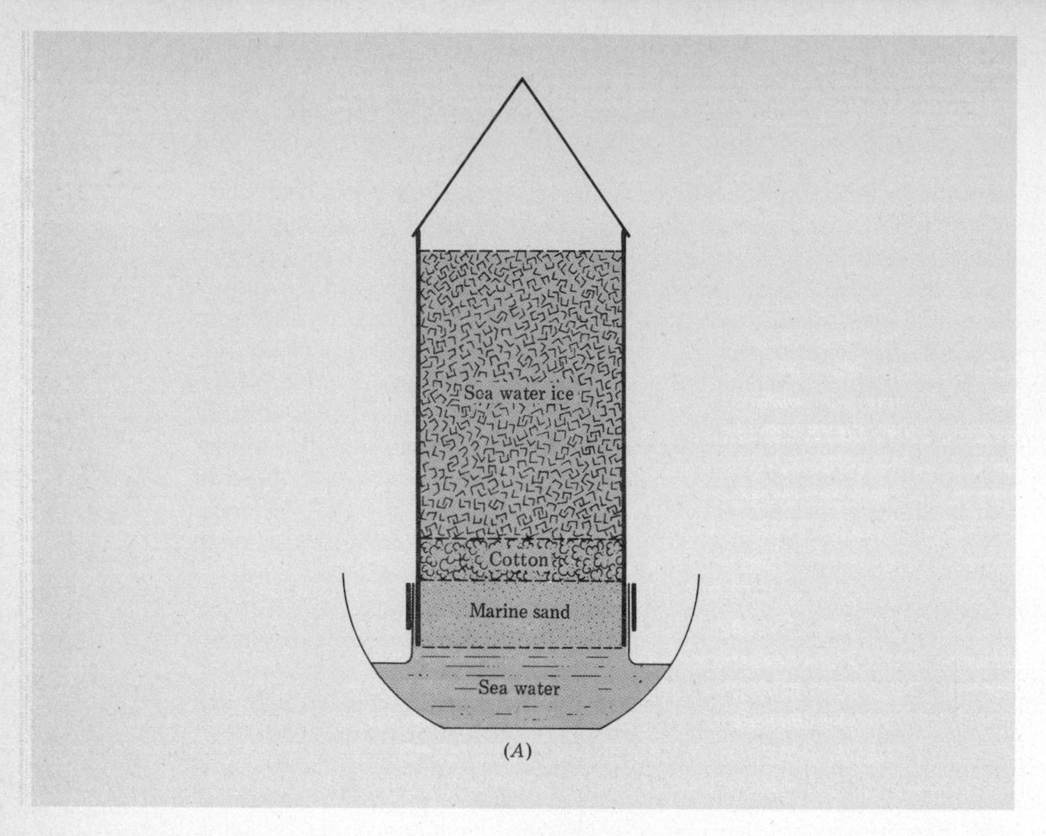

Figure 7-12 Two sets of apparatus for separating interstitial fauna from the sand. *(A)* Uhlig seawater ice method. *(B)* Boisseau elutriation method. (*A,* after G. Uhlig, 1968, Trans. Amer. Microsc. Soc., vol. 87, no. 2; *B,* photograph by the author.)

example, in the polychaetes, the complex pharyngeal apparatus found in macrofaunal polychaetes often disappears, as do the parapodial gills and even the kidneys (nephridia). In the interstitial hydroids, the number of tentacles is reduced from that observed in the forms not living interstitially.

One might expect that organisms living in an environment in which the substrate was often rearranged would be at risk of being crushed as sand grains banged into each other. As a response to that, many psammon animals have developed various types of reinforcement to their body walls. Skeletons of spicules are found in ciliates such as *Remanella*, turbellarian flatworms such as *Acanthomacrostomum spiculiferum*, and the acochlidiacean mollusks *Rhodope* and *Hedylopsis*. Gastrotrichs, on the other hand, have an armor of scales, while the nematodes depend on a thick, heavy cuticle. Those delicate animals lacking a protective armor seem to have developed great abilities to quickly extend and contract the body, which also permits escape from crushing.

Perhaps because the environment is so dynamic, we find that virtually all the inhabitants are free moving. This is true even for groups that are normally all sessile, such as the Bryozoa and Tunicata. The interstitial representatives of these two groups are free moving. It is possible, however, that since all the common ways of extracting the interstitial fauna depend on the fauna leaving the substrate, there may be other sessile forms existing that we have not yet observed.

Two final types of adaptation may be noted, and both are in response to the necessity of the interstitial organisms to remain in the sand and not be thrown out during the resuspension process. Many species of various phyla, such as flatworms and gastrotrichs, have adhesive organs to fix them to the sand grains. Crustaceans and tardigrades effect the same ability using hooks or claws on their limbs. Stratocysts, organs that detect gravity and hence differentiate up and down, are also common. Such sense organs enable any organism to quickly determine which way to move during resuspension and thus return to the sand rather than up into the open water.

LIFE HISTORY

As yet, we know relatively little of the life histories of many psammon. There are really two groups of interstitial organisms, a *temporary psammon*, which consists of the newly settled juveniles of the macrofauna organisms, and a *permanent psammon*, consisting of those organisms that spend their entire lives in the interstices of sand. The temporary psammon are not considered here.

Since the permanent interstitial fauna are so small in size, the number of gametes that they can produce at any one time is also limited. In contrast to the macrofauna, which often produce thousands or hundreds of thousands of eggs, the production of eggs by thalassopsammon is almost always below 100 per individual and usually ranges between one and ten!

When the number of eggs is that small, a species cannot afford to lose many or there will be no next generation. As a result, these animals have evolved a number of adaptations to ensure survival of the few propagules. First, there are adaptations that ensure that fertilization will occur. One way is to have copulation in which the sperm are directly transferred to the female. Such is the case with harpacticoid copepods. Another way is to package all the sperm in a single unit called a *spermatophore* and attach it to the female where it will be available to fertilize the eggs as they issue. Such a situation occurs in certain of the polychaete worms and the acochlidiacean mollusks. Finally, some species are hermaphroditic, having both male and female systems in the same individual and thus ensuring fertilization. This situation occurs in gastrotrichs, some hydroids (*Otohydra*), and some polychaetes (*Protodrilus*).

Once fertilization has occurred, it is important that the embryos be protected from loss. Usually, benthic organisms put their embryos into the plankton in order to obtain maximum dispersal. However, the plankton is a notoriously dangerous place, with the result that most of the embryos are destroyed via predation. As a consequence, to compensate, the benthic animals using this route must produce large numbers of larvae. Since the thalassopsammon cannot produce large numbers, it follows that it would not be a good survival strategy to put larvae in the plankton. In fact, this is what is observed. Very few interstitial animals have planktonic larvae, and those that do are the ones that produce the largest numbers of eggs. Most interstitial organisms have instead brood protection. In such a case, the eggs are kept with the female until the young have reached a sufficient size to be self-sustaining. That selection pressure for brood protection is high in this environment can be deduced by observing that brood protection occurs here in groups of invertebrates in which it is otherwise virtually unknown, such as cnidarians (*Otohydra*) and gastrotrichs (*Urodasys viviparus*).

For those few that do not have brood protection, additional safeguards are found. Eggs that are spawned are either sticky so as to quickly attach to sand, or else are confined in cocoons that are sticky. In any case, the development is usually direct; that is, the egg hatches as a juvenile and does not have a free-swimming larva or a larva is produced that is nonpelagic and remains in the interstices of the sand.

Despite the low reproductive potential of many species, seasonal reproductive periods have been observed in many species.

Studies by Gerlach (1971) suggest that generation time of interstitial organisms range from a few days to over a year. The average number of generations per year is three.

ECOLOGY

The composition of the interstitial fauna varies both vertically and horizontally in the substrate. The factors creating this zonation are the aforementioned grain size differences and physicochemical factors, particularly temperature, oxygen, and salinity. Characteristic patterns of species assemblages are correlated with these physical factors.

Most of the thalassopsammon seems confined to the uppermost strata, usually the top 5 cm. In investigations in Europe, McIntyre (1969) reported between 80–90 percent of the dominant nematodes, copepods, and ostracods in this upper layer. At the same time, interstitial organisms have been found as deep as 70 cm in some beaches, and there is often a definite vertical zonation (Fig. 7-13).

There is a specialized faunal assemblage that is confined to the anoxic layers. Characteristic of this assemblage are ciliates of the orders Trichostomatida, Heterotrichida, and Odontostomatida, of which the latter seems to contain only anaerobic species and is confined to this area. Zooflagellates, certain nematodes, turbellarians, gnathostomulids, rotifers, and gastrotrichs are also present, but it is not known whether these latter forms are permanent members of the community or transients from the upper oxidized zones.

Vertical migration of the organisms also occurs. This migration is usually triggered by temperature or salinity changes and is especially prominent in beaches in the temperate zone. Thus, the same organisms are often found deeper during the winter months and shallower during the summer. Strong tidal or wave action following storms may also trigger a migration, which may occur on a much shorter time scale, on the order of a few days or a tidal cycle.

There are also seasonal changes in abundance of the interstitial organisms. Certain species are more abundant at one season than another (Fig. 7-14). For example, on Danish beaches, Muus (1967) found that copepods were at a minimum in winter and peaked in numbers in the summer. Fenchel (1978) has noted that in temperate shallow water areas, there is an annual change in the densities of the dominant groups, with copepods succeeded by oligochaetes, nematodes, ostracods, and turbellarians. This temporal succession appears to be regulated by primary productivity and predation. Not only do abundances vary seasonally, but so does the species composition, with different species becoming abundant at different times of the year. This is probably correlated with changes in the physicochemical factors mentioned above, but may also be due to migration.

Trophic relationships among the interstitial fauna are not well known. It is known that certain flatfishes such as *Pleuronectes platessa, Platichthys flesus, Microchirus boscanion,* and *Limanda limanda* in European waters actively prey upon meiofauna when they are young and recently metamorphosed. McIntyre (1969) reports that their stomachs have been found to contain numbers of harpacticoid copepods, ostracods, oligochaetes, and nematodes. Other macrofaunal predators on thalassopsammon include shrimps, larger polychaetes, and hydroids. Macrofaunal deposit feeders such as polychaetes and holothurians must also destroy numbers of meiofauna existing on the particles that they consume.

In addition to macrofaunal predators, there are interstitial predators, includ-

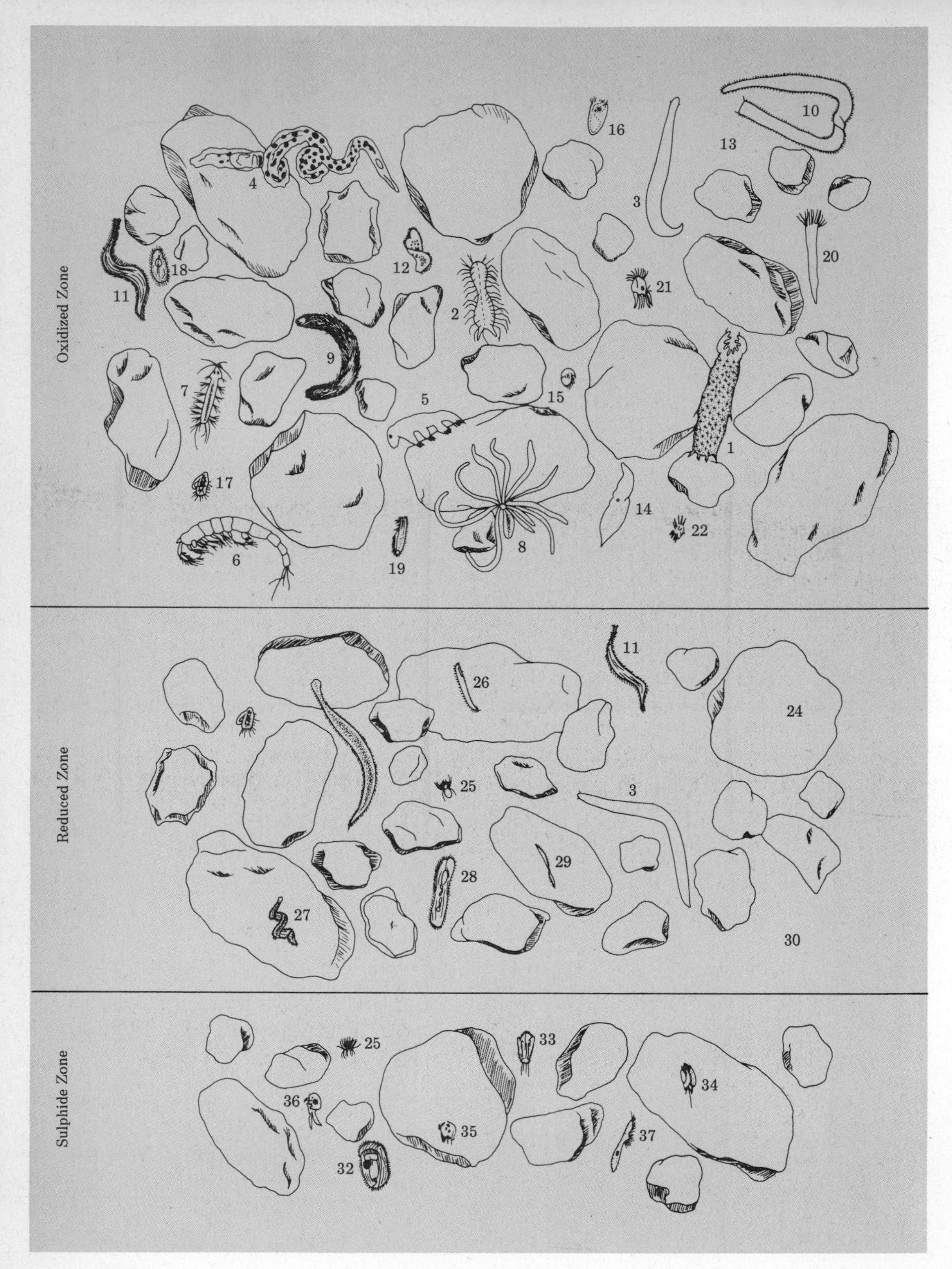
Oxidized Zone
Reduced Zone
Sulphide Zone
1
2
3
4
5
6
7
8
9
10
11
12
13
14
15
16
17
18
19
20
21
22
11
24
25
26
27
28
29
3
30
25
32
33
34
35
36
37

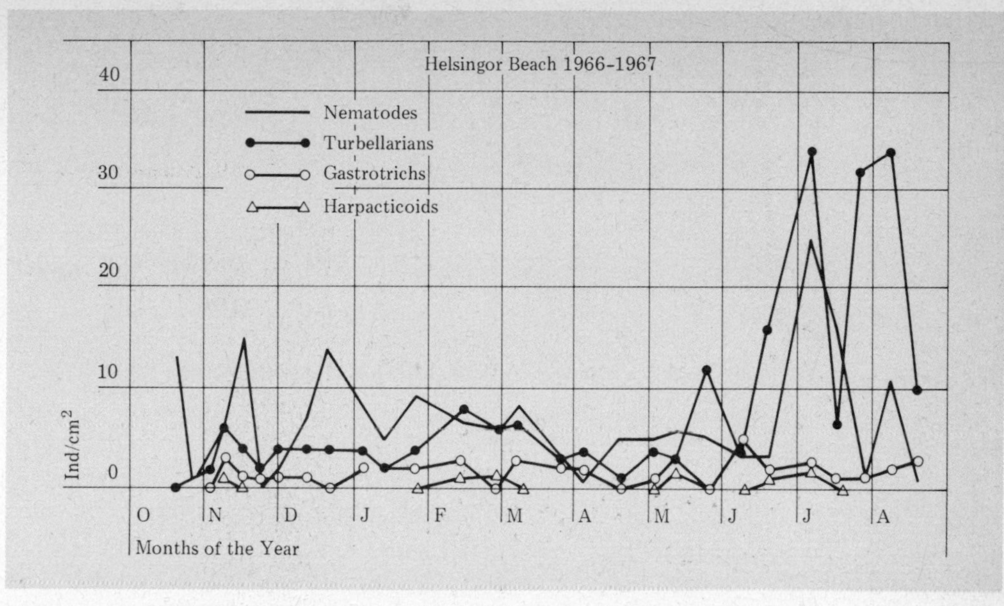

Figure 7-14 Changes in abundance of the major interstitial organisms with season in a Danish beach. (From T. Fenchel, The ecology of marine microbenthos IV, Ophelia, vol. 6, 1969.)

ing certain nematodes, turbellarians, some tardigrades, and hydroids such as *Protohydra*. The latter is a particularly voracious feeder on copepods and nematodes, according to Muus (1966).

The result of predation by both meio- and macrofaunal predators is to reduce the populations of other interstitial organisms, but just how extensive this reduction is, is not known.

In addition to predators, the interstitial fauna contains three other feeding types: herbivores, detritus feeders, and suspension feeders. While some turbellarians have been shown to be predators, others appear to be scavengers, feeding on dead meio- and macrofaunal organisms. Harpacticoid copepods seem to be primarily herbivores, feeding on the few attached diatoms or else feeding upon detritus. Ostracods also ingest diatoms. Detritus feeders include gastrotrichs, some nematodes, and archiannelid polychaetes. Suspension feeders are, as noted before, quite rare, due to lack of an interstitial plankton.

A considerable degree of food specialization appears to exist among the

---

Figure 7-13 Vertical zonation of interstitial animals in a Scandanavian beach. (1) Microdasyoid, *Urodasys*, (2) Gastrotrich, *Chaetonotus*, (3) Nematode, (4) Turbellarian, *Coelogynopora*, (5) Tardigrades, (6) Harpacticoid, *Cylindropsyllis*, (7) Archiannelid, *Nemillidion*, (8) Hydra, *Halammohydra*. Cilates: (9) *Pseudo prorodon*, (10) *Helicoprorodon*, (11) *Tracheloraphis*, (12) *Loxophyllum*, (13) *Litonotus*, (14) *Dilepthus*, (15) *Lynchella*, (16) *Chlamydodon*, (17) *Pleuronema*, (18) *Frontonia*, (19) *Blepharisma*, (20) *Condylostoma*, (21) *Diophrys*, (22) *Strombidium*, (23) *Aspidisca*, (24) *Paraspathidum*, (25) *Mesodinium*, (26) *Remanella*, (27) *Kentrophorus*, (28) *Cardiostomum*, (29) *Homalozoon*, (30) *Lacrymaria*, (31) *Geleia*, (32) *Sonderia*, (33) *Metopus*, (34) *Caenomorpha*, (35) *Saprodinium*, (36) *Myelostoma*, (37) *Parablepnarism*.

thalassopsammon. Fenchel (1968) studied the interstitial ciliates and demonstrated a surprising degree of food specificity for various bacterial and algal taxa. In addition, he showed that in certain habitats, whole groups of species subdivided their food resource according to size. A similar food niche separation has been reported by Tietjen and Lee (1977) for nematodes.

SPECIES RICHNESS AND DISTRIBUTION

One of the more remarkable features of the interstitial fauna is the species richness. Fenchel et al. (1967) found as many as 70 species of psammon organisms in 50 cm^2 of sand, and they regularly obtained 30–50 species of ciliate protozoan alone. Even more striking is the discovery by Gerlach (1965) that the Arctic interstitial fauna is rich in contrast to the impoverished macrofauna at that latitude. Based on the limited studies available now, it appears that there is no gradient in species richness in this fauna moving from polar regions to the tropics. This is in direct contrast to the common distribution pattern of most macrofaunal organisms, where lower latitudes tend to have greater richness than higher latitudes.

Zoogeographically, the psammon have a pattern in which the common genera are often cosmopolitan, but the species are not. As a result, several different zoogeographical regions or provinces can be recognized.

References

Fenchel, T. 1978. Ecology of micro- and meiobenthos. Ann. Rev. Ecol. and Systematics 9:99–121.

Hulings, N. C. (ed.). 1971. Proceedings of the first international conference on meiofauna. Smithsonian Contributions to Zoology no. 76. 205 pp.

Hulings, N. C., and J. S. Gray (eds.). 1971. A manual for the study of meiofauna. Smithsonian Contributions to Zoology no. 78. 84 pp.

Swedmark, B. 1964. The interstitial fauna of marine sand. Biol. Reviews 39:1–42.

Chapter Eight
ESTUARIES AND SALT MARSHES

Whereas the intertidal is a place of meeting of the terrestrial and the marine, estuaries are the meeting of fresh water and salt water. They act as a transition zone between the two aquatic ecosystems on this planet. Estuaries have been and remain in close association with humans, because many of the major cities of the world are established on estuaries. In contrast to other ecotones or transitional areas, estuaries are depauperate in terms of numbers of species of permanent organisms. In this chapter, we shall discuss the various special physical and chemical conditions in an estuary and how these act to create a rigorous environment in which relatively few species can persist.

TYPES OF ESTUARIES

There are an unusually large and varied number of definitions for an estuary. This great diversity of definitions is due in part to the fact that several geomorphological features of coastlines, such as lagoons, sloughs, fjords, and other shallow embayments, are often considered estuaries. A simple definition is that an *estuary* is a partially enclosed coastal embayment where fresh and sea water meet and mix. This definition implies the free connection of the sea with the fresh water source, at least during a part of the year, thus eliminating from consideration permanently isolated coastal water impoundments, as well as isolated brackish or saline bodies of water such as the Caspian Sea, Sea of Azov, and the Great Salt Lake.

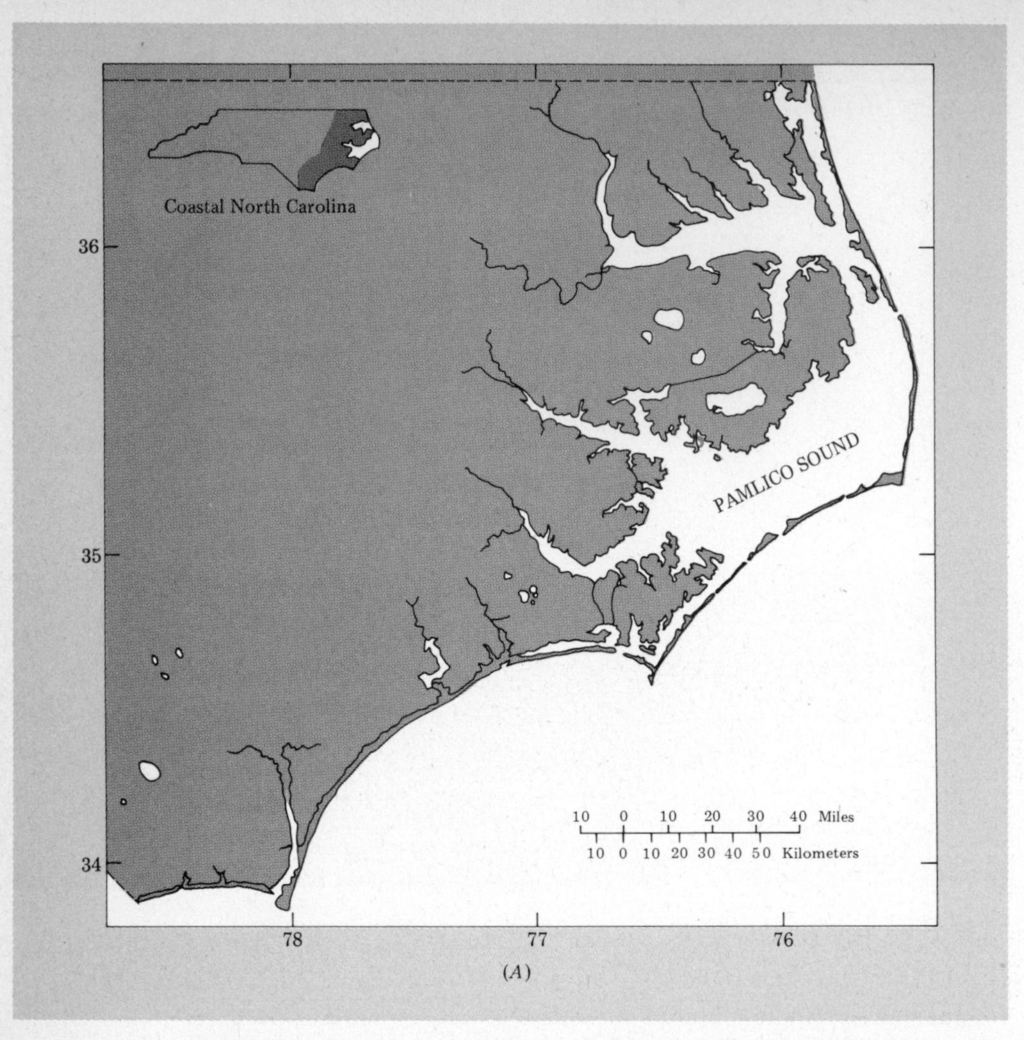

Figure 8-1 Types of estuaries. *(A)* Coastal lagoon lying behind a barrier beach, the major lagoons of the North Carolina coast. *(B)* Tectonic estuary, San Francisco Bay. *(C)* Fjord, Naeroy Fjord in Norway. (*A*, after C. Peterson, and N. M. Peterson, 1979, The ecology of intertidal flats of North Carolina: A community profile. U.S. Fish and Wildlife Service, Office of Biological Services, FWS/OBS-79/39. *B*, after T. J. Conomos, editor, San Francisco Bay, the urbanized estuary, 1979, courtesy Pacific Division—American Association for the Advancement of Science. *C*, photo by the author.)

As a result of the geomorphology of an estuary, the geological history of the area, and the prevailing climatic conditions, there may be different estuarine types, each displaying somewhat different physical and chemical conditions. These may be grouped into a few basic types (Fig. 8-1). Perhaps the most common type of estuary is the *coastal plain estuary*. Coastal plain estuaries were formed at the end of the last ice age when the rising sea level invaded low-lying coastal river valleys. Estuaries such as the Chesapeake Bay and the mouths of the Delware and Hudson rivers are examples of this type. Similar to this is the *tectonic estuary*.

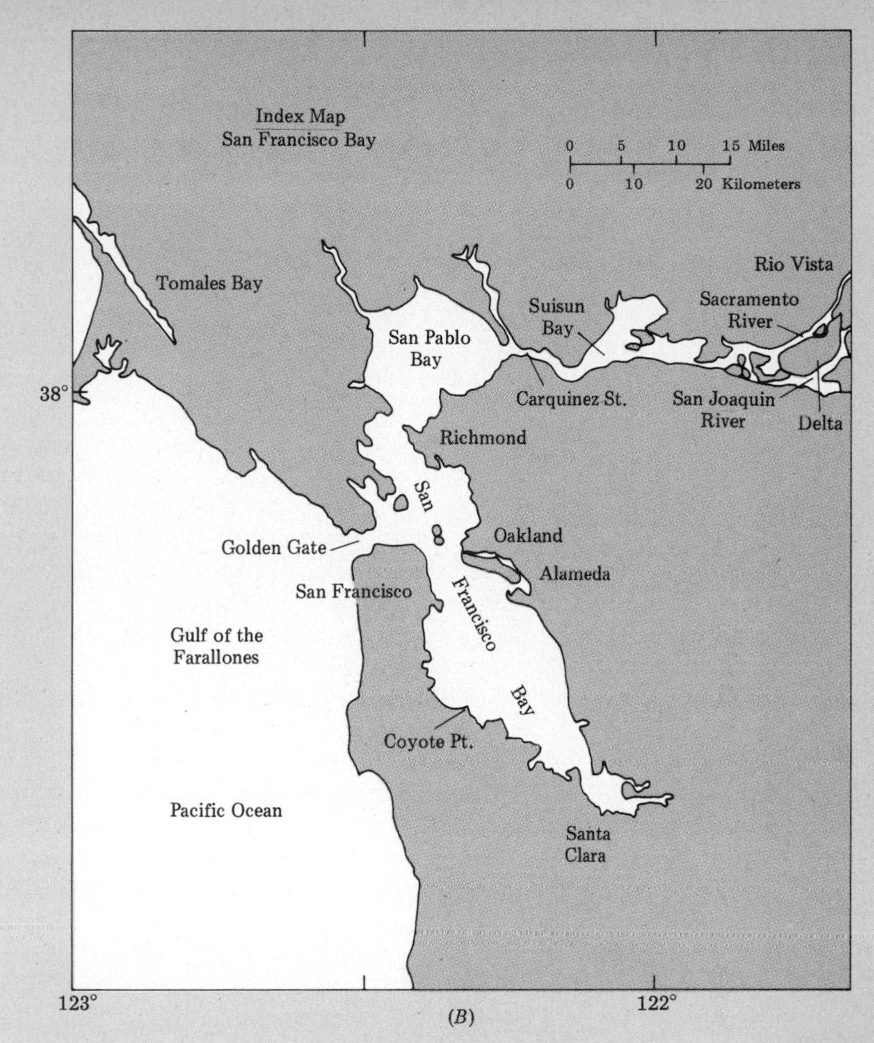
Index Map
San Francisco Bay
0 5 10 15 Miles
0 10 20 Kilometers
Tomales Bay
San Pablo Bay
Suisun Bay
Rio Vista
Sacramento River
Carquinez St.
San Joaquin River
Delta
38°
Richmond
San Francisco Bay
Oakland
Alameda
Golden Gate
San Francisco
Gulf of the Farallones
Coyote Pt.
Pacific Ocean
Santa Clara
123°
122°

(B)

(C)

In this class of estuary, the sea reinvades the land due to subsidence of the land, not as a result of a rising sea level. A good example is San Francisco Bay. A third type of estuary is the *semienclosed bay* or *lagoon*. Here, sandbars build up parallel to the coastline and partially cut off the waters behind them from the sea. This creates a shallow lagoon behind the sandbars, which collects the fresh water discharge from the land. The water in such lagoons varies in salinity, depending on the climatic conditions, whether or not any major river flows into the lagoon, and the extent to which the bars restrict sea water access. Such estuaries are common along the Texas and Florida Gulf coasts and in northwestern Europe (The Netherlands). A final type category of estuary is the *fjord*. These are valleys that have been deepened by glacial action and are then invaded by the sea. They are characterized by a shallow sill at the mouth that greatly restricts water interchange between the deeper waters of the fjord and the sea. Often, these deeper waters are stagnant because of lack of circulation. Fjords are abundant on the coasts of Norway, Chile, and British Columbia.

Estuaries may be classified in yet another way, depending upon the way the salinity gradients are formed. In most estuaries, there is a gradient in salinity from full sea water (33–37 ‰) at the mouth to fresh water at the upper reaches. Fresh water is less dense than sea water and where the two meet, the fresh water will "float" on the sea water. Mixing will occur where the two come in contact, but the extent of the mixing varies with many other environmental factors, including basin shape, tide, and river flow. In estuaries where there is substantial fresh water outflow and reduced evaporation (typical temperate zone estuaries), the fresh water will move out over the top of the salt water, mixing with it near the surface and reducing the salinity, leaving the deeper waters more saline. In such a situation, a cross section of the estuary will show isohalines (lines of equal salinity), which extend upstream at the bottom (Fig. 8-2). At any given point on the estuary, a vertical column of water will have highest salinity at or near the bottom and lowest at or near the surface. This is a *positive estuary* or *salt wedge estuary*. Such salt wedge estuaries form a continuum from those with little mixing and very prominent salt wedges, through those with partial mixing and lesser salt wedges, to homogeneous estuaries where complete mixing gives similar salinities vertically from surface to bottom at any point. Where an estuary fits in this continuum depends not only on the amount of mixing of the water masses, but also on the tidal regime, the geometry of the estuary basin, and the river flow. The tidal regime and river flow may be further altered seasonally. In desert climates where the amount of fresh water input to the estuary is small and the rate of evaporation high, a *negative estuary* results. In a negative estuary, the incoming salt water enters at the surface and is somewhat diluted by mixing with the small amount of fresh water. The high evaporation rate, however, causes this surface water to become hypersaline. Hypersaline water is more dense than sea water, sinks to the bottom, and moves out of the estuary as a bottom current. A salinity profile of such an estuary is the reverse of the positive estuary with highest values at the bottom and lowest at the top (Fig. 8-2).

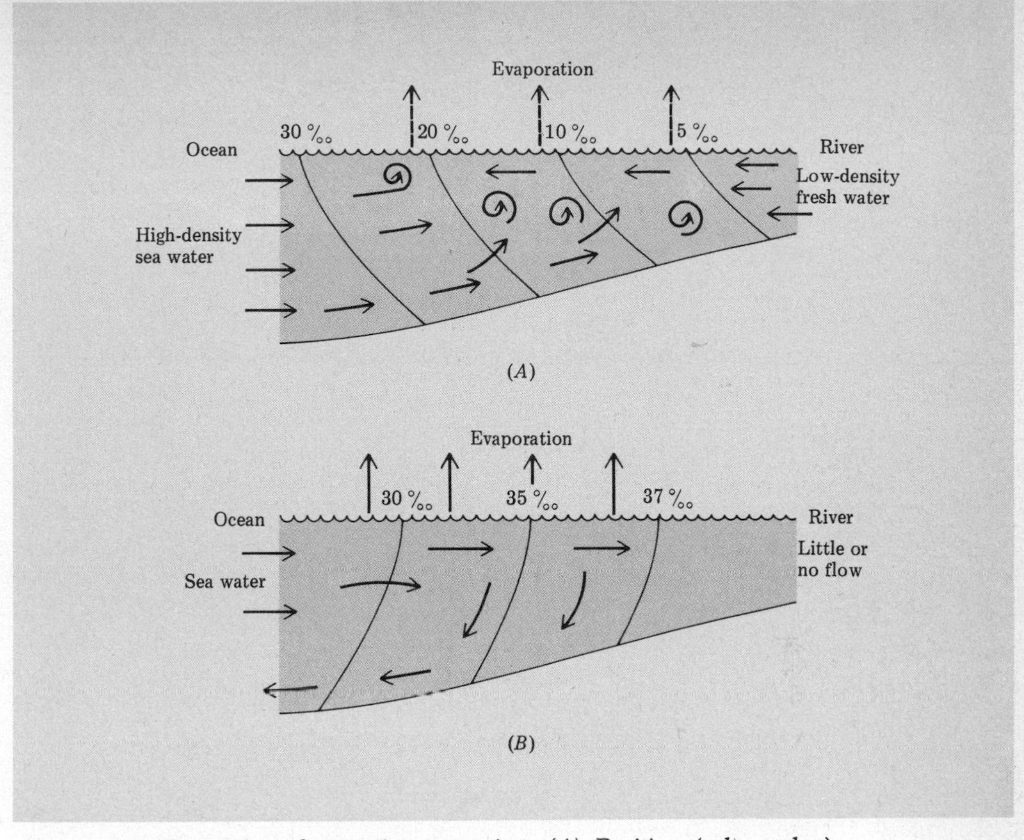

Figure 8-2 Positive and negative estuaries. *(A)* Positive (salt wedge) estuary. Fresh water runoff is significant throughout the year and greater than evaporation. Sea water enters along the bottom and gradually mixes with outward flowing fresh water. *(B)* Negative estuary. Fresh water flow is diminished or absent during part of the year. Sea water enters along the surface. Evaporation is greater than runoff so the salinity increases as one moves up the estuary. The hypersaline water sinks and flows out below the incoming sea water.

PHYSICAL CHARACTERISTICS OF ESTUARIES

The physical-chemical regime of estuaries is one with large variations in many parameters, which often create a stressful environment for organisms. It is probably due to such stresses that the number of species living in an estuarine area is small in comparison to other marine habitats.

SALINITY

The dominant feature of the estuarine environment is the fluctuation in salinity. By definition, a salinity gradient exists at some time in an estuary, but the pattern of that gradient varies seasonally, with the topography of the estuary, with the tides, and with the amount of fresh water. We have already discussed the pattern of salinity distribution in salt wedges, partially mixed, and negative estuaries (see above). There are, however, yet other factors that act to alter salinity patterns. The tide itself is one such force. Where the tidal range is significant, high tide acts to drive the salt water further up the estuary, displacing the isohalines upstream. Low tides, by contrast, displace the isohalines downstream. The result is that there is a certain area of the estuary that is subject to a salinity regime that changes with each tide (Fig. 8-3). This is the area of an estuary having maximum salinity fluctuation. The time scale of these salinity fluctuations is such that the salinity range over a 6–12 hour period equals or exceeds the entire annual range of salinities for some areas, even within an estuary.

A second force is the *Coriolis force*. The rotation of the earth has the effect of deflecting flowing water. In the Northern Hemisphere, this force deflects outflowing fresh water to the right as one looks down the estuary toward the sea. Salt water flowing into the estuary from the ocean is also displaced to the right looking from the sea toward the estuary. The opposite is true in the Southern Hemisphere. As a result, two points, each on opposite sides of the

Figure 8-3 Change of salinity in an estuary with change in tide level and/or change in river discharge. Points A and B remain at the same position, but B is inter-tidal and covered with water only at high tide. Because B is covered only at high tide, it is inundated only with high salinity water and is unaffected by low salinity water at low tide. Point A, in contrast, is covered by water of different salinity at different tidal levels.

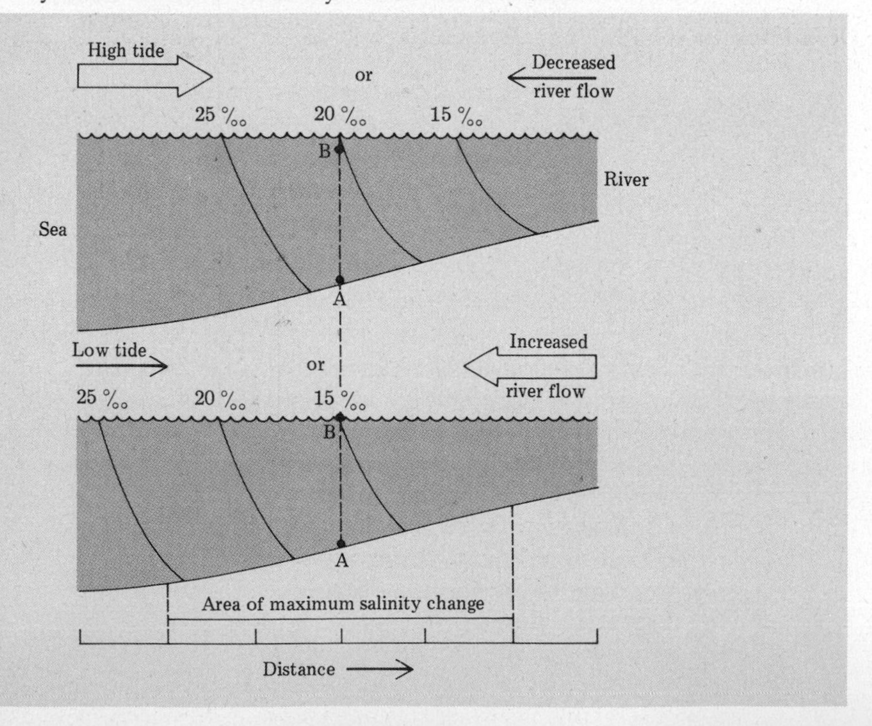

estuary equidistant from the mouth, may well have predictably different salinities (Fig. 8-4).

Seasonal changes in salinity in the estuary are usually the result of seasonal changes in evaporation and/or in fresh water flow. In areas where fresh water discharge is reduced or absent for a part of the year, higher salinities (= salt wedge) may be found further upstream. With the onset of increased fresh water flow, the salinity gradients are moved downstream toward the mouth. Therefore, at different seasons of the year, a given point in the estuary may experience different salinities.

Thus far, we have considered only the salinity changes in the water column itself. In the substrate, a different situation may prevail. Since estuarine substrates are sand or mud, water is held in the interstices between the particles. This "interstitial" water originates from the overlying water. The interstitial

Figure 8-4 Effect of Coriolis force on salinity in the Chesapeake Bay Estuary. Note the deflection of the isohalines to the right in this Northern Hemisphere estuary. Circled numbers refer to salinity values. (After D. S. McLusky, 1971, Ecology of estuaries, Heinemann Educational Books, Ltd.)

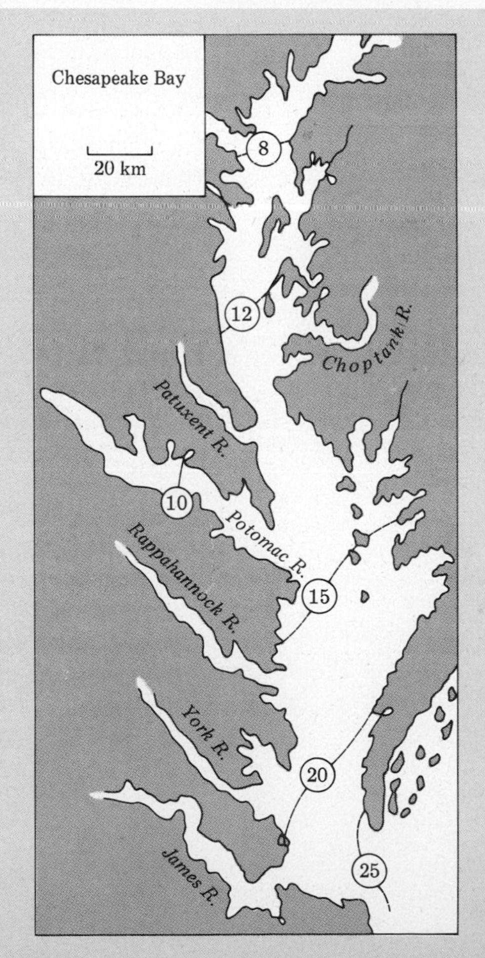

water changes in salinity much more slowly than the overlying water because of the slow interchange between the two. The interstitial water and the surrounding mud and sand are therefore "buffered" with respect to the overlying water. Organisms dwelling within the substrate are also subject to less drastic salinity changes than organisms in the water column at the same point. Since the interstitial water reflects the salinity of the overlying water, we can find a curious situation in the intertidal zones of estuaries. Here, the upper reaches are inundated only by waters of high salinity, since they are covered only when the tide is in, and that means water of maximum salinity (Fig. 8-3). By contrast, the lower intertidal areas are covered by water of lower salinity, the result of increased amounts of fresh water moving down the estuary when the tide is receding (Fig. 8-3). Therefore, the interstitial salinity remains greater higher in the intertidal than lower. This often permits marine animals to penetrate further up an estuary in the high intertidal than in the low (Fig. 8-5).

SUBSTRATE

Most estuaries are dominated by muddy substrates, which are often very soft. These are derived from sediments carried into the estuary by both sea water and fresh water. Wind (aeolian) transport of larger sand particles into the estuary is often significant in certain areas, particularly for coastal lagoons behind barrier beaches. In the case of fresh water, rivers and streams carry silt particles in suspension. When these suspended particles reach and mix with sea water in the estuary, the presence of the various ions in the sea water causes the silt particles to flocculate, thus creating larger, heavier particles, and settle out, forming the characteristic mud bottom. Sea water also carries a considerable amount of suspended material. When it enters an estuary, the sheltered conditions reduce the water motion that has been responsible for keeping the various particles in suspension. As a result, the particles settle out, contributing

Figure 8-5 Comparison of salinity fluctuations in the water column with that interstitially in the bottom mud. Data from the Pocasset River estuary in Massachusetts. (After P. C. Mangelsdoff, Jr., 1967, pp. 71–79, *in:* G. H. Lauff, editor, Estuaries, pub. no. 83. Copyright 1967 by the American Association for the Advancement of Science.)

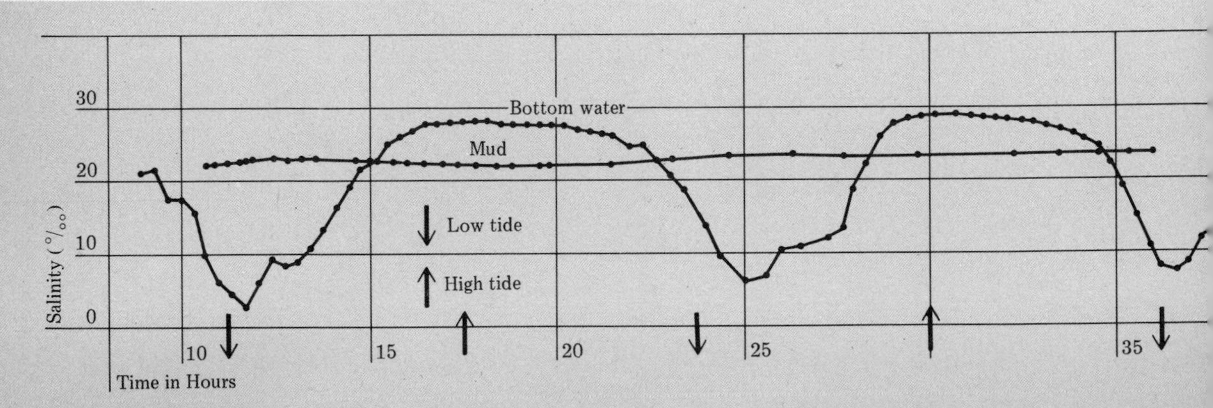

to the formation of the mud or sand substrate. The relative importance of fresh water-borne or marine-borne particles to the development of the muddy substrate varies from estuary to estuary and also geographically.

The deposition of particles is also controlled by currents and the size of the particle. Large particles settle out faster than small, and strong currents keep particles in suspension longer than slow currents. Hence, where strong currents prevail, the substrate will be coarse (sand or gravel), as only large particles are permitted to settle out; whereas where waters are calm and currents weak, fine silt will settle out. Thus, both sea water and fresh water tend to drop their coarse sediments first, the former at the mouth of the estuary and the latter in the upper reaches or the river itself, leaving the area of mixing to be dominated by fine silt (= mud), resulting from decreased water movement and flocculation from the intermixing of the two water masses.

The ecological conditions within the mud are similar to those already noted for muddy shores (see pp. 244–245). Indeed, many of the mud shores throughout the world occur in estuarine situations.

Among the particles that settle out in the estuary are many of organic origin. As a result, the substrate that accumulates is very rich in organic material. This material in turn serves as a large food reservoir for estuarine organisms. The large surface area relative to volume of the very small particles means that there is a very large area for bacteria to inhabit.

TEMPERATURE

Water temperatures in estuaries are more variable than in the nearby coastal waters. In part, this is due to the fact that there is usually a smaller volume of water in an estuary and a larger surface area, and therefore, it heats up and cools down more rapidly under prevailing atmospheric conditions (fjords, being deep and with a large volume, do not show this). Another reason for the variation is fresh water input. Fresh water in rivers and streams is much more subject to seasonal temperature change than sea water. Rivers in the temperate zones are colder in winter and warmer in summer than adjacent sea water. When these fresh waters enter the estuary and mix with the sea water, they alter the temperature. As a result, estuarine waters are colder in winter and warmer in summer than surrounding coastal waters. The time scale is of interest in that with the changing tides, a given point in the estuary will show large temperature variation as a function of the difference between sea water and river water temperature.

Because the fresh waters show the greatest temperature ranges, it follows that as one progresses up the estuary, the temperature range on an annual basis becomes greater. Similarly, the temperature range is least at the entrance to the estuary where mixing with fresh water is minimal. On a short term basis, the central area of an estuary may show the greatest change with changing tide.

Temperature also varies vertically. The surface waters have the greatest range, and the deeper waters the smallest temperature range. In salt wedge estuaries, this vertical temperature difference also reflects the fact that surface waters are dominated by fresh water, whereas the deeper waters may be completely or predominantly marine.

WAVE ACTION AND CURRENTS

Estuaries are surrounded by land on three sides. This means that the water distance over which a wind can blow to create waves is minimal, at least with respect to the oceans. Since the height of waves depends on the "fetch" or the open water distance over which the wind can blow, it follows that a small area of water can generate only small waves. The shallow depth of the water in most estuaries also precludes development of large waves. The narrowness of the mouth of the estuary, coupled with the shallow bottom, combine to dissipate rapidly the effects of any waves entering the estuary from the sea. As a result of these processes, estuaries are generally places of calm water.

Currents in estuaries are caused primarily by tidal action and river flow. Currents are generally confined to channels, but within these channels, velocities up to several knots can occur. The highest velocities occur in the middle of channels, where the frictional resistance from the bottom and side banks is lowest. Although the estuary is an overall place of sediment deposition as noted above, the channels where currents are concentrated are often areas of marked erosion. Whenever currents change position, new channels may be quickly eroded and old channels filled. This is particularly true in the intertidal areas where a natural cycle of erosion and deposition occurs. However, in most estuaries, deposition exceeds erosion so there is a net accumulation of silt.

For most estuaries, there is a continual input of fresh water at the head. A given amount of this fresh water moves down the estuary, mixing to a greater or lesser degree with sea water. A volume of this water of sufficient size is eventually discharged from the estuary or evaporated to compensate for the next volume introduced at the head. The time interval that is required for a given water mass of fresh water to be discharged from the estuary is the *flushing time*. This time interval can be a measure of the stability of the estuarine system. Long flushing times are important to the maintenance of estuarine plankton communities.

TURBIDITY

Because of the great number of particles in suspension in the water of estuaries, at least at certain times of the year, the turbidity of the water is high. Highest turbidities occur during the times of maximum river flow. Turbidity is generally at a minimum near the mouth, where full sea water occurs, and increases with distance inland. The more an estuary approaches a lagoon type, the more the turbidity is a function of the plankton concentration and/or wind speed (producing resuspension).

The major ecological effect of turbidity is to markedly decrease the penetration of light. This, in turn, decreases photosynthesis by phytoplankton and benthic plants, which reduces productivity. Under conditions of severe turbidity, phytoplankton production may be negligible, and the major production of organic matter is by emergent marsh plants (see pp. 304–305).

OXYGEN

The regular influx of fresh and salt water into the estuary coupled with the shallowness, turbulence, and wind mixing usually means that there is an ample supply of oxygen in the water column. Since the solubility of oxygen in water decreases with increased temperature and salinity, the precise amount of oxygen in the water will vary as those parameters vary. In salt wedge estuaries, or any deep estuary during the summer when a thermocline can develop and where there is a vertical salinity stratification, there is often little interchange between the oxygen-rich surface waters and the deeper layers. This isolation of the deep waters from interchange with an oxygen source, coupled with high biological activity and slow renewal or flushing rate, may deplete the oxygen of these bottom waters.

Oxygen is severely depleted in the substrate. The high organic content and high bacterial populations of the sediments exert a large oxygen demand on the interstitial water. Since the fine particle size of the sediments restricts the interchange of interstitial water with the water column above, oxygen is quickly depleted. Estuarine sediments are, therefore, anoxic below the first few centimeters unless they have large particle size and/or large numbers of burrowing animals such as *Callianassa* and *Balanoglossus* which by their activities oxygenate lower sediment layers.

THE BIOTA OF ESTUARIES

FAUNAL COMPOSITION

There are three components of the fauna of estuaries: marine, fresh water, and brackish water or estuarine.

The marine component of the fauna is the largest in terms of numbers of species and includes two subgroups. The *stenohaline* marine animals are typical marine forms that either are unable, or have very limited abilities, to tolerate salinity changes. This component is usually restricted to the mouths of estuaries where salinity is generally 30 ‰ or above. These animals are often the same species found in the open sea. *Euryhaline* marine animals compose the second subgroup. These are typical marine animals that are capable of tolerating varying amounts of salinity reduction below 30 ‰. Such species are capable of penetrating varying distances up the estuary. Most tolerate salinities down to 15 ‰, with a few hardy species tolerating levels down to 3 ‰ (Fig. 8-6).

The brackish water or true estuarine component comprises those species that are found in the middle reaches of the estuary in salinities between 5 ‰ and 30 ‰, but that are not found in fresh water or in full sea water. Examples of this component include the polychaete *Nereis diversicolor*, various oysters (*Crassostrea*, *Ostrea*) and clams (*Scrobicularia plana*, *Macoma balthica*, *Rangia flexuosa*), certain small gastropods (*Hydrobia*), and shrimps (*Palaemonetes*) (Fig. 8-7). It may be, however, that certain of these estuarine genera are limited in a seaward direction not by physiological tolerances but by biological interactions such as predation. Hence this component may not be easily defined.

The final component is derived from fresh water. These animals usually cannot tolerate salinities above 5 ‰ and are restricted to the upper reaches of the estuary (Fig. 8-6).

There is also what we might term a transitional component. This includes those organisms such as migratory fishes that pass through the estuary on their way to breeding grounds, either in fresh or salt water. Common examples are the salmon (*Salmo*, *Oncorhynchus*) and the eels (*Anguilla*). Also included here are those forms that spend only a part of their life histories in the estuary. Usually it is the juvenile stage, the adults being found at sea. Good examples of this latter group are the various shrimps of the family Penaeidae (*Penaeus setiferus*, *P. aztecus*, *P. duorarum*), which form the basis of the Gulf of Mexico shrimp fishery. The young occur in estuaries. The transitional fauna also includes those forms that enter the estuary only to feed and includes many birds and fishes.

Figure 8-6 Numbers of species in each of the three major components of the fauna of estuaries and their distribution with salinity. (After D. McLusky, 1971, Ecology of estuaries, Heinemann Books, Ltd.)

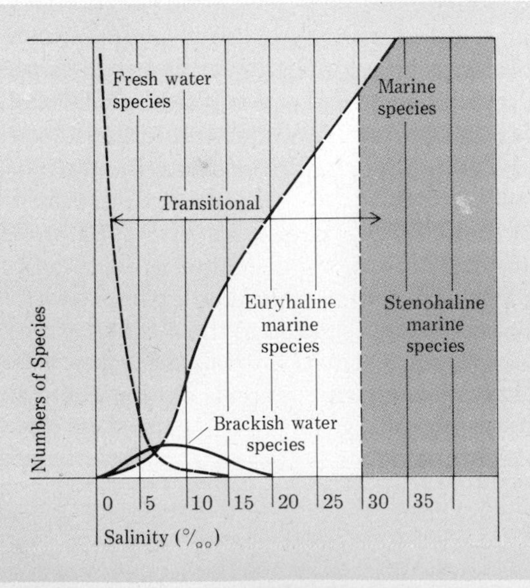

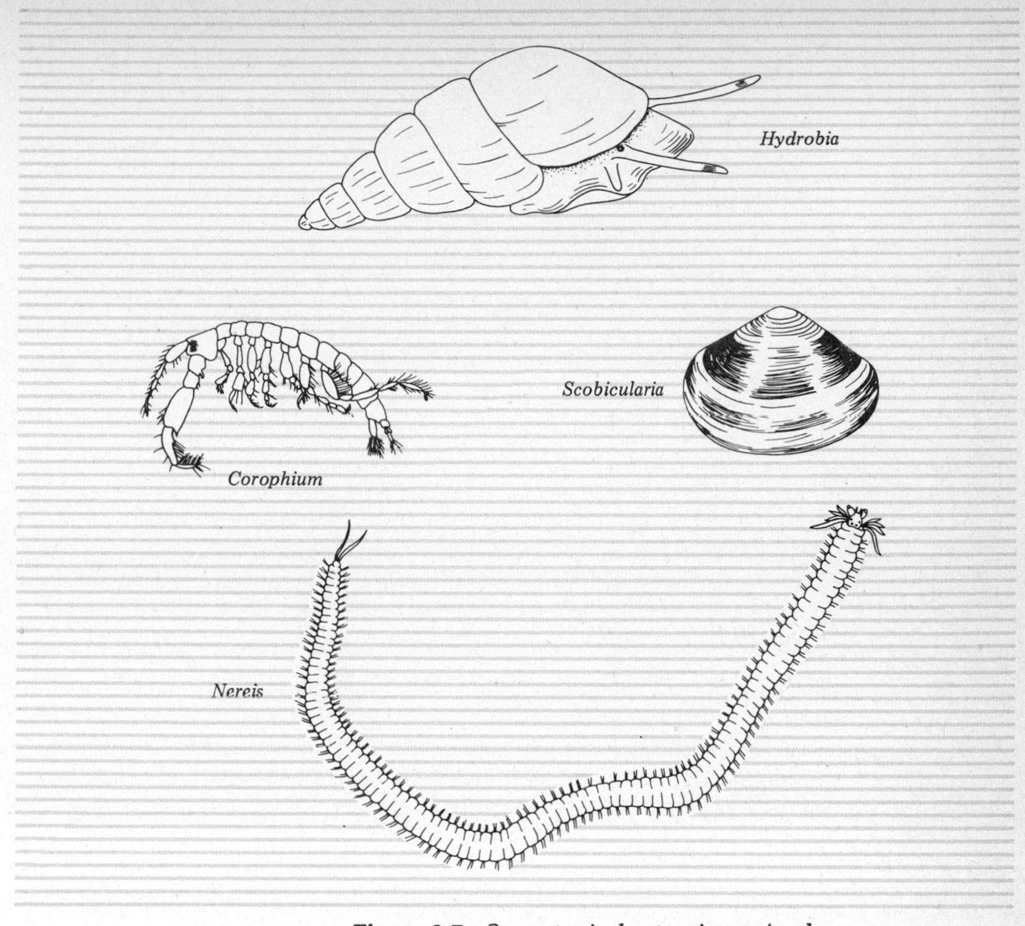

Figure 8-7 Some typical estuarine animals.

The number of species of organisms inhabiting estuarine systems, as noted by Barnes (1974), is generally significantly lower than the number inhabiting nearby marine or fresh water habitats. This is probably partly due to the inability of fresh water organisms to tolerate the increased salinities and marine organisms to tolerate the decreased salinities of the estuary.

The true estuarine organisms are derived primarily from marine stocks and not fresh water. This is similar to the situation in the other transitional zone, the intertidal, which is also populated mainly with marine organisms, not terrestrial forms. In contrast to the intertidal, however, the number of true estuarine species is very small, and the middle reaches of estuaries are depauperate (Fig. 8-6).

Because the marine animals are able to tolerate a greater reduction in salinity than the fresh water species are able to endure salinity increases, and because the true estuarine organisms are primarily derived from marine stocks, the majority of the estuary is inhabited by marine animals.

Why are there so few estuarine species? The commonest explanation is that the fluctuating environmental conditions, mainly salinity, are of such magnitude that only a few species have been able to evolve the necessary physiological specializations to enable them to exist there. Another explanation is that estuaries have not existed long enough in geological time to permit a complete estuarine fauna to develop. Yet a final reason may be that estuarine areas have little topographic diversity, being mainly broad expanses of mud. There are fewer niches and therefore, fewer species. It is not possible at this time to say if one, all, or none of the above are responsible for the depauperate condition, but apparently euryhalinity is a trait not easily acquired.

ESTUARINE VEGETATION

The flora of estuaries is also depauperate. Most of the permanently submerged portions of estuaries consist of mud substrates unsuitable for macroalgal attachment. In addition, the highly turbid water restricts light penetration to a narrow upper layer. As a result, the deeper layers of the estuary are often barren of plant life. The uppermost layers of water and the intertidal zone have a limited number of plants. In the lower reaches of the estuary at and below mean low water, there may be beds of sea grasses (*Zostera*, *Thalassia*, *Cymodocea*). These beds are considered as subtidal communities in Chapter Five.

The intertidal mud flats are inhabited by a limited number of green algal species. Common genera include *Ulva*, *Enteromorpha*, *Chaetomorpha*, and *Cladophora*. These are often seasonally abundant, disappearing during certain seasons (Fig. 8-8).

Estuarine mud flats often have an abundant diatom flora. In fact, as Lackey (1967) noted, the benthic diatoms are more abundant in estuaries than their planktonic relatives. Many are motile and undergo a rhythmic migratory pattern, moving up to the surface or down into the mud depending upon the illumination.

The highly turbid water in estuaries means that by far the dominant vegetation in terms of biomass is emergent plants. These are generally long-lived flowering plants that root in the upper intertidal and form the characteristic *salt marsh* communities that fringe estuaries throughout the temperate zones of the world. The dominant genera include *Spartina* and *Salicornia* (see subsequent sections of this chapter for discussion). In the tropics, salt marshes are replaced by mangrove forests (see Chapter Nine for a discussion).

Salt marshes are highly productive regions, in some cases responsible for most of the primary productivity in the estuary. In addition, they act as a trap for sediments, accelerating the rate of deposition around their bases. As this deposition continues, the height of the mud flat increases, establishing conditions that are less and less marine and leading to replacement of the typical salt marsh plants with more terrestrial species. This succession gives rise to a series of vegetative zones progressing away from the water. In this way, salt marshes tend to reduce aquatic habitat by converting it into dry land.

A final component to mention is the bacteria. Both the water and the mud of

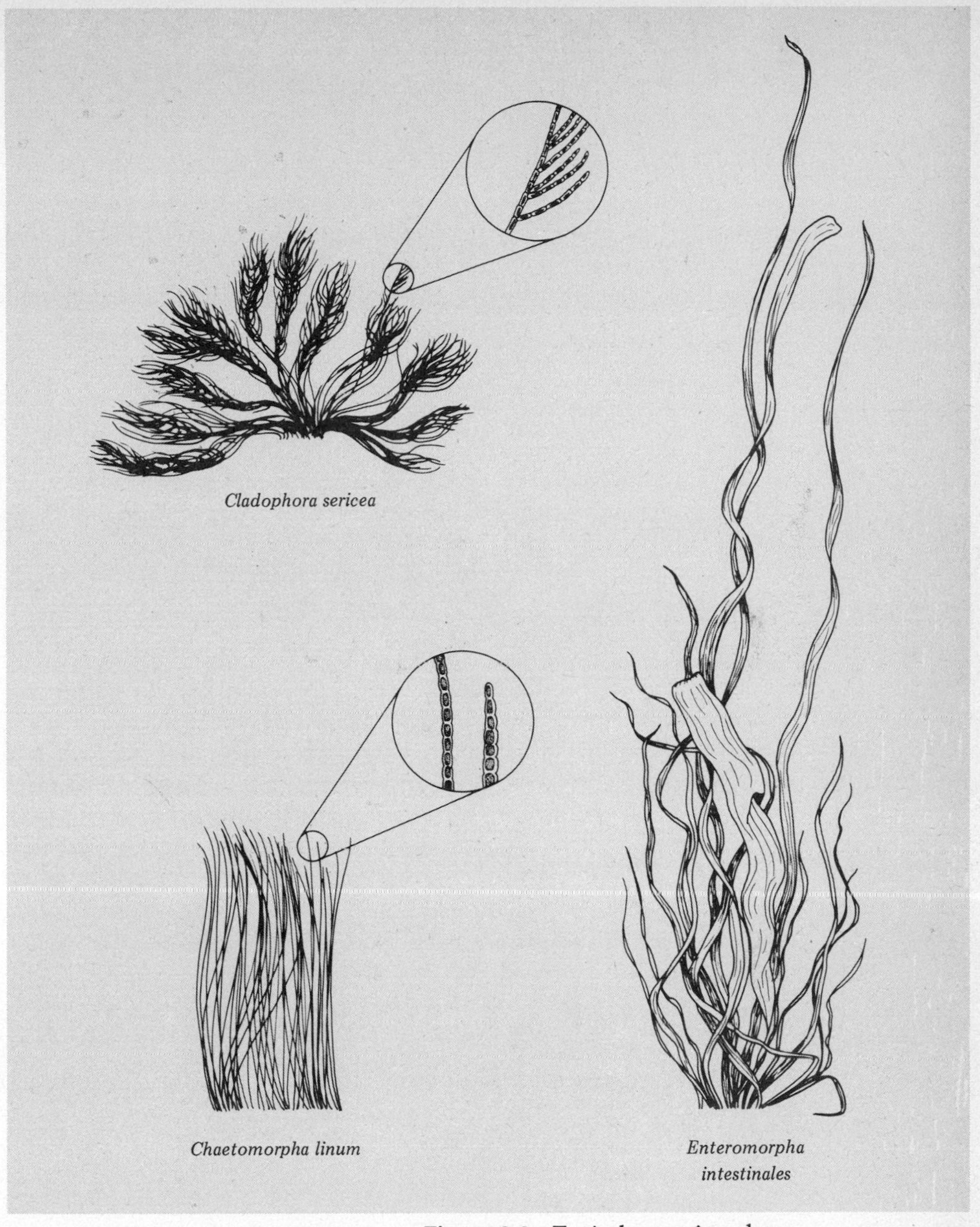

Figure 8-8 Typical estuarine algae.

estuaries are extremely rich in bacteria, due to the abundance of organic matter to decompose. Estuarine waters have been shown by Zobell and Feltham (1942) to contain hundreds of times more bacteria than sea water, and the upper layers of mud more than a thousand times more bacteria than the overlying water. Densities of bacteria in estuarine muds of 100–400 million per gram have been reported.

ESTUARINE PLANKTON

The estuarine plankton is depauperate in number of species. It thus follows the same trend as observed for the macrofauna and macrovegetation. Diatoms frequently dominate the phytoplankton, but dinoflagellates may achieve dominance during the warmer months and may remain dominant in some estuaries at all times. Dominant diatom genera include *Skeletonema*, *Asterionella*, *Chaetoceros*, *Nitzchia*, *Thalassionema*, and *Melosira*. Abundant dinoflagellate genera include *Gymnodinium*, *Gonyaulax*, *Peridinium*, and *Ceratium*. The phytoplankton may also be temporarily enriched by resuspension of characteristic bottom-dwelling diatoms. High turbidity and rapid flushing may restrict phytoplankton numbers and productivity in some estuaries. However, where turbidity is low and flushing time long, diverse populations and relatively high productivity may result. Consequently, depending on prevailing conditions, estuaries may differ considerably among themselves with respect to phytoplankton numbers and productivity.

Zooplankton in estuaries mirror the phytoplankton in being limited in species composition. The species composition also varies, both seasonally and with salinity gradients up the estuary. The few true estuarine zooplankters occur in larger, more stable estuaries, where salinity gradients are less variable; shallow, rapidly flushed estuaries are inhabited mainly by a typical marine zooplankton assemblage carried in and out with the tide. Characteristic estuarine zooplankters include species of the copepod genera *Eurytemora*, *Acartia*, *Pseudodiaptomus*, and *Centropages*; certain mysids such as species of the genera *Neomysis*, *Praunus*, and *Mesopodopsis*; and certain amphipods such as species of *Gammarus*. The estuarine zooplankton average about 1 ml/m^3 displacement volume, or somewhat greater than the concentration in adjacent coastal waters.

ADAPTATIONS OF ESTUARINE ORGANISMS

The variable nature of the estuarine habitat, especially defined by fluctuating salinities and temperatures, makes this a particularly stressful and rigorous habitat. In order for organisms to survive and successfully colonize this area, they must possess certain adaptations. In this section, we shall discuss some of these adaptations.

MORPHOLOGICAL ADAPTATIONS

Few morphological adaptations can be recognized among estuarine organisms that can be attributed simply to living under conditions of fluctuating temperature and salinity. Most are simply the result of adaptations to a given habitat such as burrowing into mud. (See Chapter Six for a discussion of adaptations of organisms living in mud.) Mud-dwelling organisms whether estuarine or not often have, for example, fine fringes of hair or setae, which guard the entrances to respiratory chambers so as to prevent the clogging of those respiratory surfaces by the silt particles. Such a situation prevails for estuarine crabs and many bivalve mollusks.

Other morphological changes in estuarine organisms, reported by Remane and Schlieper (1971), include a generally smaller body size than relatives living in full sea water and a reduced number of vertebrae among fishes. Reproduction is also affected. Marine species often have a lowered reproductive rate and lowered fecundity; fresh water species may be partially sterile.

PHYSIOLOGICAL ADAPTATIONS

The dominant adaptations that are required for continued estuarine life are those associated with maintaining the ionic balance of the body fluids in the face of fluctuating external salinities. *Osmosis* is the name given to the physical process in which water passes through a semipermeable membrane, separating two fluids of different salt concentration, moving from the area of lower to higher salt concentration. The ability to control the concentration of salts or water in internal fluids is called *osmoregulation*. Most marine organisms do not have the ability to control their internal salt content and are *osmoconformers*. Their ability to penetrate estuaries is thus limited to their tolerance for changes in their internal fluids. *Osmoregulators* are those organisms that have the physiological mechanisms to control the salt content of their internal fluids. Most estuarine animals either are osmoregulators or else possess the ability to function with fluctuating internal salt concentrations.

Penetrating into an estuary means encountering water with lowered salinity. Since the internal salt concentration of marine species is higher than that of the estuarine water, water tends to move across membranes into their bodies in an attempt to equalize the concentrations. Regulation means excreting the excess water without losing salts or excreting water and salts and replacing lost salts with active uptake of ions from the environment. For fresh water animals, which move from a more dilute to a less dilute medium when entering estuaries, the reverse is true.

Osmoregulatory ability is found in several taxa of more advanced estuarine animals, primarily polychaete worms, mollusks, and crustaceans. This field of osmoregulation has been the subject of intensive research for many years, resulting in a large literature. Detailed coverage is beyond the scope of this introductory book, but students interested in more in-depth studies may refer to the literature cited at the end of this chapter. Suffice it to say that osmoregulation can operate in three ways: (1) animals may move water; (2) they may move ions; or (3) they may adjust the internal water-ion balance. In advanced invertebrates and vertebrates, the osmoregulatory organ is generally the kidney, where excess water is excreted and needed ions resorbed. However, special cells may also exist in other parts of the body, particularly for taking up or removing certain ions.

In certain soft-bodied forms, such as polychaete worms, the osmoregulatory mechanism is developed, but relatively slow to respond. These animals are able to tolerate wide ranges of internal ion concentrations, at least for certain periods of time. *Nereis diversicolor*, for example, when placed in dilute sea water takes up water due to osmosis. After some time, however, it begins to osmoregulate and, according to McLusky (1971), the internal water content falls to near that found when it was in full sea water (Fig. 8-9). This osmoregulation not only is delayed, but does not come into effect until the salinity of the external environment falls below a certain level. That is, *N. diversicolor* remains an osmoconformer as salinity decreases and regulates only when the external salinity falls below a certain level (Fig. 8-9).

Bivalve mollusks are generally poor osmoregulators and respond to drastic salinity decreases by closing up in their shells to prevent excessive dilution of internal fluids with water. Certain gastropods, such as *Hydrobia*, do have limited osmoregulatory ability, and the members of the genus are found in enormous densities in northern European estuaries, where Muus (1963) reports they live in salinities down to 5 ‰ or lower.

Among higher crustaceans, such as crabs, osmoregulation is well developed. The combination of greatly restricted body permeability due to the exoskeleton and the pronounced ability to regulate ion concentrations of internal fluids are probable reasons for their success in estuaries. Osmoregulation is primarily through water removal by the excretory organs coupled with ion uptake from the environment to make up for unavoidable losses in water removal. In addition, the cells of crustaceans are regulated by changes in the concentration of amino acids. This "intracellular osmotic regulation" results from movement of free amino acids in and out of individual cells to equalize the osmotic concentration in the cells with the extracellular bathing fluid, thus preventing excessive water buildup and possible rupture of the cells. This ability to regulate osmotic concentrations can be seen in a graph from Barnes (1967) contrasting the salt concentration of the blood of the Australian estuarine crab, *Australoplax tridentata*, with that of the external medium (Fig. 8-10).

As Kinne (1963) has noted, there appears to be some correlation between temperature and the ability to osmoregulate. In the tropics where water temperatures are higher and the difference in temperature between fresh and salt water is minimal, there are considerably more estuarine species, and stenohaline marine species penetrate further upstream than they do in the temperate zone estuaries. The reasons for this phenomenon are not as yet clearly understood.

BEHAVIORAL ADAPTATIONS

A certain number of behavioral adaptations are common to many estuarine organisms. Among the invertebrates one such adaptation is burrowing into the mud. Although this adaptation is certainly not one which is exclusive to

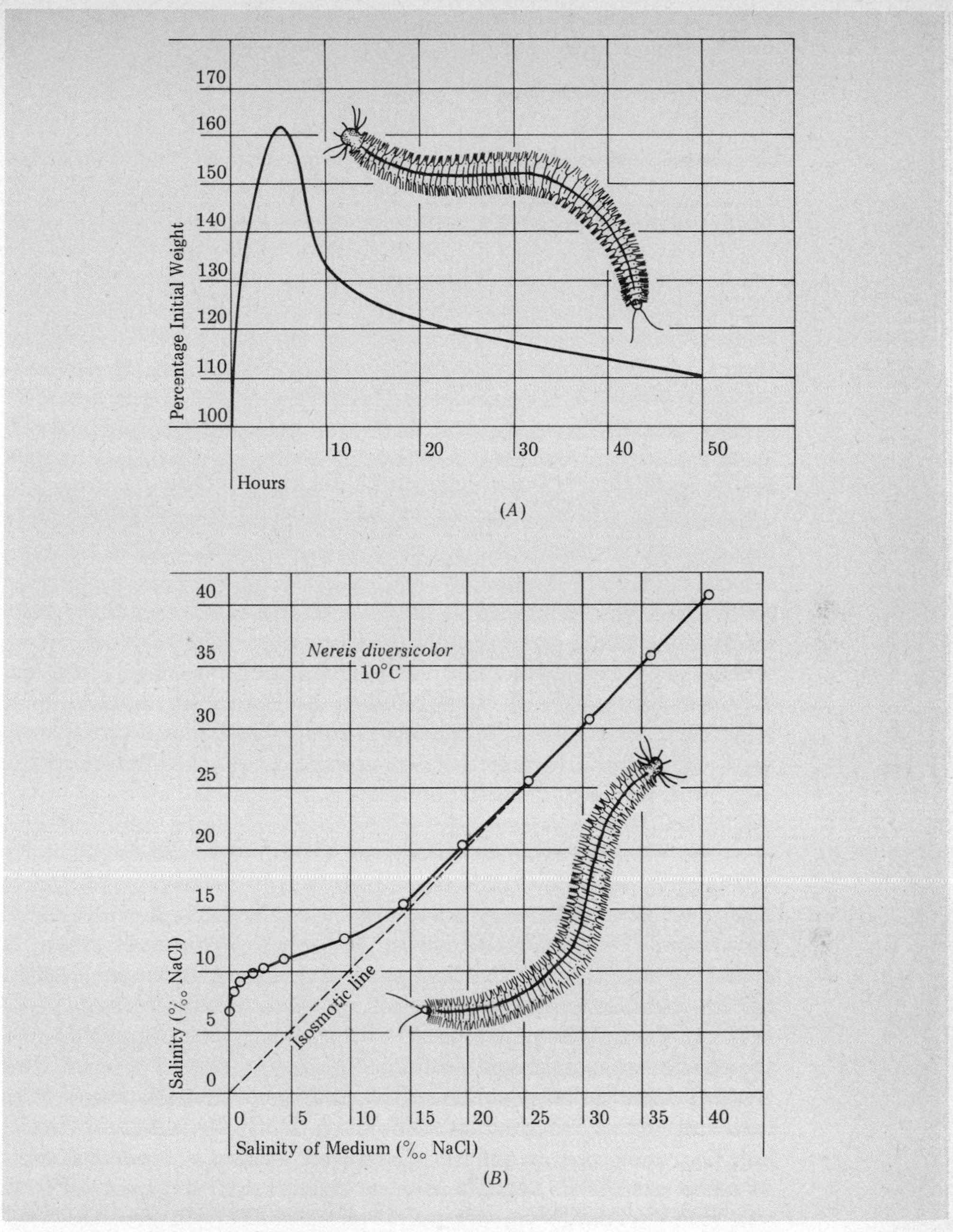

Figure 8-9 Changes in the body fluids of *Nereis diversicolor* with change in salinity. *(A)* Weight changes occurring after the animals are transferred to 20 percent sea water. Initial weight increase is due to the take-up of water by osmosis. The weight decrease occurs when the animal osmoregulates. *(B)* Osmotic concentration of the body fluids in relation to salinity change. The straight line on the graph indicates osmoconformity. (*A, B,* after D. S. McLusky, 1971, Ecology of estuaries, Heinemann Educational Books, Ltd.)

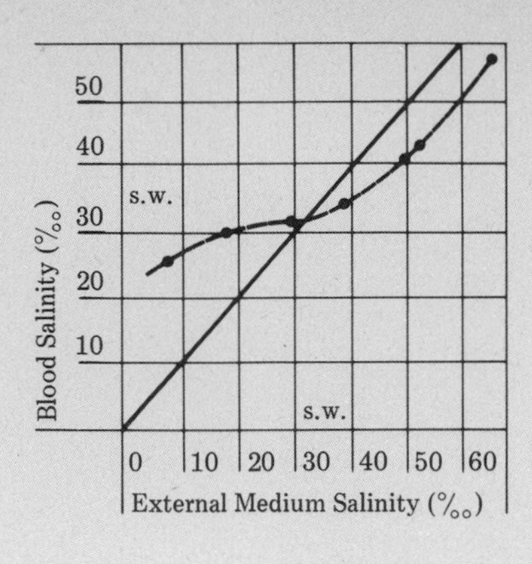

Figure 8-10 Graph showing the change in salinity of the blood of the crab *Australoplax tridentata* with change in the salinity of the external medium. (Modified from D. McLusky, 1971, Ecology of estuaries, Heinemann Books, Ltd.)

estuarine animals, being common to many invertebrates in soft sediments throughout the world's oceans, it has two beneficial effects for an estuarine animal. First, in the case of those species with limited or imperfect osmoregulatory abilities, existence in the mud means exposure to the interstitial water, which has much less salinity and temperature variation than the open water above. It is thus a means to achieve reduced salinity and temperature changes. Second, burying in the substrate means that they are less likely to be consumed by surface or water-dwelling predators such as birds, fishes, or crabs.

Another behavioral adaptation is to change position on the substrate as the species moves up or down the estuary. This behavior will have the effect of keeping the organism in an area where the salinity change is minimal. Consider, for example, a stenohaline epifaunal species that thrives best at a salinity of 25 ‰. As such a species penetrates farther up a salt wedge estuary, it remains under a salinity regime of 25 ‰ if it moves to deeper water as it progressed up the estuary (Figs. 8-2, 8-3). Such behavior means that stenohaline animals will be found deeper farther up an estuary.

Many adult estuarine crabs are able to live successfully in low salinities due to their developed osmoregulatory abilities. However, their eggs and young often lack these regulatory abilities. Therefore, many exhibit a specific migratory pattern, moving from the estuary to the adjacent sea for the breeding season. One example of this type of behavior is the crab *Eriocheir sinensis*, which has been introduced into Europe from its native China. Although adults live hundreds of miles up into rivers, *E. sinensis* migrates back to the sea to breed and to spend the early part of larval life. Similarly, the blue crab *Callinectes sapidus* lives as an adult in estuaries of the American East and Gulf coasts such as Chesapeake Bay. The females migrate to waters of higher salinity to hatch the young. Once the larval stages are passed, the young crabs migrate back up the estuary (Fig. 8-11).

On the other hand, certain other species take advantage of the great amount of food in the estuary and the relative paucity of predators to use the estuary as a

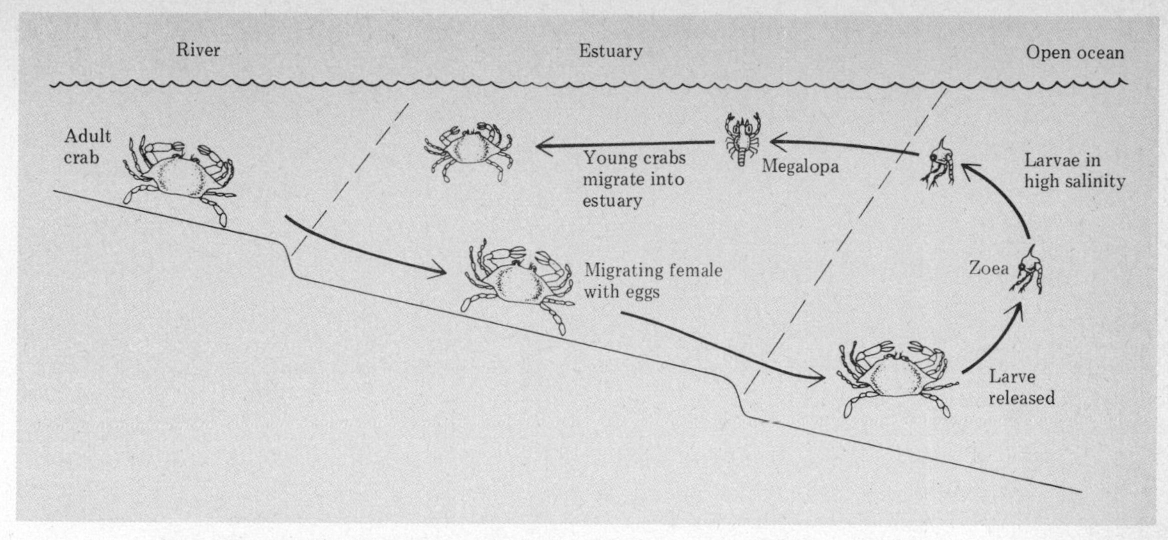

Figure 8-11 The life cycle of the blue crab *Callinectes sapidus* in estuaries of the East Coast of the United States.

nursery for their young. Most of these species are fishes. Examples include mullet (*Mugil* sp.), striped bass (*Roccus saxatilis*), and flounder (*Platichthys flesus*), which enter the estuary as juveniles and migrate back to the sea when mature.

ECOLOGY OF ESTUARIES

Having considered the major physical-chemical features of estuaries, the common organisms, and the major adaptations of the organisms, it is now possible to consider certain aspects of the ecology of these transition areas.

PRODUCTIVITY, ORGANIC MATTER, AND FOOD SOURCES

Ignoring for now the salt marsh, the primary productivity of estuaries resides in the phytoplankton, the benthic diatoms, seagrasses, and the various algal mats. Of these, the benthic diatoms and algal mats appear to be significant. Considering all these sources together, however, primary productivity by algae has traditionally been considered to be very low. Yet, estuaries are areas that have large amounts of organic material, large numbers of organisms, and high secondary productivity. If algal primary productivity is low, where does this organic material originate? Classically it has been considered that the major source of primary productivity in estuaries resided in the emergent plants of the salt marsh surrounding the estuary (see pp. 304–305). In a study by Teal (1962) on a Georgia salt marsh, it was found that there was a net productivity of 6,850 kcal/m^2/year, whereas the various algae contributed only 1,600 kcal/m^2/year. Algae may, however, be more important in some areas such as the Pacific Coast of North America, where Zedler et al. (1978) have shown they contribute significantly to primary productivity. More recently, studies by Haines (1979) in the Georgia marshes have suggested that the detritus derived from benthic and planktonic algae forms the bulk of the organic detritus. Haines found salt marsh plant detritus to be largely accumulated and consumed within the marsh itself leaving the algal detritus as the major material exported. In this scenario, the role of the marsh is primarily as a nursery for fish, crab, and shrimp, which are later exported to coastal waters.

Primary productivity in and around the estuary, however, is not the only source of organic material. As we noted previously, estuaries act as sinks for organic material brought down by rivers and in from the sea. Thus, organic matter arrives from these sources as well. It is difficult to assess the role of primary productivity within the estuarine system in contributing to the total organic production for several reasons. In the first place, there are few herbivores that feed directly on the plants themselves. Therefore, most of the plant material must be broken down to detritus before entering the various food webs. This process of breakdown involves bacterial action, which is an additional complicating factor. Also, once the plant material has become detritus, it is not easily distinguished from other organic detritus brought into the system by the river and the sea. Thus, of the great amounts of organic material that move through the estuary, some fraction is produced in the estuary itself (autochthonous), but some is transported in, the result of primary production elsewhere (allochthonous), including transport from the surrounding salt marsh. Currently the relative amount from each source is under debate. In Georgia salt marshes, for example, Teal (1962) has estimated that 45 percent of the net production is exported to the estuary, whereas Haines (1979) suggests that not only are we uncertain regarding the quantities of organic material moving through estuarine systems, but that there may be little or no export from the system.

The detritus forms a substrate for a rich bacterial and algal growth, which, in turn, serves as an important food source for various suspension and detritus feeding animals. When one refers, therefore, to the detritus food source of estuaries, detritus is used in the widest sense to refer to organic particles, bacteria, algae, and even associated protozoans. The net effect of this detritus is to form an accumulation of food for estuarine organisms. As an example of the amount of food, Odum and de la Cruz (1967) noted that open sea water contains 1–3 mg of dry organic matter per liter, whereas drainage waters in estuaries may contain up to 110 mg dry weight organic matter per liter.

PLANKTON CYCLES

Although the nutrient levels in estuaries are high, they are not well balanced; nitrogen is often low and may even occasionally be limiting to estuarine phytoplankton. As we observed in the open sea (Chapter Two), there is an annual cycle of phytoplankton in estuaries also. Low phytoplankton populations usually occur in late fall and winter due to reduced light and high turbidity, resulting from high river runoff and turbulence. This is followed by a later winter diatom bloom. This bloom is terminated in late spring, often not by zooplankton grazing, but probably by depletion of nitrogen sources leading to the accumulation of diatoms on the mud surface. Populations remain low in the summer due to low nutrient levels and grazing, but there may be occasional flowerings of dinoflagellates leading to "red tide" situations. A large fall bloom occurs in some estuaries that rivals the spring bloom, whereas in others, it is absent or reduced.

The zooplankton cycle in estuaries is quite variable and, as Deevey (1960) found, may even change drastically from year to year. It does not closely follow the phytoplankton cycle. Zooplankton peak in number later in the spring than the phytoplankton, and their numbers remain high throughout the summer, declining through the fall to winter lows. Zooplankton in estuaries appear to be consuming only 50–60 percent of the net phytoplankton production. This leaves a significant portion available to the bottom suspension feeders and indeed, in shallow estuaries, the suspension feeding benthos competes directly with the zooplankton for food, a situation not present in offshore waters.

FOOD WEBS

Low primary productivity in the water column, few herbivores, and the presence of large amounts of detritus have traditionally suggested that the basis of the food web in estuaries is detritus. This claim is currently under debate among scientists, but is a convenient starting point.

To some extent to consider detritus as the basis of estuarine food webs is misleading for it implies that detritivores actually digest the organic particles. In fact, they most likely are digesting the bacteria and other microorganisms on the particles and excreting the particles without further degradation. This should be kept in mind in the following discussion.

Detritus may be consumed directly, either while in suspension in the water mass by various suspension feeding benthic invertebrates, or more commonly by direct consumption of the material in or on the substrate. Suspension feeders, such as the clams *Cardium* and *Mya* may obtain a small portion of their food from suspended detritus along with phytoplankton and zooplankton. Direct detritus consumers include certain clams such as *Scrobicularia* and *Macoma*, which have separated siphons and feed by "vacuuming" up the detrital particles from the substrate via their inhalent siphons (Fig. 6-32).

Polychaete worms are very abundant in estuaries and are represented by many detritus feeders. Members of the families Spionidae, Terebellidae, and Ampharetidae spread out long tentacles over the surface of the mud and collect and transport the organic particles via ciliary-mucus tracts. Still others, such as the families Capitellidae and Arenicolidae, ingest the substrate directly, digesting out the organic material and bacteria during passage through their guts. Bacteria may be consumed directly by small protozoans and nematodes, both of which occur in large numbers in the upper layers of the mud.

Still other deposit feeders, such as amphipods of the genus *Corophium*, mechanically sort particles using their appendages, while small gastropods such as *Hydrobia* and *Batillaria* rasp up the surface of the mud (Fig. 8-12). Such sorting out or digesting out of the sediments leads to a reworking of the substrate.

The large numbers of detritus feeders and suspension feeders are, in turn, consumed by a number of predators, both vertebrate and invertebrate. The main invertebrate predators are various species of crabs and shrimps, certain polychaetes such as the genera *Glycera*, *Nepthys*, and perhaps some *Nereis*, and various predatory gastropod mollusks (*Polinices*, *Aglaja*, *Chelidonura*, and *Busycon*).

The dominant predators, however, are various species of fishes and birds. Many estuarine fishes are predaceous. In European estuaries, for example, Hartley (1940) found that the plaice (*Pleuronectes platessa*) feeds on polychaetes, while the flounder (*Platichthys flesus*) feeds on mysids, shrimps, and amphipods. In Louisiana estuaries, Darnell (1961) found the three most important larger invertebrate species eaten by fishes to be the clam *Rangia cuneata* and two crabs, *Rithropanopeus harissi* and *Callinectes sapidus*. Different fish species tend to concentrate on different types of animals, and feeding habits change with age. A common pattern in estuarine fishes is to progress from eating zooplankton, to detritus, to macroinvertebrates to even other fishes. Much of the productivity of estuaries that ends up in fishes is ultimately lost to the estuarine system when the fishes move offshore as adults.

Estuarine birds include many ducks and geese as well as a host of shore birds, wading birds, and gulls and terns. The majority of the shore birds feed upon the infaunal invertebrates of the mud flats, which they obtain by probing at low tide. The variety of bill lengths among these birds enables the different species to feed

Figure 8-12 Mud snails, *Batillaria attramentaria*, on the surface of the mud in a California estuary. (Photo by the author.)

at different levels in the mud (Fig. 8–13). They follow the rise and fall of the tide during daylight hours to obtain their food. Observations and analyses of stomachs of these birds suggest that they are capable of consuming large numbers of invertebrates. Studies by Drinnan (1957), Goss-Custard (1969), and Prater (1972) have reported oyster catchers (*Haematopus ostralegus*) to consume 315 *Cardium*; red shanks (*Tringa totanus*), 40,000 *Corophium*; and knots (*Calidrus canutus*), 730 *Macoma* per day. The effect of this predation on the populations of infaunal invertebrates is not well known, but various scientists have suggested that birds take between 4–20 percent of the populations.

In recent years, there has been considerable interest in conducting field experiments that exclude large predators from the soft bottom invertebrate communities of estuaries to determine the effect of predators on these communities. In summing up the results of these experimental manipulations, Peterson (1979) has noted that when such systems are freed from predation, the usual results are for the density of benthic organisms to increase, but without a corresponding tendency toward competitive exclusion by certain dominants. This is in marked contrast to the results we have seen on rocky shores where predator removal has led to a competitive dominant occupying all space, forcing other species out. Why should such a different result occur here? Peterson (1979) has suggested several reasons. The initial one is simply that such experiments

Figure 8-13 Diagram showing the bill length of some common European shore birds in relation to the depth of some common estuarine invertebrates. (After J. Green, 1968, The biology of estuarine animals, University of Washington Press.)

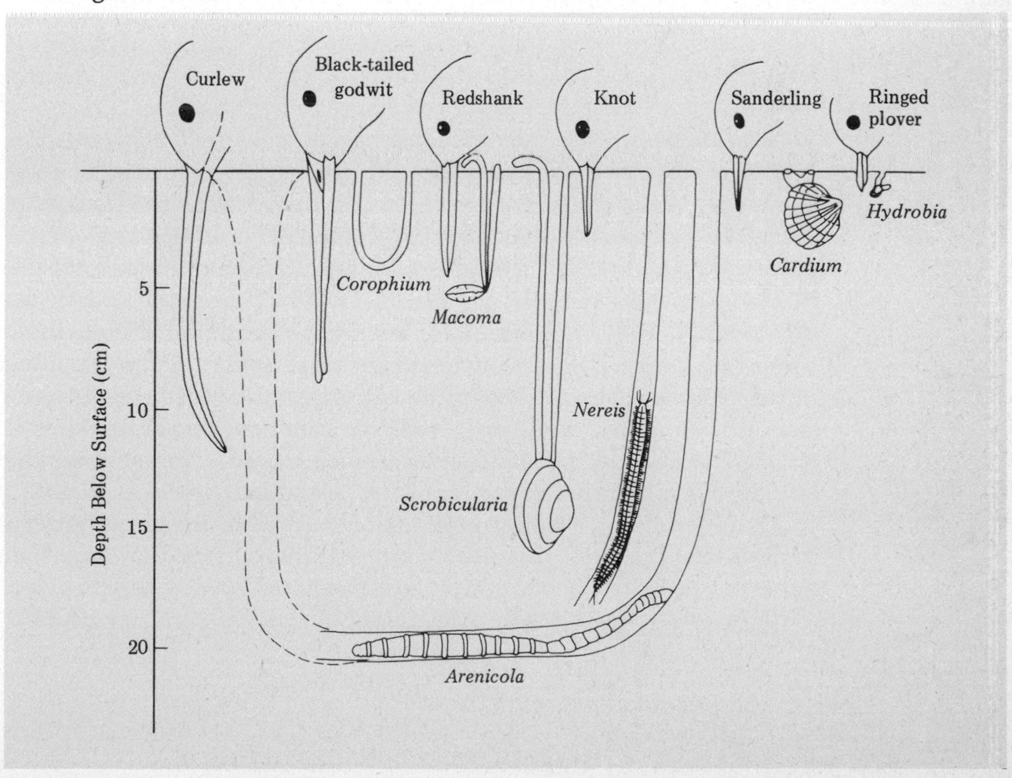

have not been conducted for a sufficient length of time. A second is that interference competition, which operates in the rocky intertidal, cannot operate in these soft muds, due to the inability of the organisms to obtain a purchase from which to push away or crush another organism and due to the ability of the organisms to exploit the third dimension (= depth) of the soft substrate. Competitive exclusion by overgrowth, common on rocky shores, is also precluded because of high sediment mobility, which excludes those species best able to employ this mechanism, namely, colonial or clonal organisms. It may also be possible that various adult-larval interactions are of sufficient intensity and development that they can hold the community at a density below its carrying capacity and hence, prevent competitive exclusion without the necessity of predation.

The estuarine food web can be summarized as having a number of pathways of energy flow generated primarily from a detrital base, and by a somewhat incomplete use of energy, allowing export to other areas primarily by means of fishes and birds. It is the abundance of food and the relative lack of predators that has made estuaries nurseries exploited by the juveniles of many animals that spend their adult lives elsewhere. It is also the reason for their use as a feeding ground by migratory adults of many species of birds and fishes. The complex interactions of the estuarine food web are summarized in simplified form in Figure 8-14.

SALT MARSHES

Bordering temperate and subpolar estuaries and protected marine shores and embayments throughout the world are special plant associations known as salt marshes (Fig. 8-15). In the tropics and subtropics, these are replaced by mangrove associations. (See Chapter Nine for discussion.) These marshes form a transition area between aquatic and terrestrial ecosystems.

DEFINITION AND CHARACTERIZATION

Salt marshes are communities of emergent vegetation rooted in soils that are alternately inundated and drained by tidal action. They occur mainly at the higher tidal levels in areas of protected water and most often in association with estuaries. Since the dominant plants are emergent flowering plants, they invade only the shallowest intertidal areas. They are *halophytes*, meaning that they grow in soils with a high salt content. The dominant plants are herbaceous angiosperms. Because the upper portions of the plants are above water even during periods of high tide, this association has both terrestrial and aquatic components.

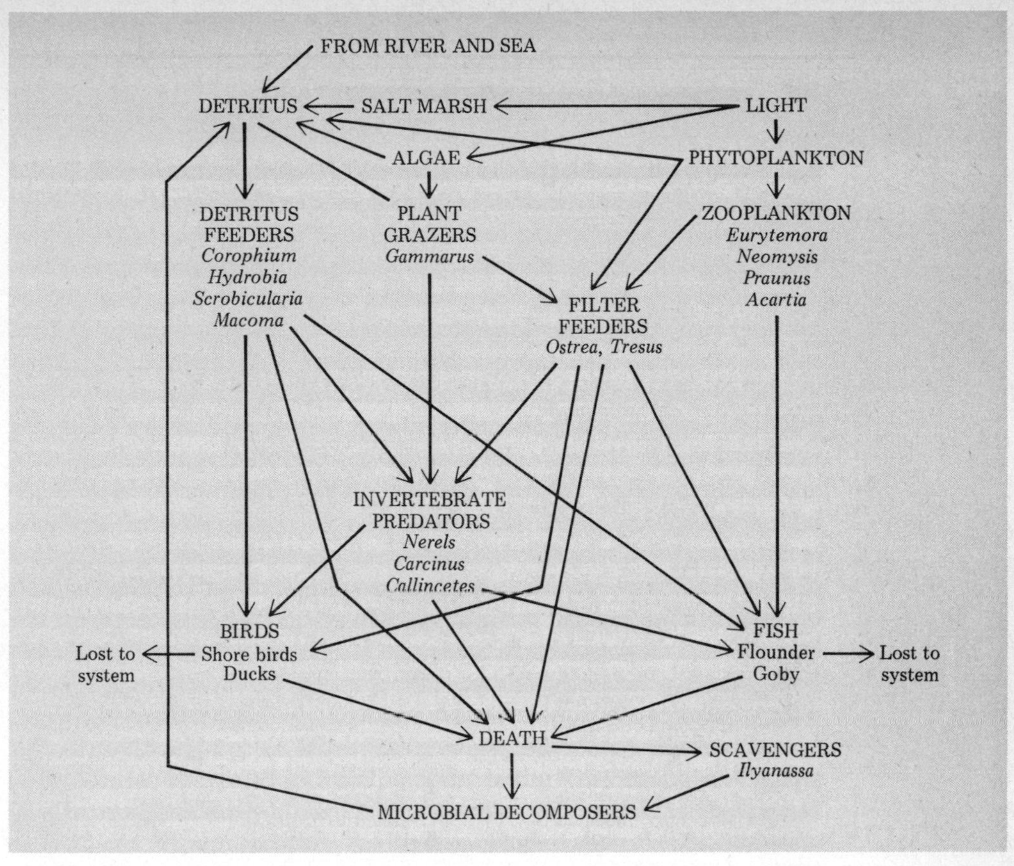

Figure 8-14 A generalized estuarine food web. (Modified from D. S. McLusky, 1971, Ecology of estuaries, Heinemann Educational Books, Ltd.)

ENVIRONMENTAL CHARACTERISTICS

The salt marsh is a rigorous environment that shows wide variations in several environmental factors. It shares with the estuary fluctuating *salinity* due to interaction of river flow and marine water, but because it is intertidal, the salinity variation may be more sudden and more extreme than that experienced in the waters of the estuary itself. A sudden rainstorm at low tide may reduce surface salinities to near zero, whereas the return of the high tide may inundate the marsh with nearly full strength sea water. It is not uncommon for marshes to experience salinities varying from 20 ‰ to 40 ‰ on a single tidal cycle.

Similarly, *temperature* undergoes wide fluctuations. Exposure at low tide opens the marsh and substratum to the extreme air temperatures of the terrestrial environment. Since these marshes occur in the temperate zone, this means air temperatures below freezing in winter and above 30°C in summer. Indeed, mud surface temperatures may vary 10°C during a single day.

The *substrate* is typically mud, essentially similar to the estuarine sediments, and with a high salt content, which results from the perfusion with salt water coupled with a high evaporation rate.

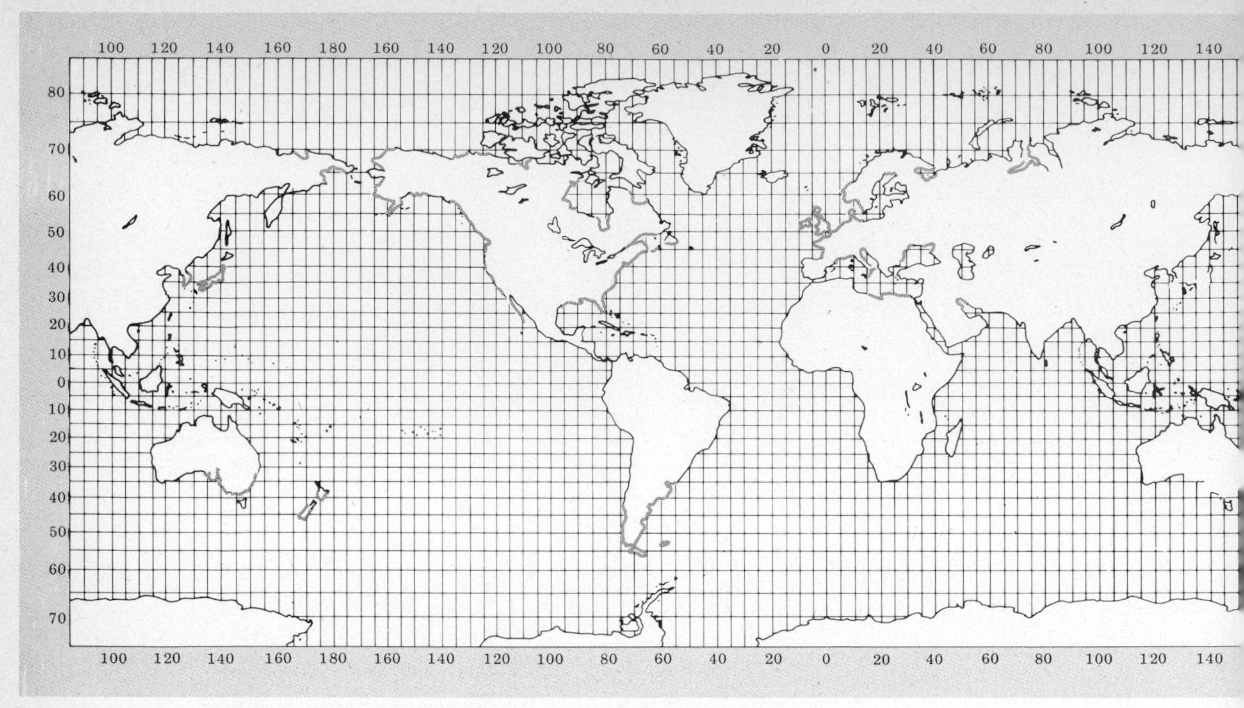

Figure 8-15 The distribution of salt marshes in the world.

COMPOSITION AND DISTRIBUTION

Because of the rigorous environmental conditions and the high soil salt content, relatively few plants and animals are able to inhabit this area. Furthermore, these tolerant organisms show a high degree of taxonomic similarity over a wide geographic area. Dominant plants of salt marshes worldwide are grasses of the genus *Spartina* and species of the genera *Juncus* and *Salicornia* (Fig. 8-16). Characteristic animals of marine origin are various species of crabs (*Uca, Hemigrapsus*), mussels (*Modiolus*), certain snails (*Littorina, Cerithidea, Melampus*) and smaller crustaceans such as amphipods.

The insects represent the major terrestrial animal component of the marsh and may live permanently on the marsh. There is also a terrestrial animal component, that enters the marsh only for feeding such as raccoons. At high tide, marine and estuarine animals enter the marsh, and at low tide, terrestrial species forage in the marsh (Fig. 8-17).

Despite the taxonomic similarity of the salt marsh species over broad geographic areas, certain combinations of topographic features and species distributions have produced several different marsh types. In the United States, there are two major groups of salt marshes, those of the Atlantic and Gulf coasts on one hand and those of the Pacific Coast on the other. East and Gulf coasts salt marshes are far more extensive than West Coast marshes and occupy large areas of gently sloping coastline surrounding the numerous broad estuaries and shallow bays associated with the extensive shallow offshore continental shelf. The Pacific Coast, by contrast, has few rivers and bays, and steep coastal

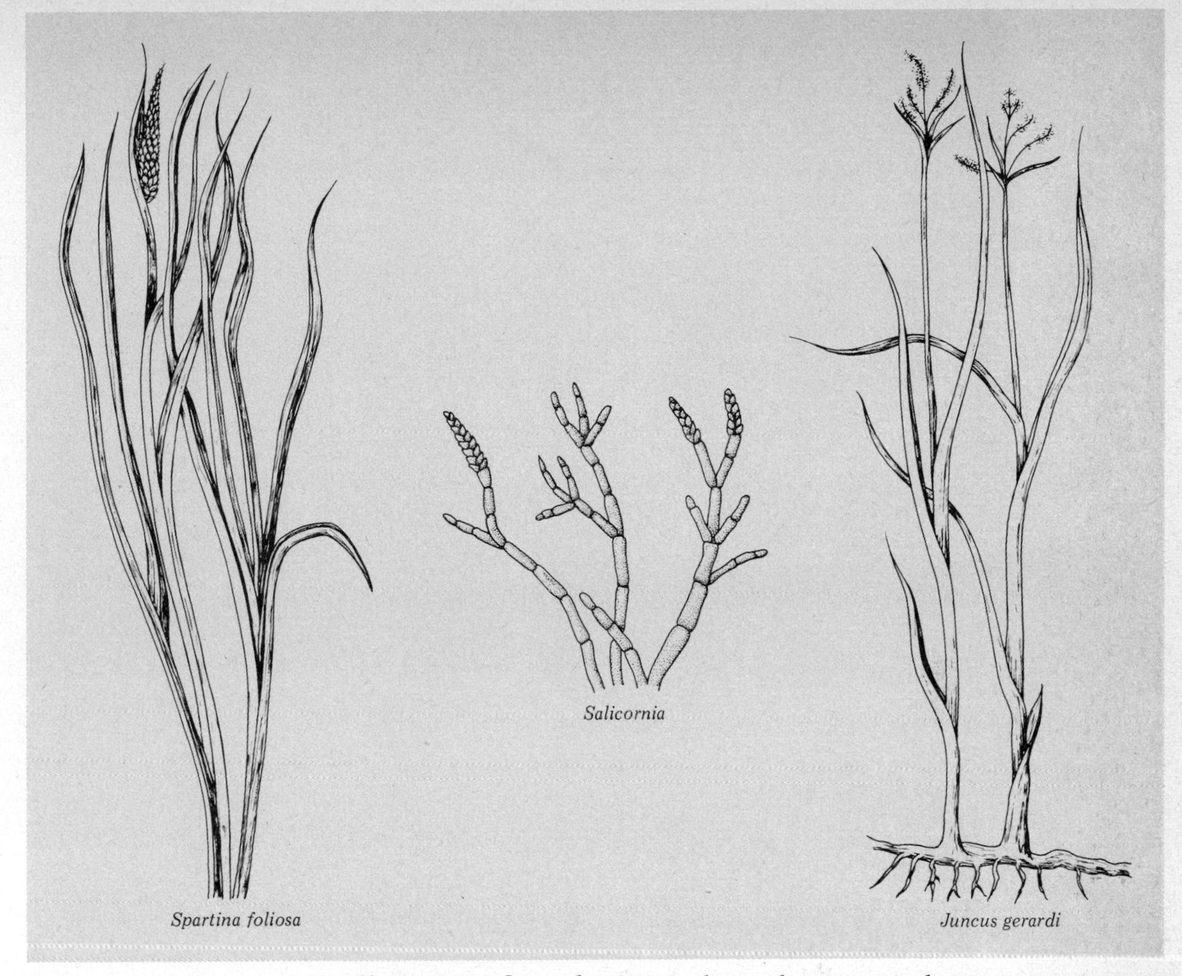

Figure 8-16 Some dominant salt marsh emergent plants.

mountains lead directly to a narrow continental shelf. Consequently, river mouths are narrow and both estuaries and salt marshes are restricted in extent. East and Gulf coasts marshes are dominated by various species of *Spartina*, which succeed each other in elevation and geographically. The more poorly developed West Coast marshes are characterized by broad bands of *Salicornia*. *Spartina* is of lesser importance or, in some cases, absent.

PHYSIOGNOMY AND ZONATION

The general physiognomy of a salt marsh is similar throughout the world wherever these associations occur. The picture is that of a broad, flat expanse of low shrubby or grassy plants, all of generally similar form, cut by a dendritic arrangement of channels leading to larger tidal creeks (Fig. 8-18). These, in turn, lead out to mud flats devoid of macrophytes or open waters of the estuary.

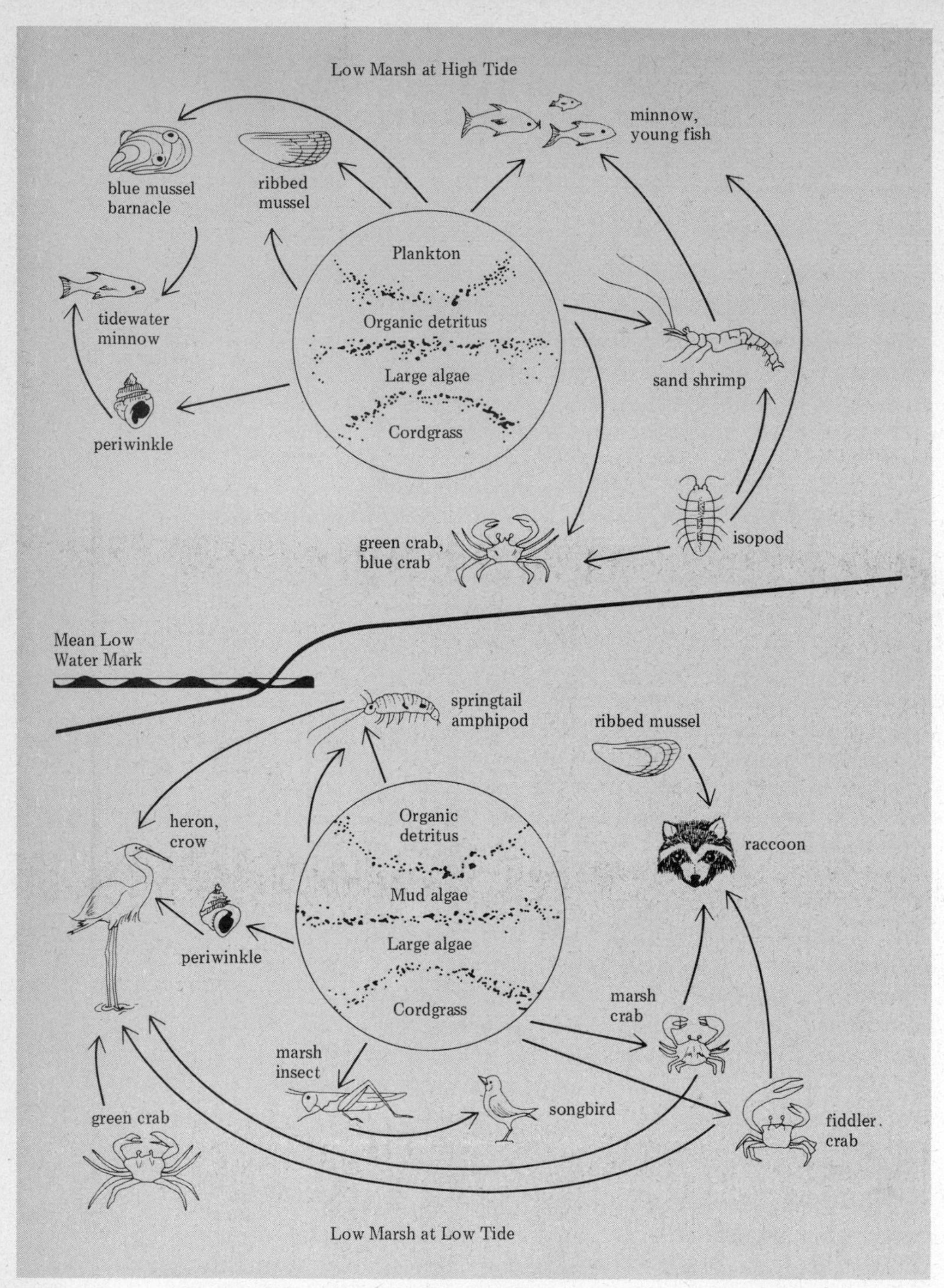

Figure 8-17 Characteristic animals present in a salt marsh at low and high tides on the Atlantic coast of North America.

Figure 8-18 A tidal creek penetrating into a coastal salt marsh on the Pacific coast of North America. (Photo by the author.)

Within the marsh are scattered open areas and shallow pools. At the upland side of the marsh, a variety of terrestrial shrubs and trees replace the halophytes.

A cross section of a typical marsh from the tidal creek to the terrestrial vegetation reveals a zonation pattern. This pattern varies in detail and species composition from one geographical area to another due to tidal regime, drainage, slope, climate, and other factors, but a basic common zonation pattern exists.

The lowest zone is that of the creek bank and bottom, usually barren of macrophytic vegetation and containing mainly infaunal estuarine and marine animals and many mud snails. Progressing upward, the next zone comprises the creekside marsh, a border zone between the creek bank and the main part of the marsh. The next zone is the main part of the marsh, which may be variously subdivided. On the East and Gulf coasts, the subdivision is into a lower, tall *Spartina* zone and an upper, short *Spartina* or *Juncus* zone. The final marsh zone is a *Salicornia* zone, often with various other plants (Fig. 8-19). West Coast

Figure 8-19 Generalized zonation pattern for Atlantic Coast salt marshes. (From several sources.)

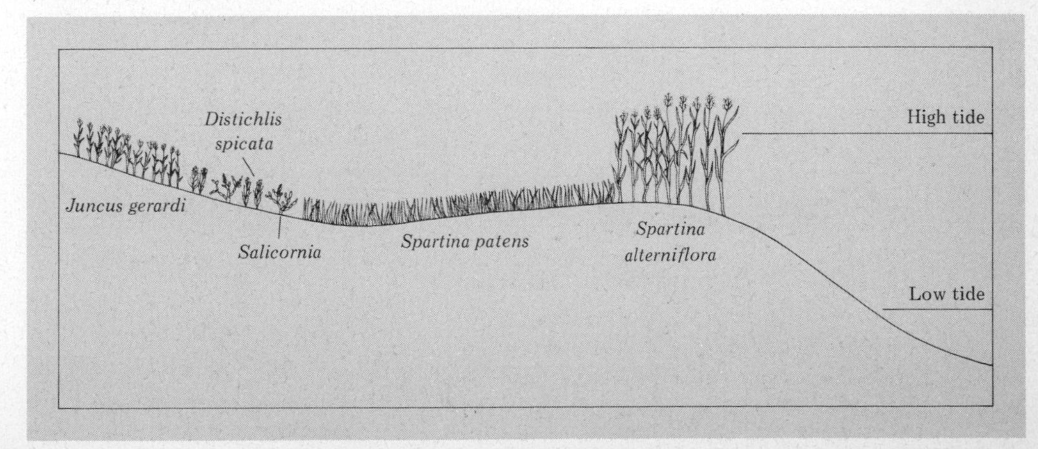

marshes tend to have a narrow fringe of *Spartina* in the lower zone followed by a broad expanse of *Salicornia* occupying most of the mid-intertidal zone. The highest intertidal area is occupied by a more diverse asemblage, of which *Jaumea*, *Distichlis*, and *Limonium* are common (Fig. 8-20).

PRODUCTIVITY

Primary productivity within the marsh is carried out by the emergent marsh plants and the various micoalgae that occur on the surface of the marsh plants and the mud. In contrast to the water and mud flats of estuaries, productivity is very high in marshes, especially for the rooted emergent plants. The most complete data on productivity exist for the East Coast marshes. The major primary producers in these marshes are the various species of *Spartina*. Teal (1962) has shown Georgia marshes to have an average annual net productivity of 1600 g/m^2/year. This decreases to the north such that in New Jersey, Good (1965) reports only 325 g/m^2/year. On the Pacific Coast, Atwater et al. (1979) report San Francisco marshes at 50–1500 g/m^2/year of net production (compare with Table 2-1, p. 70).

Due to few herbivores, most of the productivity is not consumed directly; rather, the dead plant parts are broken down by bacteria on the surface of the mud or in the water. During this process, the food value of the plant parts is actually increased as the indigestible cellulose is broken down into digestible carbohydrates. The detrital particles then form the basis of a food chain for a wide variety of marsh animals and may also be exported via tidal action to the estuary. It has been estimated that about 45 percent of the original net primary production is not used by marsh animals and is exported to the estuary. Because of low productivity in the estuary, this exported detritus has been suggested as the major basis of food for estuarine animals. However, this has recently been challenged by some scientists (see pp. 293–294) who claim most of the food produced is consumed in the marsh.

Figure 8-20 Generalized zonation of a Pacific Coast salt marsh based on the marshes of San Francisco Bay.

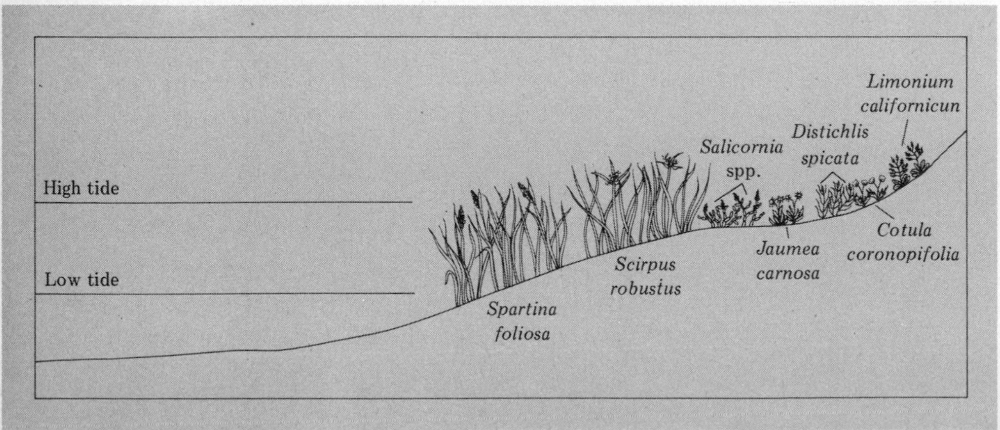

INTERACTIONS AND FOOD WEBS

The salt marsh is an ecotone between estuarine organisms and terrestrial organisms. There are numerous interactions between the animals of both areas and the marsh plants and their environment.

In general, salt marsh animals find themselves in a stressful environment of widely varying physical conditions. They must either adapt to the range of environmental conditions or else evolve habits that will minimize the effects. Few animals possess tolerance limits that include the range of temperature and salinity experienced in a marsh; therefore, marsh animals tend to evolve strategies that minimize the environmental variations to which they are exposed. As a result, there are more aquatic species found in low areas seldom exposed to the air and more terrestrial forms in areas above the effects of the highest tides. The fewest species occur in the intermediate area.

Marsh animals minimize the environmental changes in a variety of ways. Crabs commonly burrow into the mud, by which they escape desiccation and surround themselves with soil water of higher salinity. Mussels (*Modiolus*) avoid desiccation by closing up. Still others, such as coffee bean snails (*Melampus*) and certain insects, escape inundation at high tide by climbing up the marsh plants. A great many animals simply migrate in and out of the marsh with the tide, aquatic animals in at high tide and terrestrial ones at low tide.

The marsh plants themselves are beneficial to the animals in a number of ways. They may provide cover that reduces predation. The various small invertebrates, such as crabs, snails, and insects, are less subject to predation from birds and fish if they are beneath the canopy of marsh plants than if they are exposed on the open mud flat. The plants also act to stabilize the mud and provide a substrate to which certain animals may attach. Without the presence of *Spartina*, for example, the mussel *Modiolus* (*Geukensia*) would have nothing to which to attach its byssal threads; barnacles would be unable to exist due to lack of a surface to cement to; and the terrestrial insects and pulmonate snails would have no vertical substrate to climb, to enable them to escape inundation at high tide. The marsh plants also act indirectly by reducing wave and current action, which might otherwise remove or destroy some animals. They also reduce light penetration to the mud surface and ameliorate the temperatures on the mud surface, thus decreasing the range of variation to which the animals are exposed.

In turn, the animals have effects on the plants. The direct effects are minimal because only a few relatively rare grasshoppers and leafhoppers actually consume the living plants. They do not appear to have any significant effect on the plants. The burrowing activities of various crabs allow aeration of an otherwise stagnant mud and may benefit the roots of the marsh plants. On the other hand, burrowing may lead to harmful cropping of the roots. Burrowing by crabs may also lead to bank erosion and loss of parts of the marsh as they collapse into the tidal creeks. Invertebrate fecal matter may provide nutrients to the marsh plants, but this remains to be tested. Certain snails, such as *Melampus*

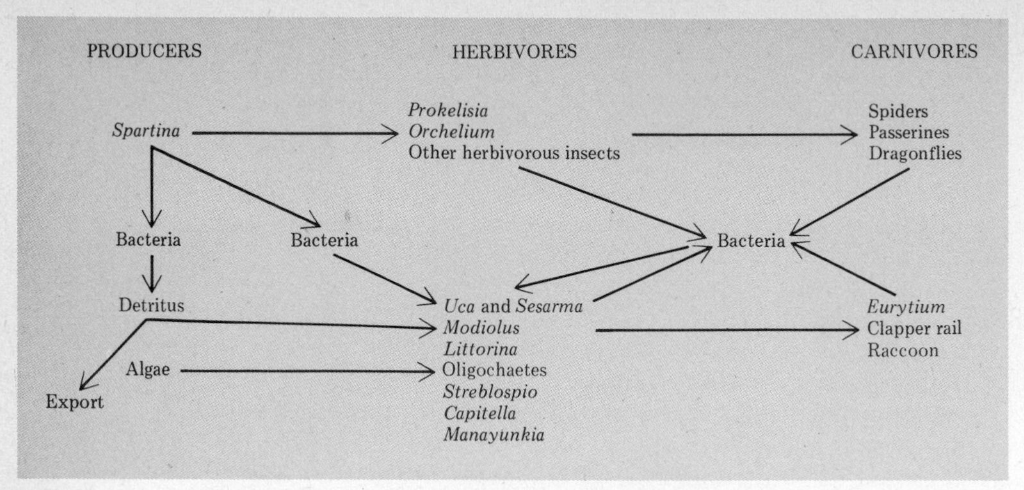

Figure 8-21 The food web of an Atlantic Coast salt marsh. (After J. Teal, 1962, Energy flow in the salt marsh ecosystem of Georgia, Ecology, vol. 43, copyright 1962, the Ecological Society of America.)

and *Littorina*, graze the algae that settle on the stems of the marsh halophytes and may enhance their growth by eliminating or reducing this algal cover.

The food webs of salt marshes are based primarily on the consumption of detritus (and associated bacteria). Since most of the detrital consumers distinguish only by particle size, a certain amount of live benthic algae is also consumed. These detrital consumers plus the few insects feeding directly on the marsh plants are in turn consumed by a variety of both aquatic and terrestrial carnivores (Fig. 8-21).

References

Barnes, R. S. K. 1974. Estuarine biology. Studies in biology no. 49. Edward Arnold, London. 76 pp. (Distributed in U.S. by Univ. Park Press, Baltimore.)

Chapman, V. J. (ed.). 1977. Ecosystems of the world. 1, West coastal ecosystems. Elsevier, N.Y. 428 pp.

Green, J. 1968. The biology of estuarine animals. Univ. of Wash. Press, Seattle. 401 pp.

Lauff, G. H. (ed.). 1967. Estuaries. Pub. no. 83. AAAS, Washington, D.C. 757 pp.

McLusky, D. S. 1971. Ecology of estuaries. Heinemann Educational Books, Ltd., London. 144 pp.

Odum, H. T., B. J. Copeland, and E. A. McMahan. 1974. Coastal ecosystems of the United States, 4 vols. Vol. I, 533 pp.; Vol. II, 521 pp.; Vol. III, 453 pp.; Vol. IV, 470 pp. Vol. II has information on marshes and estuaries. The Conservation Foundation, Washington.

Reid, G. K., and R. D. Wood. 1976. Ecology of inland waters and estuaries, 2nd ed. Van Nostrand, N.Y. 485 pp.

Chapter Nine TROPICAL COMMUNITIES

For color, sheer beauty of form and design, and tremendous variety of life, perhaps no natural area in the world can equal coral reefs. Their beauty has fascinated generations of people, both scientific and lay, down through the years. By the same token, few areas of the marine environment have less initial esthetic appeal than the dark, mud-wreathed areas known as mangrove forests, where passage is barred by a maze of tangled roots standing above a soft mud surface into which one quickly sinks. Yet these two very different associations are characteristic of vast tropical regions of the world, are unique, and must be considered if one is to understand the functioning of shallow water areas in the tropics. In this chapter, we shall take up first the distribution, zonation, and structure of these assemblages and then attempt to account for the maintenance and continuation of these communities through a consideration of the component organisms and their interactions.

CORAL REEFS

Over a vast region (millions of square miles) of the tropics, the shallow inshore waters are dominated by the formation of coral reefs and indeed, they are often used to define the limits of the tropical marine environment (Fig. 9-1). Coral reefs are unique among marine associations or communities in that they are built up entirely by biological activity. The reefs are essentially massive deposits of calcium carbonate that have been produced primarily by corals (phylum Cnidaria, class Anthozoa, order Madreporaria = Scleractinia) with minor additions from calcareous algae and other organisms that secrete calcium carbonate. Although corals are found throughout the oceans of the world in polar and temperate waters as well as in the tropics, it is only in the tropics that reefs are

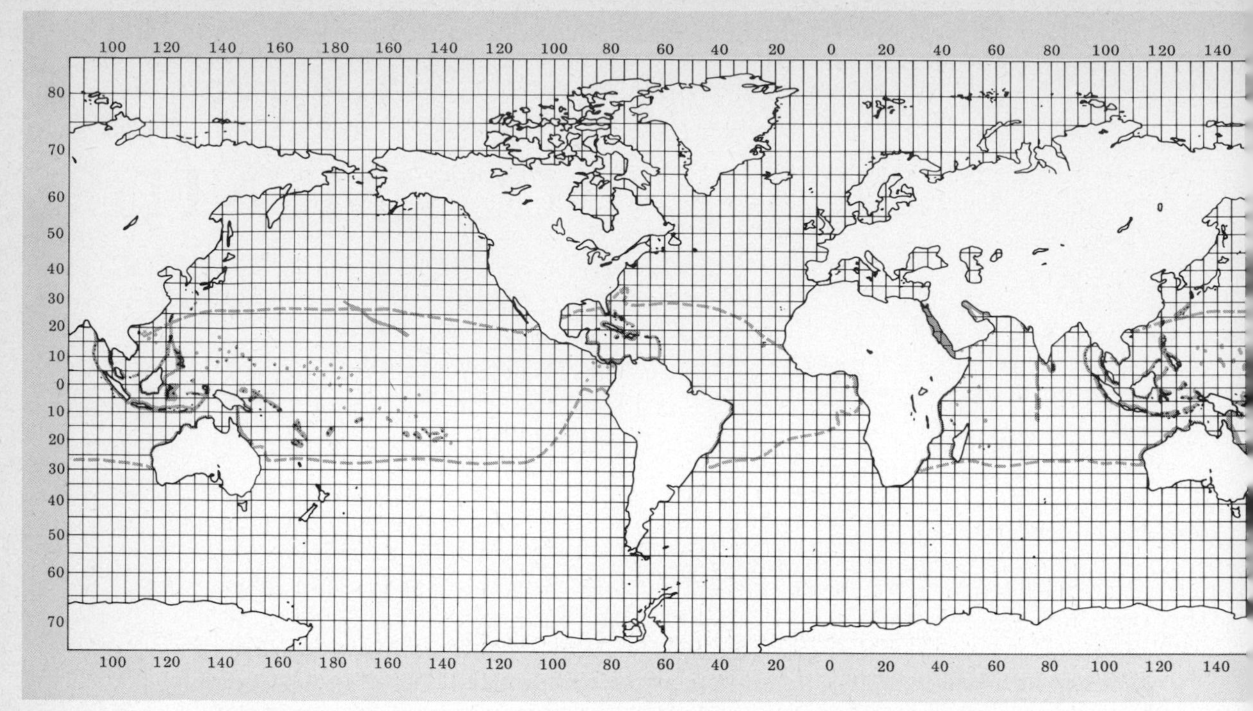

Figure 9-1 Distribution of coral reefs in the world; 20°C isotherm (dashed line).

developed. This is because there are two different groups of corals, one called *hermatypic* and the other *ahermatypic*. Hermatypic corals are those that produce reefs; ahermatypic do not. Ahermatypic corals are distributed worldwide, but hermatypic corals are found only in the tropical regions. The distinguishing feature between the two is that hermatypic corals have in their tissues small symbiotic (living together) plant cells called *zooxanthellae*, whereas ahermatypic corals do not. The role of these plants will be considered later (see pp. 322–323).

REEF DISTRIBUTION AND LIMITING FACTORS

A glance at a map of coral reef distribution in the world immediately shows that nearly all coral reefs are found only in those waters bounded by the 20°C surface isotherm (Fig. 9-1). Hermatypic corals can maintain themselves for periods at temperatures somewhat below this; however, as Wells (1957) notes, no reefs develop where the annual minimum temperature is below 18°C. Optimal reef development occurs in waters where the mean annual temperatures are about 23–25°C. Coral reefs can tolerate temperatures up to about 36–40°C. It may be noted that reefs are absent from large areas on the west coast of South and Central America and also from the west coast of Africa, both areas of which lie well within the tropical zone. The reason for the absence from these areas is that the west coasts of both of these continents are areas of strong upwelling of cold water, which reduces the temperature of the shallow inshore waters below that required for reef development. Furthermore, both of these coasts have strong, north-flowing cold currents that keep the temperature down, the Humboldt Current on the South American coast and the Benguela Current off West Africa.

Coral reefs are also limited by *depth*. Coral reefs do not develop in water that is deeper than about 50–70 m. Most reefs grow in depths of 25 m or less. This explains why these structures are restricted to the margin of continents or islands. The peculiar feature of the development of atolls, which occur rising out of water many kilometers deep, will be discussed later (see pp. 311–314). The reason for the depth restriction has to do with the hermatypic corals' requirement for light. *Light* is one of the most important factors limiting coral reefs. Why? Sufficient light must be available to allow photosynthesis by the symbiotic zooxanthellae in the coral tissue. Without sufficient light, the photosynthetic rate is reduced and with it the ability of the corals to secrete calcium carbonate and produce reefs. The compensation point for corals seems to be the depth where light intensity has been reduced to 15–20 percent of surface intensity.

Another factor that acts to restrict coral reef development is *salinity*. Hermatypic corals are true marine organisms and are intolerant of salinities deviating significantly from that of normal sea water (32–35 ‰). Wherever inshore waters are subject to continuing influxes of fresh water from river discharge so that the salinity is lowered, reefs will be absent. Such is the case along large portions of the Atlantic coast of South America, where the Amazon and Orinoco rivers discharge a huge volume of fresh water, and reefs are absent (Fig. 9-1). On a smaller scale, this occurs in many areas of the tropics where rivers and streams running out to sea cause breaks in reef development. At the other extreme, coral reefs do occur in regions of elevated salinity such as the Persian Gulf, where reefs flourish at 42 ‰.

Often correlated with fresh water runoff is the factor of *sedimentation*. Sediment, both in the water and settling out on the coral, has an adverse effect on the corals. Most hermatypic corals cannot withstand heavy sedimentation, which seems to smother them and clog their feeding structures. Sediment in the water has an additional adverse effect in that it reduces the light necessary for photosynthesis by the zooxanthellae in the coral tissue. As a result, coral reef development is reduced or eliminated in areas of high sedimentation. When this sediment is carried by rivers or streams, the combination of reduced salinity and excess sediment is responsible for the absence of reefs. Hermatypic coral species may vary, however, in their tolerance for sedimentation; few are able to tolerate rather high sedimentation rates, and these are found as isolated colonies in sedimentary areas.

In general, coral reef development is greater in areas that are subject to strong wave action. Coral colonies with their dense, massive skeletons of calcium carbonate are very resistant to damage by wave action. At the same time, the wave action provides a constant source of fresh, oxygenated sea water and prevents sediment from settling on the colony. Wave action is also responsible for renewing the plankton, which serves as food for the coral colony.

Finally, coral reefs are limited in an upward direction by immersion in air.

Most corals are killed by long exposure to air, thus limiting their upward growth to the level of the lowest tides. During an unusual and extremely low tide series over a five-day period in the Gulf of Aquaba, Loya (1976) found that coral mortality was between 80–90 percent, which suggests this factor can be of considerable importance. On very low spring tides, coral reef areas may, however, be exposed for short periods of time, a few hours, which does not seem to harm them, at least in the Indo-Pacific region.

All these factors are summarized diagrammatically in Figure 9-2.

STRUCTURE OF CORALS

Since the dominant members of coral reefs are the corals, it is necessary to have some understanding of their anatomy. Corals are members of the phylum Cnidaria, which includes such diverse forms as jellyfish, hydroids, the fresh water *Hydra*, and sea anemones. Corals and sea anemones are members of the same taxonomic class, Anthozoa, and differ primarily in that corals secrete an external calcium carbonate skeleton, whereas anemones do not. Corals may be colonial or solitary, but almost all hermatypic corals are colonial, with the various individual coral animals or *polyps* occupying little cups or *corallites* in the massive skeleton (Fig. 9-3; Plates 22 and 23). Each cup or corallite has a series of sharp, bladelike *septa* rising from the base. The pattern of these septa differs from species to species and is one basis for separating coral species. The septa of the skeletal cup alternate with the internal septa of the gastrovascular cavity of the coral polyp. Each polyp is a two-layered animal with an outer epidermis separated by a nonliving mesoglea from an internal gastrodermis. Around the mouth is a series of tentacles that have batteries of stinging capsules or nematocysts, which the animals use to capture their zooplankton food. The

Figure 9-2 Summary of physical factors acting on coral polyps and coral reefs that may act to limit their distribution.

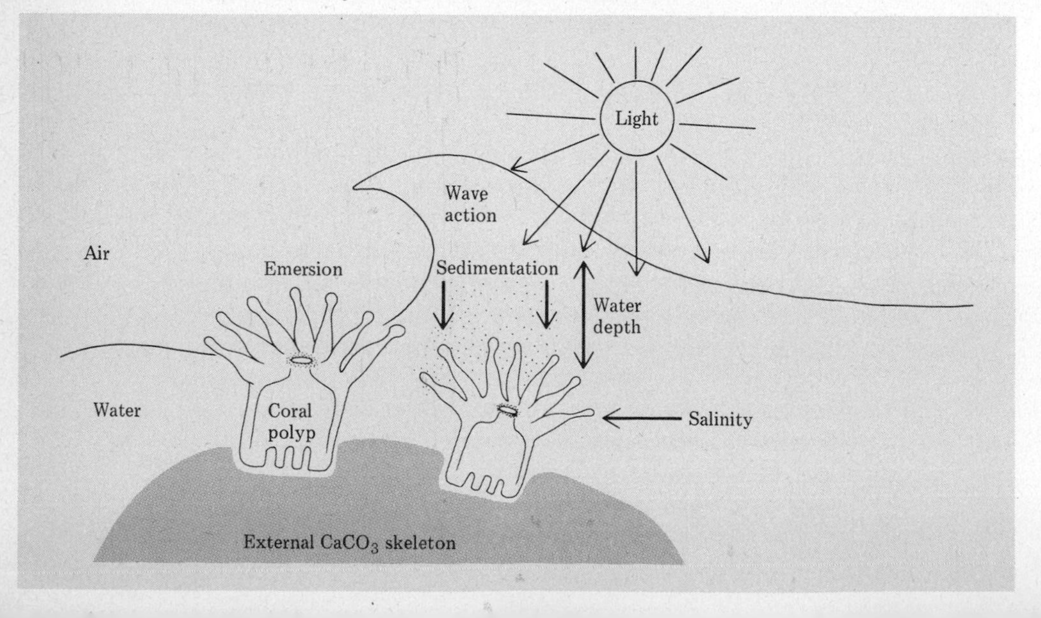

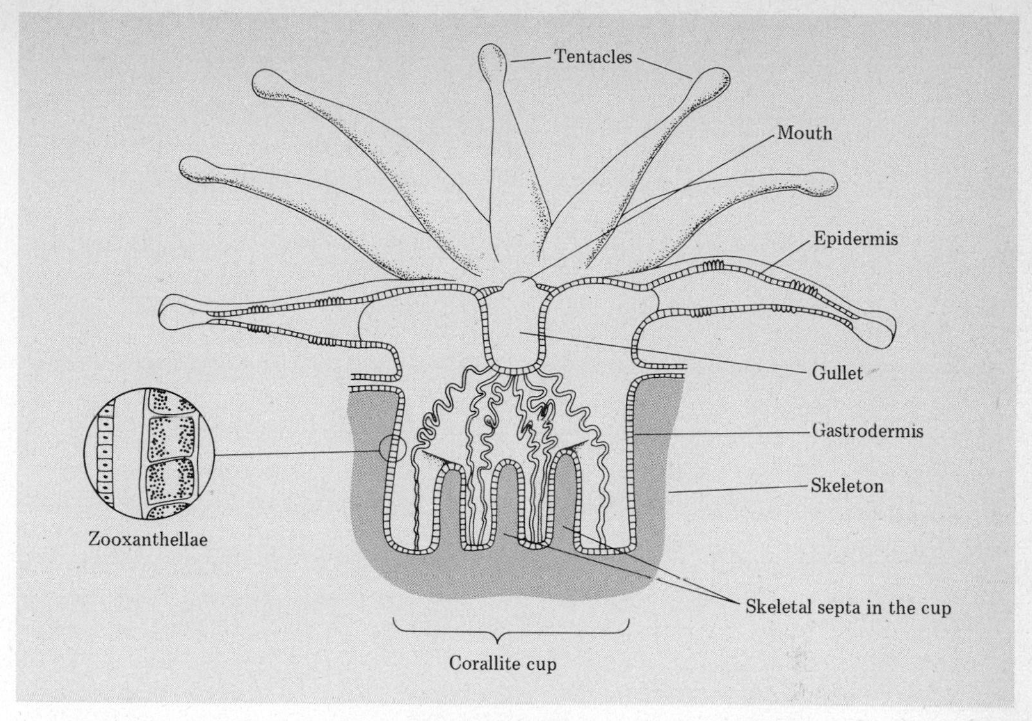

Figure 9-3 Anatomy of a coral polyp.

symbiotic zooxanthellae, actually dinoflagellates (see Chapter Ten), are found in the gastrodermis layer. Coral colonies grow by having the polyps bud off new polyps asexually. New colonies are established through the settlement of a planktonic *planula* larva, which itself is the result of sexual reproduction.

TYPES OF REEFS

Several different types of coral reefs are recognized. Most commonly, they are grouped into one of three categories: *atolls*, *barrier reefs*, and *fringing reefs*. Atolls are usually quite distinguishable as they are modified ring-shaped reefs that arise out of very deep water far from land and enclose a lagoon, which itself may contain *lagoon reefs* or *patch reefs* (Fig. 9-4). With few exceptions they are found only in the Indo-Pacific area and not in the Atlantic. Barrier reefs and fringing reefs, on the other hand, tend to grade into each other and are not readily separable. Some scientists would prefer to group them into a single category. Both types occur adjacent to a land mass, with a barrier reef being separated from the land mass by a greater distance and deeper water channel than the fringing reef (Fig. 9-4). Fringing reefs and barrier reefs are common throughout the coral reef zones in all oceans. The largest barrier reef is the Great Barrier Reef of Australia, which stretches for almost 2,000 km along the eastern coast of Australia from near New Guinea to just north of Brisbane.

ORIGIN OF REEFS

Different types of reef and reefs in different oceans may have diverse origins and histories. The greatest interest in the origin of reefs has been with respect to

(*A*)

(*B*)

Figure 9-4 Aerial photo of an atoll *(A)* and a barrier reef *(B)*. (Photos by Dr. Richard Mariscal.)

atolls. For many years, humans speculated as to how such reefs could develop out of such deep water, miles from the nearest emergent land. This interest was heightened when it was discovered that reef corals could not live deeper than 50–70 m. This led to the development of several theories concerning the origin of atolls. Only one need be discussed here; that is the *subsidence theory*. This explanation for the origin of atolls was first promulgated by Charles Darwin following his five-year voyage on the *Beagle*, during which time he had the opportunity to study reefs in several areas. According to Darwin's subsidence theory, atolls have their origin when fringing reefs begin to grow on the shores of newly formed volcanic islands, which have pushed to the surface from deep water. These islands then often begin to subside and, if the subsidence is not too fast, reef growth will keep up with the subsidence, forming next a barrier reef and finally an atoll as the island disappears beneath the sea (Fig. 9-5). When the island has disappeared, continued growth of the corals on the outside keeps the reef at the surface; but on the inside, where the island used to be, quiet water conditions and high sedimentation prevail, which preclude continued vigorous coral growth, and hence, a lagoon develops. This theory thus links all three reef types in an evolutionary sequence, but is not an explanation for all fringing and barrier reef types.

Since the current surface features of atolls give no evidence of a volcanic base, in the years after the development of Darwin's theory, other explanations were offered and the whole concept of the origin of atolls became embroiled in the "coral reef problem." If Darwin's theory were correct, it must be assumed that drilling down through the current atoll reefs would yield layer after layer of reef limestone until finally, volcanic rock was encountered. Such ability to drill to the base of atoll reefs and resolve the problem had to wait until the mid-twentieth century. In 1953, Ladd et al. reported borings at Eniwetok Atoll in the Marshall Islands penetrated 1,283 m of reef limestone and then hit volcanic rock. This was the initial evidence that Darwin's theory was substantially correct. Correctness of this theory has been strengthened by the discovery of flat-topped submerged mountains or *guyots* that, at the present time, have their tops many hundreds or thousands of meters below the present ocean surface, but which have on their surface the remains of shallow water corals. Evidently, these mountains sank too fast for reef growth to keep within the lighted zone.

Although the subsidence theory links together all three reef types in a successional sequence, this does not mean that everywhere barrier reefs and fringing reefs occur today, a similar sequence is happening. Indeed, the reasons for the occurrence of barrier and fringing reef types around continental margins and around high, nonvolcanic islands is simply that these areas offer suitable environmental conditions for the growth of reefs and a suitable substrate on which to begin growth. Hence, the extensive reefs around the Indonesian

Figure 9-5 Geological evolution of a coral atoll according to the subsidence hypothesis of Darwin. (After T. Storer, R. Usinger, R. Stebbins, and J. Nybakken, General zoology, 6th ed., McGraw-Hill. Copyright 1979. Used with the permission of McGraw-Hill Book Company.)

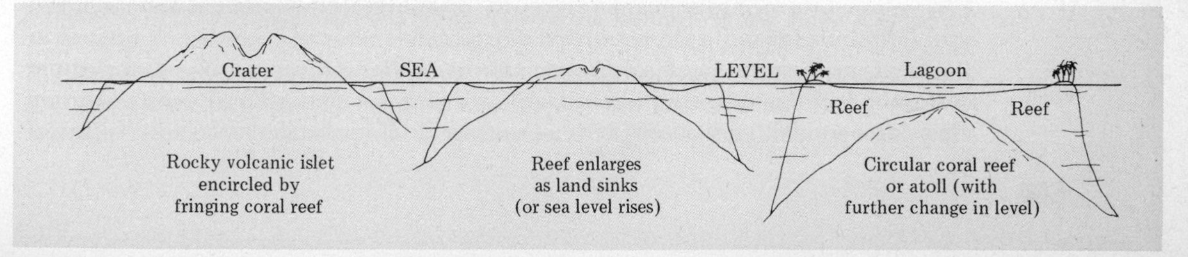

Islands, the Philippines, New Guinea, Fiji, and most of the Caribbean Islands and Florida are there because a suitable substrate in shallow water existed on which they could begin growth. In none of these areas are large land areas subsiding, nor will these reefs ultimately become atolls.

Whereas atolls are often very old structures (for example, the 1,283 m of reef limestone at Eniwetok represent reef growth for about 60 million years), other reefs may be much less old. This is particularly true for the reefs in the Atlantic Ocean. They are all geologically very recent, most dating only from the last glacial age or 10,000–15,000 years before the present.

COMPOSITION OF REEFS

Although the corals are the major organisms on coral reefs in that they form the basic reef structure, there is a bewildering array of other organisms associated with reefs such that these areas are perhaps the most diverse and species-rich areas that exist in the marine environment today. Members of practically all phyla and classes may be found on coral reefs. It is not practical here to mention all of these forms, nor do we even know the roles played by many of them, but before considering the ecological structure of reefs, it is worth pointing out some of the more important and/or abundant groups.

In addition to the stony corals (order Madreporaria), certain other cnidarians contribute strongly to the reef. These include those anthozoan relatives of the corals called gorgonians, the sea fans, and sea whips (Fig. 9-6; Plates 24 and 25). These organisms have internal skeletons of spicules. They are particularly conspicuous on the Atlantic reefs. The soft corals (Anthozoa, Octocorallia) are very common on Indo-Pacific reefs and may be more abundant than the stony corals in some areas of the reef. They are often rare on Atlantic reefs.

The *coralline algae* are an extremely important group in constructing and maintaining reefs. These algae precipitate $CaCO_3$ as do the corals, but the algae tend to be encrusting, spreading out in thin layers over the reef, literally cementing the various pieces together. By this action, the various individual calcium carbonate pieces are welded together to form a strong bulwark that resists wave destruction. Still other algae are nonencrusting, grow erect, and secrete calcium carbonate, but are not coralline. Much of the sand in coral reef systems is derived from breakup of these erect algae, usually of the green alga genus *Halimeda*. Other free-living, noncoralline algae are inconspicuous on reefs, but a significant algal flora also exists just below the surface layers of calcium carbonate in the coral colonies themselves.

Significant contributions to the calcium carbonate deposits on reefs are also made by mollusks of various types. The most conspicuous and important of these are the various giant clams (*Tridacna, Hippopus*), which Salvat has recorded occurring in numbers as great as 200 per square meter in some atoll lagoons in the Tuamotu Archipelago! Generally, McMichael (1974) records their density at $1/m^2$ or less (Plate 27). Giant clams are absent from the Atlantic reefs. Various

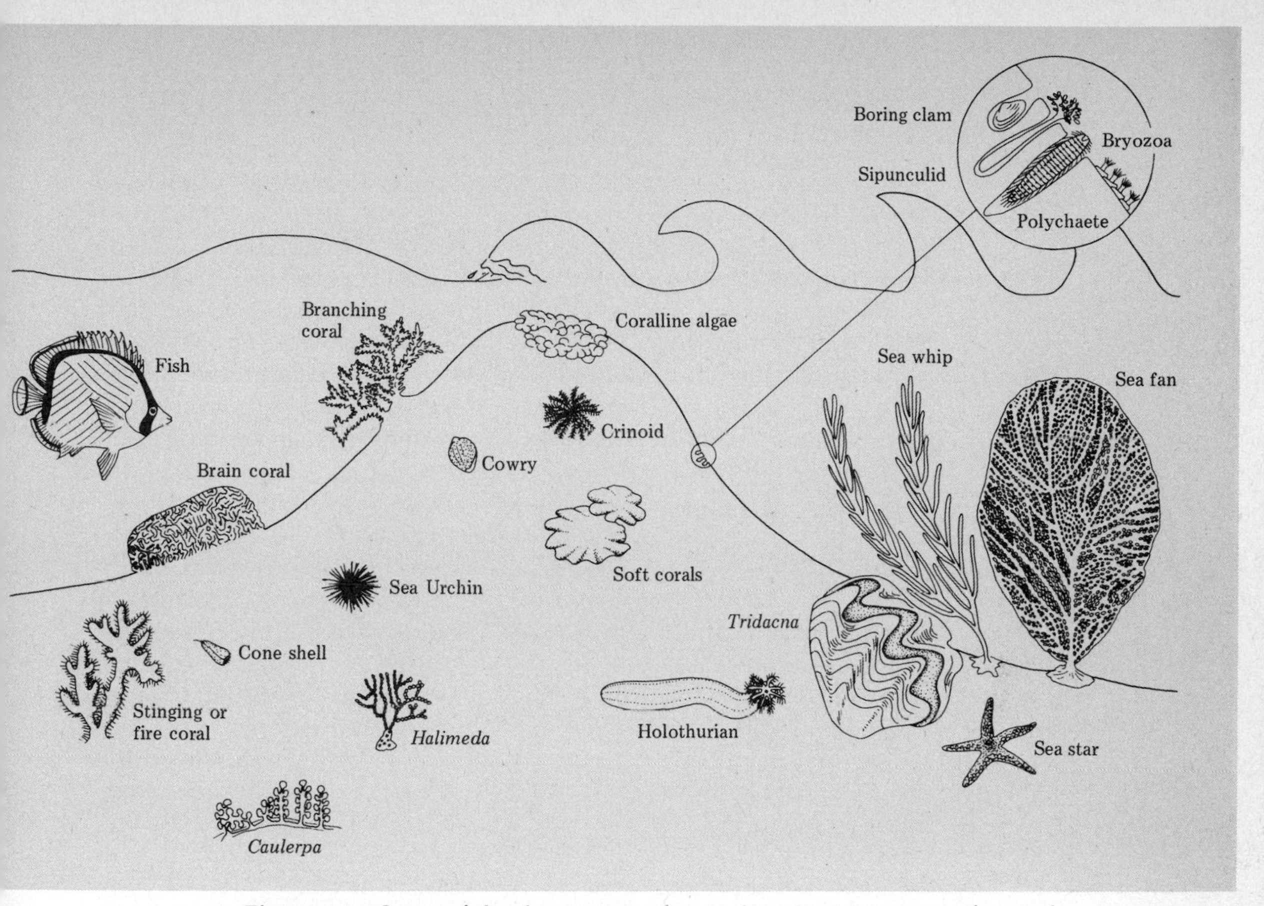

Figure 9-6 Some of the dominant and conspicuous components of a coral reef.

small gastropods are also abundant on reefs, but usually inconspicuous due to their size or because they are hidden.

Echinoderms, primarily sea urchins, sea cucumbers, starfish, and sea lilies (crinoids) are another abundant and conspicuous group on reefs, but their role in reef ecosystems (except for the sea star *Acanthaster* and the urchin *Diadema*) is not well understood (Plate 28).

Although various crustaceans and polychaete worms are very abundant on reefs, they are inconspicuous, and their various functions in the reef ecosystem unknown.

The final large group, which is very conspicuous on reefs and most certainly contributes to the structure of the reef system, is the various fishes.

Bacteria are also probably very abundant on reefs and are responsible for the decomposition and quick cycling of organic matter.

CORAL DISTRIBUTION AND REEF ZONATION

The largest number of reef coral species and genera occur in an area of the Indo-Pacific that includes the Philippine Islands, the Indonesian Archipelago,

New Guinea, and northern Australia. Within this area, Crossland (1952) and Wells (1954) list 50 genera and several hundred species. Considering the whole Indo-Pacific region, Wells (1957) lists 80 genera and 700 species. The number of genera and species declines away from this area, both north and south and east and west, but most of the central Indo-Pacific has reefs with from 20 to 40 genera of corals. This is in contrast to the Atlantic Ocean, where Goreau and Wells (1967) report only 36 genera and 62 species are found. The dominant reef-building genus in both oceans, *Acropora*, has 150 species in the Indo-Pacific, but only three in the Atlantic, and some important reef-building genera of the Pacific such as *Pocillopora*, *Pavona*, and *Goniopora* are absent from the Atlantic. Thus, the species complexity of Atlantic reefs is much less than that of Pacific reefs. Whether in the Atlantic or Pacific, as one moves away from the centers of the distribution of coral species, the number of species declines and is dramatically reduced at the edges of the coral reef zone. For example, the northernmost Pacific reefs, at Midway Island in the Hawaiian chain, have but nine genera of corals.

Coral reefs are large and complex associations of organisms that have a number of different habitat types all present in the same system. Thus, for example, there are areas of hard substrate upon which many sessile organisms attach and which are analogous to rocky shores; at the same time, there are areas of sand, which require a different set of adaptations, much as we saw for soft bottoms. Similarly, there are areas of heavy wave action and strong currents and areas of virtual calm where water movement is minimal. Still other areas may have a lush growth of calcareous green algae (*Halimeda*) mimicking sea grass beds. All of these different habitat types are found in a relatively small area. Therefore, one of the reasons for the great diversity of life in coral reefs is the great diversity of habitats. As might be expected, these different habitat types, as well as others not mentioned here, vary in their extent or presence on reefs in different geographic areas; hence, it is difficult to give a pattern of zonation for reefs which will be generally applicable. The pattern of zonation, though most complex, is most consistent for atolls, and it can be given as a general guide.

Reef development and zonation on atolls are strongly influenced by wave action, a product of the prevailing wind, and, therefore, even though these reefs are essentially circular, a cross section shows different zonation, depending on whether the reefs face the prevailing wind and waves or lie in the lee of them. Figure 9-7 is a diagrammatic cross section of a typical atoll, which indicates the major features of zonation.

Beginning on the windward side, the first zone on a reef is the *outer seaward slope*, where living coral begins to become abundant at a depth of about 50 m. Corals in this area are few and often delicate. At about 15 m, there is often a terrace or step in the otherwise steep outer slope; from here to the surface, coral growth is lush. This is also the area where wave action is the most severe and, hence, we know little of this area because of its inaccessibility to humans. At the surface of the water is the *windward reef margin*, which is characterized by the development of coral-algal spurs or buttresses extending out into the waves and

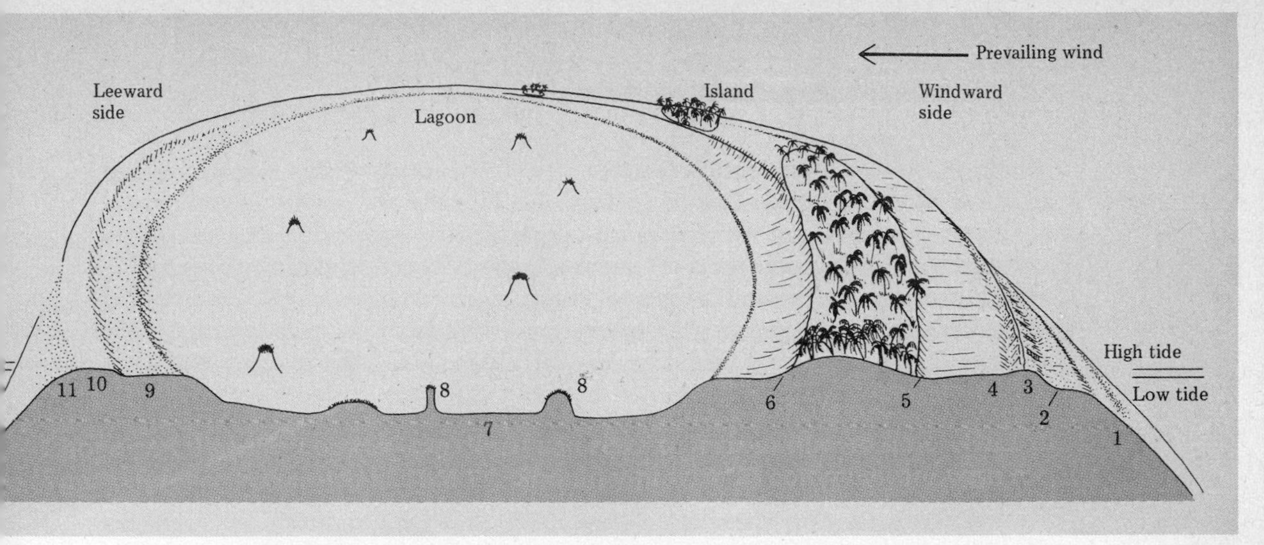

Figure 9-7 Diagrammatic cross section of a typical atoll: (1) outer seaward slope, (2) windward reef margin with spur and buttress zone, (3) algal ridge, (4) the reef flat, (5) seaward beach of the island, (6) lagoon beach of the island, (7) lagoon floor, (8) lagoon reefs, (9) leeward reef flat, (10) leeward reef margin, (11) leeward reef slope. (Modified from C. M. Yonge, 1963, The biology of coral reefs, Advances in marine biology, vol. 1.)

down in depth to the level of the terrace. These spurs or buttresses alternate irregularly with deep channels or grooves into which water surges onto the reef itself and through which water and associated debris leave the reef (Fig. 9-8). This is the so-called *spur and groove* or *buttress zone*. The sides of the grooves or channels and the spurs themselves support a lush growth of the dominant reef-forming corals such as *Acropora*, and it is here that coral reef accretion occurs most rapidly. This is because environmental conditions are optimal. Immediate-

Figure 9-8 Generalized sketch of the spur and buttress zone of a windward reef. *AR*, algal ridge; *B*, buttresses or spurs; *G*, grooves; *SC*, surge channels; *T*, 10-fathom terrace. (After C. M. Yonge, 1963, The biology of coral reefs, Advances in marine biology, vol. 1.)

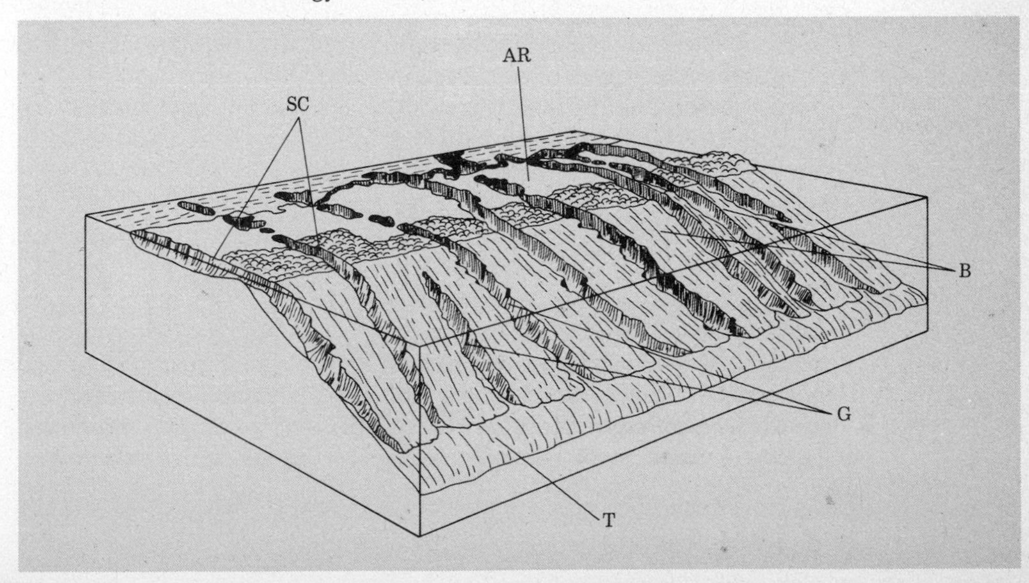

ly behind the buttress zone is a smooth, coral-free ridge of encrusting coralline algae called the *algal ridge*. This ridge receives the full impact of the wave energy and hence, is virtually free of any organisms except the encrusting coralline algae. The coralline algae present are usually of the genus *Lithothamnion* or related genera and hence, the origin of the alternate name for this zone, the "Lithothamnion ridge." Algal ridges are not developed on all windward reefs in the Indo-Pacific and are rare in the Atlantic. The grooves of the buttress zone, often called "surge channels," penetrate behind the algal ridge as deep channels or often as "blow holes" where the channel has been roofed over by algal and/or coral growth, leaving only one or a few openings on the inside of the ridge through which water may enter or leave. These "blow holes" may offer a spectacular display under certain wave conditions, as the water is forced into the tubes at the outer slope of the reef and explodes as a geyser on the inside. This whole area, algal ridge and buttress zone, is an extremely dangerous area, since wave action is violent and a step into a blow hole or surge channel could mean disaster.

Behind the algal ridge is a very shallow *reef flat* zone. This is a complex area with gradients in environmental factors such as temperature, turbidity, and exposure across it from the algal ridge to the shore of the atoll island. These gradients, coupled with differences in depth and substrate type (coral rock, sand), provide a great number of habitats that have caused this zone to be variously subdivided. It is also one of the most species-rich areas of the reef. Near the algal ridge, conditions for coral growth are excellent and a great variety of corals is found. This is another area where active accumulation of reef material occurs. Conditions for coral growth depart more from optimal as one approaches the island and as a consequence, corals become less and less abundant. Deeper areas of the middle reef flat often experience the development of micro-atolls, which are simply large colonies of a massive coral such as *Porites*, in which the uppermost part of the colony has been killed, either by exposure to air or excess sedimentation. This is followed by erosion of the skeleton so that it becomes a small basin filled with water and in which other organisms may grow (Fig. 9-9). These micro-atolls support a variety of organisms. Also present in this reef flat area are the giant clams (*Tridacna*, *Hippopus*).

The reef flat zone ends at the *seaward beach* of the atoll island. On the lagoon side of the island is a narrow beach, followed by the quiet waters of the lagoon. Two major zones can be recognized in the lagoon, the *lagoon reefs* and the *lagoon floor*. Lagoon reefs are found around the perimeter of the lagoon and also exist as patch reefs or pinnacles rising from the lagoon floor to the surface. The depth of the lagoon is usually less than 50 m and therefore within the depth range of coral growth. Conditions within the lagoon are not as good for coral growth as on the windward side, because wave action and circulation are not as great and sedimentation is greater. Hence, coral growth is often restricted to areas where conditions are better, leading to the existence of patch reefs and pinnacles and leaving large areas of the lagoon floor free of coral growth. Sedimentation leads

Figure 9-9 Photo of a micro-atoll on the Great Barrier Reef.

to large areas of sand, and the lagoon floor may have extensive beds of sea grasses (*Thalassia*, *Cymodocea*) or of green algae such as *Caulerpa* and *Halimeda*.

The leeward side of the lagoon is margined by the *leeward reef flat*, which is generally narrower than the seaward reef flat and separated from the lagoon reefs by a barren boulder zone or by an area of poor coral growth dominated by isolated colonies of *Porites*. The seaward margin of this flat is similar to the windward flat. At the leeward reef margin, the algal ridge is usually poorly developed or absent and there are no surge channels or buttresses. At the seaward edge, extending down to 15–20 m, coral growth is lush and diverse, dominated by the branching fronds of *Acropora*. In addition to the massive corals seen on the seaward reef margin, numerous branching species are found here, due to the absence of violent wave action. Below this, the leeward reef slope is similar to the outer seaward slope.

Fringing reefs and barrier reefs lack the complexity of zones of atolls and, in effect, can be considered as truncated portions of the atoll zonation scheme. Fringing reefs and barrier reefs on the windward sides of islands have a zonation like that of the windward side of atolls and those on the leeward side similar to the leeward reef of atolls. The major difference between the two is that barrier reefs, especially large ones such as the Great Barrier Reef, may have a very extensive lagoon between the reef flat and the shore of the continent or island and that lagoon may itself develop additional coral reefs.

ATLANTIC AND INDO-PACIFIC REEFS

As we have already indicated, there are certain differences between the Atlantic and Indo-Pacific reefs. These will be detailed at this time. They include not only structure and zonation features, but also presence, absence, and abundance of various faunal and floral components.

The physical conditions under which reefs grow in both the Atlantic and Pacific are quite similar with respect to temperature, salinity, and turbidity, but the number of coral genera and species is very different between the two areas as described in the preceding section. This reduction in number of species between Pacific and Atlantic reefs is not limited to corals, but also occurs among most of the reef components, including mollusks, fishes, and crustaceans. Atlantic reefs commonly have large numbers of sea fans and whips (gorgonians), whereas these are much less in evidence in Indo-Pacific reefs. Indo-Pacific reefs have large numbers of soft or alcyonarian coral such as *Sarcophyton* and *Lobophyton*, whereas these are inconspicuous in the Atlantic. Algal ridges are rare on Atlantic reefs and calcareous and coralline algae tend to play a lesser role on the reef. They also lack the giant clams (*Tridacna, Hippopus*) and the alcyonarian corals *Heliopora* and *Tubipora*.

Not only are there differences in species composition between these two areas, there are also differences in age and reef morphology (= construction). Atlantic reefs usually rest on shallow banks or platforms, which are probably the result of erosion by wave action during the Pleistocene period, when sea level was much lower than at present. Therefore, borings through Atlantic reefs do not show the great depth of reef limestone found in the Pacific which indicate the great age of the Pacific reefs. In fact, as noted above, most Atlantic reefs are apparently of very recent origin, dating only from the end of the last ice age. Why this great discrepancy in age? This is not known for certain, but two explanations may be offered. In the first place, the Atlantic is a geologically more recent ocean, and hence, there has been less time for reefs to develop. Secondly, in the ice age, the Atlantic may have become too cold for reefs to survive, and hence they died out and were replaced only when the seas warmed up following the retreat of the glaciers. Glynn (1973) has stated that Atlantic reefs, although of more recent origin, do develop massive structural frameworks similar to Pacific reefs and show similar vigorous growth.

Zonation is somewhat different. Atlantic reefs have the buttress zone deeper than the Pacific reefs, and active coral growth extends to 100 m in the Atlantic, but only to 60 m in the Pacific. Atoll islands are often absent in the Atlantic, but reef flats and shallow lagoons are common.

PRODUCTIVITY

Coral reefs are truly oases in a watery desert. As we noted in Chapter Two, tropical marine waters tend to be extremely poor in nutrients, and hence, very low in productivity (see pp. 66–70). Yet, coral reefs abound with life and all studies done on them have indicated a very high productivity. Most studies, such as Kohn and Helfrich (1957) and Odum and Odum (1955), have given primary productivity estimates of 1500–3500 g C/m^2/yr. This is in contrast to a productivity in the open tropical oceans of 18–50 g C/m^2/yr. How can coral reefs maintain such production in such nutrient-poor areas? This is less hard to explain for reefs bordering high islands or continents, where much nutrient material is available from the land and from the shallow inshore waters, than it is for atolls isolated far from land. In the latter case, we do not as yet know all the reasons for the high productivity, but at least some can be suggested. In the first place, the amount of plant tissue present on a reef and capable of photosynthesis is very large, much larger than in a similar area of the open ocean. Zooxanthellae are extremely efficient autotrophic organisms, and since they are found in virtually all corals on the reef, they form a significant biomass. In addition, there are the algae in the coral skeleton itself, the numerous coralline algae and the various green and brown algae. Taken together, the plant biomass, according to Odum and Odum (1955), is greater than the animal biomass, even if less conspicuous. Even given the potential productivity of this large plant biomass on reefs and the optimal physical conditions, it would not be realized if there were not enough nutrients, the lack of which is the very reason for the low productivity of the open ocean immediately offshore. The secret, apparently, of reefs is their ability to tenaciously hold on to all nutrients in the system and to act as a sink for any brought in from the outside. This ensures that nutrients are cycled within the reef system and are not lost to deeper offshore waters. It also means that any plankton from the open sea offshore that impinges on the reef remains there, as do the nutrients brought with them.

As an example of the cycling that prevents loss, we can point to the zooxanthellae in the coral tissue. Fixing the plants in the coral tissue means that the algae will not be washed from the reef. It also means that any nutrients produced by the coral (NO_3, PO_4) as a result of its metabolism will be available directly to the plant without cycling it outside into the water first, where it might be lost. Similarly, most of the free-living algae on the reef contain calcium carbonate. They are too heavy to be moved easily off the reef; hence, their nutrients remain on the reef to nourish additional generations. Rapid cycling is also aided by the large bacteria populations in coral reefs. These bacteria act quickly on dead material, breaking it down and making available enclosed nutrients. For all these reasons and probably for others yet to be discovered, coral atolls have very high productivity, as do fringing and barrier reefs. However, even given the above, it is as yet far from understood just how such systems are maintained. Much of this explanation, however, can be forthcoming from a more complete understanding of how the various components in the reef ecosystem interact. Of these components, the first and most significant are the corals themselves. They are the subject of our next consideration.

BIOLOGY OF HERMATYPIC CORALS

Hermatypic corals are the dominant group involved in reef formation and maintenance; therefore, a knowledge of certain of the biological aspects of the group has considerable relevance to our understanding of the ecology of reefs themselves.

Nutrition. Reef corals have long been known to be carnivorous animals, similar in that respect to most other members of their phylum. They have tentacles studded with stinging capsules, nematocysts, which are used to sting and capture small plankton organisms. Since they occur in colonies of hundreds or thousands of individuals, and the number of colonies is also huge, the total area for feeding is enormous. In addition to the tentacles and their nematocysts, the outer epidermis of corals is ciliated and produces mucus. This ciliary mucous mechanism is generally used by corals to rid themselves of sediment settling on the surface; in some corals, it has been modified to be used in feeding as well. Those species employing mucous mechanisms to capture plankton tend to have smaller, shorter tentacles than those that feed directly on plankton by snaring them with their tentacles. Food organisms are detected by means of chemoreception.

Since corals exist in such numbers on reefs, and the surrounding ocean is notoriously poor in plankton, how is it possible for these animals to obtain enough food? This is precisely the question that has puzzled generations of coral reef biologists and is only now becoming clear. Rigorous plankton sampling on reefs and just offshore by Porter (1974), Johannes et al. (1970), and others, both during the day and at night, has revealed an abundant plankton population. However, a large fraction of this plankton population is not from the open ocean area, but is a population indigenous to the reef itself, a fact not previously known. This indigenous plankton population is composed primarily of meroplankton, which seems to spend the daylight hours on the bottom, rising into the waters over the reef only at dusk. This may also explain why many corals are closed up during the daylight hours, expanding to feed only during the hours of darkness. In spite of this discovery of enhanced abundances of plankton over the reefs, Johannes et al. (1970) have determined that the amount of plankton available to the corals is only sufficient to satisfy 5–10 percent of their total food requirements. Where, then, does the remainder of their food come from? Only one source seems possible, and that is the zooxanthellae found in the tissues.

The role of zooxanthellae in coral biology has long puzzled biologists. It was early learned that corals were carnivores and that they could not digest plant tissue. Furthermore, early experimental studies showed that, if corals were starved or kept in the dark, they would expel their zooxanthellae, but could continue to live for a short time without them. All of these suggested that the zooxanthellae were really unnecessary to the coral. More recently, however, Muscatine and Cernichiari (1969) have employed radioactive tracers to prove that organic compounds fixed by the zooxanthellae in photosynthesis are transferred to the coral, where they undoubtedly serve as food for the coral. Franzisket (1969) has also demonstrated that, if corals are not fed but kept in the light, they gain weight. This could only happen if the zooxanthellae furnished food to them. Whereas the role of zooxanthellae in nutrition thus seems proved, we do not know as yet how important this role is for all coral species. At present, it appears that corals with large polyps are less dependent on zooxanthellae for nutrition than other species with smaller polyps.

Growth and Calcification. Although there has been much interest in the growth rate of both corals and coral reefs for many years, it is only in recent years that we have come to have a reasonable understanding of growth in individual corals and reefs. The first requirement for active coral growth is light. As Connell (1973) has shown, if corals are shaded or in some way prevented from obtaining access to the light, they cease growth and, if light is withheld long enough, they will die. This requirement for light is undoubtedly due to the needs of the zooxanthellae, and it suggests the other important role of zooxanthellae in the life of corals. Zooxanthellae have been found by Goreau (1961) to significantly increase the calcification rate of corals and hence the rate of growth of the coral colonies. Exactly how the zooxanthellae act to increase this rate of skeletal accretion is not understood as yet, but again points up the important role of the zooxanthellae in the coral reef ecosystem, where rapid calcification is needed to maintain the reef against the various destructive forces acting against it.

Growth rates for individual colonies are generally obtained by continued measurement of marked individual colonies over a period of time. A rapid and simple way of doing this has been suggested by Connell (1973). It is to take photographs of an area of reef, then measure the colonies later from the developed print (Fig. 9-10).

The rate at which coral colonies grow differs with different species, age of the colonies, and in different areas of a reef, but some generalities can be noted. Young, small colonies tend to grow more rapidly than older, larger colonies, and branched or foliaceous corals tend to grow more rapidly than massive (= brain) corals. *Acropora*, for example, is a foliaceous genus; measurements of the Atlantic species by Vaughn (1915) suggest that the species could grow 5–10 cm in diameter and 2–5 cm in height per year. *Montastrea annularis*, a massive type, grew only 0.5–2 cm in diameter and 0.25–0.75 cm in height per year. Despite much recent work that has given us good estimates of growth rates in some species, the fact that there is growth throughout the life of the colony and that growth rates in the larger, older colonies are variable and difficult to estimate with any precision, means that we still do not have any good feeling for the age structure of the colonies seen on a reef, nor do we know how long corals can live. What little information there is, primarily that of Connell (1973), suggests that most coral colonies on reefs are ten years of age or younger. However, some of the very large massive corals, such as those forming the aforementioned micro-atolls, may be 100 years old or older, suggesting that corals can live much longer than the ages generally realized by most colonies on the reef.

It is also difficult to translate the growth rates of individual species into growth rates of a whole reef system. As a result, most estimates of the rate of reef

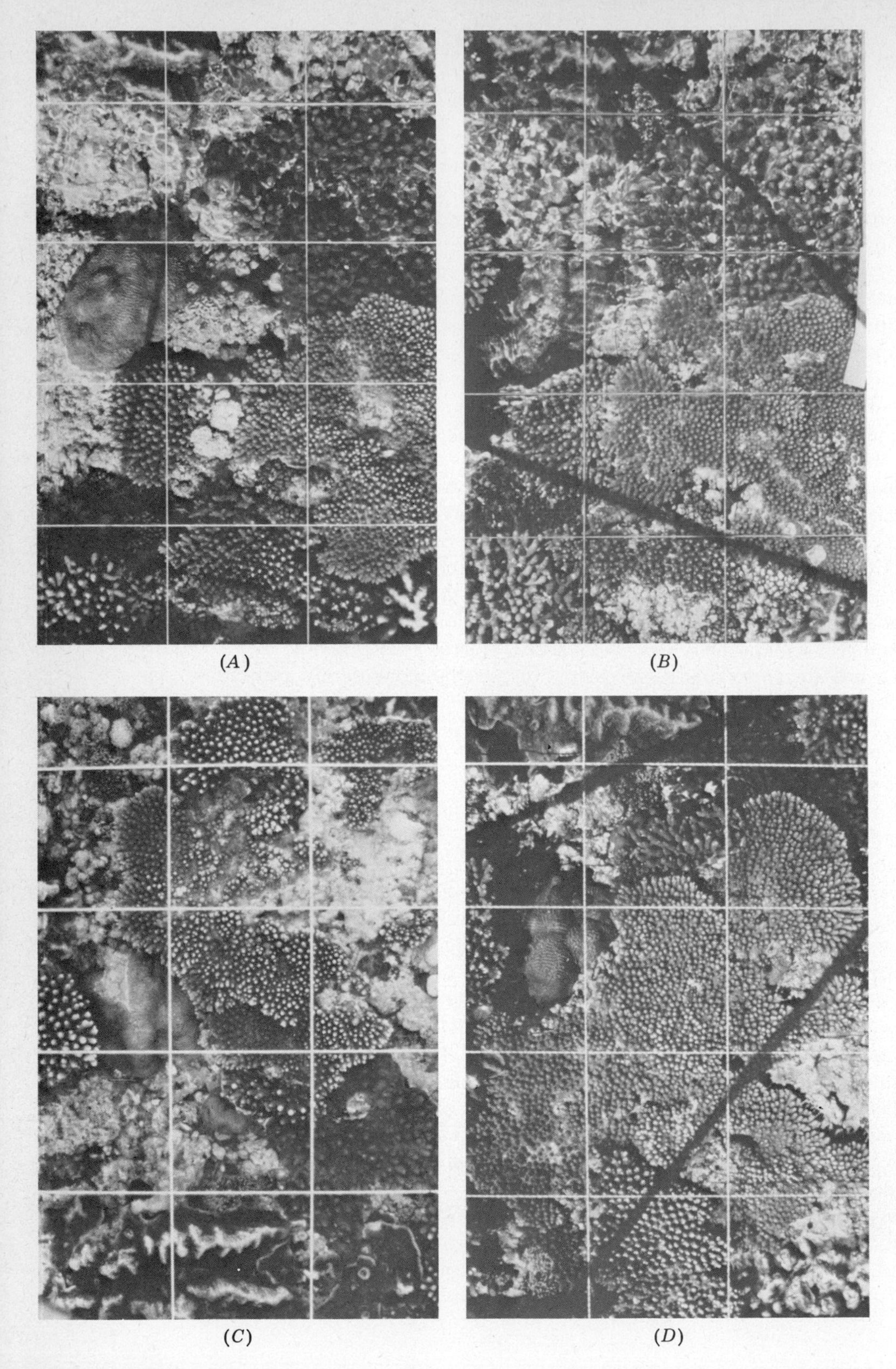
(A) (B) (C) (D)

growth have been made by extrapolating from changes in reef topography over a period of years or from age of the reef as determined from geological data such as thickness of reef limestone deposits. These estimates vary quite widely and also vary for reefs in different areas of the world. The range reported by Stoddard (1969) is from 0.2 mm of upward growth per year to perhaps 8 mm.

The growth form of a given coral species may also vary considerably, depending on its location on a reef. In contrast to the same species in shallow water, a coral species at greater depth is thinner and more spindly, probably as a result of less calcification. Wave action tends to force branched species into shapes in which the branches are shorter and stubbier, and currents cause branching forms to have a definite alignment of the branches. This ability of coral species to have different skeletal shapes has resulted in much confusion among coral reef taxonomists, such that different growth forms of the same species have often been named as different species.

Sexual Maturity, Reproduction, and Recruitment. Corals perform both sexual and asexual reproduction. Asexual reproduction is generally accomplished by budding off a new individual from the parent, and continued budding is the mechanism of increasing the size of the colony, but not for producing new colonies. Sexual reproduction results in the production of a free-swimming *planula* larva, which, when it settles down, initiates development of a new colony (Fig. 9-11).

From our very limited data, it appears that most corals reach sexual maturity at between 7–10 years of age. Corals may be hermaphroditic or dioecious. Fertilization usually occurs in the gastrovascular cavity of the female, the sperm being released into the water and drawn into the gastrovascular cavity. Fertilized eggs are usually retained until development has proceeded to the planula larvae stage. The planulae are released and swim in the open water for an undetermined time, but perhaps for only a few days, before settling down and starting a new colony. Since adult corals are fixed in place, the planula larvae are the means of dispersal of the various coral species.

Except for two studies, little information exists on the rate of recruitment of new coral colonies into reef systems. The study by Connell (1973) has indicated that the rate of recruitment on the Great Barrier Reef was between 0 and 13 new colonies per square meter per year. The commoner coral species were the most abundant of the new colonizers, but there were great differences in species that

---

Figure 9-10 Photographs of the same one-meter quadrat showing the changes in growth and occurrence of coral colonies over an eight-year period on the Great Barrier Reef at Heron Island. For each photograph the square meter frame, divided into 20 × 20 cm squares, was positioned on permanent stakes. Note the great changes which have occurred in growth and in presence of corals over short time intervals. *(A)* 1963. *(B)* 1965. *(C)* 1969. *(D)* 1971. (Photographs courtesy of Dr. Joseph Connell, Dept. of Biological Sciences, University of California, Santa Barbara.)

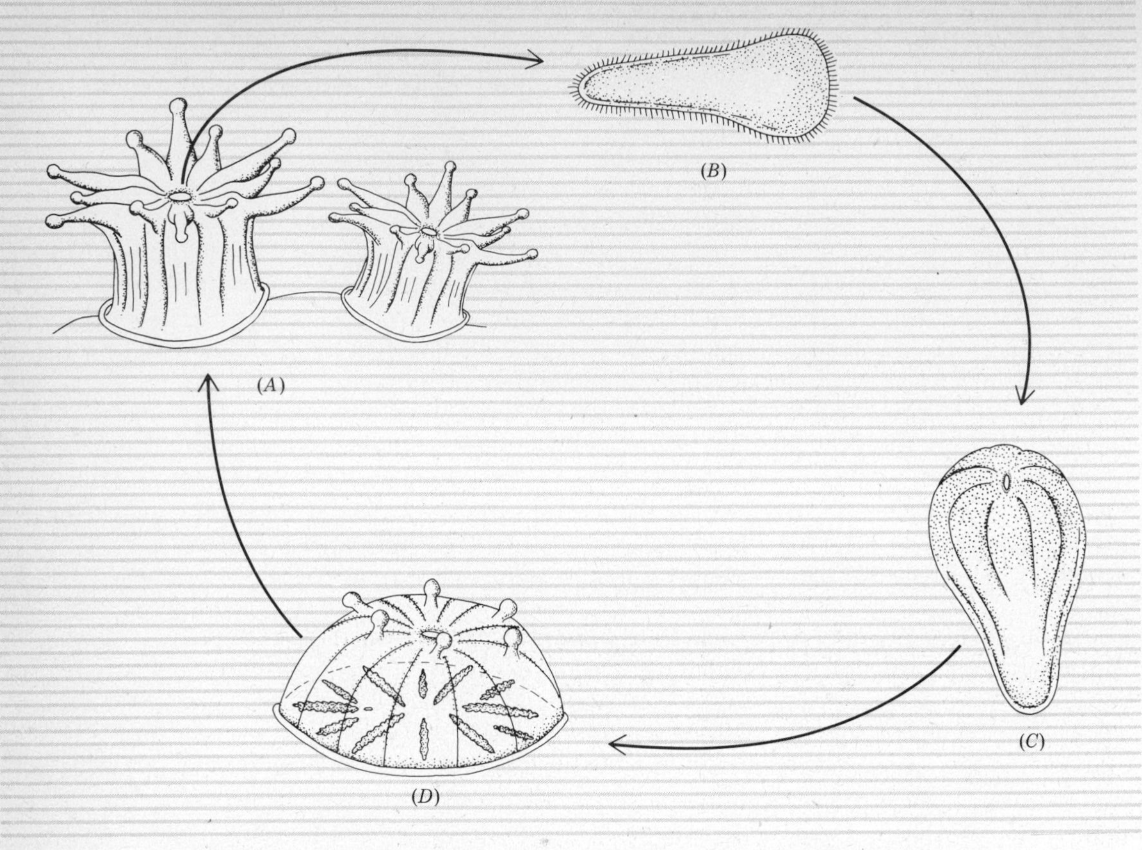

Figure 9-11 Reproduction of corals. *(A)* Adult polyp. *(B)* Planula larva. *(C)* Later planula with developing septa. *(D)* Young polyp after attachment. (Modified from L. Hyman, 1940, The invertebrates: vol. 1, Protozoa through Ctenophora. McGraw-Hill. Copyright 1940. Used with the permission of McGraw-Hill Book Company.)

settled among different areas. It also appeared that mortality was high among the newly settled colonies.

SPECIES INTERACTIONS AND ECOLOGY OF REEFS

At this point, we have covered the basic biology of the dominant organisms of the reefs, namely, the scleractinian corals, and have outlined the general zonation pattern of reefs and the physical conditions acting on them. Reefs, however, are much more than simple static systems. They are living systems that increase or decrease in size as a result of the complex interactions among various biological and physical forces. We can now consider what we know about how these interacting factors produce the reef as we view it, and also how a balance is maintained to ensure the persistence of a reef system through time. Since coral reefs are perhaps the most complex systems in the marine environment and since they have not received as much detailed ecological study as, say, rocky shores, our knowledge of this aspect of the coral reef system is not as complete as we would like. However, it is still possible to suggest how some factors interact to maintain the system. The general picture that is beginning to emerge is that the interaction that produces the reef system is similar to that which is responsible for the patterns we observed on rocky shores.

Competition. One of the most obvious features of a coral reef in an area of active coral growth is that there is virtually no open space. The primary space is completely covered by the dominant space occupiers, namely the corals themselves. Since all space is occupied and since corals must have access to the light to survive, we might expect competition to occur among corals for space and light.

Such competition does occur among coral colonies. Upright, branching coral colonies grow more rapidly than encrusting or massive corals, and they often extend themselves up and over the encrusting forms, shutting them off from the light. Where this occurs, the part of the encrusting colony in the shade dies. This is indirect competition. Since these branching corals are able to grow faster and shade out the slower growing massive and encrusting corals, how then do these slow-growing forms manage to persist? There are a number of reasons, but the most intriguing is the discovery by Lang (1973) that the slow-growing coral species have the ability to extend out filaments from their gastrovascular cavities that, when they encounter the living tissue of an adjacent coral colony of a different species, digest that tissue, thereby killing that portion of the adjacent colony (Fig. 9-12). This is direct competition and is reminiscent of the activities of anemone clones (see pp. 226–227). Thus, the slow-growing corals can prevent the faster-growing corals from overshadowing them: they maintain space on the reef by simply killing off that portion of the colony about to overgrow them. Coral species may be ranked in an "aggressive pecking order" in which each species is capable of attacking and killing those below it and, in turn, is attacked by those

Figure 9-12 Photograph showing the killing of portions of a tabular *Acropora* colony (left) by the brain coral *Ctenella chagius* (right) thereby preventing overgrowth by the faster growing *Acropora*. (Photograph from C. R. C. Sheppard, Interspecific aggression between reef corals with reference to their distribution, Marine ecology progress series, vol. 1.)

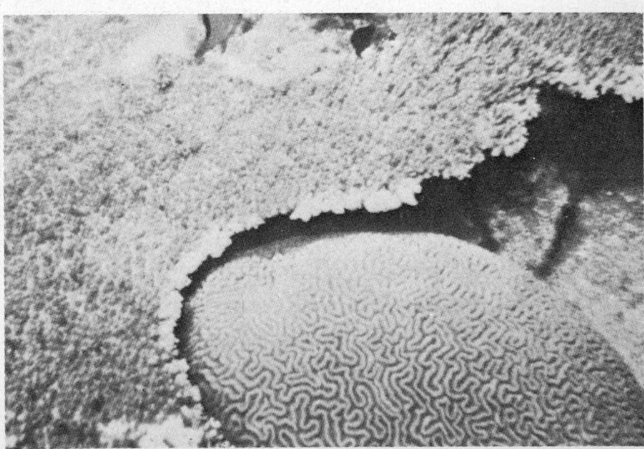

above it. In this hierarchy, those corals that are fast-growing and branching and thus capable of shading out others are lower in rank than the slow-growing forms. Hence, this interspecific aggression is one mechanism that acts to prevent the monopolization of space and thereby maintains coral diversity. One effect of the intense space competition on reefs is the tendency of various organisms to utilize the same space in association. This is *commensalism*. Commensal relationships are much more common in reefs than elsewhere (see Chapter Ten for complete discussion).

Predation. Although coral reefs have huge numbers of species of various other invertebrates, swimming over a reef, one sees virtually no invertebrates except for a few large echinoderms (sea cucumbers, urchins, feather stars) and large mollusks (*Tridacna* and others). The reef seems dominated by the corals and by the abundant fishes. This is because the other invertebrates are all hidden from view. It has been suggested by Bakus (1964) that the reason for this cryptic pattern is the intense predation pressure on the reef. Any soft-bodied invertebrate in the open would be quickly consumed. This can be verified by turning over coral heads and exposing the fauna, which immediately attracts fishes to the area to consume the exposed organisms. Presumably, the larger echinoderms and mollusks are more immune to predation.

The role of predation in determining the structure and composition of coral reefs is not as well studied or understood as it is for rocky intertidal areas, but we do have some evidence of its importance for corals.

A surprising number of animals feed on coral and hence can be classified as predators. Most of these predators, however, are small in relation to the coral colony. The process of predation in these forms resembles the process of grazing by herbivores, whereby patches of coral polyps are removed but the entire colony is not destroyed. Thus, if not too much tissue is removed, the coral may regrow polyps to cover the area which has been "grazed" very much as grass regrows after grazing. These small predators include a number of gastropod mollusks (families Architectonidae, Epitoniidae, Ovulidae, Muricidae, and Coralliophilidae) at least one nudibranch mollusk (*Phestilla*), amphinomid polychaete worms (*Hermodice*), certain barnacles (*Pyrgoma*), and several crabs (*Mithraculus*, *Trapezia*, *Tetralia*). Taken together, these predators do not have a significant effect on coral colonies, nor do they seem able to affect the community structure.

Two other taxa of predators are, however, capable of destroying coral colonies and thus of modifying the structure of reefs. These are the starfish *Acanthaster planci* and various fishes. *Acanthaster planci* is a very large, multi-armed starfish that feeds only on the tissue of living coral (Fig. 9-13). Because of its size, it is capable of destroying an entire colony during feeding. Normally, *Acanthaster* is quite rare on coral reefs in the Indo-Pacific (not found in the Atlantic), but can exert an influence on the composition of reefs. As Porter (1972) has reported, it does this first through its preference for feeding upon the fast-growing, space-monopolizing coral species. Thus, by selectively reducing or removing the

Figure 9-13 Photo showing the large numbers of *Acanthaster planci* on a heavily infested reef in Palau in 1979. (Photo courtesy of Dr. Charles Birkeland.)

faster-growing corals, *Acanthaster* promotes diversity of corals and helps ensure survival of the slower-growing species. The second influence on the reef structure by *Acanthaster* comes when conditions bring about very high starfish densities. In recent years, *Acanthaster planci* has undergone a population explosion on certain reefs in the western Pacific such that the normal population density reported by Endean (1973) of one to three starfish observed in four to five hours of search has gone to more than 100 starfish seen in 20 minutes of search. The latter figure represents tens of thousands of starfish on each reef (Fig. 9-13). In these cases, the entire reef is destroyed by the voracious starfish (see pp. 342–343) and nearly all corals are consumed.

Two groups of fishes actively graze over coral colonies: those species that consume the coral polyps themselves, such as puffers (Tetraodontidae), filefish (Monacanthidae), triggerfish (Balistidae), and butterfly fish (Chaetodontidae); and a group of multivores (= omnivores) that remove the coral polyps to obtain either the algae in the coral skeleton or various invertebrates that have bored into the skeleton (Acanthuridae, Scaridae) (Fig. 9-14). It has been demonstrated by Motoda (1940) that the grazing of coral colonies by fishes may, in certain cases, be sufficiently intense to kill the colony. Corals that are restricted to certain regions of the reef when transplanted to other areas have been killed by fish grazing. The extent to which this activity restricts corals to certain regions or is responsible for coral distribution patterns is, however, unknown at present. What is certain is that the two families of fishes, Acanthuridae (surgeon fishes) and Scaridae (parrot fishes), that do consume living and dead coral are among the more abundant fishes on reefs; by passing coral skeletons through their guts, they are significant producers of sand and sediment on reefs.

Grazing. Aside from the abundant and conspicuous coralline algae on reefs, there is little evidence of other large algae or other plants. If you look closely,

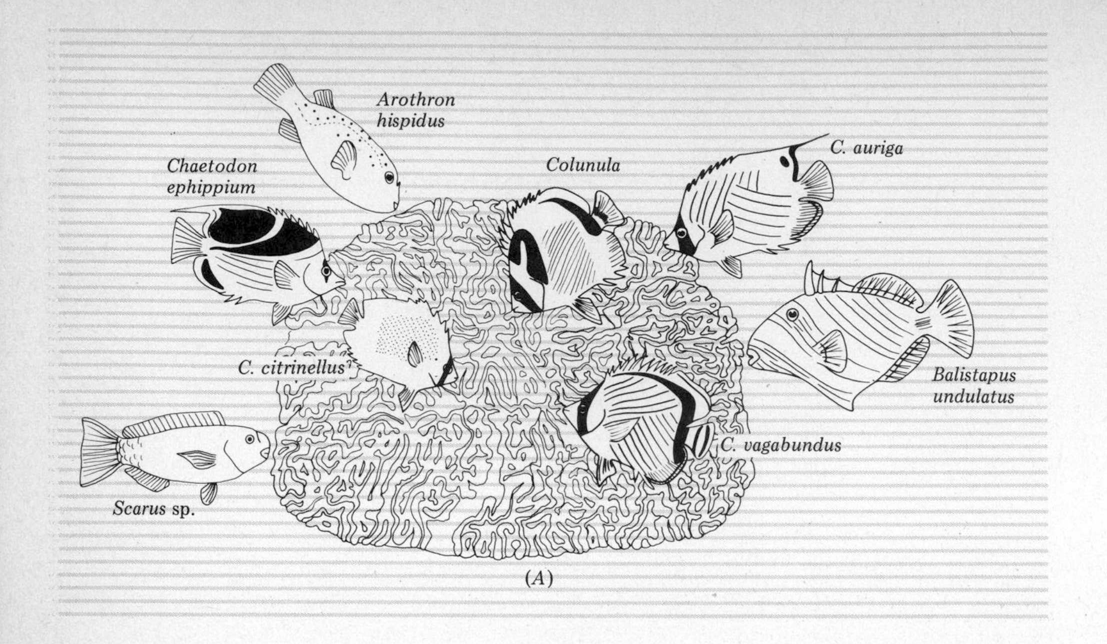
Arothron hispidus
Chaetodon ephippium
Colunula
C. auriga
Balistapus undulatus
C. citrinellus
C. vagabundus
Scarus sp.

(A)

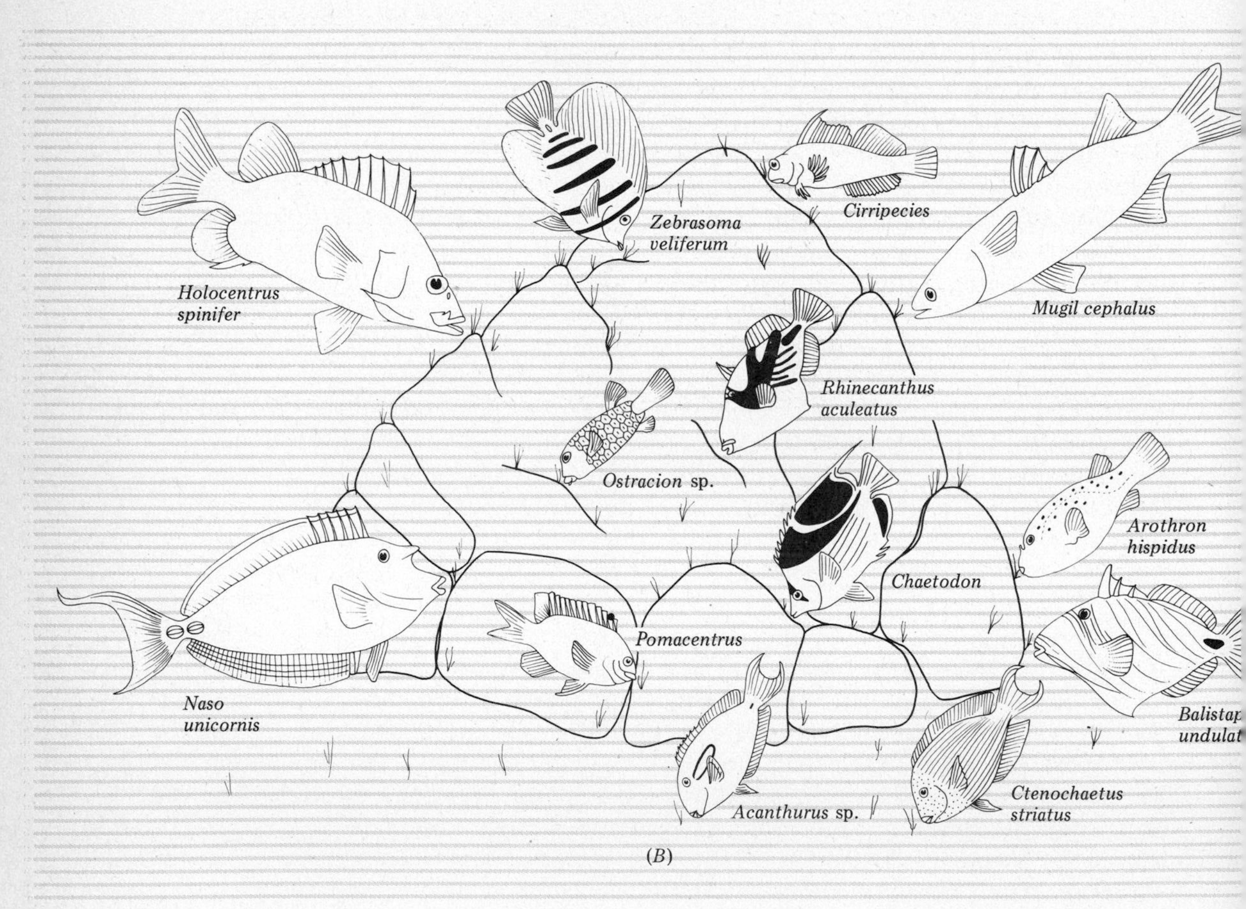
Zebrasoma veliferum
Cirripecies
Holocentrus spinifer
Mugil cephalus
Rhinecanthus aculeatus
Ostracion sp.
Arothron hispidus
Chaetodon
Pomacentrus
Naso unicornis
Acanthurus sp.
Ctenochaetus striatus

(B)

however, you will find in many parts of the reef a short algal turf or small clumps of algae. Algae can be major space competitors with corals on reefs, invading and establishing themselves more quickly than corals. Since the algae are not dominant on the reef and obviously do not outcompete the corals, what restrains them? The answer appears to be intense grazing pressure by fishes and also certain invertebrates, primarily sea urchins. Many families of fishes contain herbivorous grazers, including Siganididae, Pomacentridae, as well as the aforementioned Acanthuridae and Scaridae. Continued grazing by these fishes, plus the grazing by abundant sea urchins such as the ubiquitous genus *Diadema*, keep the algae reduced to a minimum and enhance the survival of coral recruits (Fig. 9-14).

Algal grazing by fishes, however, may also indirectly retard coral growth. For example, damselfish (Pomacentridae) are territorial fishes that graze either selectively or nonselectively upon the algae composing the algal mat within their territory, but keep other grazing fishes out. The result of this territorial foraging, according to Lobel (1980), is that corals and coralline algae are excluded from the territory due to overgrowth by the algae, and the territory serves as a refuge for juvenile invertebrates and demersal plankton.

The grazing effect of urchins such as *Diadema* and *Eucidaris* may have even more significant effects. The studies done by Sammarco, Levinton, and Ogden (1974) with *Diadema* suggest that, at very high densities, this urchin will graze off all organisms, not just algae, and thus prevent coral growth as well. At lower densities, the urchin effectively removes the algae and permits coral colonies to become established, while at very low densities, algae take over the area by outcompeting the corals in the absence of significant grazing. Presumably, the grazing by moderate densities of *Diadema* clears areas for coral planulae to settle into and thus indirectly contributes to maintenance of reef corals.

In the Galapagos, Glynn, Wellington, and Birkeland (1979) have demonstrated that the urchin *Eucidaris thouarsii* grazes on the dominant coral, thus interfering with the establishment of the reef framework, and may be responsible for the lack of reefs in these islands.

Figure 9-14 *(A)* Fishes which actively graze coral colonies: butterfly fish (Chaetodontidae), *Chaetodon, Colunula;* parrot fish (Scaridae), *Scarus;* triggerfish (Balistidae), *Balistapus;* puffers (Tetraodontidae), *Arothron.* *(B)* Herbivorous fishes of the reef: triggerfish (Balistidae), *Balistapus, Rhinecanthus;* squirrel fishes (Holocentridae), *Holocentrus;* surgeon fish (Acanthuridae), *Acanthurus, Ctenochaetus, Zebrasoma, Naso;* damselfishes (Pomacentridae), *Pomacentrus;* butterfly fishes (Chaetodontidae), *Chaetodon,* puffers (Tetraodontidae), *Arothron;* mullet (Mugilidae), *Mugil;* boxfishes (Ostraciontidae), *Ostracion;* blennies (Blenniidae), *Cirripecies.* (Not to scale.) (Modified from R. W. Hiatt and D. W. Strasburg, 1960, Ecological relationships of the fish fauna on coral reefs of the Marshall Islands, Ecological monographs, vol. 30. Copyright 1960, the Ecological Society of America.)

ROLE OF ALGAE IN REEF SYSTEMS

As we have noted, other than corallines, algae are very inconspicuous on coral reefs, in marked contrast to the great kelp forests of temperate shores. Yet, algae are found in reef systems and their importance to the entire reef system can be summarized here.

First, the encrusting red coralline algae such as *Lithothamnion* are important in maintaining the integrity of the reef by constantly cementing together various pieces of calcium carbonate, thus reinforcing the reef framework against destruction by wave action and preventing individual pieces being carried off the reef. The algal ridge is further responsible for breaking the velocity of the waves and producing calmer conditions, allowing growth of other organisms in the reef flat behind it. The second group of calcareous algae are the greens, of which the genus *Halimeda* is dominant. These algae are significant contributors to the sand found in reefs, particularly in the lagoon area. They thus create a special habitat type.

Certain algae also bore into coral skeletons and thus are responsible for the breakdown of reef structure. Algae also create habitats in and around themselves and may furnish necessary shade from the hot tropical sun for certain organisms.

They, of course, are important as primary producers in the reef system and as a food for various herbivores. These various roles are summarized in Figure 9-15.

ECOLOGY OF REEF FISHES

It is impossible to conceive of a coral reef without the presence of myriads of brilliantly colored fishes gracefully moving among the various corals and associated organisms. Indeed, fishes are certainly the most abundant and conspicuous large organisms that one encounters on a reef (Plates 26 and 30). Because of their abundance and all-pervading presence in all areas of the reef, it

Figure 9-15 The various roles of free-living algae in different zones of a reef. (Modified from M. Doty, 1974, *in:* Proceedings of the second international symposium on coral reefs, vol. 1, Great Barrier Reef Committee, Brisbane. Used by permission of the Great Barrier Reef Committee.)

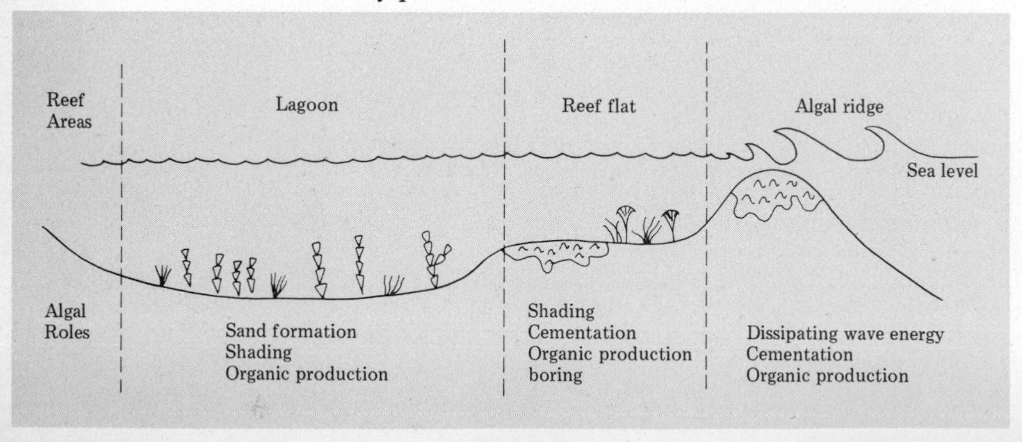

seems obvious that they must contribute in significant ways to the reef ecosystem. We have noted some of these relationships in previous sections and will in this section review other aspects of their ecology.

The species richness of coral reef fishes is very similar to that observed for corals. That is to say, the central Indo-Pacific area of the Philippine and Indonesian archipelagos have the greatest number of species and the number decreases in all directions away from this center (Table 9-1). The Atlantic reefs are also relatively species-poor in this comparison. The number of species that have been found on a single reef is really quite remarkable, 500 for one reef in the Great Barrier Reef system. How does this great diversity of species come about, and how is it maintained? This is the question that has intrigued coral reef biologists for many years and is still generating debate. We can only suggest some of the reasons here; more remain to be discovered.

One of the reasons for the high diversity of species on reefs is the great variety of habitats that exists on reefs. Coral reefs encompass not only coral, but also areas of sand, various caves and crevices, areas of algae, as well as shallow and deep water and different zones progressing across the reef. This great diversity of habitats alone can go far to explain the increased number of fishes. These various habitats and their associated fish inhabitants are summarized diagrammatically in Figures 9-16 and 9-17.

The numerous habitats, however, are not enough to explain the high diversity of coral reef fishes, particularly in local areas. High local fish diversity has led recently to numerous experimental studies attempting to explain how so many species can continue to coexist in a given area. As a result two opposing theories of reef-fish diversity and community structure have arisen. The more classical view is that coexistence is the result of a high degree of specialization such that each species has a specific set of adaptations that give it the competitive edge in at least one situation on a reef. That is to say, these fishes have narrower ecological niches and hence more species can be accommodated in a given area. The opposing view, promulgated by Sale (1977), is the "lottery" hypothesis. This hypothesis states that fishes are not specialized, many similar species having the same requirements, and that there is active competition among the species. Local success and persistence result from chance as to which species

TABLE 9-1 Number of Fish Species in Several Coral Reef Areas

Geographical Area	Number of Fish Species
Philippine Islands	2,177
New Guinea	1,700
Great Barrier Reef	1,500
Seychelle Islands	880
Marshall and Mariana Islands	669
Bahama Islands	507
Hawaiian Islands	448

After Goldman and Talbot, 1976.

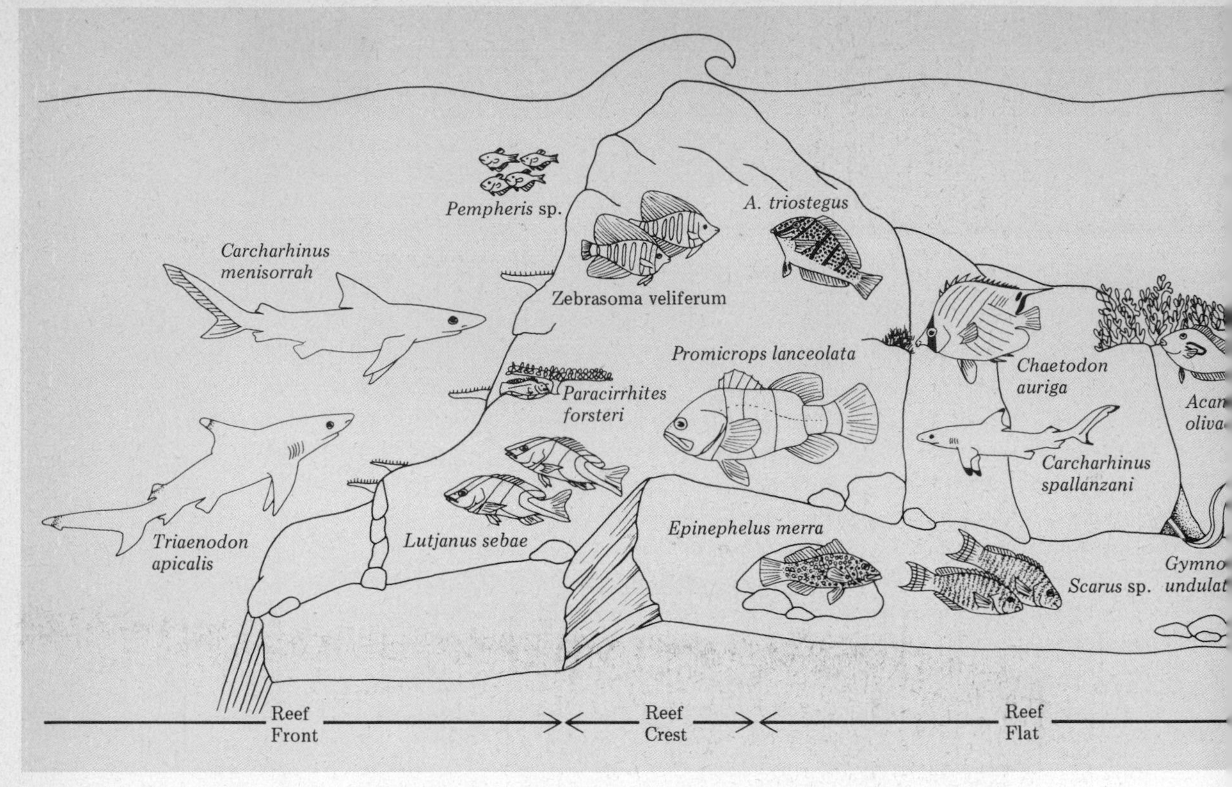

Figure 9-16 Generalized picture of the characteristic fish species and their habitats on a seaward coral reef in the Marshall Islands. Requiem sharks (Carcharinidae), *Triaenodon; Carcharinus;* moray eels (Muraenidae), *Gymnothorax;* hawkfishes (Cirrhitidae), *Paracirrhites;* groupers (Serranidae), *Epinephalus, Promicrops;* snappers (Lutjanidae), *Lutjanus;* surgeon fishes (Acantharidae), *Acanthurus, Zebrasoma;* parrot fishes (Scaridae) *Scarus;* butterfly fishes (Chaetodontidae), *Chaetodon,* sweepers (Pempheridae), *Pempheris.* (Modified from R. W. Hiatt and D. W. Strasburg, 1960, *Ecological monographs,* vol. 30. Copyright 1960, the Ecological Society of America.)

occupies the vacant space. Evidence is currently inconclusive as to which of these views, or neither, are correct.

Perhaps as a result of the large numbers of species and the partitioning of the habitat, we find that most reef fishes, despite their obvious mobility, are

Figure 9-17 Fishes associated with individual coral colonies of the *(A)* branching and *(B)* plate type. Butterfly fishes (Chaetodontidae), *Chaetodon, Forcipeger;* damsel fishes (Pomacentridae), *Chromis, Dascyllus;* wrasses (Labridae), *Epibulus, Thalassoma, Stethojulis, Gomphosus;* velvet fishes (Caracanthidae), *Caracanthus;* gobies (Gobiidae), *Paragobiodon;* puffers (Tetraodontidae), *Arothron;* file fishes (Monocanthidae), *Oxymonocanthus;* hawk fishes (Cirrhitidae), *Paracirrhites;* trigger fish (Balistidae), *Balistapus, Rhinecanthus, Abalistes;* squirrel fishes (Holocentridae), *Holocentrus.* (Modified from R. W. Hiatt and D. W. Strasburg, 1960, *Ecological monographs,* vol. 30. Copyright 1960, the Ecological Society of America.)

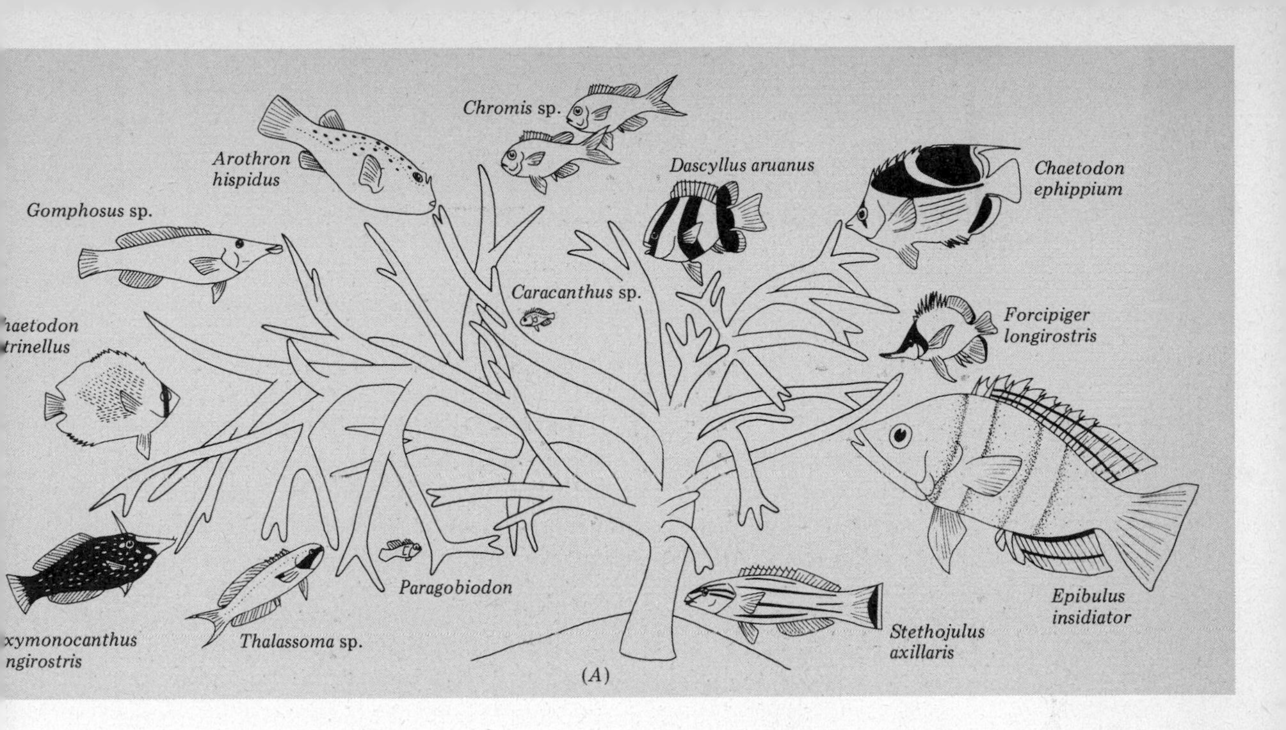
Chromis sp.
Arothron hispidus
Dascyllus aruanus
Chaetodon ephippium
Gomphosus sp.
Caracanthus sp.
haetodon
trinellus
Forcipiger longirostris
Paragobiodon
Epibulus insidiator
xymonocanthus
ngirostris
Thalassoma sp.
Stethojulus axillaris
(A)

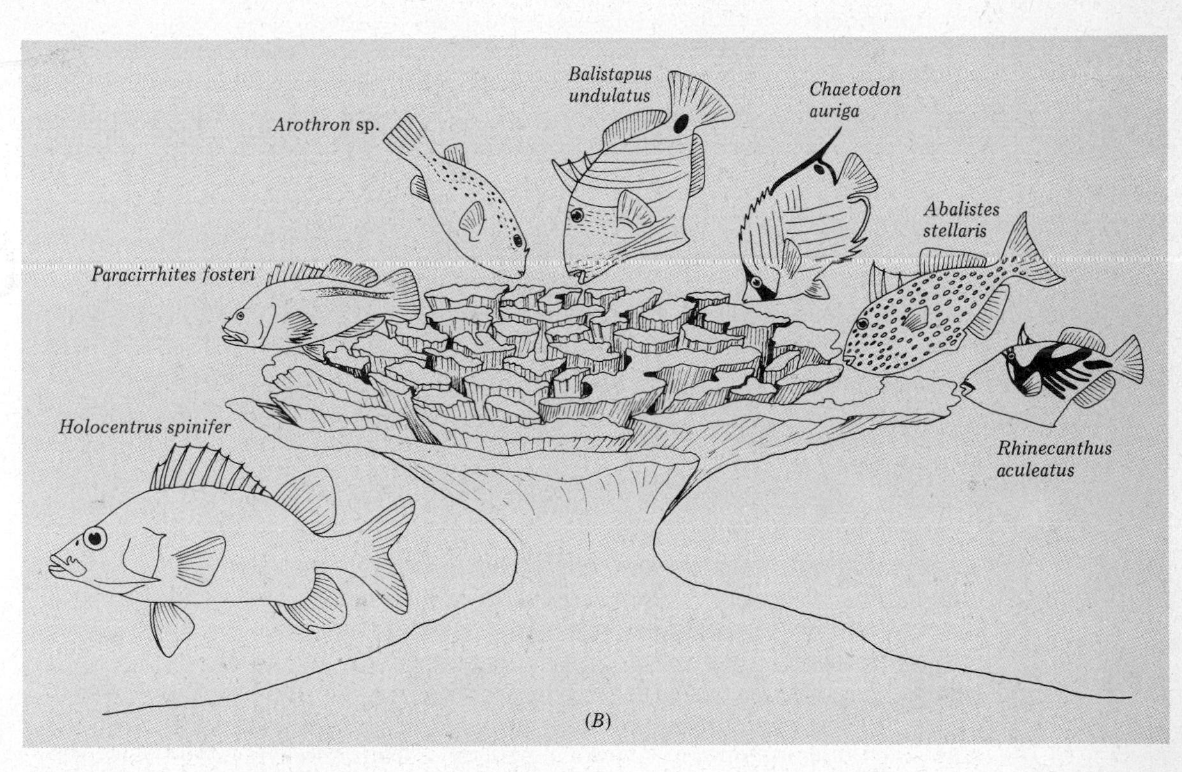
Balistapus undulatus
Chaetodon auriga
Arothron sp.
Abalistes stellaris
Paracirrhites fosteri
Holocentrus spinifer
Rhinecanthus aculeatus
(B)

restricted to certain areas of the reef and are very localized. They also do not migrate, and many of the smaller species such as gobies, blennies, and damselfishes are known to defend territories.

One of the most interesting discoveries concerning the fishes of coral reefs is the differences in fishes between day and night. Most people see reefs during the day when the great majority of the fish species are visible. At night, however, these diurnal fishes seek shelter in the reef and are replaced by a smaller number of nocturnal species not seen at all during the day (Fig. 9-18 Plate 29). Since some of the nocturnal species are ecologically similar to certain diurnal species (Apogonidae, for example, replace Pomacentridae), this is another way of permitting a greater number of species to exist on the reef without competing directly. According to Hobson (1968), all nocturnal species are predaceous.

The feeding relationships of reef fishes are of some interest. The most abundant feeding type on reefs are the carnivores, which may constitute 50–70 percent of the fish species. Goldman and Talbot (1976) say many of these carnivores appear not to be highly specialized to a given food source, but instead are opportunistic, taking whatever is available to them. They also feed upon different prey at different stages in their life cycles. The fact that the majority of the fishes on reefs are generalized carnivores seems to argue against the aforementioned explanation for the great species richness of reefs as being due to narrower niches (= more specialization). Indeed, this lack of specialization makes it difficult to satisfactorily explain the species richness. This is not to say that no specialized carnivores exist; there are a few. Correlated with the generalized feeding habits of most carnivores, the numbers of scavenger fishes are few because the carnivores simply pick up any recently dead organisms.

Herbivores and coral grazers make up the next largest group of fishes (about 15 percent of the species), and the most important of these are the families Scaridae and Acanthuridae (Fig. 9-19). The remainder of the fishes are generally classed as omnivores or multivores and include representatives from virtually all families of fishes on the reef (Pomacentridae, Chaetodontidae, Pomocanthidae, Monocanthidae, Ostractiontidae, Tetraodontidae). Only a few fishes are zooplankton feeders, and these are mainly small, schooling fishes of the families Clupeidae and Atherinidae.

Considering that the preponderance of fishes on the reef are unspecialized carnivores, it is easy to understand why most invertebrate organisms are hidden from view, since to expose a soft body on a reef would certainly invite instantaneous attack by some carnivore. How, then, do a few invertebrates, notably the large sea cucumbers, remain abundant lying unprotected on reefs? The answer apparently is that those few that are visible have evolved mechanisms to deter potential predators. The most common is a toxin or poison. According to Bakus (1973), tropical sea cucumbers produce toxic substances that can kill fishes, and, in addition, have viscous, sticky strands (cuvierian organs), which can be extruded to literally tie up any potential predator.

The development of distasteful or toxic substances is not limited to invertebrates. A great many of the fishes also produce toxic substances. These may take the form of venom associated with various spines, of poisonous material simply extruded onto the body surface (*crinotoxin*), or the flesh and internal organs may be toxic. Truly venomous fishes are rare on reefs and are confined mainly to the stonefishes (Synanceiidae) and scorpion fishes (Scorpaenidae), but a large number of fishes have toxic secretions on their outer surfaces, including the abundant parrot fish (Scaridae), wrasses (Labridae), and surgeon fish (Acanthuridae). One explanation for the prevalence of these substances is that they deter predation by the abundant carnivores. It seems a most reasonable explanation.

The final category, namely, toxic flesh or internal organs, has caused the most interest among humans, because eating tropical fishes with toxins in the body flesh or organs produces a serious disease called *ciguatera*. Ciguatera is one of the most mysterious and puzzling diseases found in the tropics. As noted by Banner (1973), the symptoms in humans are primarily neurological, including exhaustion, visual disturbance, "inversion" of senses (hot feels cold and vice versa), paralysis, loss of reflexes, and finally, death by respiratory failure. The mystery comes from several sources: (1) The symptoms are not consistent. (2) The fishes that have the toxin can include virtually any or all of the large reef fishes, ranging from carnivores to herbivores, but the fishes vary in toxicity, both in time and in space, so that a given species may be toxic on one reef but not on another, or the same species may be toxic at one time of the year on a given reef and not at another. (3) There seems to be no repeatable pattern to the occurrence of the toxin. All that is currently known about ciguatera is that the toxin originates on the coral reef and is transmitted through the food web. It remains of considerable importance to the peoples of the tropics, since the fisheries of many areas cannot be developed until this toxicity problem is solved.

Another of the outstanding characteristics of reef fishes is their color. Why, especially given the great predation pressure, should they be so brightly colored? Biologists cannot agree as yet, and it is likely that color and pattern serve several different functions. One explanation is that the bright colors serve to advertise that the species is toxic or otherwise distasteful so predators will ignore it (= warning coloration). Another explanation of the colors is that they serve for species recognition. They may also serve to camouflage the species, either by breaking up the shape of the fish or else making it appear as something else. This latter explanation has been especially invoked to explain the dramatic banding patterns seen in some reef fishes (Fig. 9-20).

A final aspect of reef fish ecology concerns the phenomenon of *cleaning behavior*. Cleaning behavior is a specialized form of predation in which certain small fishes (*Labroides* spp.) or shrimps remove various ectoparasites from other, usually larger, fish species. Cleaning behavior is extremely widespread and apparently occurs on all reefs. In this process, the cleaner fishes (or shrimps) often set up "cleaning stations" where they appear to advertise their presence

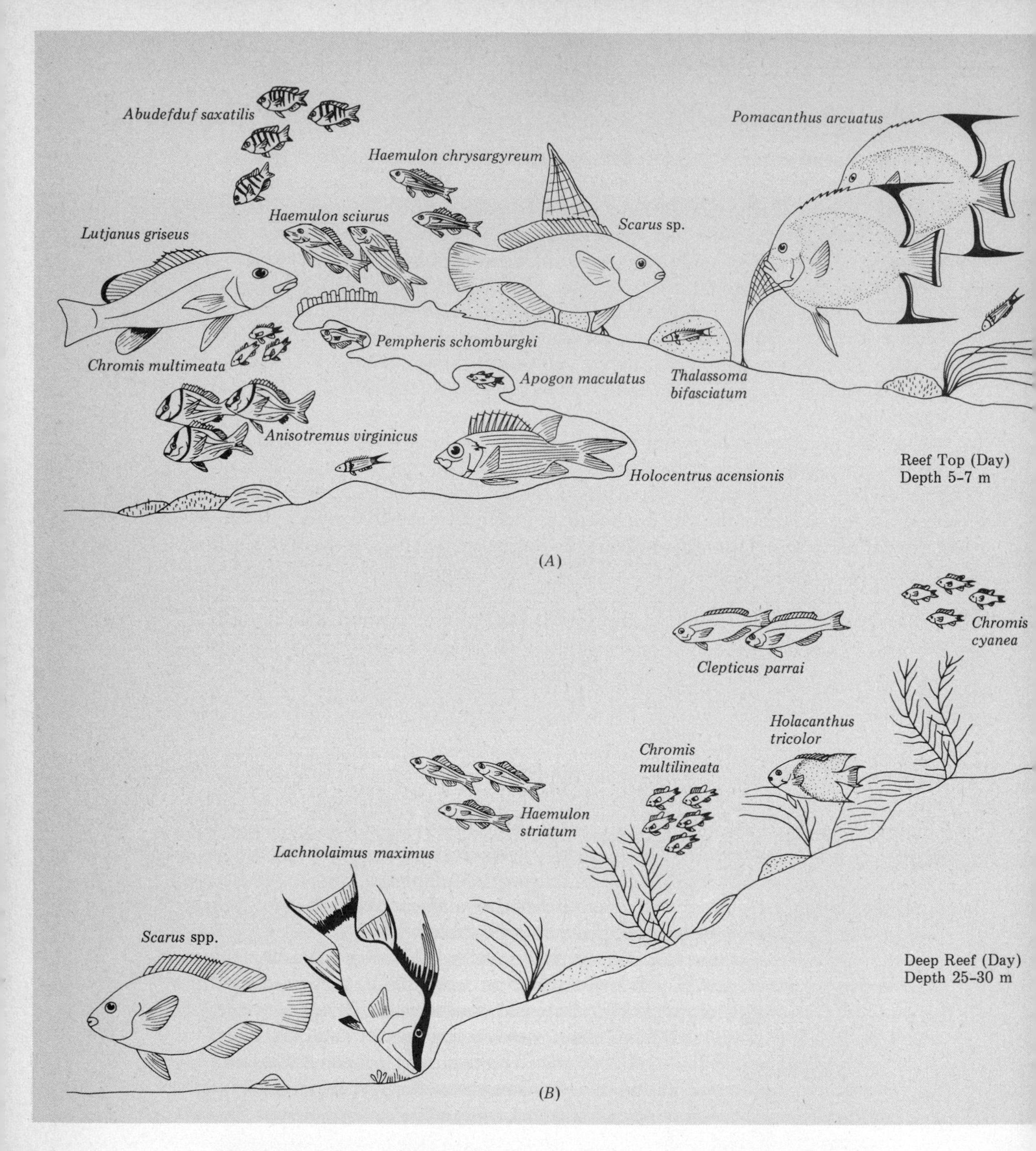

Figure 9-18 Day and night fish distributions on a Florida coral reef at two different depths. *(A)* Reef top at 5–7 m. *(B)* Deep reef at 25–30 m. Damsel fishes (Pomacentridae) *Abedefduf, Chromis;* wrasses (Labridae), *Ciepticus, Thalassoma, Lachnolaimus;* parrot fishes (Scaridae) *Scarus;* angel fishes (Chaetodontidae), *Pomacanthus, Holocanthus;* grunts (Pomadasyidae), *Haem-*

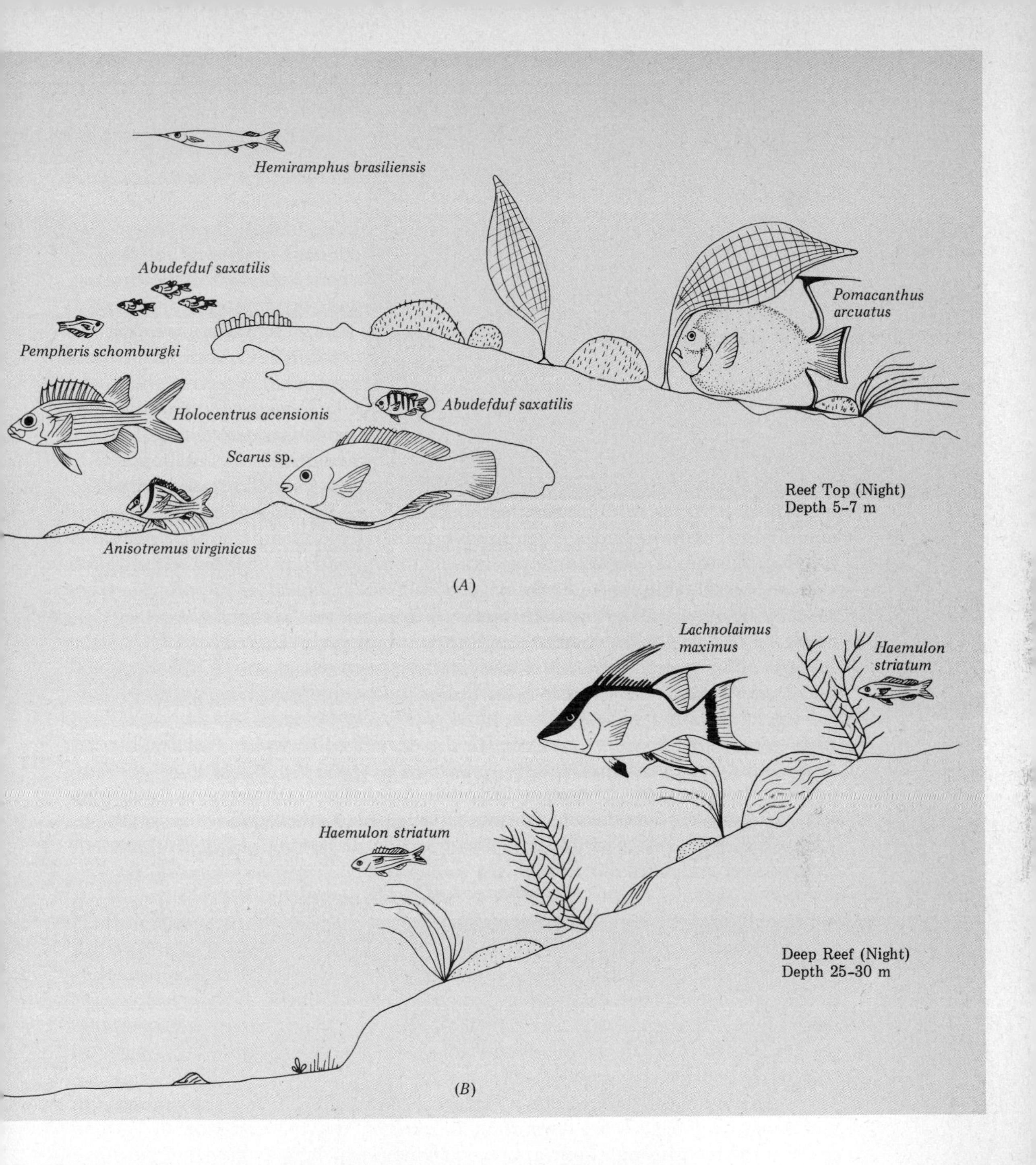

ulon, Anisotremus; cardinal fishes (Apogonidae), *Apogon;* squirrel fishes (Holocentridae), *Holocentrus,* sweepers (Pempheridae), *Pempheris;* snappers (Lutjanidae) *Lutjanus;* halfbeaks (Hemiramphidae) *Hemiramphus.* (Data for the drawing based on W. A. Starck and W. P. Davis, 1966, Night habits of fishes of Alligator Reef, Florida, Ichthyologica, vol. 38, no. 4.)

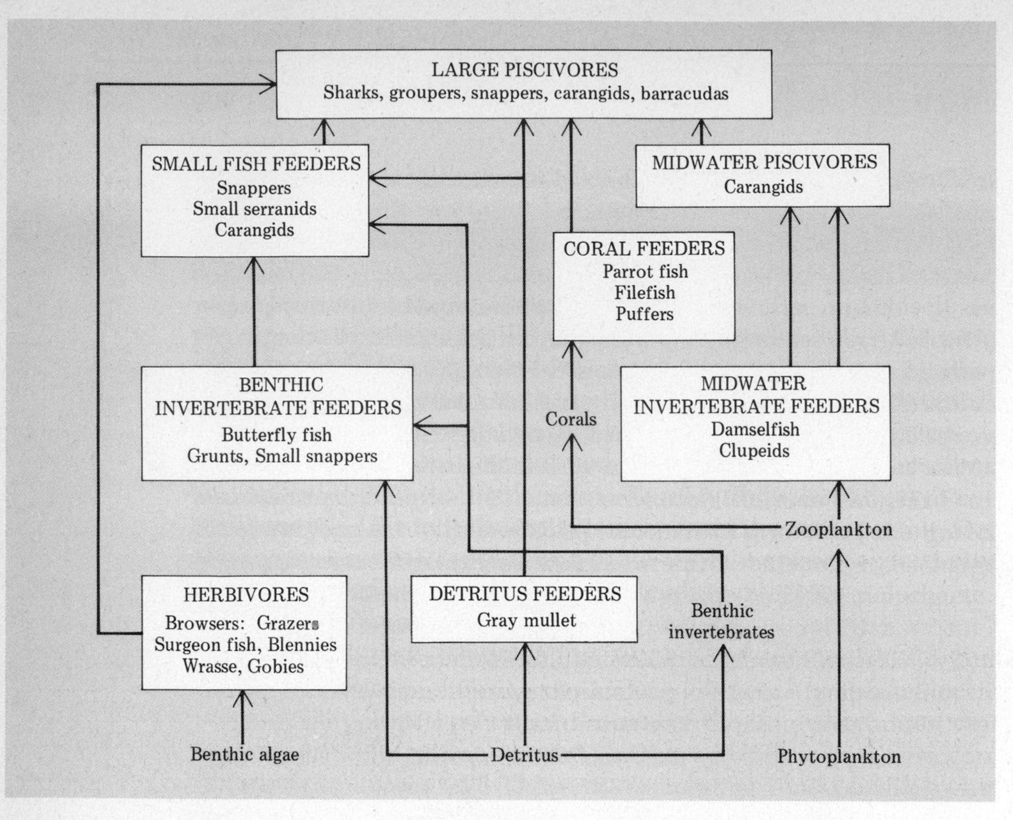

Figure 9-19 Trophic relationships of coral reef fishes. (After R. H. Lowe-McConnell, 1977, Ecology of fishes in tropical waters, Edward Arnold.)

through bright, contrasting colors. The fish to be cleaned comes to the cleaning station area (often a prominent coral head or boulder) and remains motionless as the cleaner moves over its body removing the parasites (Fig. 9-21). The cleaners may even enter the mouth and gill chambers of the fish. A sort of "truce" among predators prevails at the cleaning stations, as the fishes being cleaned are vulnerable to predation. Fishes even "line up" at these stations, awaiting their turn for cleaning.

The importance of this cleaning behavior to the fish population and economy of the reef has not been established. Some biologists believe that, if cleaners are removed from a reef, the fish fauna decreases, moving away or otherwise showing signs of distress. Yet, other biologists have removed cleaners with no noticeable effect on the reef or reef fishes.

We can summarize this section by stating again the importance of fishes to the reef ecosystem. Fishes are important in determining the zonation of various parts of the reef through their grazing activities, which probably keep the algae reduced to a short mat and prevent them from outcompeting the corals. At the same time, their grazing on corals is responsible for the exclusion of certain corals from areas of the reef, which also contributes to the zonation of reefs. Consumption of coral by fishes also leads to the breakdown of coral and contributes to the sediment budget of a reef. The tremendous number of

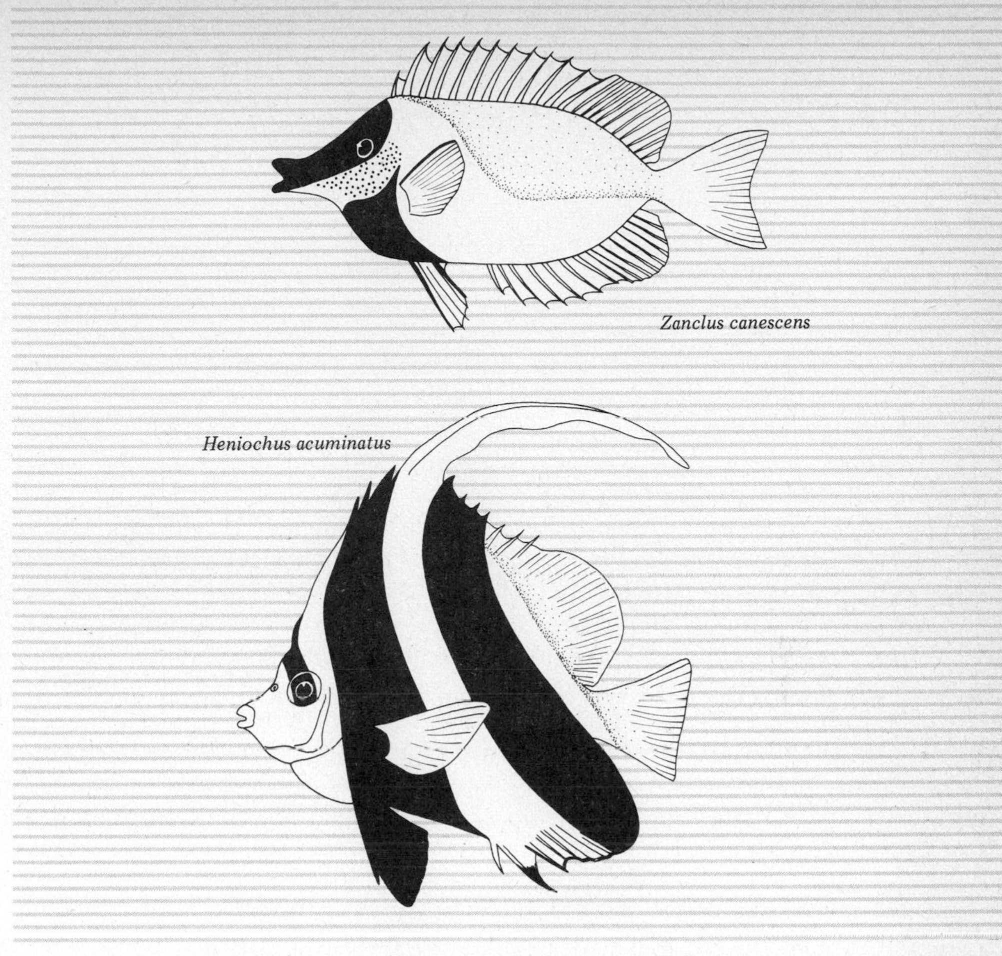

Figure 9-20 Reef fishes with dramatic banding patterns.

carnivores among the fishes undoubtedly exerts a great predation pressure on all reef organisms and is the reason for the cryptic nature of most soft-bodied invertebrates and for the development of various toxic passive defense mechanisms among fishes and invertebrates.

Figure 9-21 A cleaner wrasse *(Labroides)* cleaning the mouth of a large grouper.

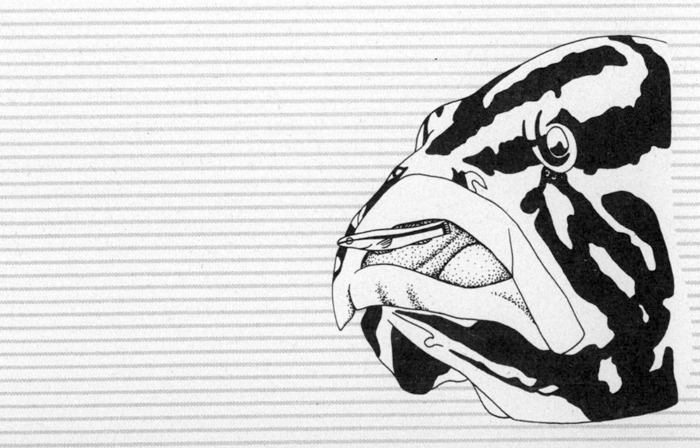

CATASTROPHIC MORTALITY AND RECOVERY OF REEFS

Although reefs appear to be large and very stable systems continuing, in the case of atolls, for millions of years, in fact they do suffer large-scale destruction from various forces. Perhaps the major source of massive reef mortality is mechanical destruction by severe tropical storms. Hurricanes or typhoons are sufficiently intense that their passage over a reef area often means destruction of large areas of reef. Since most coral reefs lie within the zone most frequently traversed by typhoons or hurricanes, the chance that a reef or a section of a reef will be destroyed or severely damaged is great. Damage in such storms usually is due to coral colonies being uprooted and carried off the reef, thus freeing space for new occupants.

The second major source of catastrophic mortality of reefs, at least in recent times, is the population explosion of the sea star *Acanthaster planci*. Since 1957, when the population explosions were first observed, *A. planci* has caused catastrophic mortality of reefs in a number of places in the western Pacific. The ability of this starfish to destroy large areas of reef is tremendous. In Guam, Chesher (1969) estimated that 90 percent of the coral reef along 38 km of shoreline was destroyed in 2-½ years, and on the Great Barrier Reef, Endean (1973) indicates that the bulk of the corals on an 8 km^2 reef were destroyed in 12 months. When these rates are multiplied by the number of reefs in the Pacific that were found to be infested in the 1960s (including 60 percent of the reefs of the Great Barrier Reef between 16–19°S), one can understand the concern for the survival of reefs.

What causes these population explosions and are they a natural phenomenon? These are the two major questions that biologists have asked regarding the "Acanthaster problem," and their answers point up how little we really know about how coral reefs function. There is no consensus concerning the answer to either question. It is possible that similar outbreaks have occurred in the past, because recent work on the Great Barrier Reef by Frankel (1978) has provided evidence for such population explosions on the reefs, either as independent episodic events or cyclic events at 200–300 year intervals. Whether such explosions are natural to other reef systems is not known. Even if these earlier outbreaks were natural, the recent ones could still have been caused by human factors. Whether natural or human-induced, what is the trigger? Examination of the infested reefs has produced no consistent pattern of human-induced or natural activity that would allow us to isolate a single cause. Many suggested reasons have been proffered, such as various dredging activities opening new space for the *A. planci* juveniles; that some chemical (e.g., pesticides) somehow freed *A. planci* from its normal population controls; that humans removed an important predator, allowing increased survival of juveniles; or that severe tropical storms opened space for juvenile settlement or caused adults to aggregate.

Although no one knows for certain what the cause is, of the above suggestions the most plausible are the latter two involving predators and the effects of storms. The predator removal theory necessitates the presence of predators on *Acanthaster* that normally keeps the population in check. Is there such a predator? Yes, there is a large gastropod, *Charonia tritonis*, which does feed on *Acanthaster*. It is also in demand by shell collectors. Endean (1977) has argued that it is possible that the removal of this relatively uncommon gastropod from the reefs by collectors could have allowed the population explosion to begin, but currently there is no evidence to support this hypothesis. If this explanation is correct, then *Charonia tritonis* is a "keystone" species in the same sense that *Pisaster ochraceus* was in the rocky intertidal. However, we know so little about the interactions on reefs that this certainly could not have been predicted ahead of time. Another, more acceptable, theory by Pearson (1975) suggests that juvenile recruitment of *A. planci* is enhanced by a combination of unusually low salinities and high temperatures. In this hypothesis, the occurrence of a year or years of abnormally heavy rainfall in an area, especially if coupled with human destruction of the native vegetation on adjacent land areas causing increased runoff, could trigger an explosion. This model invokes basically a natural explanation that fits the pattern of recent outbreaks, but currently has little evidence to support it. Correlated with this theory is the adult aggregation theory of Dana et al. (1972), which says that the destruction of reef areas by typhoons causes the starfish to condense into aggregations, which then proceed to attack the remaining living coral en masse.

The activities of humans can directly cause catastrophic mortality on reefs through dredging and/or pollution. The reefs in Kaneohe Bay, Hawaii, for example, were destroyed through sewage pollution. During World War II, the military closed off the lagoon at Palmyra Atoll, which resulted in the death of all lagoon reefs. In certain areas of India, coral is actively mined for use in building and this is also causing catastrophic mortality. In the Philippines, dynamiting for fishes is a major cause of destruction.

When catastrophic mortality occurs, how long does it take a reef to recover? Again, we are ignorant, but the little information we have on recolonization of reefs destroyed by hurricanes suggests that it may take as long as 25–30 years before the reef has completely recovered. Estimates of time for recover from *Acanthaster* devastation range from 7–40 years. The reasons for the widely varying estimates of recovery time are due to the fact that recovery depends on several factors, including the extent of initial destruction, the nearness of a source of larvae for recolonization, and favorable water currents and conditions for establishment of the larvae. Recovery of some reefs may be inhibited by the mats of algae and soft corals, which quickly take over the dead coral skeletons and thus do not offer suitable substrates for the recolonizing reef corals.

MANGROVE FORESTS

Mangrove forest or *mangal* is a general term used to describe a variety of tropical inshore communities dominated by several species of peculiar trees or shrubs that have the ability to grow in salt water. These "mangroves" are flowering terrestrial plants that have reinvaded the fringes of the sea. The term mangrove refers to the individual plants, whereas mangal refers to the whole community or association dominated by these plants. Walsh (1974) reports that from 60–75 percent of the coastline of the earth's tropical regions are lined with mangroves, so their importance is clear. Recognizable descriptions of mangal associations were made by the Greeks as early as 325 B.C. In this section, we shall cover the basic ecological factors governing the existence of these communities.

STRUCTURE AND ADAPTATIONS

The mangal comprises trees and shrubs belonging to some 12 genera of flowering plants in eight different families (Table 9-2). The most important or dominant genera appear to be *Rhizophora*, *Avicennia*, *Bruguiera*, and *Sonneratia*. Mangroves share a number of characteristic features that allow them to live in shallow marine waters. They are, first of all, shallow rooted, with their roots spreading widely or else with peculiar prop roots from the trunk and/or branches (Fig. 9-22). The shallow roots often send up extensions, called pneumatophores, to the surface of the substrate that allow the roots to receive oxygen in the otherwise anoxic mud in which these trees grow (Fig. 9-23). The leaves are tough and succulent and have internal water storage tissue and elevated salt concentrations. Some mangroves have salt glands, which help to maintain the osmotic balance by secreting salt (see also salt glands in sea-birds in Chapter Three). Certain mangroves (*Bruguiera*, *Rhizophora*) that establish themselves in marine waters have developed a peculiar form of seed germination and dispersal. The seed, while still on the parent plant, germinates and begins to grow into a seedling without any intervening resting stage. During this time, the seedling elongates and the weight distribution changes, so that it becomes heavier at the outer, free end (Fig. 9-24). Eventually, this seedling is dropped from the parent plant and, because of the weight distribution, floats upright in

TABLE 9-2 The Genera of Mangroves

Avicennia
Suaeda
Laguncularia
Lumnitzera
Conocarpus
Xylocarpus
Aegiceras
Aegialitis
Rhizophora
Bruguiera
Ceriops
Sonneratia

After Lugo and Snedaker, 1974.

Figure 9-22 Mangrove forest showing typical root system and roots descending from the branches. (Photo by Dr. Richard Mariscal.)

the water. It then is carried by the water currents until it enters water shallow enough for its root end to strike the bottom. When this happens, it puts out roots to anchor itself and continues growth into a tree. The advantages to this system of reproduction are obvious for a plant living on the fringe of the sea. Having seeds able to float means it can be dispersed by water currents, while the fact that it floats upright with much of the seed below the water means that when water shallow enough for the mangrove to grow in is reached, the seed will ground itself (Plate 32).

DISTRIBUTION

Mangal associations are distributed throughout the tropical and subtropical oceans of the world. They are able to grow only on shores that are sheltered from wave action; otherwise, the seeds would never be able to ground properly and put down roots. These shores may be directly along the lee sides of islands or island chains or on islands or land masses behind protecting offshore coral reefs.

Figure 9-23 Pneumatophores on the roots of the mangrove tree. (Photo courtesy of Wards Natural Science Establishment.)

They are particularly well developed in estuarine areas of the tropics, where they reach their greatest areal extent.

Mangroves occur over a larger geographic area than coral reefs and may be found in areas well outside the tropics, such as the northern shores of the Gulf of Mexico; all along the west coast of central and northern South America and Africa, where reefs are rare or absent; and as far south as the north island of New Zealand (Fig. 9-25).

These mangrove forests may also penetrate some distance upstream along the banks of rivers (as far as 300 km along the Fly River in New Guinea, according to McNae [1968]).

Mangroves are usually absent from atolls and isolated high islands like Hawaii. (Mangroves were introduced to Hawaii in 1902.)

PHYSICAL CONDITIONS OF MANGROVE FORESTS

Since mangroves can establish themselves only where there is no significant wave action, the first physical condition noted in mangrove areas is that the water motion is minimal. This lack of vigorous water motion has in its turn significant effects. Slow water movement means that fine sediment particles tend to settle out and accumulate on the bottom. The result is an accumulation of mud, and hence, the substrate in mangrove swamps is usually mud. In this sense, then, these areas are much like the previously discussed muddy shores, where poor interstitial circulation and high bacterial numbers lead to anoxic conditions. Perhaps this is also why mangroves have such shallow roots and/or pneumatophores.

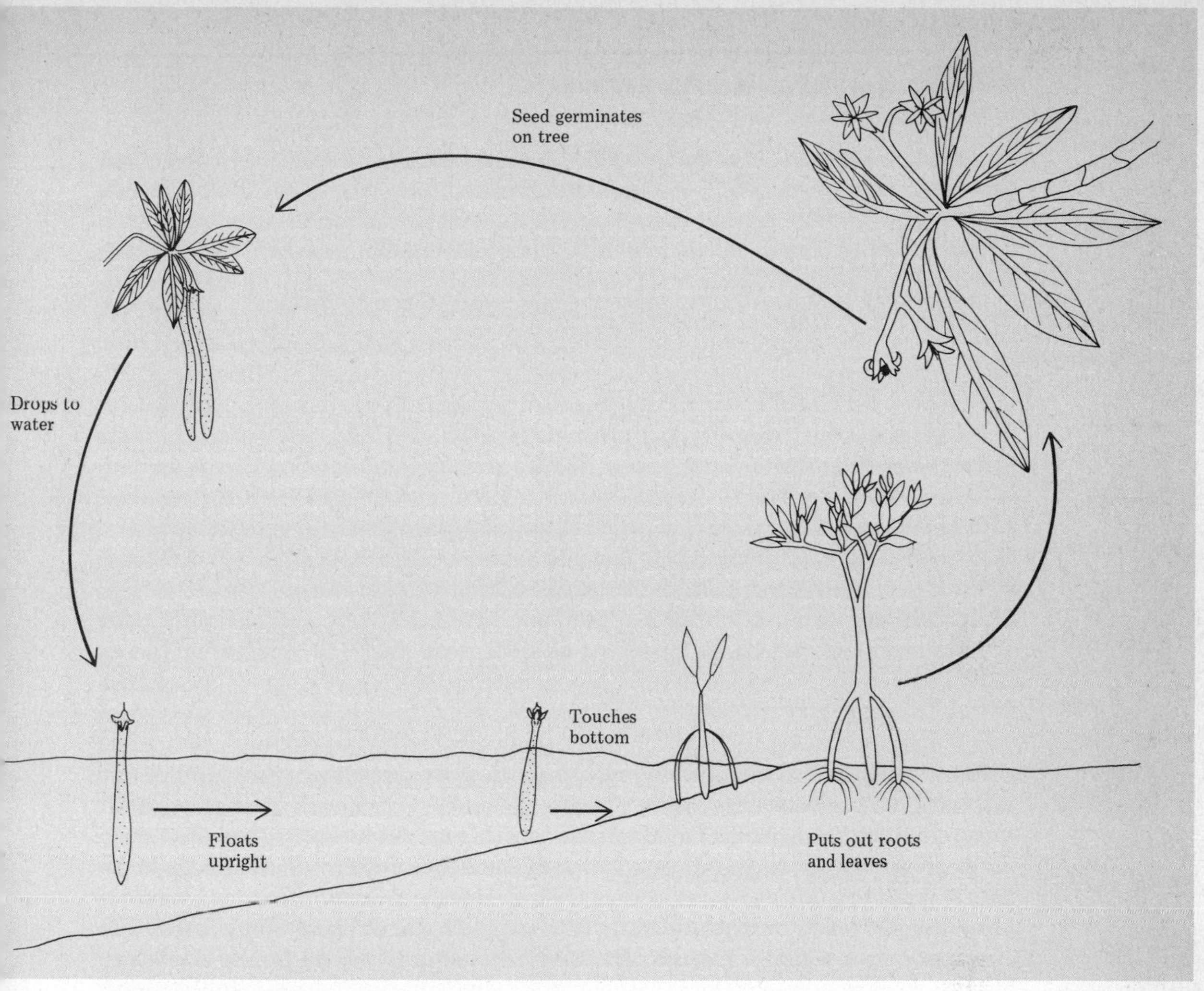

Figure 9-24 The life cycle of a typical mangrove tree.

The initial slow water movement in mangrove forests is further enhanced by the mangroves themselves. As we noted above, many mangroves have peculiar prop (stilt) roots, which extend downward from the trunk and branches. These roots are often so numerous that they form an impenetrable tangle between the surface of the mud and the surface of the water. The result of the presence of these dense root systems is to further decrease the water movement, so that the very finest particles settle out around the roots of the mangroves, creating ever accumulating layers of sediment. Once dropped, the sediment is usually not picked up again and transported out. The result of this trapping is the rather rapid accumulation of sediment within the mangrove association. Mangrove forests are very significant producers of new coastal land at the expense of the sea. The rate at which this occurs is really quite remarkable. McNae (1968) notes, for example, that the great Venetian traveler Marco Polo visited a port city called Palembang in Sumatra in 1292. At that time, the city was directly on the sea. At

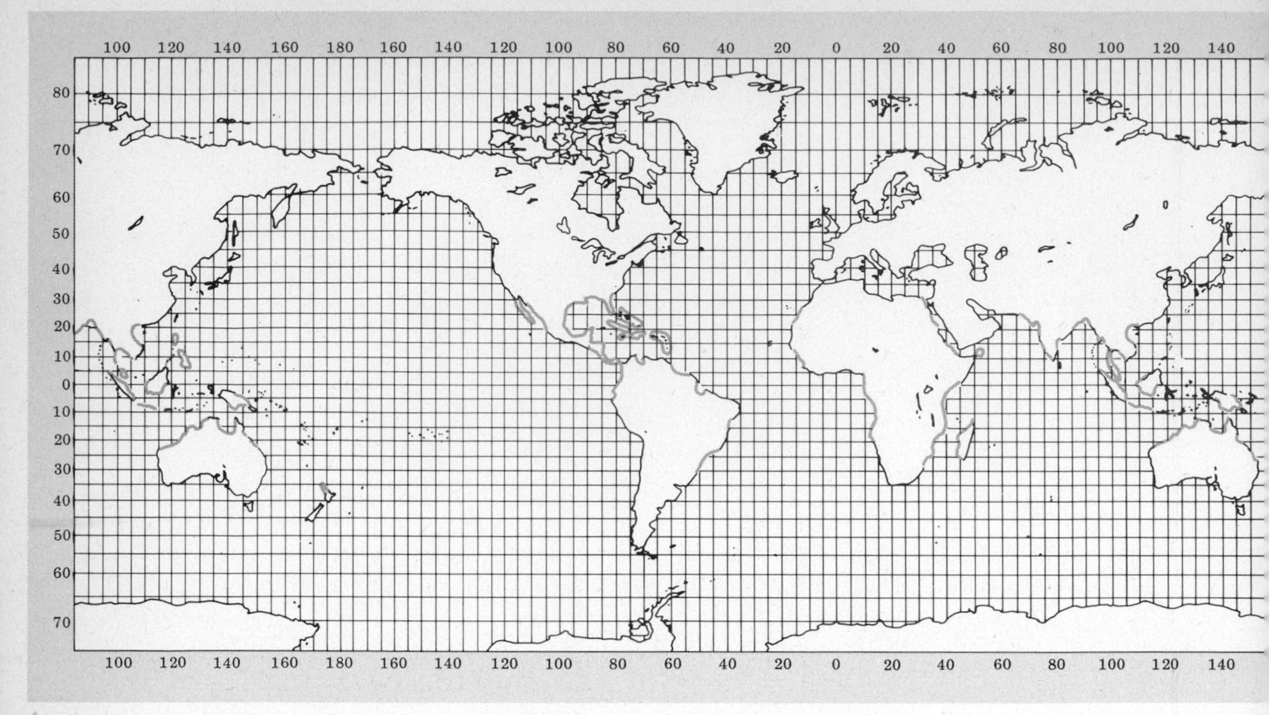

Figure 9-25 Distribution of mangrove forests in the world.

present it lies 50 km inland! This translates into an accumulation of land at the rate of 73 m/year over the last 675 years. In other areas, the rate may be even faster (200 m/year in Bodri Delta of east Java). This creation of land means that, as one progresses through a mangrove association from seaward side to landward edge, one will see a progressive change from marine to terrestrial conditions. This is probably also the main reason for the zonation observed in mangrove forests (see below). The soils of mangrove forests are characterized by low oxygen and high salt content. They are fine-grained, with a high organic content.

The final physical factor to be considered is the tide. The tidal range and type of tide vary across the geographical range of the mangroves. Mangals develop only in shallow water and intertidal areas and are thus strongly influenced by the tides. It is the tide and its vertical range that determine the periodicity of the inundation of the forests. This periodicity of inundation seems to be of considerable importance in determining what kinds of mangrove associations develop in an area and may be responsible for the different types of zonation observed.

ZONATION

To the present, most studies on mangrove forests have been descriptive works in which the emphasis has been on describing the changes in vegetation across the associations from seaward edge to true terrestrial communities. As might be expected, each of these studies has produced differing schemes that reflect the differing underlying conditions (such as the aforementioned tidal factors). No one has produced a truly universal scheme such as has been proposed for rocky shores (see Chapter Six), but a general scheme of wide applicability for mangals of the Indo-Pacific region may be suggested.

The seaward area of most Pacific mangals is dominated by one or more species of *Avicennia*. This *Avicennia* fringe is usually narrow, because *Avicennia* seedlings do not grow well under the conditions of shade or heavy siltation prevailing in the interior of the forest. Associates in this zone and growing seaward of it are trees of the genus *Sonneratia*, which grow where they experience daily wetting (Fig. 9-26).

Behind the *Avicennia* fringe lies the *Rhizophora zone*, dominated by one or more species of *Rhizophora*. These are the trees that most characterize mangal communities, because they have the arching stilt roots that make these areas such an impenetrable maze for humans. Species of *Rhizophora* are often quite tall and develop across a broad area of the intertidal, from levels flooded at all high tides to areas flooded only on highest spring tides.

Progressing landward, the next zone is the *Bruguiera* zone. Trees of the genus *Bruguiera* develop on more heavy sediment (clays) at the level of high water of spring tides.

The final zone of the mangal, which is sometimes present, is the *Ceriops* zone, an association of small shrubs. When present, this is a variable zone and may, in fact, be merged with the trees of the *Bruguiera* zone.

In the Americas, the zonation of mangal associations is less pronounced, perhaps in part due to the lesser number of mangrove species. Zonation in the Americas, according to West (1977), places the *Rhizophora* mangal zone at the seaward edge. This zone is followed inland by a broad band of *Avicennia* and then a belt of *Laguncularia* or *Conocarpetum* (Fig. 9-26).

It should also be remembered that the above are general schemes and not all mangrove forests will correspond to them. In fact, in some areas the association may be considerably abbreviated or represented only by a few individuals. This is particularly true at the limits of the distribution of the mangroves. The zonation may also be interrupted where local conditions are such that evaporation of water from the soils makes them hypersaline. Hypersalinity tends to kill off mangroves, creating bare areas. Full development of mangrove forests is thus found in areas of high rainfall or in areas where rivers furnish enough fresh water to preclude the development of hypersaline conditions.

Zonation may also be limited by tidal action. Wherever the tidal range is small, the intertidal zone is also restricted, as are the mangrove forests. The most extensive forests are developed on shores that have a substantial vertical tidal range.

ASSOCIATED ORGANISMS

Mangal communities are unique: due to the vertical extent of the trees, true terrestrial organisms can occupy the upper levels, while simultaneously true marine animals occupy the bases. Mangrove forests, then, form a strange

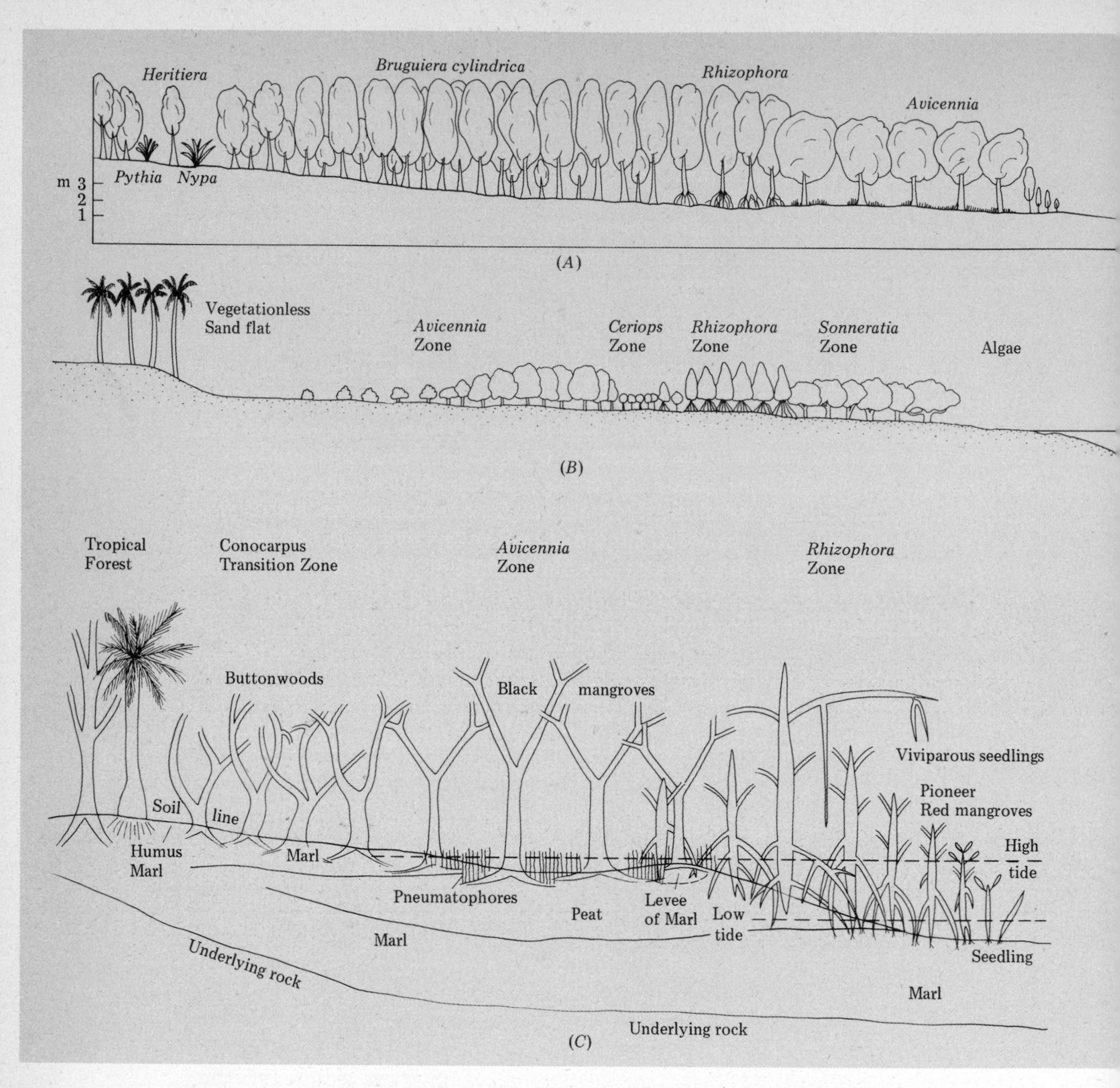

Figure 9-26 Diagrams of the zonation of mangroves in several areas. *(A)* Malaya. *(B)* East Africa. *(C)* Florida. (*A, B,* after V. J. Chapman, editor, 1977, Ecosystems of the World, I. Wet Coastal Ecosystems, Elsevier Scientific Publishing Co. *C,* after J. H. Davis, 1940, The ecology and geologic role of mangroves in Florida, Carnegie Institution of Washington, no. 517.)

mixture of marine and terrestrial organisms and have been suggested as a pathway from land to sea and vice versa.

Terrestrial organisms seem to show no special adaptation for life in the mangal, since they spend their lives out of reach of the marine waters in the upper reaches of the trees, though they may forage on the marine animals at low tide.

The marine organisms are of two types: those that inhabit the hard substrate represented by the numerous stilt roots of the mangroves, and those that occupy the mud. Mangal associations are different from muddy shores primarily because of the vast hard surface area of roots available to organisms, which is absent from typical mud shores (see pp. 243–245).

The dominant groups of marine animals in the mangrove forests are mollusks, certain crustaceans, and some peculiar fishes.

Mollusks are represented by a number of snails, one group of which generally lives on the roots and trunks of the mangrove trees (Littorinidae) and the other on the mud at the base of the roots, comprising mainly detritus feeders (Ellobiidae and Potamididae). Little is known of the contribution of these snails to the mangal. A second group of mollusks includes the bivalves. Oysters are the dominant bivalves. They attach to the roots of the mangroves, where they form a significant biomass (Fig. 9-27).

Mangrove forests are inhabited by numerous large-sized crabs and shrimps. These animals excavate burrows in the soft substrate and include such common genera as *Uca*, the fiddler crabs; *Cardisoma*, the tropical land crabs; and various ghost crabs (*Dotilla, Cleistostoma*). These crabs usually are specialized to feed upon the detrital particles found in the mud. Generally, they separate the organic detrital particles from the nonorganic matter by filtering the substrate through a set of fine hairs around the mouth. These crabs also show varying degrees of adaptation toward a more terrestrial mode of life. This usually expresses itself in a vascularization of the walls of the gill chambers so that they become more "lunglike." As we noted previously, mangrove associations are transition areas between land and sea, and the crab fauna certainly reflect this in their partial adaptation to air breathing.

The burrows of these crabs, as well as those of certain shrimps such as *Upogebia* and *Thalassina*, serve several functions. They serve as refuges from predation, as a breeding place, and as an aid to feeding. However, they also serve the mangal community in that they allow oxygen to enter more deeply into the substrate, thus ameliorating anoxic conditions.

Mangrove areas also serve as nursery grounds for penaeid shrimps and fishes such as mullet, which may spend the early part of their life cycles in these areas before moving offshore.

Conspicuous by their size and abundance in the water and on the mud are the small, big-eyed fishes of the genus *Periophthalmus* and relatives. These fishes, collectively called "mud-skippers," are remarkable because they spend most of their time out of water, crawling around on the mud flat or even climbing the mangrove roots (Fig. 9-28)! They act like frogs or toads. These fishes create burrows in the mud, which serve for refuge and for breeding. They are quite fast, moving over the exposed mud flats by "walking" on their strong pectoral fins or else via a series of "skipping" motions or "bounds," in which the tail and caudal fins provide the thrust.

Aside from their peculiar unfishlike habit of crawling around on the mud flats, the other feature of these fishes is the modification of the eyes. The eyes are set

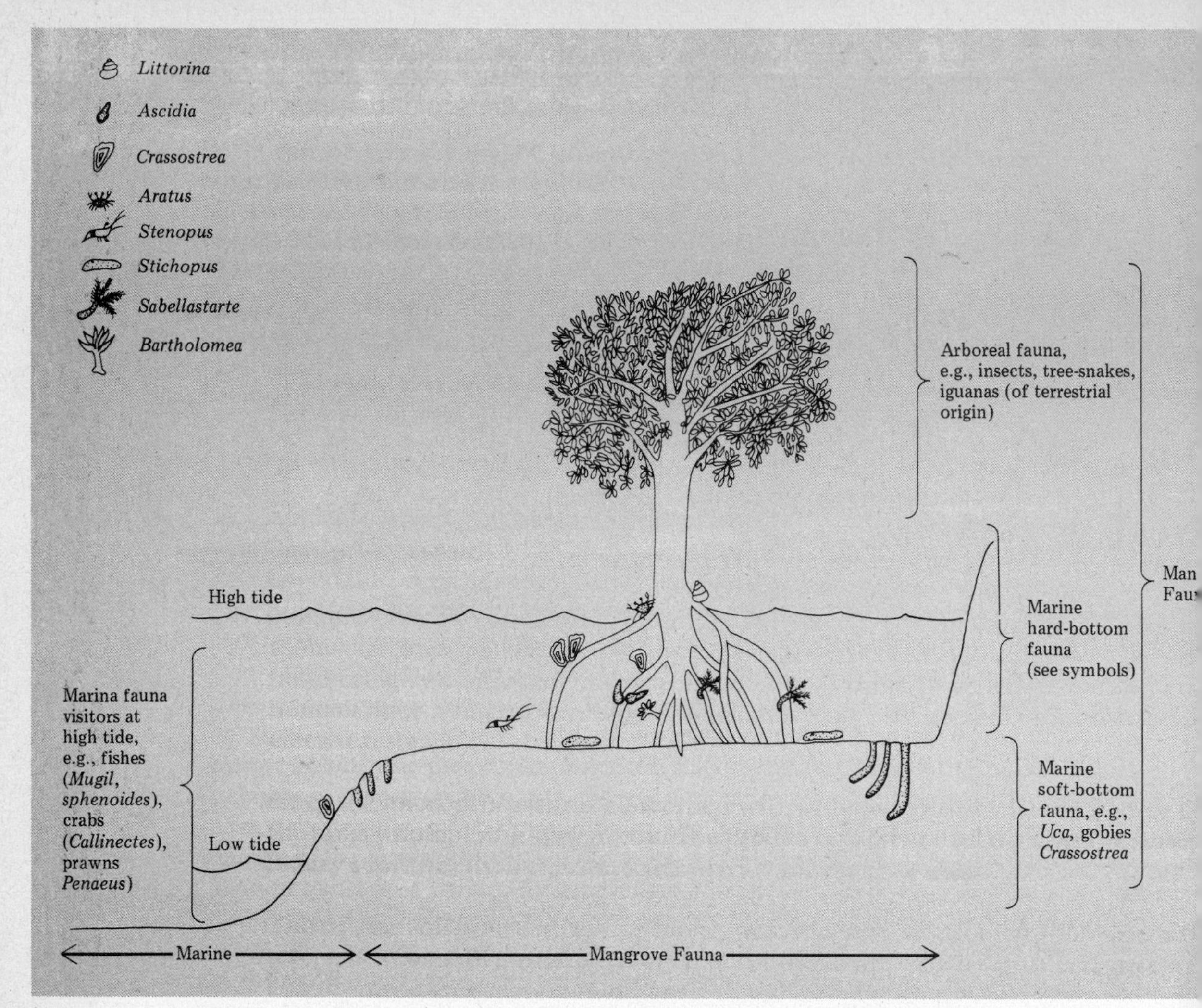

Figure 9-27 Representation of the macrofauna of a mangrove forest showing the vertical distribution and ecological relationships. (Modified from H. Friedrich, 1969, Marine biology, University of Washington Press.)

high on the head and are arranged so that they focus best in air, not in the water. When swimming, then, the eyes protrude on the surface and these fishes look for all the world like frogs. A final noteworthy adaptation is in the respiratory system. The gills are often reduced, and aerial respiration is accomplished by vascularized sacs in the mouth cavity and gill chambers.

SUCCESSION AND MORTALITY

Mangal associations are subject to mortality from a number of different natural and human-induced causes. They exist under rather delicately balanced conditions involving a rather predictable steady sedimentation rate, minimal water movement, a certain tidal regime, and water and soil of a certain salinity. Any change that upsets this balance produces corresponding changes in the mangal

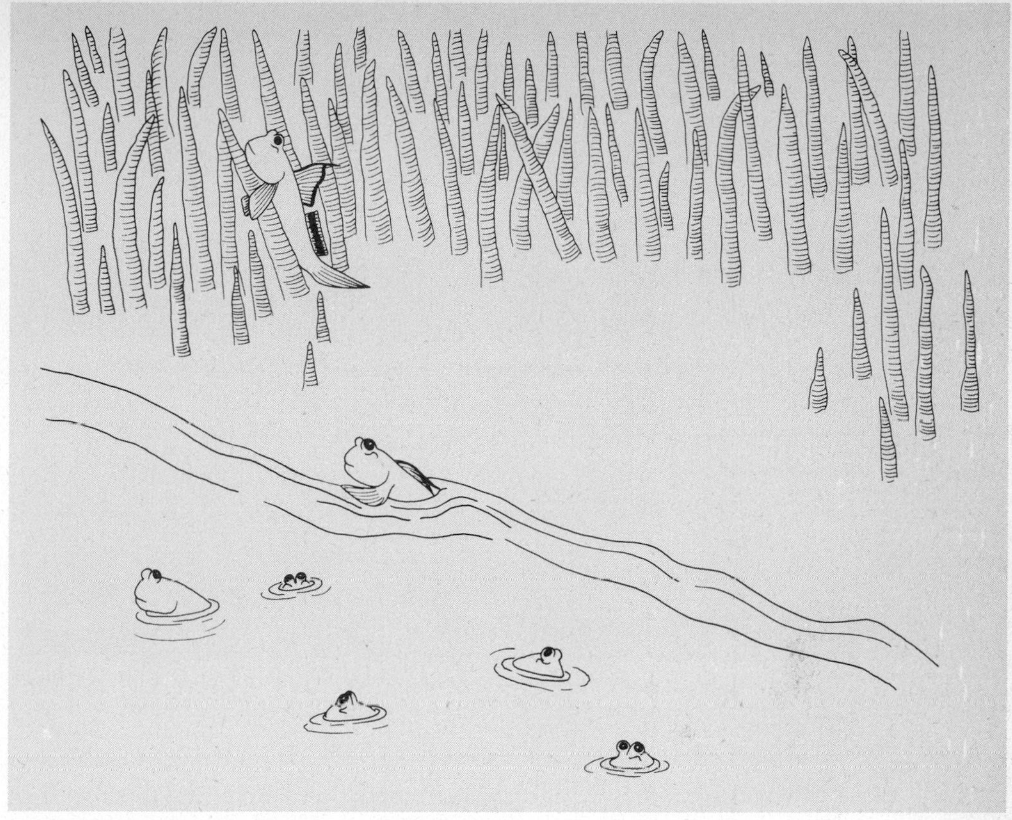

Figure 9-28 Fishes of the genus *Periophthalmus* in the mud and on the roots of the the mangroves. (From various sources.)

community. If the changes are slow enough, there is a gradual change or *succession*. Thus, as sediment builds seaward, the forest changes from *Avicennia* to *Rhizophora* to *Bruguiera* and finally to true terrestrial communities. When the changes are rapid, however, entire areas of the mangrove association may be destroyed.

Perhaps the greatest cause of large-scale mortality is typhoons and hurricanes. These violent storms destroy large areas of mangrove forests by uprooting of trees or by massive sedimentation or altered salinity of water and soil. Hurricane Donna, which struck Everglade National Park in 1960, caused a mortality ranging from 25–75 percent of 100,000 acres of mangroves in the park. Recovery from such massive mortality has a similar time span for mangroves as for coral reefs, and estimates of recovery time are about 20–25 years (Fig. 9-29).

Another mortality source is the small isopod *Sphaeroma terebrans*. This isopod destroys the prop roots of mangroves by boring into them. This leaves the underlying substrate free to erode away, thus allowing the trees to fall into the water. In this way, the whole forest is slowly cut down and destroyed. The importance of this organism as a destructive force has not been fully analyzed.

Most other massive mortality is due to human activity. Perhaps the greatest recent destruction has been as a result of herbicide spraying on the mangrove forests of Vietnam during the war. According to Tschirley (1969), perhaps

Figure 9-29 Photo of dead mangroves destroyed in Everglades National Park by the hurricane of 1960. (Photo by the author.)

100,000 hectares of these forests were defoliated and destroyed by herbicides during the war. It is not known how long recovery of these forests will take, or if they will regenerate, due to the extreme sensitivity of mangroves to residual herbicides.

Mangrove forests have also been destroyed by filling, by dredging, and by channelizing the waters. Historically, mangrove forests probably occurred along the Persian Gulf, from which they are now absent due to cutting by early man for firewood and boat building. In the tropics today, the mangroves are exploited by humans mainly for firewood, but may also be partially destroyed in order to build ponds for fish and shrimp culture or for salt production, or filled for construction.

References

Chapman, V. J. (ed.). 1977. Ecosystems of the world. Vol. I, Wet coastal ecosystems. Elsevier, N.Y. 428 pp.

Goreau, T. F., N. I. Goreau, and T. J. Goreau. 1979. Corals and coral reefs. Sci. Amer. 241(2):124–136.

Jones, O. A., and R. Endean (eds.). 1973. Biology and geology of coral reefs. Vol. II: Biology 1. Academic Press, N.Y. 480 pp.

Jones, O. A., and R. Endean (eds.). 1976. Biology and geology of coral reefs. Vol. III: Biology 2. Academic Press, N.Y. 435 pp.

Lowe-McConnell, R. H. 1977. Ecology of fishes in tropical waters. Studies in biology no. 76, Edward Arnold, London. 64 pp.
Lugo, A. E., and S. C. Snedaker. 1974. The ecology of mangroves. Annual Review of Ecology and Systematics 5:39–64.
McNae, W. 1968. A general account of the fauna and flora of mangrove swamps and forests in the Indo-West Pacific region. Advances in Marine Biology 6:73–270.
Newell, N. D. 1972. The evolution of reefs. Sci. Amer. 226(6):54–65.
Odum, H. T., B. J. Copeland, and E. A. McMahan (eds.). 1974. Coastal ecological systems of the United States. Vol. I: B, Natural tropical ecosystems of high diversity. Conservation Foundation, Washington, pp. 346–514.
Proceedings of the second international symposium on coral reefs. Vol. 1, 630 pp.; Vol. 2, 753 pp. 1974. Great Barrier Reef Committee, Brisbane, Australia.
Proceedings of the third international coral reef symposium. Vol. 1, Biology, 656 pp; Vol. 2, Geology, 628 pp. University of Miami Rosentiel School of Marine and Atmospheric Science.
Stoddard, D. R. 1969. Ecology and morphology of recent coral reefs. Biol. Rev. 44(4):433–498.
Walsh, G. E. 1974. Mangroves, a review, pp. 51–174. *In*: Reimold, R. J., and W. J. Queen (eds.), Ecology of Halophytes. Academic Press, N.Y.
Wells, J. W. 1957. Coral reefs, pp. 609–631. *In*: The treatise on marine ecology and paleoecology. Vol. I, Ecology. Memoir 67, Geol. Soc. of Amer.
Yonge, C. M. 1963. The biology of coral reefs. Advances in Marine Biology 1:209–260.

Chapter Ten
SYMBIOTIC RELATIONSHIPS

One of the most striking features of marine organisms is the large number of close associations that can be observed among many different species. These are a series of associations that are not predator-prey or herbivore-plant relationships in which one member consumes the other. These close relationships between unlike species generally seem to be either unharmful to either member or, more likely, to be beneficial to one or both. *Symbiosis* is the name given to such associations and means an interrelationship between two different species. Symbiotic relationships are found in the terrestrial environment as well as in the aquatic, but appear to be disproportionately common, extensive, and well developed in the marine environment, such that they merit separate coverage in a book such as this.

DEFINITIONS AND COVERAGE

Symbiotic relationships cover a broad spectrum of associations from random, casual, or facultative associations through more and more obligatory groupings that benefit one or both members to finally those that are parasitic. Such differences in the degree of association have led to the subdivision of symbiosis into more narrowly defined groupings. The term *commensalism* is often used to refer to an association that is clearly to the advantage of one member while not harming the other member. *Inquilinism* is a special subdivision of commensalism, in which an animal lives in the home of another, or in its digestive tract, without being parasitic. *Mutualism* is that form of symbiosis in which two species associate together for their mutual benefit. In a mutualism relationship, the partners are often called *symbionts*. In a commensal or inquiline relationship, however, the partner gaining advantage is called the *commensal* and the other the *host*.

Parasitism generally refers to an association in which one species lives in or upon another and draws nourishment from that species at the expense of, or to the detriment of, the other. In other words, it is an association in which the advantage is solely to one member at the expense of the other. Parasitism is an association that seems to be extremely common, in both marine and terrestrial environments. It will not be discussed in this chapter, since it contains few features unique to the marine environment.

Basically, there are two broad groupings of nonparasitic symbiotic associations in the sea: those between algal cells and various invertebrate animals and those between various animals, both vertebrate and invertebrate. We shall consider these in that order.

SYMBIOSIS OF ALGAE AND ANIMALS

All symbiotic relationships in the sea between plants and animals are between unicellular algae or their parts and a wide variety of marine invertebrate animals. These symbiotic relationships have been shown to be most common in tropical waters, but they also are prevalent in temperate oceans. They do, however, appear to be absent, or virtually so, from polar waters. Furthermore, the associations are, for obvious reasons, restricted to the very shallow subtidal or intertidal areas or to the uppermost layers of the pelagic realm where sufficient light is present.

TYPES AND COMPOSITION OF THE ASSOCIATIONS

There are basically two types of symbiotic association between algae and invertebrates. The more common is to have the entire functioning algal cell associated with the invertebrate animal. The second is to have only the functioning chloroplasts from the algal cells incorporated into the tissues of the invertebrate body.

The algal cell symbionts have been typically classified into groups on the basis of their color. *Zooxanthellae* is the name given to those cells appearing brown, golden, or brownish yellow in color, and *Zoochlorellae* to those that appear green. A third, smaller group appears blue or bluish green and has been called *Cyanellae* (Table 10-1). These color groups, however, do not distinguish among the actual algal species involved. Zooxanthellae, the most common in the seas of the world, are primarily species of dinoflagellates (see pp. 38–39), but include a few diatoms and cryptomonads (Fig. 10-1). As Schoenberg and Trench (1976) have noted, the most common zooxanthellae species is the dinoflagellate *Symbiodinium* (= *Gymnodinium*) *microadriaticum.* The zoochlorellae are much less common in the sea, but dominate symbiotic associations in fresh water. Among the marine associations, zoochlorellae include the alga *Platymonas convolutae* (Pyramimonadales) and unknown species of Chlorophyceae. Cyanellae are all

TABLE 10-1 Summary of the Types of Associations Between Algae and Marine Invertebrates

Algal Symbiont Group	Algal Taxon	Invertebrate Animal Host Taxa
Zooxanthellae	Dinoflagellata	Protozoa, Porifera Cnidaria, Platyhelminthes Mollusca
	Bacillariophyceae	Platyhelminthes *(Convoluta convoluta)*
	Cryptomonadida	Protozoa
Zoochlorellae	Pyramimonadales	Platyhelminthes *(Convoluta roscoffensis*
	(?) Chlorophyceae	Cnidaria *(Anthopleura* sp.)
Cyanellae	Cyanophyceae	Porifera, Protozoa
Chloroplasts	Chlorophyceae	Mollusca (Sacoglossa)

Modified from D. C. Smith, Symbiosis of algae with invertebrates, Oxford Biology Reader #43. 16 pp. Copyright Oxford University Press.

blue-green algae, but at present, the identity of the species involved remains unknown (Table 10-1). They are most common as symbionts in sponges and in planktonic diatoms.

The algae all occur inside the bodies of the animals and most are found either within vacuoles inside individual tissue cells or else in various body spaces between or within tissue layers. Generally, the algal cells are restricted to certain tissues or areas of the host, and they grow and reproduce within the invertebrate without undergoing digestion by the host.

Figure 10-1 Photomicrograph of symbiotic zooxanthellae from the anemone *Anthopleura xanthogrammica.* (Photo by the author.)

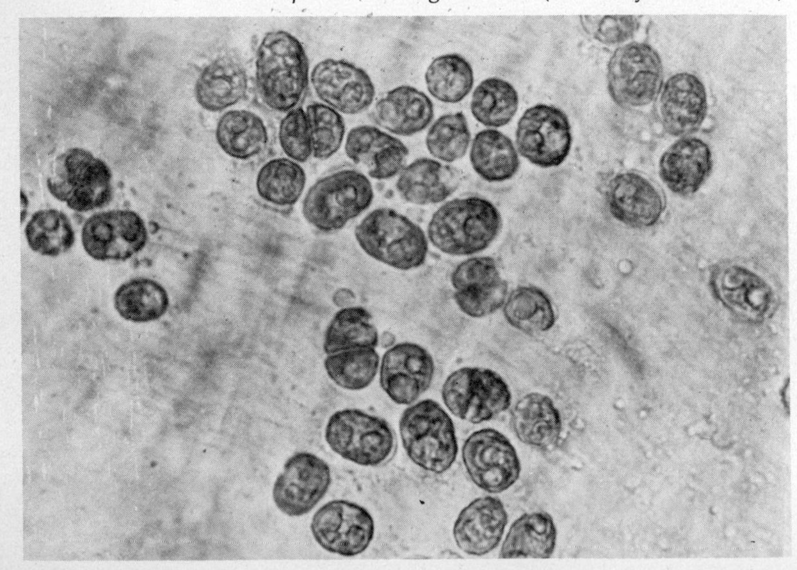

Among those associations that involve only the chloroplasts and the invertebrate body, the chloroplasts are usually derived from the cells of larger green algae *(Codium, Caulerpa, Cladophora, Bryopsis)*, which are first ingested by the animal in feeding and subsequently are transferred from the digestive tract of the animal to other tissues.

Algal symbiosis, either cellular or chloroplast, is widespread among various invertebrate groups. Symbiotic associations occur in Protozoa, Porifera, Cnidaria, Platyhelminthes, Mollusca, and Echiura. Within these phyla, McLaughlin and Zahl (1966) report about 130 genera having algal symbionts. It is most common among Protozoa and Cnidaria.

ORIGIN OF THE ASSOCIATION

Although we will probably never know the exact origin of algal-invertebrate symbiotic relationships, it is possible to outline some suggestions. In all the phyla mentioned above that contain symbiotic algae or chloroplasts, the final phase of digestion is intracellular, where individual cells take up particles from the stomach. It is thus possible to conceive of the origin of this association through the ingestion by cells of the digestive tract of either intact algal cells or chloroplasts, taken in by the animal during feeding. If, then, the algal cells were resistant to digestive action or the animal lacked enzymes to digest plant cellulose, the basis for a new association would be laid. This association could then evolve into a more obligatory relationship, provided that the new association conveyed an enhanced survival value to both symbionts. Since these associations, in fact, are common, they must have some selective value (see pp. 365–366).

DISTRIBUTION OF ALGAE-INVERTEBRATE ASSOCIATIONS

Although there is not space or time to list and discuss all the species in which algal symbionts occur, it is of use to discuss briefly the various groups in which this association is prevalent.

Beginning with Protozoa, symbiotic relationships are found in all of the epipelagic planktonic Radiolaria where the zooxanthellae occur in the outermost frothy layer (Fig. 10-2). They also are found in a number of planktonic Foraminifera *(Globigerinoides)* and even in marine ciliates *(Paraeuplotes, Trichodina)*.

Symbiosis is not common among sponges (Porifera), but has been recorded in a few species, zooxanthellae in *Cliona* and cyanellae in *Demospongia*.

Among the Cnidaria, symbiosis becomes extremely common, especially among tropical cnidarians. Virtually all tropical, shallow water anemones, soft corals, sea fans, whips, and stony corals have symbiotic zooxanthellae in their tissues. Even certain of the tropical jellyfish *(Cassiopeia)* contain zooxanthellae. In

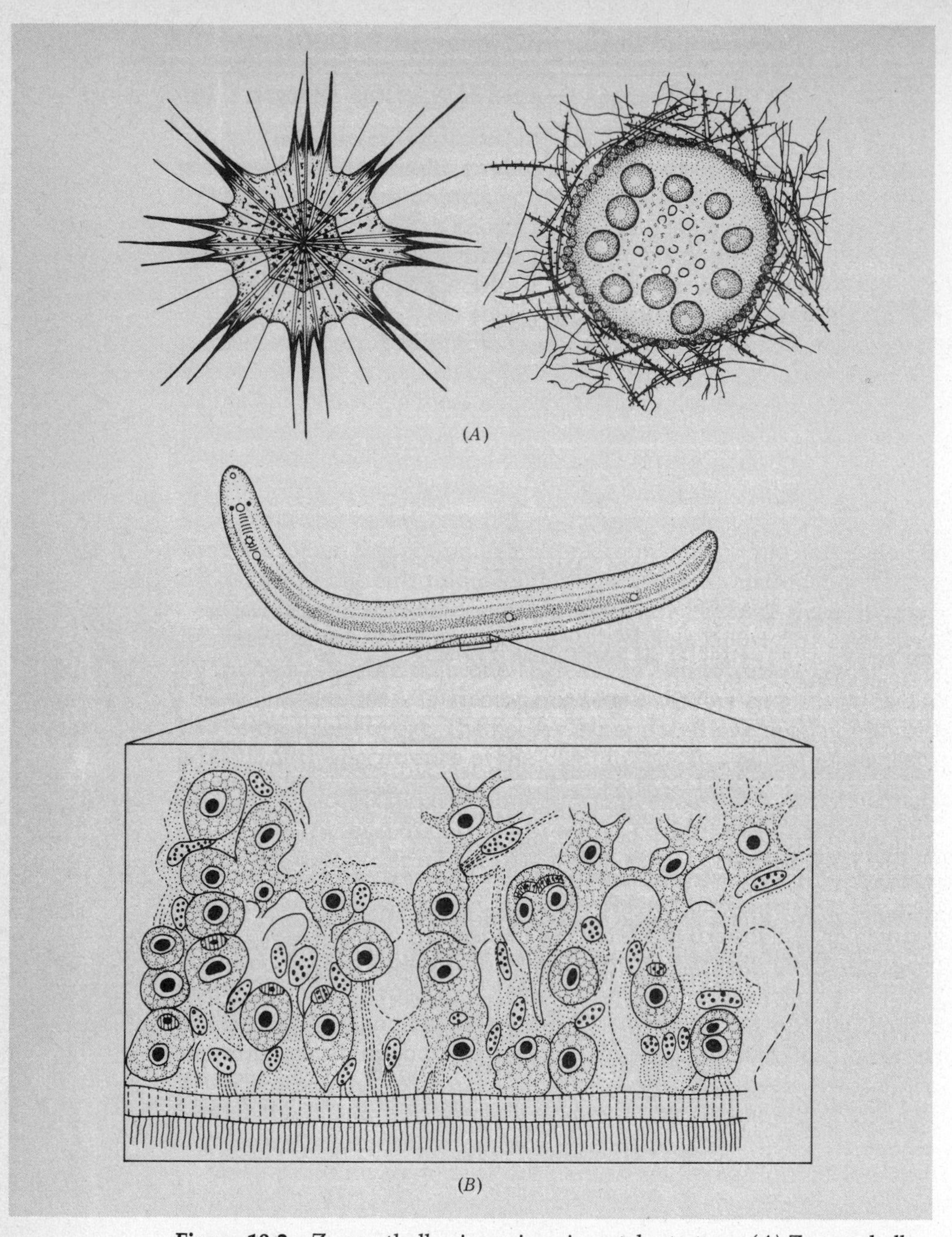

Figure 10-2 Zooxanthellae in various invertebrate taxa. *(A)* Zooxanthellae in the radiolarian *Acanthometra pellucida* (left) and *Sphaerozoum acuferum* (right). *(B)* Zooxanthellae in the acoel flatworm *Convoluta roscoffensis.* (*A, B,* after J. A. C. Nicol, 1960, The biology of marine animals, Wiley. © J. A. Colin Nicol, Pitman Publishing Ltd.)

temperate seas, the incidence is not as great, though many anemones such as *Anthopleura* retain the symbionts.

Whereas a few symbiotic relationships occur in Ctenophora *(Beröe)*, Annelida *(Eunice)*, Echinodermata *(Ophioglypha)*, and Tunicata *(Didemnum, Trididemnum)*, the remaining major reservoirs of symbiotic relationships are found in the Platyhelminthes and Mollusca. Among marine flatworms, the classic case is that of the genus *Convoluta*, which has been extensively investigated. Symbiotic relationships in the Mollusca are generally restricted to certain marine gastropods of the order Sacoglossa, where the animals retain only the chloroplasts, and to the bivalve family Tridacnidae.

MODIFICATIONS RESULTING FROM THE ASSOCIATION

The symbiotic association between the algae and the invertebrates is generally a very close one (mutualistic), which has resulted in rather significant anatomical and physiological changes in both the algal cells and the various invertebrate hosts.

Perhaps the most profound changes that occur as a result of the association are found in the algal cells. Most marine symbiotic algae are dinoflagellates. The symbionts, however, have lost their locomotory flagellae and the characteristic grooves around the body. Furthermore, they have cell walls that are much reduced in thickness. The zoochlorellae in the flatworm *Convoluta* have lost even more, in that the cell wall disappears, as does the light-sensitive stigmata. These cells become little more than bags containing chloroplasts.

If the above algal cells are removed from the host and grown outside the animal in culture, they will develop the characteristic flagellae, cell walls, and other organs of a typical free-living form. It is thus apparent that all the changes observed are a direct result of the symbiotic association (Fig. 10-3).

In the case of the animals, the changes vary with the type of organism and with the degree of interdependence established between the symbionts. The single universal modification is that all of these invertebrates live in very shallow water, where they can obtain adequate light so that the algae can carry on photosynthesis.

Figure 10-3 Free-living *(B)* and symbiotic *(A)* zooxanthellae cells.

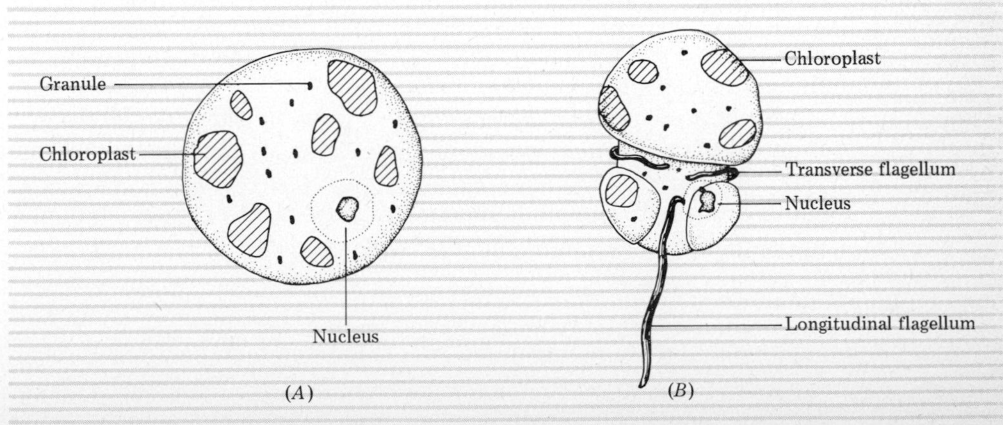

Perhaps the least amount of modification occurs in the lower invertebrates, such as Protozoa and Porifera, where the algae occur as simple inclusions in the cytoplasm of the animal. No special anatomical modifications are observed.

Among the Cnidaria, definite modifications begin to become apparent. The algae occur in marine Cnidaria in the innermost of the two cell layers, the gastrodermis (endodermis). The numbers of zooxanthellae in different corals vary, and those that have the most seem to have reduced tentacle size, indicating that they are less able to capture zooplankton food. Among certain soft corals of the family Xeniidae the digestive regions of the animal are reduced and the animals are not responsive to animal food. In the jellyfish *Cassiopeia,* a striking modification is behavioral. These animals, rather than swimming in open waters as do most jellyfishes, lie upside down on the bottom in shallow tropical waters, exposing their oral arms to the light in order to illuminate their algae. The oral arms are also much enlarged and expanded to provide more area of habitation by the algal cells. Truly a strange way of life for a jellyfish!

Among flatworms, the most studied case of symbiosis is that of *Convoluta roscoffensis,* a small animal inhabiting sand beaches on the Brittany coast of France. The alga inhabiting *Convoluta* is *Platymonas convolutae,* which also occurs free-living in the same area. *Convoluta* exhibit perhaps the most profound behavioral and life history modifications, but little in the way of anatomical changes (Fig. 10-2).

Convoluta roscoffensis live burrowed into the sand of beaches near the upper reaches of the tidal zone. Whenever the tide is in, the animals are found buried. When the tide recedes, the animals move up onto the surface of the sand, where they spread out to expose their symbionts to the light. When the tide begins to return, the vibrations trigger a reburrowing response so that the animals are safely under the sand again before the rising tidal waters are able to sweep them out. Even more significantly, as Smith (1973) notes, these animals apparently do not feed as adults. Furthermore, if the young worms do not ingest the algal symbionts upon hatching, they will not complete development and will die even if they feed.

It is among the mollusks, however, that we find the most dramatic changes of all resulting from the association. Although there are thousands of species of mollusks in the seas of the world (the second largest phylum after arthropods), symbiotic associations with algal cells have developed in only seven species. All are bivalve mollusks in a single family, the Tridacnidae, which include the giant clams. Six of the species are in the genus *Tridacna* itself and the remaining one in the genus *Hippopus.*

All tridacnid clams are distributed only in the Old World tropics of the Indo-Pacific. They also include the largest bivalve mollusk in existence: *Tridacna gigas,* which has been recorded by Rosewater (1965) to reach 4 ft in length and 580 lb in weight. The other species are more modest in size, but all are still large in comparison with most other bivalve species.

These tridacnid clams are inhabitants of coral reef areas, where they are found abundantly in the shallow, sunlit water (Fig. 10-4). They usually have a very uncharacteristic position for a bivalve mollusk, in that they lie either on the surface of the bottom or bored into coral or coral rock with the opening between the valves facing up toward the surface of the water. The valves usually gape widely, and within this opening can be seen an extensive, brightly colored tissue layer. It is in this tissue layer that the symbiotic zooxanthellae are found. The brightly colored tissue exposed within the gape is siphonal tissue; the color results from the interaction of various pigments deposited there (Plate 27). The reason for the bright colors is that they act to protect the tissues of the clam from the damaging effects of sunlight, while passing enough light to allow the zooxanthellae to photosynthesize.

What is truly remarkable about tridacnids, however, is the tremendous change that their bodies have undergone from that of a typical clam in order to accommodate this symbiotic association. Bivalves generally rest with their foot either embedded in the substrate or held against it. In this position, a normal clam has its hinge uppermost. Obviously, this would not do for a tridacnid, since such a position would not permit the tissue and zooxanthellae to be illuminated. The result, as Yonge (1975) has described, is that the entire tridacnid has undergone a tremendous rotation with respect to the foot, so that the hinge comes to lie on the underside next to the foot, and the opening of the shell faces upward (Fig. 10-5). At the same time, the siphons and siphonal tissue underwent an expansion and grew and extended themselves, covering the length of the upward-facing opening and providing the expanded area for occupation by the zooxanthellae. As a result of this rotation and expansion, one of the tridacnid's shell-closing muscles was lost. Because of these profound anatomical changes, the tridacnids are markedly different in body orientation from any

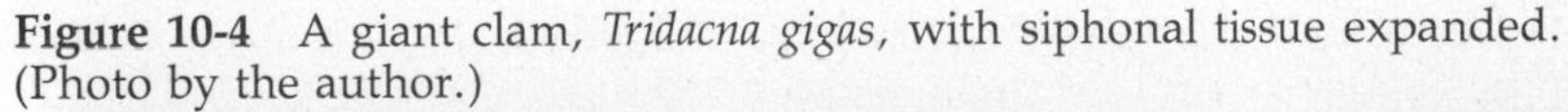
Figure 10-4 A giant clam, *Tridacna gigas,* with siphonal tissue expanded. (Photo by the author.)

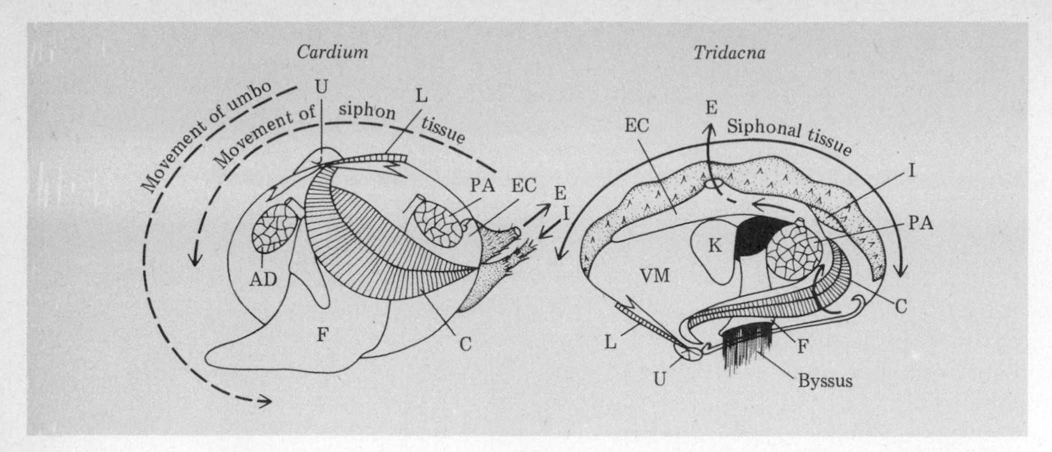

Figure 10-5 Comparison of the body orientation and internal anatomy of a typical clam, *Cardium*, with that of *Tridacna*. *AD*, anterior adductor; *C*, ctenidium; *E*, exhalent current; *EC*, exhalent chamber; *F*, foot; *I*, inhalent current entering inhalent aperture; *K*, kidney; *L*, ligament; *PA*, posterior adductor; *U*, umbo; *VM*, visceral mass. (Modified from C. M. Yonge, 1957, Symbiosis, *in:* J. C. Hedgpeth, ed. The treatise on marine ecology and paleoecology, vol. I, Ecology, Geological Society of America.)

other bivalves. These changes can only be attributed to the association with the symbiotic algae.

Symbiotic relationships with chloroplasts have been reported by Greene (1970) in the marine gastropods of the order Sacoglossa *(Elysia, Placida, Placobranchus)* (Fig. 10-6). Animals of this order regularly consume the contents of algal cells, and some species have evolved the ability to retain the chloroplasts in a functioning position on the dorsal surface. The only modifications so far observed are that they do not feed very often and the tropical forms provide a screen of light-absorbing material above the chloroplasts to cut down the light intensity.

Figure 10-6 *Elysia hedgpethi*, a sacoglossan mollusk which contains chloroplasts obtained from the green algae *Bryopsis corticulans* or *Codium fragile*. (Photo by the author.)

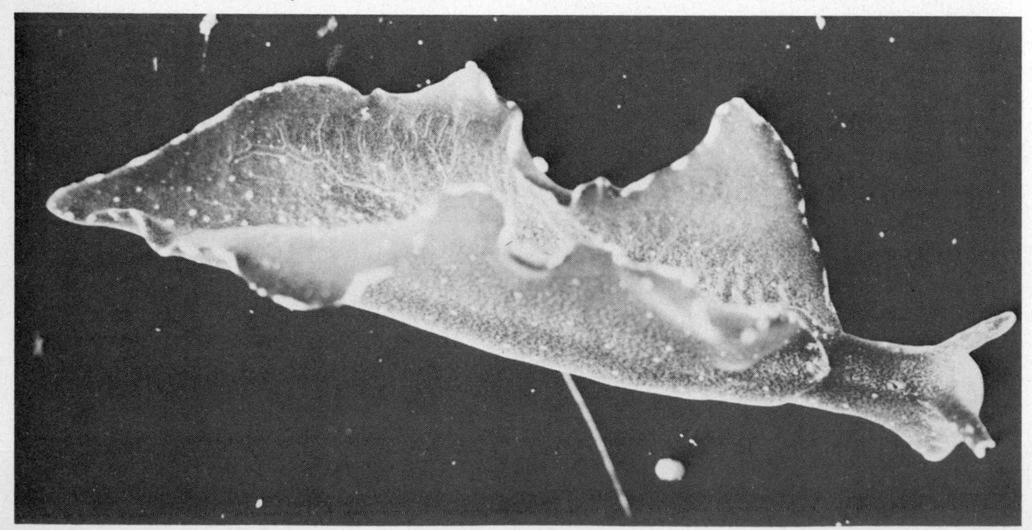

VALUE OF THE ASSOCIATION

Since symbiotic associations with algae are so common among invertebrates, and since they often result in profound anatomical or behavioral changes in the partners, it seems only natural to assume that they must have some positive value to each of the partners. What might be the value of such associations?

We have already partially answered this question with respect to corals in Chapter Nine, where we saw that the corals obtained food materials (see pp. 322–323) from the zooxanthellae and also that the zooxanthellae enhanced the ability of corals to lay down calcium carbonate (see pp. 323–324). In turn, the zooxanthellae in corals received nutrients in the form of nitrates and phosphates produced in the metabolic processes of the coral, but rare in external waters.

Among most invertebrates that have a symbiotic association with algal cells, the cells retain their integrity and are not digested by the animal to obtain nutrients. Thus, the nutrients that pass are in the form of chemical compounds. Energy-containing molecules, such as glycerol, produced by the zooxanthellae in photosynthesis, pass to the animal, and nitrates and phosphates, needed nutrients for the algae, pass from the invertebrate to the algal cell.

A similar situation also occurs between the sacoglossan mollusks and the symbiotic chloroplasts where Green (1970) has shown translocation of organic material from chloroplast to animal. However, in those sacoglossans thus far investigated, the duration of life of the chloroplasts appears to be much shorter and the animals must periodically replenish the supply by ingesting the contents of algal cells.

Among the giant clams, however, Yonge (1975) has suggested that, in addition to reciprocal transference of nutrients, the animals may also actively digest the zooxanthellae cells. This would seem to be counterproductive, especially considering the great anatomical changes that these animals have undergone in order to properly provide for their algal guests. Apparently, however, the clams can discern between healthy zooxanthellae cells and those that are degenerating or senile. As a result, the clam transports only the senile or degenerating cells from the outer mantle blood spaces, where they photosynthesize, into the deeper tissues, where they are consumed by the blood cells. In this manner, then, the clams retain the symbiotic relationship, and at the same time, they cull the unfit algal cells to obtain additional nutrients otherwise lost when the cell dies.

Other values may result from the association. As we have noted, most of these associations are tropical; these waters are classically low in plant nutrients. It may well be that the algae benefit by the association through access to a larger and more reliable source of nutrients in the form of the metabolic products of the animal (NO_2, PO_4, CO_2) than they would obtain from the open water.

Tropical waters also are lower in oxygen than temperate waters, because water at higher temperatures holds less oxygen. Since the photosynthetic process produces oxygen, it may be that the animals, especially in the crowded conditions of the shallow waters of the tropics, gain through additional amounts of oxygen produced by the symbiotic algae.

ESTABLISHMENT AND TRANSMISSION OF ZOOXANTHELLAE

Whereas each generation of sacoglossans must actually feed on algae to obtain their supply of symbiotic chloroplasts, perpetuation of the symbiotic association with zooxanthellae from generation to generation of hosts is accomplished differently among the different groups. Basically, two major routes are available. Either the zooxanthellae cells must be passed on directly from the parent to the eggs or larvae, or each new generation must reinfect itself anew from algal cells in the surrounding environment.

Passage of algal symbionts directly to the next generation via the egg or larvae seems to be the method used by most Cnidaria. In most corals, for example, the zooxanthellae enter the eggs at some time prior to their release from the parent. It is also assumed, though without much evidence, that the giant clams also transmit the zooxanthellae through the eggs. It has also been suggested the clams receive them from the surrounding corals.

For other hosts, the perpetuation of a symbiotic relationship depends upon the reinfection of each generation. In such cases, the algal symbionts must be obtained anew from the environment. Among Protozoa and Porifera, this is the common means of transmission. Each new generation obtains the algal cells through ingestion of the free-living form of the symbiont. Surprisingly, this is also the case in *Convoluta roscoffensis,* where the association is a highly dependent one. In this case, the algae-free larvae ingest the free-living *Platymonas convolutae.* In order to ensure that the larvae will be infected, however, there has evolved a chemical that attracts the alga to the egg cases of *Convoluta,* so that when the young hatch and begin to feed, the algae are present. Under such conditions, the new generation is virtually assured of reinfection.

A final possibility for transmission is the ingestion by the potential host of food organisms themselves with symbiotic algae. Although this has not as yet been shown to be important, it remains an area requiring study.

SYMBIOSIS AMONG ANIMALS

We are concerned here with those special associations in which members of different species are regularly associated with each other in nonparasitic relationships. Such symbiotic associations are widespread in the sea, primarily in the crowded reaches of the epipelagic and shallow subtidal zones. As with algal symbiosis, such relationships appear to be somewhat more common or spectacular in the tropics, but are certainly also common in temperate seas.

TYPES OF ASSOCIATIONS

Symbiotic relationships among marine animals cover a broader spectrum than the strictly mutualistic associations we have seen among plants and chloroplasts and marine invertebrates. The simplest type of associations are commensal, whereon the "guest" lives on another organism "host," or in or upon some construction of that organism, such as a tube or burrow. Such associations are similar to epiphytic relationships seen among terrestrial plants, wherein the epiphyte simply uses the other plant as a substrate without actually taking sustenance from it. In these cases, the commensal usually gains in some measurable way from the association and the host is not seriously inconvenienced.

Marine commensals that live on or upon other invertebrates are called *epizoites*. Those that live inside other animals but are not parasites are called *endozoites*. Epizoites are extremely abundant in marine waters and many are probably not true commensals; the relationship is the result of organisms that normally settle on the substrate settling at random on the outside of a slow-moving or sessile invertebrate. These will not be considered further here. Other epizoites are highly specific, and a symbiotic relationship is certainly the case, with the "guest" somehow seeking out the correct "host." These symbiotic epizoites are the most abundant group of commensals and are spread among the phyla Protozoa, Cnidaria, Entoprocta, Annelida, Arthropoda, and Mollusca.

Many invertebrates harbor specialized ciliate protozoans, either on their external surfaces, internally in the digestive tract, or among the gills. For example, the vorticelled ciliate *Ellobiophyra donacis* is restricted to the gills of the clam *Donax vittatus*, while the collared ciliate *Lobochona prorates* is found on the telson of the isopod crustacean *Limnoria tripunctata* (Fig. 10-7). Many other examples are also known. Endozoic commensal ciliates are also common, especially in the digestive tracts of larger animals. Thus, for example, Beers (1961) found that the large green sea urchin of the cold temperate waters of North America, *Strongylocentrotus drobachiensis*, has as many as seven species of ciliate protozoans in its gut, which are found nowhere else and die if removed.

Among the Cnidaria, the class Hydrozoa furnishes most of the cases of epizoites. The best studied is the case of two species of *Probisodactyla*, two-tentacled hydroids, which always occur on the rim of the tubes of polychaete worms of the genus *Pseudopotamilla* in the North Pacific Ocean. Similarly, along the Pacific and Gulf coasts, the hydroid genus *Clytia* occurs on the clams of the genus *Donax*. Other hydroids are epizoic on gorgonians, pennatulids, and ascidians, but the closeness of the association is not always known.

Turbellarian flatworms are often endozoic in the digestive tracts of larger marine invertebrates and in the mantle cavity of various mollusks. Thus, the polyclad *Notoplana ovalis* occurs in the mantle cavity of the limpet *Patella oculis*,

(A)

(B)

while the triclad *Nexillis epichitonius* occurs in the mantle cavity of the common Pacific coast chiton, *Mopalia hindsii.* The rhabdocoels *Syndesmis dendrastorum* and *Syndisyrinx franciscanus* are endozoites in the guts of the sand dollar, *Dendraster excentricus,* and the sea urchins *Strongylocentrotus franciscanus* and *S. purpuratus,* respectively.

The phylum Entoprocta is a small, little-known group, most of which, as Nielsen (1964) has documented, are epizoic on other marine invertebrates, particularly polychaete worms, where they inhabit respiratory current areas (Fig. 10-8).

Among the Annelida, a number of polychaete worms are epizoites with definite host specificity. The best studied is the hesionid *Ophiodromus pugettensis,* which occurs on the ambulacral grooves of the bat star, *Patiria miniata,* on the Pacific coast (Fig. 10-9). However, the whole group of so-called "scale worms" (= Polynoidae) are all very common epizoites on various other echinoderms and mollusks.

Among the class Crustacea, the epizoic forms occur mainly within three groups: copepods, amphipods, and decapod crabs. Copepods associated with other marine animals are usually highly evolved and modified parasites, but there are a few that are simply commensal, scampering over the outer surfaces of various invertebrates. Such an example are various species of the genus *Hemicyclops,* which move on the surface of other, larger crustaceans. Still others, like *Paranthessius,* are found in the mantle cavity of various bivalves *(Tresus, Protothaca, Saxidomus).* One of the most specialized epizoic associations is that among several species of barnacles and various large whales. Two families of barnacles are found only on whales and several are species-specific. Thus, *Cryptolepas rachianecti* occurs only on the California gray whale, *Eschrichtius gibbosus,* and *Conchoderma auritum* attaches to *Coronula diadema,* which, in turn, is attached to the skin of whales (Fig. 10-10; Plate 8). A very curious form of epizoic relationship occurs between the amphipods of the suborder Hyperiida and the large jellyfish. Hyperiids are almost always epizoic or endozoic on jellyfish. Also epizoic on jellyfish are certain juveniles of various benthic crab species (e.g., *Cancer jordani;* Plate 1).

Most mollusk epizoites are found among certain groups of clams. Generally, these are small clams that attach via a foot or byssus to the outside of various other invertebrates. Thus, *Montacuta ferruginosa* is found on the heart urchin, *Echinocardium cordatum; Orobitella rugifera,* attached via a byssus on the underside of the burrowing shrimp *Upogebia stellata;* and *Mysella pedroana* on the legs of the

---

Figure 10-7 Some epizooites. *(A) Elliobiophyra donacis* on the gills of the bivalve mollusk *Donax. (B) Proboscidactyla stellata* on the tube of *Branchiomma vesiculosum.* (*A,* after S. M. Henry, ed., 1966, Symbiosis, vol. 1, Associations of microorganisms, plants and marine organisms, Academic Press. *B,* after J. A. Colin Nicol, 1960, The biology of marine animals, Wiley. © J. A. Colin Nicol, Pitman Publishing Ltd.)

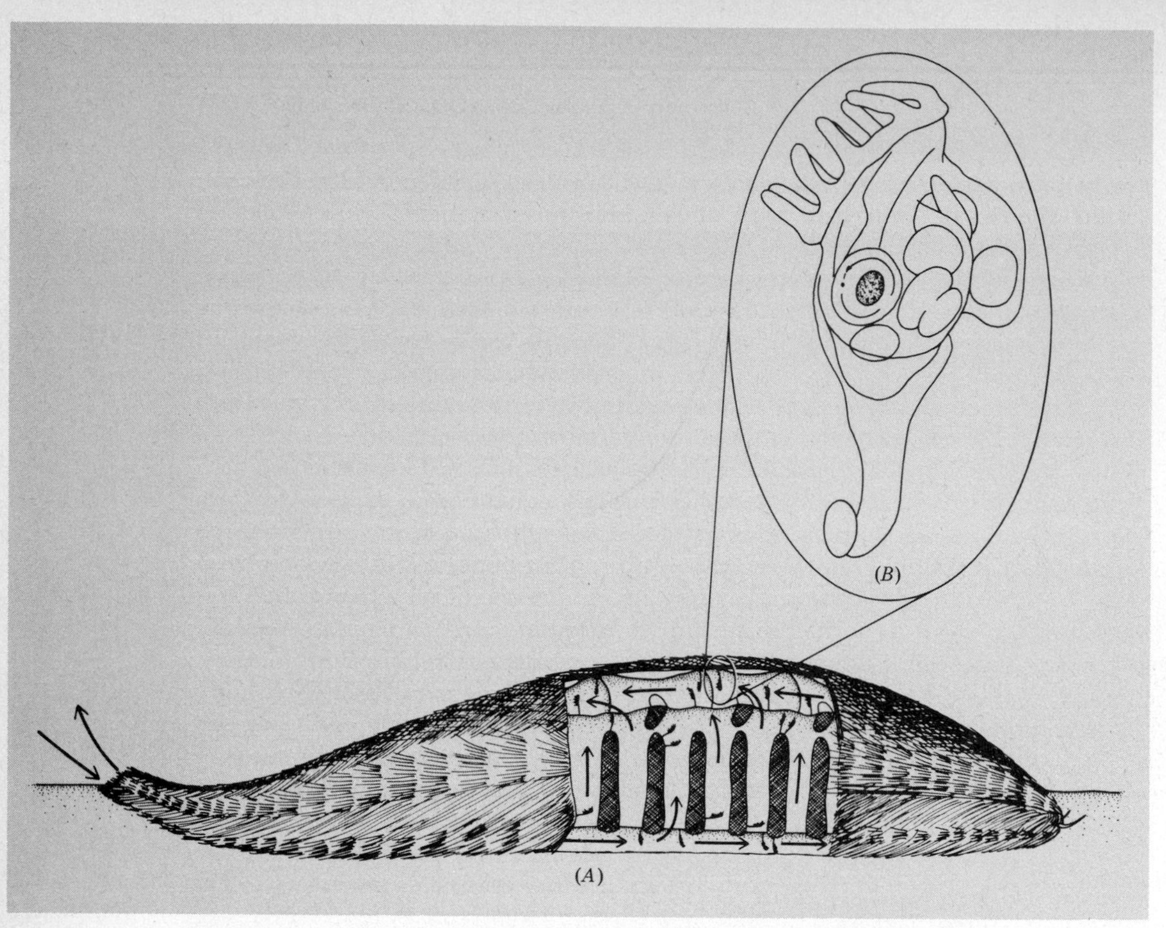

Figure 10-8 Epizoic entoprocta, *Loxosomella fauveli,* on the polychaete *Aphrodite aculeata. (A)* Position of the entoprocts in the respiratory channels of the worm. *(B)* Enlarged view of a single *Loxosomella fauveli* individual. (Modified from C. Nielsen, 1964, Studies on Danish Entoprocta, Ophelia, vol. 1, no. 1.)

sand crab *Blepharipoda occidentalis.* Among the gastropods found on other invertebrates the genus *Crepidula* is one of the most common (Plate 20).

The above is certainly not an exhaustive list of the various epizoic and endozoic symbiotic associations, but gives an idea of the extent and diversity of such associations. Many, many more could be listed, but would not serve the purpose of this text.

A second type of symbiotic association is that of organisms associated with animals forming a tube or burrow. In this case, the commensal or guest occupies the tube or burrow constructed by the host, but is not necessarily in close association with the body of the host. These commensals use the tube or burrow as a refuge from predation. They may or may not also tap the food source of the host. This type of symbiotic relationship ranges from one simply fortuitous, as in the case of an animal diving into any convenient burrow to avoid a predator, to examples of obligatory associations, where the commensals are not found other

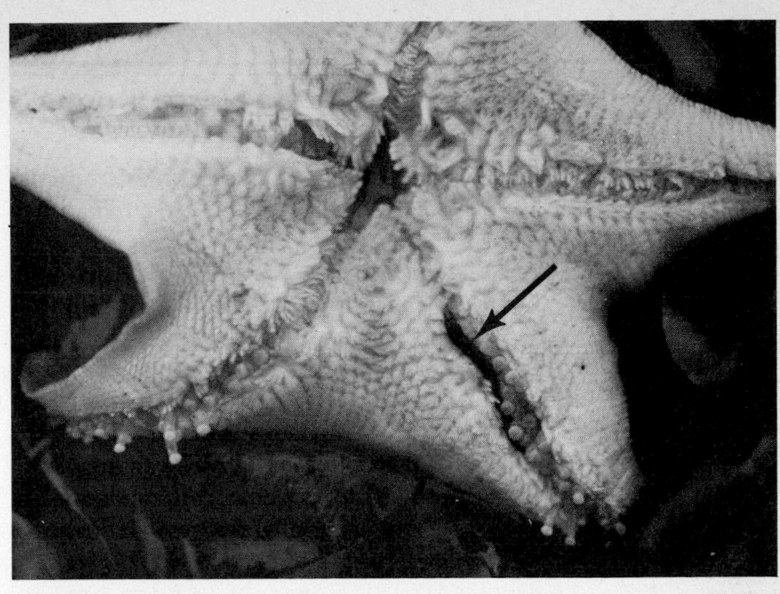

Figure 10-9 The scale worm *Ophiodromus pugettensis* on the bat star *Patiria miniata*. Arrow points to worm. (Photo by the author.)

than in such tubes or burrows and are dependent upon them for continued existence. In the former category are certain small gobiid fishes such as *Clevelandia ios*, which take refuge in any available tube or burrow but emerge to feed. Obligate tube dwellers include many small crabs of the family Pinnotheridae, known as "pea crabs." Many species of this family are known throughout the world, where they inhabit the tubes and burrows of polychaete annelids and other invertebrates, sometimes in species-specific associations.

Figure 10-10 The whale barnacle *Cryptolepas rachianectes* and the cyamid amphipod *Cyamus scammoni* on the skin of the California gray whale, *Eschrichtius gibbosus*. (Photo by the author.)

One of the best known examples of symbiosis with tube dwellers is that of the echiurid worm *Urechis caupo* on the Pacific coast of North America. This animal creates a permanent U-shaped burrow in which can be found four or more commensals (Fig. 10-11). At the upper end is the fish *Clevelandia ios,* which merely uses the tube as a refuge. Further in the tube and closely associated with the worm are the scale worm *Hesperonöe adventor* and the pinnotherid crab *Scleroplax granulata. Hesperonöe adventor* seems restricted to *Urechis* burrows and lives against the worm, snatching food from it. Although the crabs are restricted to tubes, they may be found in the burrows of other animals as well. Another commensal is the small clam *Cryptomya californica,* which has very short siphons and inserts them into the burrow of *Urechis,* thus enabling it to live lower in the substrate than the length of its siphons would normally permit. Occasionally the shrimp *Betaeus longidactylus* may occur in the burrow.

A similar situation prevails with the tube-building polychaete *Chaetopterus.* This genus is virtually cosmopolitan, building tough parchment tubes. These tubes are also inhabited by various species of pinnotherid crabs and scale worms, the species differing depending on the geographical locality.

Another type of association concerns those invertebrates that live in the mantle cavities of various mollusks. This appears to be a very common site for commensals to occupy. Thus, the nemertine *Malacobdella grossi* lives in the mantle cavity of various clams on the Pacific coast. Pinnotherid crabs occupy the mantle cavity of numerous bivalves, including oysters, where they are considered a pest. Polychaete worms inhabit mantle cavities of chitons and limpets. In the tropics, shrimps of the family Palaemonidae are often found in mantle cavities of large bivalves such as *Pinna, Atrina,* and *Tridacna.*

Figure 10-11 The echiuran worm *Urechis caupo,* to the left, and some of the symbionts which inhabit its burrow. Shown are the pea crab *Scleroplax granulata,* the scale worm *Hesperonoe adventor,* and the long-fingered shrimp *Betaeus longidactylus.* (Photo by Dr. Richard Mariscal.)

The remaining types of associations are those that appear to be mainly mutualistic rather than strictly commensal, as the aforementioned. The organisms display varying levels of behavioral and physiological modifications for the association.

One of the most obvious but not well understood associations is that which occurs between various crabs and sea anemones. Many hermit crabs all over the world have anemones attached to their shells, and in some genera, such as *Dardanus*, the presence of anemones seems to be universal. Still other crabs, such as *Hepatus, Munidopagurus macrocheles,* and *Stenocionops furcata,* are found with anemones attached directly to their backs. The ultimate symbiotic relationship appears to be that evolved among a few crabs *(Lybia, Polydactylus),* which carry anemones on their chelae and actively use them for defense and/or food capture (Fig. 10-12). Interestingly, most of the anemones involved in these associations are of three genera, *Calliactis, Paracalliactis,* and *Adamsia,* species of which are rarely found elsewhere than with crabs. These associations must be mutualistic, because whenever the crab changes shells or molts its carapace, the anemone is usually also transferred. This is effected by the manipulatory movements of the crab, but requires a degree of cooperation on the part of the anemone in loosening its grip on the shell or carapace and allowing itself to be physically moved.

Another type of association is that between other Cnidaria and fishes. Throughout the Indo-West Pacific in the coral reef areas, there is a striking association of small fishes of the genera *Amphiprion, Premnas,* and *Dascyllus* with various large anemones. These fishes are able to live nestled among the tentacles of the anemones by preventing discharge of the formidable nematocysts of the anemone tentacles (Fig. 10-13 Plate 31). Other fishes are not able to prevent this discharge and the anemones can kill and eat fishes of a similar size. A similar association seems to exist between the dangerous siphonophore, the Portuguese man-of-war *(Physalia),* and the little fish *Nomeus gronovii,* which swims among the tentacles bearing the powerful nematocysts. Still other juvenile fishes often

Figure 10-12 Symbiotic relationships among crabs and anemones. *(A)* The anemones *Calliactis tricolor* perched on a hermit crab shell. *(B)* The crab *Lybia tessellata* carrying two anemones of the genus *Triactis* in its chelae. (Redrawn from photographs.)

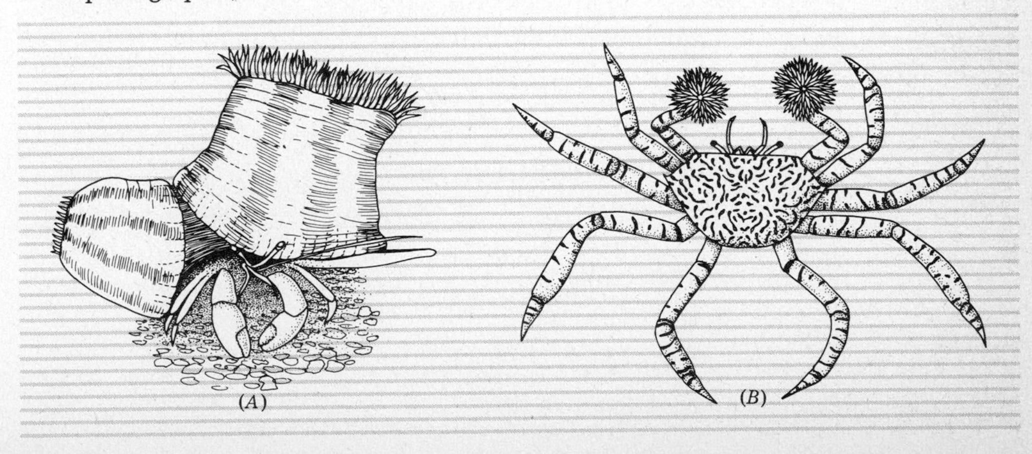

Figure 10-13 The anemone fish *Amphiprion* sp. and its host anemone *Radianthus* sp. (Redrawn from photographs.)

congregate under the bells of the large scyphozoan jellyfish *(Cyanea, Chrysaora),* where they presumably obtain protection from predation.

A similar situation prevails between fishes and the long-spined tropical sea urchin, *Diadema.* These urchins are circumtropical in shallow water and have enormously long, thin spines, which easily penetrate flesh. At least two tropical fishes, *Aeoliscus strigatus* and *Diademichthys deversor,* have adapted themselves to live among these spines, presumably for protection (Fig. 10-14). They also look like sea urchin spines and so are camouflaged.

There is also the association of fishes with other fishes, usually large, predaceous fishes. Pilot fishes *(Naucrates ductor)* and remoras *(Echeneis remora)* are always found with other, larger fishes, other marine vertebrates such as turtles, or even any moving, inanimate objects (Fig. 10-15).

The final type of association is that known as "cleaning behavior." As noted in Chapter Nine this is an association in which various species of fishes and shrimps actively attract large fishes to themselves for the purpose of cleaning them of various ectoparasites (see pp. 337–340 for a discussion).

ORIGIN OF THE ASSOCIATION

With such a diverse array of relationships as we have noted, it does not seem likely that there is any single cause for the origin of all the different types of associations. However, one factor common to all associations is that they have

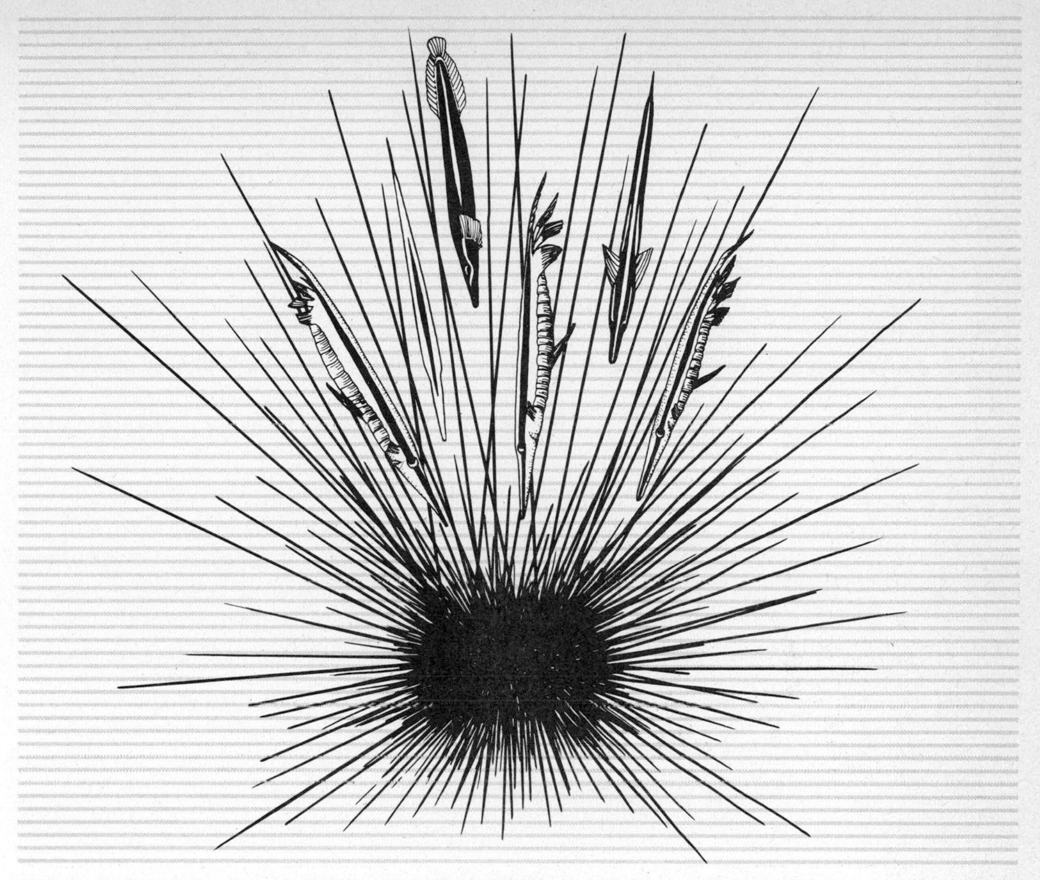

Figure 10-14 The long-spined sea urchin *Diadema* with three shrimp fish, *Aeoliscus strigatus*, and two cling fish, *Diademichthys deversor*, sheltering among the spines. (After R. V. Gotto, 1969, Marine animals, partnerships and other Associations, Elsevier.)

arisen in those areas of the ocean that are the most crowded with life. It would, therefore, appear reasonable that epizoites in particular could have become established due to the great competition for space among the settling invertebrate larvae. As a result, many settled upon other invertebrates and this then later evolved, in some cases, into more obligatory relationships in which those individuals that were commensal gained some advantage over the free-living forms. This is borne out by noting that many epizoites or burrow-inhabiting

Figure 10-15 A remora, *Echenesis* sp., showing the suckerlike dorsal fin. *(A)* Dorsal view. *(B)* Side view. (Photos by the author.)

(A)

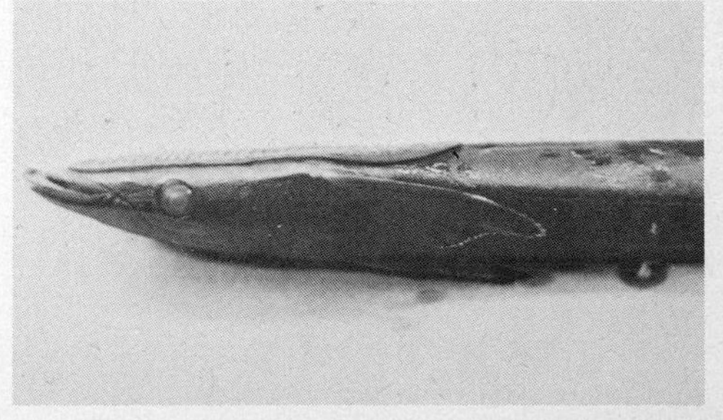

(B)

species are not strongly attracted to any given host species, while others are highly specific. Similarly, many of the tube and burrow inhabitants, and those found in the mantle cavity of various mollusks, also probably had their origin in the search for space. After all, tubes, burrows, and mantle cavities really represent a certain amount of potentially occupiable, semienclosed space.

The initial association of crabs and anemones may also well have begun when anemones began to take up residence on shells that were already occupied by hermit crabs. Alternatively, the association could have arisen out of the habit of various crabs of picking up materials from the bottom and applying them to their carapaces to aid in camouflage.

Many of the commensals that are epizoites or burrow- or tube-dwellers, especially the polychaete worms, tend to be from groups that show strong *thigmotaxis;* that is, they respond positively to stimulation by contact with other, solid bodies. Such a behavioral response would naturally preadapt them to move into tubes or burrows or into various cavities in the bodies of larger animals. Once so established, selection would favor the development of chemosensory clues enabling these animals to find such burrows, tubes, or cavities, and more specific associations would follow.

Among the fishes and Cnidarians, it seems that the various types of symbiotic relationships have their origin in the protection from larger predators that such relationships confer. It does not appear that this relationship stems from crowding, especially between epipelagic fishes and Cnidarians. Provided the symbionts could avoid having the nematocysts discharged, the tentacles of anemones and jellyfishes with their nematocysts represent a secure, formidable barrier to a large predator. However, in the case of large sharks, pilot fishes, and remoras, the association probably originated in the search for food in the epipelagic. Pilot fishes and remoras probably were originally attracted by the amounts of food that were dropped in the feeding by the larger fishes, and they could have managed to establish themselves either by judiciously avoiding falling prey themselves to the sharks, or by being small enough to not be recognized as food by the larger fishes.

Cleaning behavior is a very complex relationship involving many fish species, and its origin and ecological functions have been confounded by various alternate hypotheses and factual inaccuracies as Gorlick et al. (1978) have discussed.

MODIFICATIONS DUE TO THE ASSOCIATION

As with the algal-animal symbiosis, the animal-animal relationships can result in anatomical, physiological, and behavioral modifications to one or both of the partners.

The fewest modifications occur among the epizoites, especially those that are facultative epizoites and are found also in nonsymbiotic situations. Oftentimes, these animals have no anatomical modifications attributable to the symbiotic association, but those that are found with specific hosts usually have developed the means for *recognition* of the host from a distance. Such ability to recognize hosts is even more highly developed among obligate symbionts of various types. What is the basis for such recognition? Investigators such as Davenport (1950) have shown that it is most likely a chemical unique to the host, which is released into the water. The commensal has, in turn, developed chemical sensory receptors, which detect the chemical and enable it to "home" in on its host. If such symbionts are tested in the laboratory in an apparatus designed to give them a "choice," they show a remarkable ability to choose the "host" on which they are found from among other closely related forms. Such attraction by chemical means has been demonstrated for several scale worms (Polynoidae) commensal with various starfish species; for the peculiar fishes, *Carapus,* which live in sea cucumbers; and for the polynoid *Hesperonoe adventor* and its host, *Urechis caupo.*

Yet another modification of commensals, and one that also aids in host recognition, is the ability of the larvae or young to select the appropriate substrate. In many epizoite species that settle onto their hosts from the plankton, metamorphosis and subsequent development do not occur unless the larva is in contact with the appropriate host. The substrate discrimination abilities of some of these symbionts are truly remarkable. Some can apparently recognize substrate surface texture at the molecular level! It is not surprising, then, that whale barnacles can settle not only on whale skin, but only on certain species of whales. However, in this case, it is likely that a chemical unique to the host is also present. Similarly, the clam *Modiolaria* is able to recognize the structure of tunicin, the major chemical component of the external layer of sea squirts with which it lives.

A basic modification of tube-dwelling symbionts is that they are often smaller or thinner and flatter than their free-living relatives. Thus, the pinnotherid, or "pea" crabs, are among the smallest of the various marine crabs. The shrimps that inhabit the mantle cavities of bivalve mollusks are also among the smallest of the shrimps. Similarly, the scale worms are often very flattened in comparison with other polychaetes. The fishes, such as *Clevelandia ios,* that are commensal in tubes or in the internal spaces of clams and sea cucumbers, *Apogonichthys* and *Carapus,* respectively, are very thin in order that they may fit into these restricted areas. The pearl fishes, *Carapus,* show perhaps the greatest modifications in that they have lost their scales and pelvic fins and have shifted the anal opening far forward under the head. This latter change apparently is to ensure that defecation will occur outside the body of the sea cucumber. Furthermore, as Arnold (1957) has shown, these fishes show behavioral modification in that they enter into the cucumber tail first (Fig. 10-16)! Such extensive modifications suggest a long history of adaptation.

Among those anemones found associated with crabs, a number of behavioral modifications occur. In the first place, the anemones must respond positively to manipulations by the crab when it rushes to transfer the anemone to another

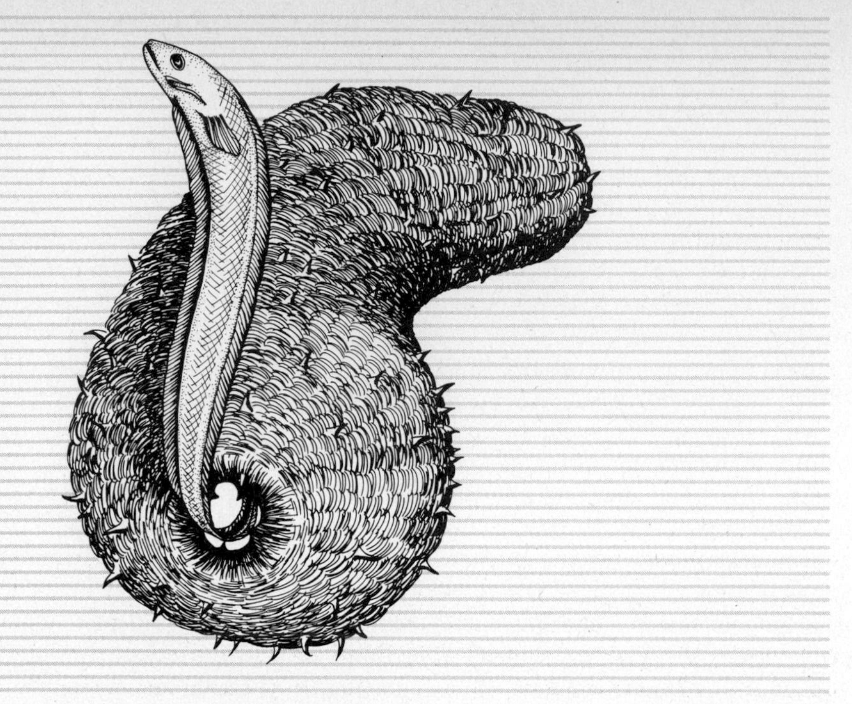

Figure 10-16 A pearl fish, *Carapus,* entering tail first into the anal opening of its holothurian host. (After R. V. Gotto, 1969, Marine animals, partnerships and other associations, Elsevier.)

shell or from its shed carapace to a new one. This it does by releasing its grip on the old shell so that the crab can remove it with its claws. This requires a higher degree of nervous integration than we have come to believe could be accomplished by the extremely primitive, brainless nervous system of the anemone! We are thus currently ignorant of an understanding of this process among those crabs that carry anemones in their chelae. The anemones again show a behavioral adaptation in that they allow themselves to be carried in the chelae and do not close up. In these cases, however, the crab also shows changes in that the chelae are modified to carry the anemones rather than gather food. Behaviorally, the crab extends the anemones when threatened or when gathering food; it uses its second pair of legs to transport food to the mouth, rather than the chelae, since the latter carry the anemones. Perhaps the most drastic changes undergone by the anemone member of these crab-anemone systems is that exhibited by *Adamsia palliata,* which lives with the hermit crab, *Eupagurus prideauxi.* Here, the anemone steadily expands its basal disc to completely encircle the vulnerable abdomen of the hermit crab, thus eliminating the need for the crab to change its shell as it grows. The anemone then expands to allow for the growth of the crab, meanwhile orienting itself so that the oral disc and mouth are ventral to the crab, thus enabling it to obtain food from the crab.

The greatest morphological changes among fishes associated with other fishes occur in the remoras, where the animals have the dorsal fin modified to form a large sucker, which enables them to remain attached to their hosts (Fig. 10-15). Pilot fishes *(Naucrates)* and the anemone fishes *(Amphiprion)* have no special morphological modifications. Among anemone fishes, however, there are definite behavioral and biochemical modifications. Studies have shown that anemone fishes of the genus *Amphiprion* must undergo a period of "acclimatization" to an anemone before they can, with impunity, dive into its tentacles. During this period, the fish is not capable of preventing nematocyst discharge by the anemone. During the acclimatization, Schlichter (1976) has demonstrated that the fish goes through a series of first brief and then longer encounters with the anemone, during which time, the fish coats itself with anemone mucus, thus tricking the anemone to discern the fish as itself. Hence, there is no nematocyst discharge from the anemone. When this coating is complete, the acclimatization period is over and the fish is able to dive in among the tentacles without causing nematocyst discharge.

In addition to this fascinating coating process, anemone fishes have additional behavioral modifications. In contrast to most fishes, which swim away when approached by humans or other large potential predators, anemone fishes swim out of their anemones toward the potential predator! This serves to attract the predator to them, at which time, they do an abrupt about-face and dive into the anemone. Presumably, the behavior would tend to make an unwary potential predator chase them and subsequently be caught and consumed by the anemone. In order to further attract attention to themselves, anemone fish also have vivid color patterns, contrasting markedly with their background (Plate 31).

In contrast to the anemone fishes, which develop immunity to the nematocyst of their host, the fishes *(Nomeus gronovii)* associated with the Portuguese man-of-war do not acclimate, nor do they ever obtain complete immunity from the nematocysts (Fig. 10-17). They instead appear to spend their lives playing a perpetual game of Russian roulette with the tentacles of the siphonophore, constantly moving so as to avoid contact with the lethal nematocyst batteries! The same is true for the juveniles of various fishes that shelter below the umbrellas of the large jellyfish.

Among cleaner fishes, certain morphological modifications can be noted, such as pointed, narrow snouts and forcepslike teeth, adaptations for picking up the small ectoparasites. Both cleaning shrimps and fishes also tend to have very bright and contrasting colors, which make them stand out against the background and advertise their presence. Certain cleaning shrimps, notably the Pederson shrimp, *Periclimenes pedersoni,* have extraordinarily long antennae, which they wave about to further advertise their presence to the fishes.

Cleaning behavior relationships also impose behavioral modifications on the fishes cleaned. These animals must seek out and present themselves at the "cleaning station" and then remain motionless while the cleaner moves over their bodies. They must also open their mouths and spread their gills and opercula to allow the cleaner to enter the mouth and gill area. Of course, at the same time, the larger fish being cleaned must refrain from sudden inhalations, which might accidentally ingest the small cleaner!

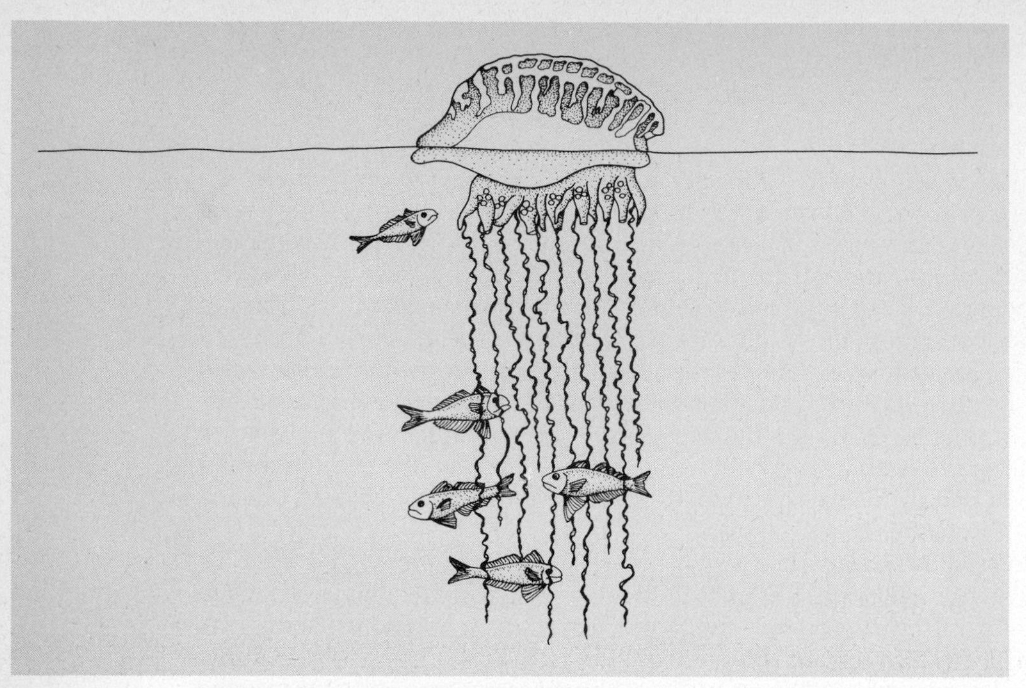

Figure 10-17 Portuguese man of war, *Physalia,* with the fish *Nomeus gronovii* sheltering among the tentacles. (After J. A. Colin Nicol, 1960, The biology of marine animals, Wiley. © J. A. Colin Nicol, Pitman Publishing Ltd.)

VALUE OF THE ASSOCIATION

These symbiotic associations have obvious value to one or both of the members, and the relative value of the relationship to each varies depending upon the association.

The major value of the symbiotic relationship to the epizoites would seem to be with the epizoite rather than the host organism. For some epizoites, the value of the association lies in the fact that the host represents the only available suitable substrate in an otherwise unsuitable area. For example, the hydroid *Clytia bakeri,* which is epizoic on the clam *Donax gouldi* in sand areas in southern California, could not live in the area unless it had the clam to settle upon, as it cannot exist on sand. There is, however, no obvious value to the clam. Other epizoites presumably fix themselves on the host to take advantage of feeding or respiratory currents produced by the host. This seems to be particularly common. The two-tentacled hydroids of the genus *Proboscidactyla* are fixed on the rims of the tubes of their annelid hosts to take advantage of the feeding tentacles and to remove food particles from the host. Similarly, the various epizoitic entoprocts are usually found where there are respiratory currents. In this latter case, they do not steal food directly from the host, however, but filter out the food from the moving current. Perhaps most epizoites benefit by the protection afforded them by the larger host. Thus, the various flatworms occurring in the guts and mantle cavities of their larger hosts are protected from predation and, in the case of intertidal forms, from possible desiccation. Surely, the hyperiid amphipods found on the large scyphozoan jellyfish would be much more subject to predation if free living in the plankton. Finally, the epizoite may also gain the advantage of movement. This advantage may be of particular significance if the epizoite is itself sessile and the host freely moving.

Commensals that inhabit tubes gain primarily protection from predation. This, for example, is the main value to the goby *Clevelandia ios* in the tubes of *Urechis caupo.* Other commensals, however, such as the scale worms (Polynoidae) gain not only protection but also food as they snatch morsels from their tube-dwelling hosts. Commensals in the mantle cavities of bivalve mollusks benefit from the protection of the bivalve shells, but also are able to employ the feeding and respiratory currents of the clam for their own needs. In a few cases, these symbionts may tend toward parasitism in that they may elect to consume pieces of the gills or mantle of their bivalve host.

Among the fishes associated with large jellyfishes, the Portuguese man-of-war, and the large sharks, it seems apparent that the whole value of the association is to the fishes in the form of protection from predation in an area (the epipelagic zone) notorious for great predation pressure. The pilot fish and remoras are also protected from predation by their association with the large predaceous sharks, but they may also obtain food when the shark is feeding. The fishes associated with sea urchins and living in the respiratory trees of sea cucumbers are protected by their association, but their presence presumably has no correlative advantage to their host.

Among the mutualistic associations, the values of the association are more pronounced and, as the term applies, extend to the host as well. In the various crab and anemone associations, the crabs, as hosts, gain the advantage of the nematocyst batteries of the anemones, either as protection against predators or else as an offensive weapon to capture food. The anemones, on the other hand, while still sessile, gain the advantage of movement; some, such as those on hermit crabs, may also be able to obtain food fragments lost during the feeding process of the crab. A similar situation prevails between anemone fish and their anemones. The fishes gain mainly protection, but they benefit their partners by attracting prey to within reach of the tentacles and also by picking up pieces of food outside the reach of the anemone and depositing them in the anemone's mouth.

In the case of "cleaning behavior," the cleaner organisms' gain is in the form of food, whereas the large fishes cleaned presumably enjoy an increased measure of health due to the removal of the various ectoparasites.

LUMINESCENT BACTERIA

A final category of symbiosis remains that does not fit well with the others and hence, its consideration here. That is the curious relationship that has developed between various marine animals and luminescent bacteria. In these relationships, the bacteria are usually confined to a cavity in the body of the larger animal near the outer surface. This cavity is usually connected to the outside through an opening of some sort.

Such symbiotic relationships are confined to two groups of marine animals, fishes and squids, and are not common. The relationship is most common among fishes and squids inhabiting the mesopelagic zone (see pp. 142–144), but is still less common among these groups than is bioluminescence produced intrinsically without bacteria. The various suggested values for the production of light by mesopelagic animals have been discussed in Chapter Four.

The relationship of bacteria and the fishes or squids is a mutualistic one in which the bacteria obtain food from the larger animal. In turn, the fishes or squid use the light produced by the bacteria for various defensive and/or offensive purposes, as discussed in Chapter Four.

The light produced by bacteria is usually continuous, and as a result, the fishes and squids often develop rather elaborate modifications to control the light. These anatomical developments can include the production of a reflecting layer behind the cavity to reflect the light outward, often lenses to focus and concentrate the light, and most often, a screen or shade on the outside which can be raised or lowered to either "turn on" or "turn off" the light. The bacteria show no corresponding changes in behavior or morphology as a result of the association.

The origin of these elaborate associations is not well understood, and the bacteria apparently are not passed from one generation of fishes or squid through the eggs. It would appear that each generation somehow must reinfect its light organs from the external environment.

References

Cheng, T. (ed.). 1971. Aspects of the biology of symbiosis. Univ. Park Press, Baltimore. 327 pp.

Dales, R. P. 1957. Commensalism, Ch. 15. *In:* Hedgpeth, J. E. (ed.). The treatise on marine ecology and paleoecology. Vol. I, Ecology. Memoir 67, Geol. Soc. of Amer.

Gotto, R. V. 1969. Marine animals, partnerships and other associations. Elsevier, N.Y. 96 pp.

Henry, S. M. 1966. Symbiosis. Vol. I, Associations of microorganisms, plants and marine organisms. Academic Press, N.Y. 478 pp.

Nicol, J. A. C. 1960. The biology of marine animals. Wiley, N.Y. 707 pp.

Smith, D. C. 1973. Symbiosis of algae with invertebrates. Oxford biology reader no. 43. Oxford Univ. Press, N.Y. 16 pp.

Vernberg, W. B. 1974. Symbiosis in the sea. The Belle W. Baruch Library in Marine Science no. 2. Univ. of South Carolina Press, Columbia. 276 pp.

Yonge, C. M. 1957. Symbiosis. Ch. 15. *In:* Hedgpeth, J. E. (ed.). The treatise on marine ecology and paleoecology. Vol. I, Ecology. Memoir 67, Geol. Soc. of Amer.

Yonge, C. M. 1975. Giant clams. Sci. Amer. 232: 96–105.

Chapter Eleven
HUMAN IMPACT ON THE SEA

The oceans are truly the last frontier for human exploration and exploitation on this planet. For years, the vastness and volume of the oceans, coupled with the fact that humans are terrestrial organisms ill-equipped naturally to enter into water more than superficially, kept the oceans and their communities safe from significant human interference or effects. As the human population of the world increased through the centuries, large land areas were significantly altered, but the oceans remained relatively untouched. The major usage of the sea during these centuries was as a source of food, but the primitive nature of gear and the limited range of vessels combined to make the effects of fishing insignificant.

In the twentieth century, all this has changed. The explosion of technology has enabled human beings to permeate all parts of the oceans. The rapidly increasing human population, coupled with sophisticated technology, have, in a few short decades, had significant effects on the ecology of the world's oceans. The purpose of this chapter is to discuss and evaluate some of these effects in light of our knowledge of the functioning of the oceans' ecosystems.

FISHERIES

The earliest use of the oceans by humans was likely for food. Early human populations living along the oceans captured various shore fishes and shellfish for consumption. This is recorded in the various large "shell mounds" excavated by archaeologists in different parts of the world. With the advent of vessels to venture onto the sea and the development of more refined nets, fishing became an important source of food to people living near the oceans. However, the gear and the vessels were sufficiently inefficient through much of human history, so that even as late as the 1880s, scientists such as T. H. Huxley believed that the major fisheries were inexhaustible. In the decades of this century, however, the old ships and gear have been replaced by much larger and more powerful vessels, more effective nets and traps, and electronic devices for detecting fish schools. The result has been a significant reduction in many fish populations and the disappearance or overexploitation of others at a time when increasing human populations are demanding more food. Since many believe the oceans are the major food reservoir for upcoming generations of humans, the impact of fishing and its potential to sustain future generations bear consideration here.

MAJOR FISHING AREAS

Geographically, the major fisheries are concentrated in the waters overlying the continental shelves around the world. The only major fisheries that operate in the open oceanic regions of the world are those for tuna and whales. There are several reasons for the concentration of fisheries in neritic waters. In the first place, the inshore waters have a much higher primary productivity than most open-ocean waters (see pp. 69–70) and therefore support larger populations of fishes at all trophic levels (Fig. 11-1). Secondly, the bottom is fairly shallow on the shelf and is accessible to the various nets and traps used by humans to capture fishes. The deep ocean floor, on the other hand, is so far removed from the surface that even with our current sophisticated capture mechanisms, we cannot commercially fish there. Finally, the great lack of food in abyssal depths precludes any large fish population from existing there. Any fishes there could likely not sustain a fishery for any length of time.

Although the fisheries are concentrated in the shelf regions, there is a very unequal distribution of the tonnage caught in various areas. The shelf areas of northwest Europe, along the upwelling coast of western South America, and off Japan produce the largest catches of fishes (Fig. 11-2). The southern oceans and the tropics, with the exception of the west coast of South America, contribute much less to world fisheries.

MAJOR COMMERCIAL SPECIES

The commercially important marine animals are drawn from four groups: bony and cartilaginous fishes, marine mammals, mollusks, and crustaceans. Of these groups, the various fishes constitute by far the greatest tonnage. Among the thousands of species of marine fishes, only a very few make up the majority of fisheries catches throughout the world. These can be assembled into a few major groups (Table 11-1). The herrings, sardines, and anchovies account for the largest tonnage of fishes. Of these, a single species, the Peruvian anchoveta *(Engraulis ringens)* accounts for almost half the catch and is the basis of the largest single fishery in the world, off Peru (Fig. 11-3). Significant other fisheries for this group are for menhaden off the Atlantic and Gulf coasts of the United States and the herring fishery of northwestern Europe. Before its collapse in the 1940s and

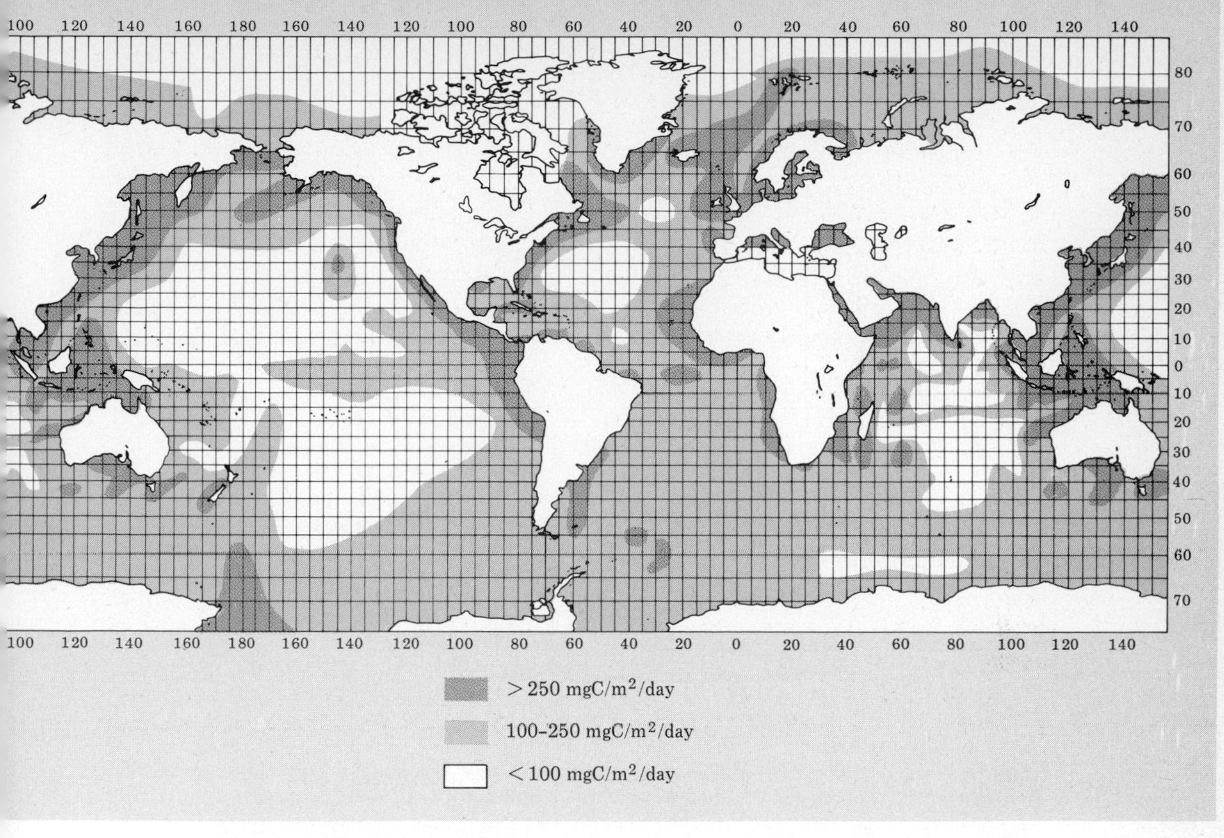

Figure 11-1 Geographical variations in the primary productivity of the world's oceans. White areas are areas of lowest productivity (less than 100 mg C/m²/day); light stipple indicates medium productivity (100–250 mg C/m²/day); heavy stipple indicates areas of highest productivity (more than 250 mg C/m²/day). (After FAO atlas of the living resources of the sea, Food and Agriculture Organization of the United Nations.)

1950s, the California sardine fishery was an important contributor to tonnage in this group. Herrings, anchovies, and anchovetas, commonly called clupeiod fishes, are small, pelagic, and feed on a low trophic level, either directly on phytoplankton *(Engraulis)* or on herbivores such as copepods. Most are not used as food for humans, but are reduced to protein meal for animal food.

The second largest group of fishes landed are the gadoids, comprising various cod, haddock, pollock, and hake. These are bottom-dwelling (demersal) fishes, and the major fisheries are on the shallow banks of the North Atlantic and North Pacific oceans (Fig. 11-3).

The third largest group are the scombroid fishes, generally known by the common name "mackerel." Fishes of this group are common in both temperate and tropical waters, where they are fast-moving pelagic carnivores in shallow water.

Closely related to mackerels are the various tunas. Tunas are among the largest fishes in commercial fisheries. They are also the basis of the only major open-ocean fishery. They are widespread, fast-swimming carnivores of tropical

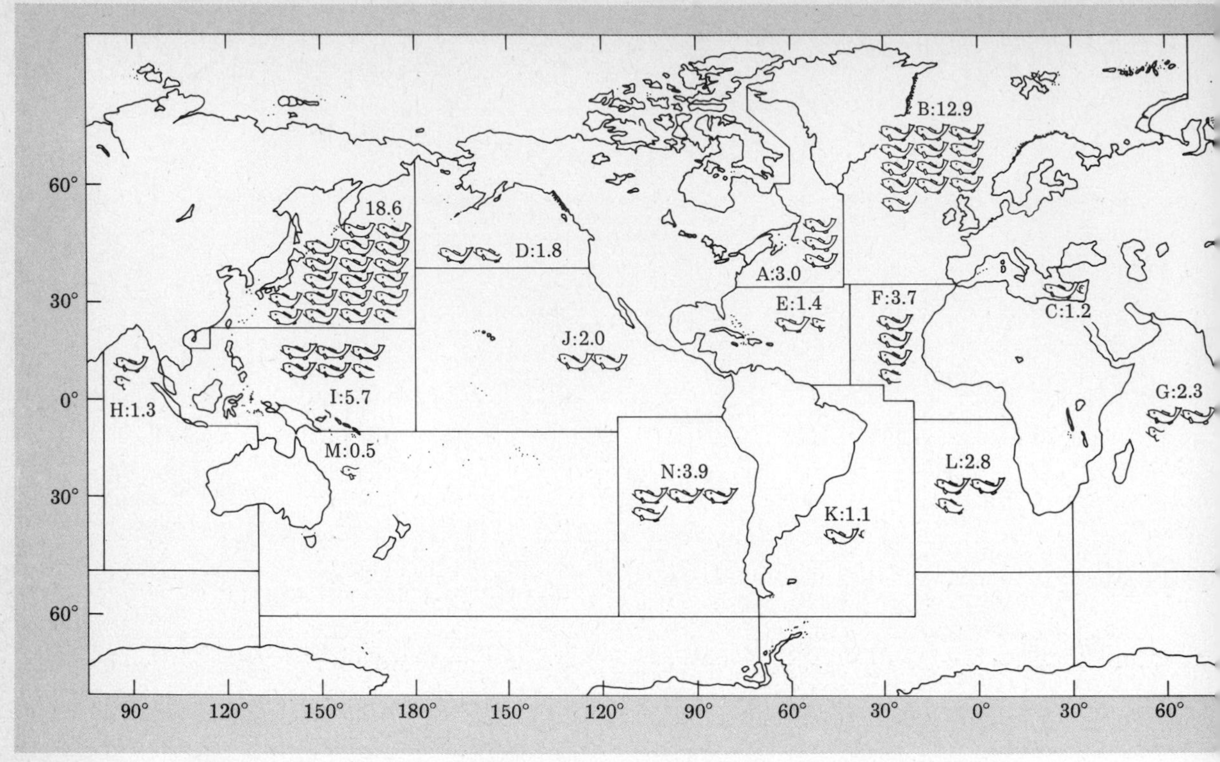

Figure 11-2 The major marine fishing regions of the world and their respective 1977 landings. Each symbol represents 1 million metric tons. The letters refer to the region in the FAO division of the oceans. (Modified from the FAO 1977 yearbook of fishery statistics, Food and Agriculture Organization of the United Nations.)

and warm temperate seas (see Chapter Three for a discussion of their biology). Although the tonnage is not as great as for other fishes, they are significant because they command a high price as a canned fish product.

Redfishes, rockfishes, and sea basses are demersal, cold-water fishes which are primarily used as human food.

The flatfishes such as halibut, sole, plaice, and flounder, are all well-known fishes that are of considerable economic importance, because they are premium eating fishes. All are caught in fairly shallow waters, and many have been heavily exploited for years.

The sharks and other cartilaginous fishes are generally not consumed directly, due to the high urea content of the flesh.

One group not separated in Table 11-1 is the salmonids. They form a very minor part of the world landings, too small to merit separate listing, but are of great economic importance in North America and Europe because of their fine taste.

The major crustacean group in world fisheries is shrimp, various species being caught in both warm and cool waters in all oceans. The other significant crustaceans are various species of crabs and lobsters. All are economically valuable and command high prices for direct human consumption.

TABLE 11-1 The Catches of Different Groups of Fishes from 1971 to 1977 (numbers are in millions of metric tons)

	1971	1972	1973	1974	1975	1976	1977
Flounders, halibut, sole	1.39	1.29	1.24	1.18	1.14	1.12	1.08
Cod, hake, haddock	10.66	11.44	11.96	12.68	11.85	12.14	10.69
Redfish, basses, congers	4.07	4.16	4.26	4.79	4.97	4.86	5.14
Jacks, mullets, sauries	4.65	4.95	5.64	5.35	5.85	7.25	8.68
Herrings, sardines, anchovies	19.81	13.36	11.41	14.04	13.75	15.30	12.96
Tunas, bonitos, billfish	1.81	1.96	2.09	2.25	2.09	2.29	2.33
Mackerels, snocks, cutlassfishes	3.27	3.18	3.43	3.61	3.61	3.30	3.56
Sharks, rays, chimaeras, etc.	0.52	0.53	0.60	0.56	0.59	0.56	0.59
Misc. marine fish	8.19	8.65	9.13	8.81	8.55	8.92	8.92
Shrimps, prawns, etc.	1.04	1.11	1.26	1.32	1.29	1.41	1.44
Krill, planktonic crustacea	0	0	0	0	0	0.005	0.0001
Crabs	0.36	0.37	0.39	0.41	0.38	0.40	0.45
Lobsters	0.095	0.094	0.096	0.093	0.095	0.099	0.096
Oysters	0.73	0.78	0.79	0.74	0.84	0.90	0.88
Mussels	0.64	0.52	0.44	0.38	0.46	0.48	0.54
Squid, octopus, etc.	0.91	1.15	1.07	1.07	1.17	1.17	1.16
Clams, cockles, arkshells	0.66	0.68	0.68	0.73	0.70	0.76	0.83
Scallops, pectens	0.16	0.19	0.20	0.24	0.28	0.36	0.41

The major mollusk group in terms of tonnage is the squid. Significant numbers of squids are caught in Japan, Europe, and California. The other major mollusk group commercially harvested includes the bivalves, such as oysters and clams.

The major fishery for marine mammals has been that for various species of whales. Persistent overexploitation of the stocks in this century has, however, virtually doomed this fishery. Other harvestable mammals are mainly seals and sea lions.

Other minor fisheries exist for various species of algae. Some are taken in Japan for human consumption, but most are collected to extract various products for use in industry (algin derivatives).

SUSTAINABLE YIELD AND THE FUTURE

In 1977, the total world fishery catch was 73.5 million metric tons, of which 50.9 million metric tons were for human consumption. Of the 73.5 million metric tons, 62.7 million metric tons were from marine fisheries. The total fishes caught for human consumption represents perhaps 1 percent of all human food, but a

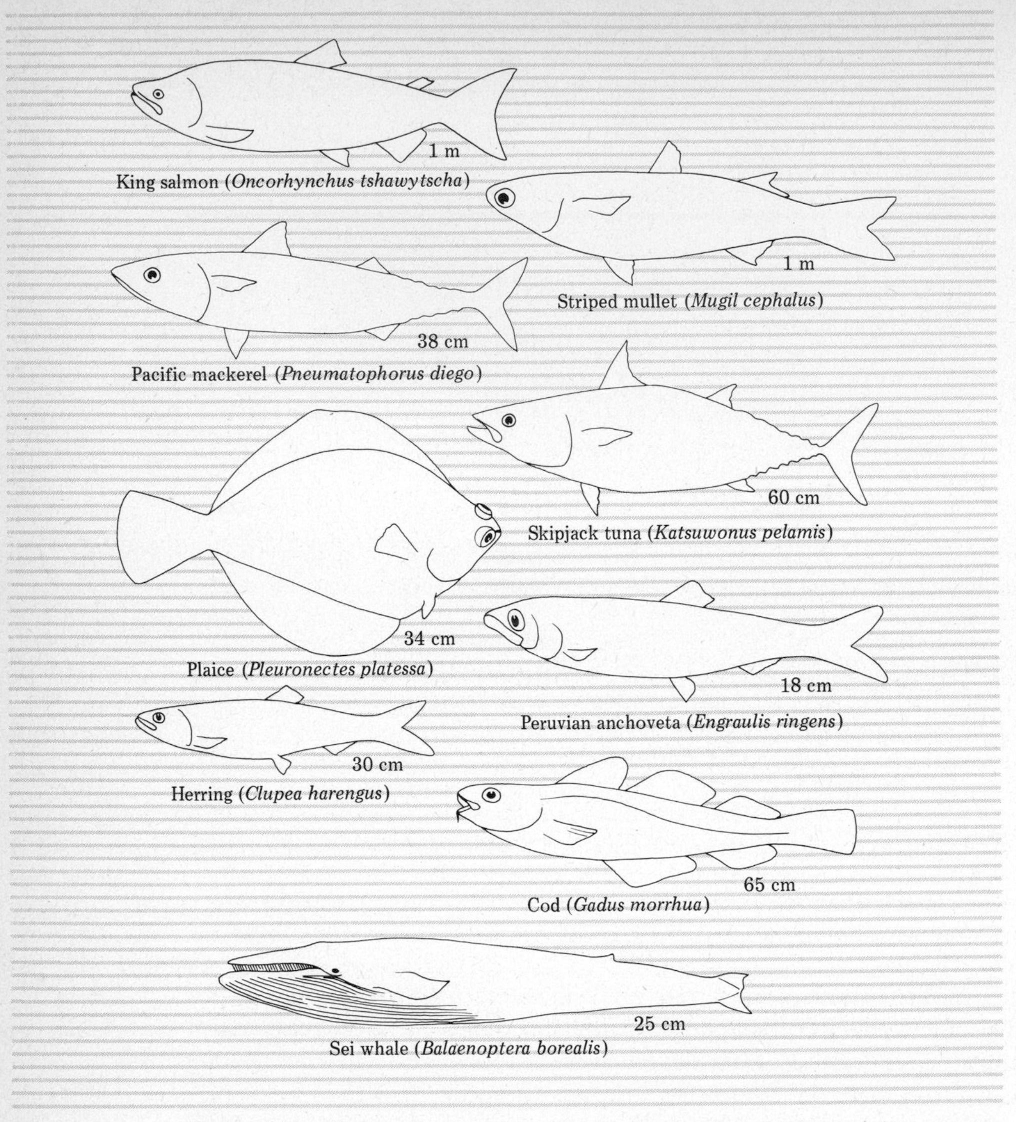

Figure 11-3 Examples of some of the major commercial fishes of the world.

significant 10 percent of the protein intake; hence its importance. The total marine tonnage in 1977 was down from 1976, when landings were a record 64.1 million metric tons. In fact, the recent landings have all been relatively stable compared to the rapid growth in the 1950s and 1960s (Fig. 11-4). Since the human population is still increasing and is expected to reach near 7 billion by the end of the century, and since many are anticipating that much of the food needed to feed these increased numbers will come from the sea, does this leveling off mean we have reached the limits? What are the facts with respect to current and potential amounts of food from the sea?

In order to answer these questions, it is first necessary to look briefly at the current fisheries situation. Of the present fishing effort, 90 percent is concen-

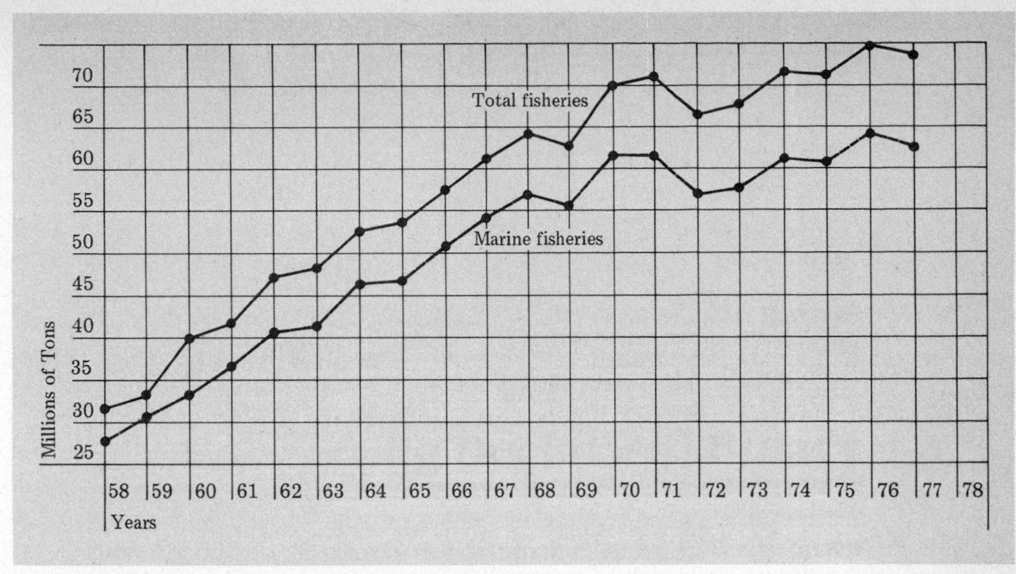

Figure 11-4 Changes in the world annual fish catch in recent years.

trated on the 7 percent of the world ocean area represented by the continental shelves. These shallow areas are the most productive ocean areas.

The great growth of the world fisheries in the 1950s and 1960s was due to two factors. First was the discovery and exploitation of the anchoveta fishery of Peru, which sent Peru from having less than 1 percent of the world fisheries landings in 1950 to 17 percent in 1967 and a position as the top fishing nation in the world. The second factor was the introduction of modern fish-capturing and handling gear, including factory ships, larger nets, and more accurate methods of locating fishes. Most of this effort was expended on the continental shelf regions around the world with the result that, at the present time, perhaps most of this limited area has been fully exploited for its existing fish stocks and many have been overfished (see pp. 391–394). It would therefore appear that little increase can be expected from these areas.

What about the remaining 93 percent of the world's oceans? Currently, the major oceanic fisheries of the world are for various tunas, swordfish, bill fish, and so forth. In 1977, the landings of this group were 2.3 million metric tons, or only 3 percent of total fishery landings for 1977. However, this 2.3 million tons represented a significant increase in landings over previous years and is probably partly due to the change from long line and pole fishing to the advent of purse seines (Table 11-1). Whereas 15 years ago the tuna stocks were considered nearly impossible to overfish, the advent of purse seines dramatically changed that fact in a few years. Today there is considerable concern about the continued ability of these fish stocks to sustain the current fishing pressure.

Bottom fishing on the abyssal plains under the oceanic area is not feasible, since the diminished amount of energy reaching the abyssal floor would not allow the existence of substantial fish stocks (see pp. 132–133).

Furthermore, as we noted in Chapter One, the surface waters of these open ocean areas are impoverished with respect to nutrients and phytoplankton.

They cannot sustain the populations of fishes, even in the surface waters, that the continental shelf waters are able to. They are "watery deserts." This suggests that, even though this is a vast area, it cannot match the production of the shallow coastal waters. In fact, we are approaching the limit of what we can obtain from the sea.

What, then, is a reasonable estimate for the *maximum sustainable yield* of food that we can expect from the seas? By maximum sustainable yield, we mean the largest number of fishes that can be harvested year after year without diminishing the stocks.

Perhaps the best estimate is based upon total primary productivity of the oceans. As we have already discussed, all food chains and trophic levels are ultimately dependent upon photosynthesis. The energy fixed in photosynthesis is finite in the world's oceans. By knowing the trophic level of the various fishes taken as food and the efficiency of energy transfer from level to level, it is thus possible to obtain a figure for maximum production. Such was the approach taken by Ryther (1969) to estimate potential yield.

In this estimate, he first divided the ocean into three areas, each representing waters with significantly different levels of primary productivity. The largest area (90 percent) was that of the open oceans of the world, but it had the lowest primary productivity (Table 11-2). The second area in size (9.9 percent) was the neritic zone, and the final—highest in productivity but smallest in size (0.1 percent)—was the restricted area where upwelling occurs (Table 11-2). Next, he estimated the efficiency of energy transfer among trophic levels in each division and finally, the trophic level of the major fishes captured in each (Table 11-3). Combining the data, he arrived at a total world fish production of 24×10^7 tons (= 240 million tons).

This figure, however, does not represent the total amount that may be taken. A significant fraction of this 240 million tons must be left as breeding stock to sustain future yields. Secondly, another large fraction is consumed by other carnivores in the sea. Ryther estimated these latter fractions to account for

TABLE 11-2 Division of the Oceans into Provinces Based on Level of Primary Productivity

Province	Percentage of the Ocean	Area (km^2)	Mean Productivity (grams of carbon per m^2 per year)	Total Productivity (10^9 tons of carbon per year)
Open ocean	90.0	326×10^6	50	16.3
Coastal zone	9.9	36×10^6	100	3.6
Upwelling areas	0.1	3.6×10^5	300	0.1
Total				20.0

After Ryther, 1969.

TABLE 11-3 Estimated Total Fish Production of the World's Oceans Based on the Three Provinces of Table 11-2

Province	Primary Production (tons of organic carbon)	Trophic Levels	Efficiency	Fish Production (tons, fresh weight)
Oceanic	16.3×10^9	5	10	16×10^5
Coastal	3.6×10^9	3	15	12×10^7
Upwelling	0.1×10^9	1½	20	12×10^7
Total				24×10^7

After Ryther, 1969.

somewhat more than half of the total production, thus leaving about 100 million tons as the sustained yield for the world's oceans.

Other scientists, such as Cushing (1969) and Shaefer (1965), have also estimated the total harvestable amounts of fishes in the sea. These estimates have been made on the basis of primary productivity, on the basis of extrapolating from current landings, and on the basis of trophic level and ecological efficiencies. Because scientists disagree on the proper figures to use for primary productivity, ecological efficiency, and average trophic level, these estimates have varied from a low of 60 million tons to a high of 290 million tons.

Given the current world marine fisheries situation in which, despite intensive effort and the wide use of modern gear, the catch appears to be leveling off in the vicinity of 60–65 million metric tons (Fig. 11-4), it would seem prudent to accept limits somewhere around 100 million metric tons as the maximum sustainable yield.

That we may be reaching the limit on the number of fishes that may be taken is manifested in the decline of various fisheries.

OVEREXPLOITATION

In recent years, there have been abundant examples of the decline of fish stocks of all types in all areas of the world. Some of these declines are clearly due to instances of overexploitation, but others have more complex origins. All indicate that we are approaching or exceeding the maximum amount of fishes we can take from the oceans, but also that we have often mismanaged many fisheries so that we have not achieved a maximum sustainable yield. The signs of overfishing are normally a decline in average size of fishes and an increase in effort needed to land the same amount of fishes. Age structure also changes. In some cases, overexploitation has been compounded by certain natural changes in the environment that also acted to reduce the stocks. In most situations, however, overexploitation is the direct result of human activity. Three recent examples are discussed here, one with environmental consequences affecting the fishery, one in which causes are uncertain, and one related directly to humans.

The small clupeid fish known as the Peruvian anchoveta *(Engraulis ringens)* is a small filter feeder that occurs in large schools in the upwelling water mass of the Humbolt Current off the west coast of South America, where it feeds on the abundant plankton. This small fish occurred in tremendous numbers in these cold waters before the advent of human fishing. The major predators were various species of marine birds, collectively called "guano birds," which existed by the millions along these shores and produced huge quantities of droppings or *guano,* which were harvested for fertilizer.

The anchoveta population depends on an abundant plankton, which, in turn, depends on the nutrient-rich upwelling water. In certain years, the upwelling is blocked in this area by an influx of warm surface water known as the *El Niño* Current. When this occurs, plankton disappear and the population of anchoveta drops, as do the bird numbers. Hence, there is a natural abundance cycle.

Direct human exploitation of the anchoveta began in the 1950s and the fishery increased at a phenomenal rate until the late 1960s. The catch was 10 million metric tons or more than one-sixth of the entire world's catch!

This huge catch was accompanied by a decline in bird numbers, since they had fewer fishes to eat. The fishery experienced one drop in numbers in 1965 as a result of the occurrence of the El Niño, but this temporary setback was made up the following year.

Fishery experts had estimated a maximum sustainable anchoveta yield of 9.5 million metric tons. This was exceeded in 1967 when over 10 million tons were landed. By 1970, over 12.3 million metric tons were landed, but the size of the fishes was decreasing, and it was apparent to experts that the fishery was in trouble and that too few fishes were escaping to provide the breeding stock. As could have been forecast, the crash came in 1972, aided by another El Niño, so that landings dropped to 4.7 million tons. This was followed by an even worse year in 1973, when only 2.3 million tons were landed. Subsequently, this fishery has not recovered, so that even in 1977 only 2.5 million tons were landed. Thus, a once huge resource has, due to overfishing, been reduced far below its sustainable yield, and the human population of the world has been denied a significant protein supply.

Much more mysterious is the demise of the California sardine fishery. The California sardine, *Sardinops caerulea,* is also a small, plankton-feeding clupeid fish with a range from Mexico to southern Alaska. In the 1930s, this species was the basis of the greatest fishery in North America. During the 1936–1937 season, fishermen landed 650,000 tons in California alone, the major fishing region. The decline began in the 1940s, when the age structure of captured sardines began to change to older age groups, reflecting the decimation of the younger fishes and their failure to recruit to the fishery. Despite the establishment of the California Cooperative Oceanic Fisheries Research Program in 1949, the fishery was commercially dead by the early 1950s. In Monterey, California, a center of the industry, the famed Cannery Row of Steinbeck closed up. The fishery has not recovered in the nearly 30 years since then, though a few tons of sardines are landed as "incidental" captures each year.

Although most people attributed the demise of the fishery over the 10-year period from 1940 to 1950 to overfishing, it seems apparent now that, in this case, other factors were also at work. The decline of the fishery began with the onset of a 15-year period of unusually cold water in the sardine habitat. It is felt by Marr (1960) that this cold water affected the sardine population adversely, either directly through a reduction in breeding or through an effect on the food supply. The result was a catastrophic decline in numbers. If overfishing were the only reason for the demise, one would have expected that, in the 25 years during which there has been no fishing, the stocks would have recovered somewhat or at least returned to their known breeding grounds. But this has not occurred; hence, the strong suspicion that other factors are involved.

Finally, Soutar and Isaacs (1969) have presented evidence from the examination of the scales of sardines preserved in layers of sediments in certain anoxic basins, that the California sardine naturally undergoes huge cyclic changes in population numbers, cycles which range from 500 to 1700 years. If this is correct, the California sardine would be present in numbers for a certain period of years, then vanish for a longer period before reappearing. During its absence, the niche is filled by another clupeid, the anchovy. Currently, the increase in anchovy numbers would seem to substantiate this theory. If true, Cannery Row might again bustle to the activities of fish rendering in only 500 years or so!

The final example of the demise of a fishery is much more simple, for it is the direct result of human overfishing. Here, we refer to the whaling industry. The ignominious end of the world's whale fishery will undoubtedly be observed in the next few years.

In contrast to most animals taken in various fisheries, whales are warm-blooded mammals which have a long life span and a very low reproductive rate. This means that they are very slow to build up their numbers once reduced and are easily reduced to extinction.

Large-scale whaling began in the North Atlantic in the sixteenth century. At first, methods were primitive and refuges from whalers existed for most species. However, the advent of long-range vessels, and especially the harpoon gun with explosive charge, spelled the end of whale stocks in the North Atlantic and North Pacific by 1900.

Attention then turned to the large stocks of baleen whales in the Antarctic region, where most whaling of the twentieth century has been concentrated. Coincident with the exploitation of these southern stocks was the advent of huge factory ships, more speedy catcher boats, and airplanes and helicopters to find the whales. All these technological advancements led to a greater slaughter of whales in the first 60 years of this century than had been achieved by all whaling efforts in all preceding centuries. In about 60 years of Antarctic whaling, Rounsefell (1975) notes that perhaps 1,300,000 whales were taken!

In an attempt to regulate this fishery, the International Whaling Commission (IWC) was set up in 1948. It was to obtain data on whale stocks and to set catch quotas for the various species so as to conserve the stocks. Unfortunately, because the Commission had no enforcement powers, because it had no independent means of obtaining data on the status of the various whale stocks outside of what the whalers reported, and because it was subject to intense political pressure by the whaling nations which it comprised, its history of attempts to conserve stocks has been a mockery. It has consistently set its quotas too high and refused to heed the advice of its own and outside experts. As a result, the stocks of whales have continued to decline to the point where today, many species are on the verge of extinction, and others are so reduced in numbers they are commercially extinct. Along with the demise of the whales has been the demise of the whaling industry. When the IWC was set up in 1948, its 20 member nations were whalers. By 1979, only two remained as whalers, Russia and Japan.

The dramatic decline in this fishery can be understood from a few examples. In the 1930–1931 season in the Antarctic, 25,000 blue whales *(Balaenoptera musculus)* were taken, but by the 1963–1964 season, only 112 were taken. In 1968, the IWC reluctantly gave complete protection to the species. Declining blue whale stocks forced whalers to change to fin whales *(Balaenoptera physalus)* after World War II, and from 1955 to 1961, more than 25,000 were taken each season. By 1965, this number had dropped to less than 2,500 and by 1974–1975 to less than 1,000. Sei whales *(Balaenoptera borealis)* went from 25,453 in 1964–1965 to 3,859 in 1974–1975.

Clearly, the unrealistically high quotas set by the IWC and the greed of the whaling nations have reduced the whale stocks to the point where it is, or soon will be, uneconomical to continue whaling. Even if all whaling is stopped now, as many countries have urged, it will take many species years, perhaps centuries, to recover because of their low fecundity and the long time necessary for them to reach reproductive size. For some species such as the blue whale, McVay (1966) has suggested that the numbers may now be so low that the remaining few will be unable to find mates in the vast reaches of the ocean. Extinction may be inevitable. That there is some hope for recovery of at least some of these leviathans lies in the recovery of the California gray whale *(Eschrichtius gibbosus)*. This species was reduced to only a few individuals in the early decades of this century, but since complete protection in 1938, they have now increased in numbers to more than 8,000.

TRAGEDY OF THE COMMONS

In 1968, Garrett Hardin published an article in *Science* magazine entitled "The Tragedy of the Commons." In this article, he pointed out that freedom of access to a common resource ultimately brings ruin to all who enter. The core of this argument is that whenever there is a resource held in common by a large number of individuals or political entities, such as countries, each entity will try to maximize its own gain from that resource because not to do so will allow another to do so and because the negative aspects of acting selfishly do not devolve on the one acting alone, but on all who share the resource. Unfortunately, such an analysis is made by each and every entity sharing the resource. As a result, the resource is destroyed. Therein is the tragedy.

The plight of the whale fishery can certainly be likened to such a tragedy. It may also be applicable to other open-ocean fisheries, such as tuna, or potential fisheries, such as krill (see pp. 396–397).

Open-ocean fisheries lie in international waters—in other words, beyond the control of any one country—and presumably are held in "common" for the good of the world's people. Since they are owned by no one, they and their resources are open to access by all. Therefore, those countries with the means to exploit these areas do so because it serves their own interests. They do not have to answer to another nation or regulatory agency, and the tragedy is that, if they don't catch the fishes, another nation will. It is just this short-term greed that has destroyed the whale fishery and may next do so for tuna. Recent interest in mining of deep-ocean manganese nodules may also soon fall into this category.

REGULATION

The decline of many world fisheries due to the tragedy of common access and the pressures of an increased demand for food by an ever-increasing human population have led to friction among fishing nations and various attempts at the regulation of fisheries resources. As long as the fishery exists on the continental shelf where individual nations have jurisdiction by international convention, the fishery can be controlled by a single nation. In the past, the area of an individual nation's jurisdiction was 12 miles (19.3 km). In recent years, however, several countries have unilaterally acted to extend these limits out to 50 or 200 miles. This has caused several confrontations such as the "cod war" between Iceland and Britain and the seizure and fining of many United States tuna boats by Ecuador and Peru. Most recently, through the actions of the United Nations, a standard 200-mile wide fishing and economic zone was established for the waters of each coastal nation. This means that individual nations will now have full control of all fishery activity within 200 miles of their shores.

International agreements among nations for the control of fisheries have proven to be more difficult to bring about and also to make work effectively, as witness the IWC. However, there are some outstanding examples that have worked. One of the oldest is the International Fur Seal Treaty, established in 1911 among the United States, Russia, Japan, and Canada, to regulate exploitation of the northern fur seal *(Callorhinus ursinus)* (Plate 9). It has been an outstanding success in regulating fur seal numbers to achieve a sustained yield. Another success has been the International Halibut Commission, established in 1924 between Canada and the United States to regulate the then declining Pacific halibut fishery. This was the first fish species to be conserved internationally. After establishment of the Commission, the stocks of halibut recovered (Fig. 11-5). Since these early attempts, other international regulatory agencies have been established, including the Atlantic and Baltic Commissions, The International Tuna Commission, and the International North Pacific Fisheries Commission (INPFC). Many of these more recent regulatory agencies have had a beneficial effect on certain fisheries, but problems still remain. The major one is that the various commissions have limited jurisdiction geographically or are limited to only certain species. Thus, the waters around Japan are not in the jurisdiction of the INPFC and the Halibut Commission regulates only that species.

In order to effect a truly worldwide regulation of fisheries, the current commissions or others will have to expand their jurisdiction to include all species in their areas, and those areas, mainly in the Southern Hemisphere, where no regulatory agencies exist will have to be brought under control.

NEW FISHERIES

Given the current leveling off of the world fisheries catch and the decline or plateauing in many long-established fisheries, are there any prospects for

Figure 11-5 Exploitation and recovery of the Pacific halibut fishery. (Modified from D. H. Cushing, 1975, Fisheries resources of the sea and their management © OUP 1975.)

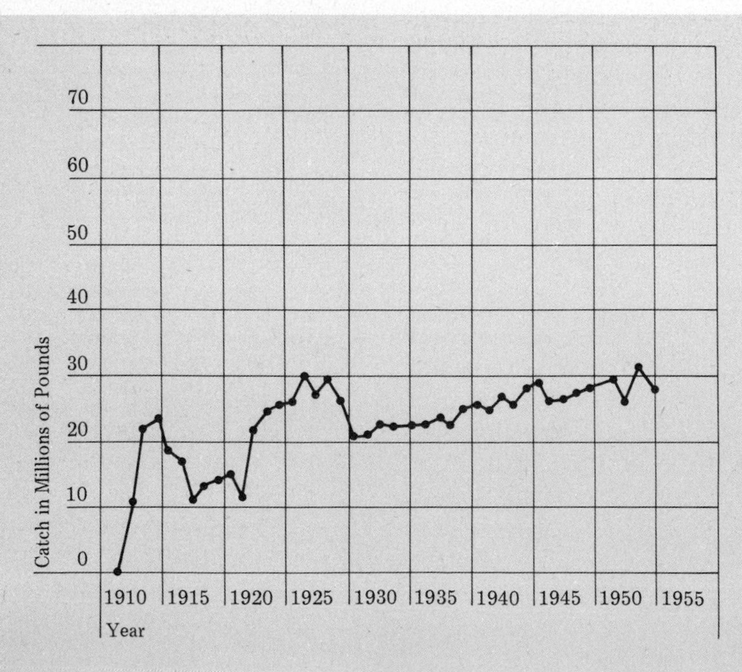

finding substantial new fisheries? The answer appears to be yes, but whether these new fisheries will be able to contribute significantly to the human food supply is much less certain.

Now that the stocks of baleen whales have been reduced to the point of extinction in Antarctic waters, a new fishery has developed for the "krill" (euphausiids) that formerly constituted the food of the whales. Krill exist in untold numbers in the seas around Antarctica, and they are large enough (2 inches) so they may be captured without resorting to very small-mesh plankton nets. The most conservative estimates of stocks of krill are 50 million tons, but other estimates range up to 100 million tons, according to Idyll (1970). Perhaps as many as 50 million tons could be harvested annually. Thus, the potential for this fishery is enormous, certainly the largest of any unexploited fishery in the world today. However, much remains to be learned about this fishery.

Another area of potential exploitation, as yet relatively untapped, is the mesopelagic zone, especially off the continental slopes in 300–1000 m. Recently, substantial stocks of blue whiting have been discovered in this zone off the British Isles. This zone also contains substantial numbers of squids and small midwater fishes such as lantern fish.

Finally, the entire Indian Ocean has a low level of fisheries development by current standards. It produces only about one-fifth the catch per unit area of the Atlantic and Pacific. It would appear that increased efforts in these waters would increase the catches without overexploiting the stocks. On the basis of primary production alone, Idyll (1970) estimated that this ocean can produce 10 million metric tons of fishes as opposed to a little over 3 million tons in 1974.

We can summarize this section by saying that current world fisheries appear to be exploited to the maximum, with many showing signs of definite overexploitation due to unregulated access and overfishing. Furthermore, there appear to be only limited amounts of increases to be expected from fishing stocks and underexploited areas, with the possible exception of the Antarctic krill. The estimate of a total sustainable yield of 100 million metric tons or less for the world's oceans appears reasonable. With catches now reaching 63 million tons, we can expect little significant increase in food from the sea. Therefore, the seas cannot and will not be the source of substantial amounts of food for an increasing human population.

MARICULTURE

As the limit is in sight for the harvesting of wild populations of marine animals and plants in fisheries, humans have begun to work with culturing various marine and estuarine species in situations that are more analogous to our terrestrial based agriculture, a field that is termed *aquaculture.* Aquaculture seeks to rear selected aquatic plants and animals under controlled conditions so as to increase the amount of food available to humans with respect to that obtained through traditional fishing. We are concerned here only with marine aquaculture, or *mariculture.*

HISTORY AND EXTENT

Despite the evidence of the occurrence of culturing of marine organisms back at least as far as ancient Rome, aquaculture through the centuries has never become a significant contributor to human food. Food from the seas remained the province of the hunters and gatherers, the fishermen. Only in this century has aquaculture received wide recognition and undergone significant development. Partly, this development is in response to the pressure for more human food, especially protein, and partly from the declining wild stocks.

Aquaculture encompasses marine organisms raised not only for human food, but also for other products, such as pearls, and terrestrial animal food.

Aquaculture is practiced throughout the world in the tropical and temperate zones, but is more technologically advanced in the industrialized nations of the Northern Hemisphere. At present, it is confined to shallow coastal embayments or artificial ponds.

SPECIES CULTIVATED

Few marine organisms are currently cultivated, and most are either the type that fetch a premium price on the market, which economically justifies the high cost of their culture, or else the type that produce large amounts of biomass under intensive culture. In the former category, we could include certain species of shrimps and abalone cultured in Japan. In the latter category are oysters and mussels. As might be anticipated, estuarine and bottom-dwelling species are more commonly cultivated than pelagic, open-ocean species.

Fishes that are successfully cultured commercially include several species of Pacific salmon *(Oncorhynchus),* mullets *(Mugil, Chanos),* yellowtail *(Seriola),* and flatfishes *(Pleuronectes, Solea).* Crustaceans cultured for market include various species of shrimps and prawns *(Penaeus, Metapenaeus, Leander).* Among the mollusks, oysters *(Crassostrea, Ostrea)* and mussels *(Mytilus)* are the subject of intensive commercial culturing (Fig. 11-6). Other mollusks raised include several clams *(Anadara, Tapes, Mactra, Mercenaria, Meretrix)* and the abalone *(Haliotis).* Compared to the total list of fishes, crustaceans, and mollusks taken for food in various fisheries, this is a very short list. Other species have been cultured experimentally, including lobsters, scallops, and crabs, but at present, no commercial industry exists. This suggests that there is considerable room for future expansion.

YIELDS

Even though aquaculture has lagged far behind agriculture in development, its potential for production of human food even in its current low state is

Figure 11-6 Oyster culture racks in a Pacific Coast estuary. (Photo by the author.)

considerable. For example, Bardach (1968) has noted that the unmanaged common oyster grounds along the east coast of the United States yield only about 10 kg of oysters per hectare, while the intensive culture of oysters in hanging racks in Japan yield 58,000 kg/hectare! Even larger yields are possible with mussels. In Spain, an intensive culture on hanging racks produces 300,000 kg/hectare.

A similar situation occurs with fishes. In Indonesia, the milk fish *(Chanos chanos)* is cultured in brackish ponds, but is not subjected to such intensive care as in Taiwan. Yield in Indonesia is 400 kg/hectare, but 2,000 kg/hectare in Taiwan.

Presently, the world yield from aquacultural practices is probably over 4 million tons, but most is from fresh water. Thus, mariculture currently contributes little human food yield. However, the science of mariculture is as yet at a low level of development. If the above examples are any indication of what can be expected, it would appear that the potential future yield could be substantial; however, it is not great for reasons discussed below.

PROBLEMS AND RESTRICTIONS TO MARICULTURE DEVELOPMENT

Mariculture is not a panacea for the problem of feeding the human race, nor is it likely to contribute significantly to food supplies in the immediate future. The reasons for this are numerous; space permits mentioning only the more significant. Marine animals require, first of all, that the water in which they live has the proper physical and chemical characteristics. This is difficult to maintain in ponds or tanks and requires complex filtering and water treatment to remove wastes and other toxic materials. Thus, attempts to raise marine organisms on a large scale mean an often economically crippling investment in equipment to maintain the water quality. Another drawback is that many marine species go through a series of larval stages, each requiring different conditions and food, before attaining adult size. To successfully rear these forms is often too costly or simply impossible in captivity. Still other species are pelagic in the open ocean and will not survive when confined in smaller spaces. Since we cannot control animals in the pelagic areas, we cannot culture there. Finally, there is the problem of various diseases and parasites, which tend to proliferate under captive or crowded conditions.

What this means then, is that present and future mariculture appears most practical for those species with fewer or no larval stages, those that are nonpelagic, and those that live and reproduce well in crowded conditions. Furthermore, our inability to control conditions or organisms in the open ocean means that future mariculture and increased yields must come from sheltered inshore bays, lagoons, and estuaries. Since the areas suitable for such culturing are a very minor part of the total world oceans, even at maximum culturing intensity, we cannot expect really substantial quantities of food from mariculture in the near future.

POLLUTION

The seas have been considered as the ultimate appropriate dumping grounds for the wastes of human societies. In the same way that fishery stocks were once considered inexhaustible, so we have felt until recently that the immense volume of the world's oceans had an infinite capacity for absorbing all of our waste materials. Dilution, in other words, was the solution to pollution. In recent years, we have come to realize that however large, the oceans are not infinite in their capacity to absorb wastes and that certain of our wastes in very small amounts may have significant effects on communities and species. It is the purpose of this section to briefly discuss a few of the more significant sources of marine pollution and their effects or potential effects on marine ecosystems. For more extensive information on pollution, the student is referred to the references at the end of this chapter.

OIL

Extensive media coverage over the last 15 years has certainly made the oil pollution problem in the seas widely known to the public. Oil pollution in the seas results primarily from the spillage of crude oil from offshore drilling platforms or from accidents involving tankers. What is less well known is that there are a few marine environments subject to natural oil contamination (Coal Oil Point in Santa Barbara, California, for example).

Crude oil released into the sea usually floats, although some components may sink, and after a time, evaporation of certain fractions may cause even more to sink. Away from land, the floating oil probably has little effect on the majority of planktonic and nektonic organisms. The exceptions are the marine birds and certain mammals such as sea otters. When birds contact the oil, they become coated, and their feathers lose their insulating qualities. As a result, most will die of exposure in the water or may be unable to feed, with the same result. Oil is therefore devastating to marine birds, causing significant losses.

Inshore, the oil may coat the shallow subtidal and intertidal and smother the communities. The devastation is nearly complete initially, but recovery usually occurs with time. More serious than the oil itself have been the various chemicals such as detergents used to break up or disperse the oil in the water. As Nelson-Smith (1973) notes, these were shown in the *Torrey Canyon* disaster in 1967 to have caused more mortality of marine organisms than the oil itself.

Oil disasters have become more numerous with the increased oil demand of the industrialized world, with the necessity of transport of the oil from distant sources, and with increased numbers of offshore drilling rigs.

The immensity of oil disasters involving the giant supertankers is revealed by the sinking of the *Amoco Cadiz* off the Brittany coast of France in March 1978. In this case, 220,000 tons of oil spilled and one-third reached shore to contaminate 320 km (192 miles) of shoreline. The spill contaminated or destroyed major commercial marine products, including oyster beds, mussel beds, and lobster holding pens. As of this writing, it is not known how long it will take the area to recover. Until recently, we had experienced no such extensive spills from offshore wells, though they are potentially more devastating, since the amount of oil gushing out is not limited by the hold of a ship. Such a disaster happened in the Gulf of Mexico where, in 1979, the Ixtoc I well blew out and spewed out 140 million gallons of oil in 295 days before being capped. It was the world's greatest oil spill. Only a few of these larger disasters per year could add up to significant destruction, particularly of shallow water communities. We can expect more such disasters in future years with increased oil shipping and increased offshore drilling.

SEWAGE AND GARBAGE

The discharge of human sewage and garbage into coastal waters is practiced throughout the world. The sewage may or may not have had some treatment before discharge. Sewage adds a large volume of small particles to the water and also large amounts of nutrients. In small volumes and with adequate diffusing pipes, it is difficult to detect any long-term effect on the communities of the open coast. In large volumes and in semienclosed embayments, the effect can be devastating. Two examples should suffice.

In southern California, the Los Angeles area discharges in excess of 330 million gallons of sewage per day at the White's Point Outfall off the Palos Verdes Peninsula. Extensive studies in and around this outfall and others in the area over a period of years by personnel of the Southern California Coastal Water Research Project (Bascom, 1978) have revealed that sewage has caused significant degradation in benthic invertebrate communities in areas near the outfall when compared to similar areas some distance away. In addition, the kelp beds in the vicinity of the outfall have disappeared; sea urchins have markedly increased; and diseased fishes are more prevalent. About 4.6 percent or 168 sq km of the 3,640 sq km southern California mainland shelf has been changed or degraded as a result of sewage discharge from four major outfalls (Fig. 11-7).

Kaneohe Bay on the Island of Oahu is a large, shallow embayment with restricted water circulation that was once known for its flourishing coral reefs. Since World War II, Banner (1974) reports that the area around the bay has been subjected to a tenfold increase in population. As a direct result of this urbanization, the bay has been subjected to massive domestic sewage discharges along with significant siltation from runoff during storms. The result has been the total destruction of the coral reef communities in two-thirds of the bay and their replacement by a massive, smothering growth of the green alga *Dictyosphaeria carvernosa* (Fig. 11-8). More recently, the sewage discharges have been eliminated from the bay and the bay appears to be recovering.

In addition to sewage, large amounts of garbage are dumped into the sea each year from shore and from ships. What effect this material may have on the marine communities is as yet unknown. Evidence of the extent of the garbage

Figure 11-7 Areas near southern California's major sewer outfalls that are "changed" or "degraded." (After W. Bascom, ed., 1978, Coastal water research project annual report, Los Angeles.)

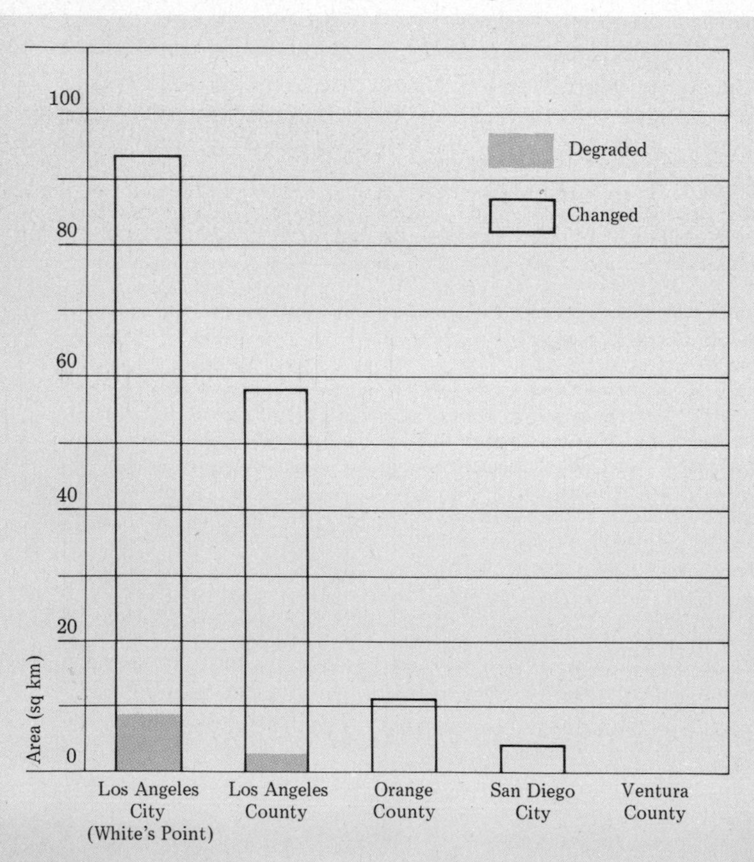

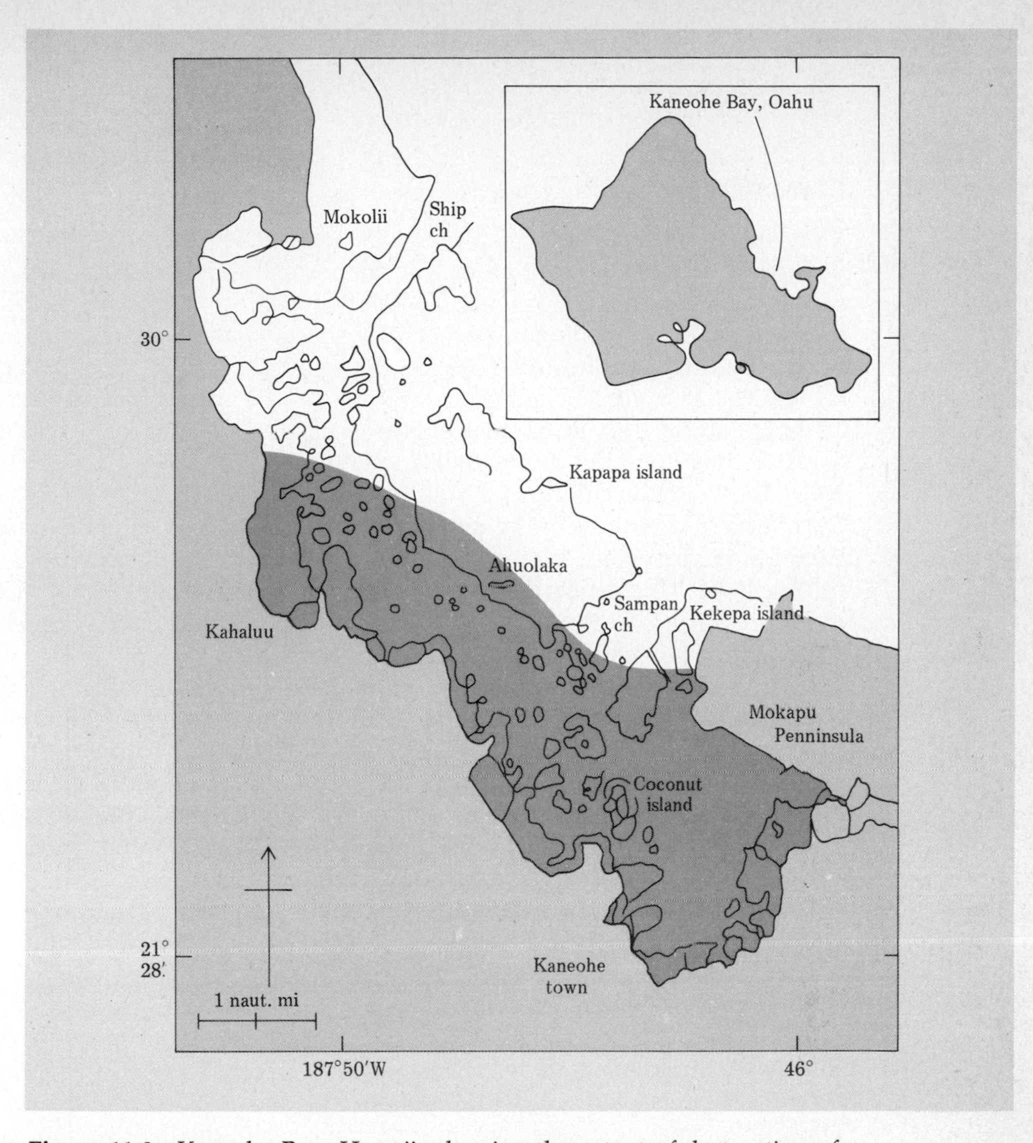

Figure 11-8 Kaneohe Bay, Hawaii, showing the extent of destruction of coral and overgrowth by the alga *Dictyosphaeria*. The shaded portion indicates the area of destruction. (Modified from A. H. Banner, 1974, Kaneohe Bay, Hawaii: Urban pollution and a coral reef ecosystem, *in:* Proceedings of the second international coral reef symposium, vol. 2, Great Barrier Reef Committee, Brisbane. Used by permission of the Great Barrier Reef Committee.)

pollution is found in the occurrence of small bits of plastic in plankton tows taken in many parts of the North Atlantic and Pacific oceans and in the digestive tracts of fishes, diving sea birds, and marine mammals. As yet, there is no evidence that these plastic bits cause mortality, though it has been suggested that they cause intestinal blockage. More ominous is that they do not decompose, and increasing use of plastics by modern human societies probably means increasing amounts in the oceans.

CHEMICALS

More insidious than oil or sewage, which are at least visible, are the various invisible toxic chemicals produced by the industrialized nations that find their way into the oceans' ecosystems. These chemicals are often transferred through the food chains in the sea and exert their effects in animals and places removed in time and in space from the source. Because of this, it is difficult to pin down the effects of a given chemical, especially if the effects turn up years later. Certain marine organisms also enhance the toxic effects of many chemicals because they have the ability to accumulate the substances in their bodies far above that found in the surrounding water. Another factor that tends to increase the effects of chemicals on living systems is *biological magnification,* in which the chemical increases in concentration in the bodies of organisms with succeeding trophic level. This results from the fact that these chemicals are not metabolized in an organism and therefore, the amount accumulated in the tissues remains there. When several such individuals are consumed by a carnivore of the next trophic level, the carnivore, in turn, gains the chemicals from all the individuals, which, in turn, increase the total concentration in its body. Continuation of this process can lead to rather significant levels in the top carnivore, if the food chains are long. As we saw in Chapter One, marine food chains are long, and so top carnivores in marine systems often have heavy loads. Humans also tend to consume marine organisms mainly from the higher trophic levels. This has already had an effect on certain of our fisheries such as swordfish and tuna, in which the levels of the element mercury have often been higher than safe levels established by the FDA for humans, causing fishes to be rejected for consumption. More tragic than that is the case of *Minimata disease,* documented by Goldberg (1974).

In the late 1930s, the Chisso Corporation of Japan established a factory on the shores of Minimata Bay to produce vinyl chloride and formaldehyde. By-products of this plant contained mercury and were discharged into the bay. Through biomagnification, the marine fishes and shellfish accumulated high concentrations of the toxic compound methyl-mercury chloride. The fishes and shellfish were in turn consumed by the inhabitants of the area. About 15 years after the dumping of mercury into the bay began, a strange, permanently disabling neurological disorder began to appear among the inhabitants, especially among the children. This disorder was called Minimata disease. The cause was diagnosed as due to mercury poisoning in 1959, but it was not until the early 1960s that the active mercury compound was identified and the link to the factory discharge established. It took even longer, until the 1970s, for Japan to halt the discharge of mercury into the ocean and for other nations to blacklist mercury dumping and establish standards for acceptable levels in food.

A second, well-documented example is the case of the pesticide DDT. Dichlorodiphenyltrichloroethane (DDT) was the first of the synthetic pesticides of a class known as chlorinated hydrocarbons. It was put into worldwide use in

1945 and was hailed as a boon to mankind for its effectiveness in destroying a wide range of insect pests while remaining relatively nontoxic to humans. It has, however, a few attributes that were ultimately to bring about serious environmental consequences. First, it is a remarkably stable compound in natural systems. It, or its first break-down product, DDE, persists for years; just how long is unknown. It is relatively insoluble in water, but quite soluble in fats or lipids, and it adheres strongly to particles.

Although DDT is not used in the marine environment, it has entered the marine food webs through runoff from land, precipitation, and dumping. As a result, by the 1960s, DDT had been found in marine organisms as remote as Antarctica. The insidious effects of DDT in marine food webs began to be apparent in the 1960s, when certain top carnivores were found to have high levels in their tissues. DDT enters the marine food chains at the level of plankton where the chemical adsorbs on the surfaces of the plankton organisms or is dissolved in the lipid. Since it is not metabolized, biomagnification concentrates it through succeeding trophic levels to the top carnivore. Concentrations of DDT in certain marine fish such as mackerel were found to exceed the permissible 5 ppm established as safe for human consumption by the FDA, and many catches had to be destroyed.

The most serious effects, however, occurred among marine birds. In birds, once DDT reached certain high levels, it had a direct effect in interfering with the calcium deposition in the egg shell. As a result, thin eggs were laid, which broke open when the birds incubated them. The greatest amount of attention was focused on the brown pelican *(Pelicanus occidentalis),* in which successful breeding virtually disappeared in the main rookeries on the islands off southern California in the late 1960s and early 1970s. As Goldberg (1976) notes, this was traced directly to the high levels of DDT in their tissues. What was curious here was that most of southern California is urbanized and had not used much DDT, certainly not as much as that used in agricultural areas. Yet, the waters around Los Angeles had concentrations of DDT averaging 370 ppm as opposed to 1 ppm off Baja California. Where did it come from? The source was discovered to be the White's Point sewer outfall. It turned out that a single chemical plant that produced most of the world's DDT was located in Los Angeles and was dumping its wastes into this outfall at the rate of 100 tons a year, a figure ten times as high as the total amount of DDT carried per year by the Mississippi River into the Gulf of Mexico. Thus, the decline of the brown pelican was linked through the marine food web to a toxic chemical discharged from a single source. Subsequent cessation of DDT dumping in Los Angeles has been followed by the slow recovery of the pelican.

The above are only two examples of the problems of toxic chemicals. They emphasize the insidious nature of the often much delayed effects and the great difficulty of tracing the source of a problem through complex food webs, about which we know relatively little. Since modern industrial states now produce thousands of actually or potentially toxic chemicals each year and many will enter the oceans, how many more of these disasters must we look forward to in the future?

MISCELLANEOUS POLLUTION PROBLEMS

In order to provide docking facilities for commercial and pleasure vessels, humans have engaged in a great deal of dredging of estuaries and bays. Such activities destroy large areas of productive habitat for marine organisms and may have effects beyond the immediate area if nursery grounds for commercial species caught offshore are destroyed. Not only are communities destroyed by physically being removed, but the increased load of silt suspended in the water reduces light, and hence photosynthesis, and clogs the feeding and respiratory surfaces of many invertebrates, leading to their deaths. Such activity has in the past destroyed many productive areas worldwide, and the expanding human population suggests that we can look forward to even more destruction in the future.

The siting of power plants along the sea coasts to take advantage of the marine waters for cooling purposes has led to the perturbation of marine communities by heated water discharge. This *thermal pollution* has been little investigated for the marine environment; indeed, thus far, the effects would appear to be restricted to the communities immediately adjacent to the discharge. The deleterious effects of thermal discharges are most pronounced in the tropics, where the organisms are living naturally at water temperatures that are close to their thermal maximum. It takes little increase in temperature to stress such communities. An example of the effects of a thermal discharge on tropical communities is described by Banus and Kolehmainen (1976) in Guayanilla on the south coast of Puerto Rico, where a power generating station discharges heated water into a lagoon. The discharged water is 8–10°C higher than the ambient sea water and produces water temperatures in the lagoon of 39°C in the summer. This is high enough to affect reproduction of the dominant mangrove trees in the bay, and no young rooted and growing mangrove seedlings have been found in the lagoon.

A final consideration concerns pollution by the introduction of *alien animals.* The devastating consequences on native fauna and flora through the introduction of alien animals and plants is well known and documented for the terrestrial environment. We know, for example, of the loss of the American chestnut through a disease imported from Europe and more recently, the demise of our native elm trees through Dutch elm disease. Equally dramatic have been the devastations of European rabbits on the flora and fauna of Australia. What about introductions in marine waters?

Our knowledge of the effects of introduction of marine species is limited. The major intentional introduction of marine species have been those of commercial or sports fishing value. Thus, the eastern *(Crassostrea virginica)* and Japanese *(Crassostrea gigas)* oysters were introduced into the bays and estuaries of the

Pacific Coast of North America to provide the basis of an oyster industry. Similarly, the striped bass (*Roccus saxatilis*) was introduced to the Pacific Coast from the Atlantic and has become the basis of a valuable sports fishery.

Along with these purposeful introductions have, however, come other, unintentional species. Other inadvertent introductions have occurred from the fouling organisms on ships and in ballast and through commercial bait and seafood importations. The number of these species and their cumulative effects on indigenous communities is very little known, but potentially great. That they may have profound effects is witnessed by recent studies of introduced species on the central California coast. Carlton (1975) has shown that between 150 and 200 marine invertebrates from various parts of the world have become established in the bays and estuaries of the Pacific Coast. In the case of San Francisco Bay, the majority of the invertebrate fauna has been introduced, and it is now nearly impossible to know what the indigenous communities of this great bay were like.

How serious these introductions are as a threat to marine communities around the world remains to be seen, but if the results we have observed in terrestrial communities are any guide, the threat could be considerable. The potential for such introductions in the future is great. If, for example, a sea level canal is put through in Central America, allowing free interchange of the tropical Atlantic and Pacific faunas, massive changes may occur. The giant offshore oil drilling rigs are another source. They accumulate a whole marine fauna in one area during drilling. Then, when they are moved, often to an entirely different area, the whole fauna is brought along, transferred intact.

CONCLUDING REMARKS

For too long, the seas of the world have been considered an inexhaustible source of food, as having an infinite capacity to absorb and purify our wastes, and as a source of all the raw materials needed to maintain an industrial society. It is now apparent that none of these assumptions is true and that the human population at the current level of technological development has the ability to inflict massive destruction on the seas, just as we have done on land. At present, the seas remain in good condition relative to the land, and we cannot afford to permit them to be degraded in the same way if we wish to continue our tenure as a species on this planet. We must employ ecological principles, such as those outlined in this text, in order to ensure that the potential of the oceans is realized—without degradation or reliving yet another tragedy of the commons.

References

Bardach, J. E., J. H. Ryther, and W. O. McLarney. 1972. Aquaculture. Wiley, N.Y. 868 pp.

Cushing, D. 1975. Fisheries resources of the sea and their management. Oxford Univ. Press. 87 pp.

Cushing, D. H. 1977. Science and the fisheries. Studies in biology no. 85. Edward Arnold. 60 pp.

Hardin, G. 1968. The tragedy of the commons. Science 162(3859):1243–1248.

Idyll, C. P. 1970. The sea against hunger. Crowell. 221 pp.

Marine pollution. 1974. A series of articles in *Oceanus* 18(1).

Rounsefell, B. A. 1975. Ecology, utilization and management of marine fisheries. C. V. Mosby, St. Louis. 516 pp.

Russell-Hunter, A. D. 1970. Aquatic productivity. Macmillan, N.Y. 306 pp.

Ryther, J. H. 1969. Photosynthesis and fish production in the sea. Science 166:72–76.

Glossary

abyssal The bottom zone of the oceans between 4,000 and 6,000 m
abyssal gigantism Phenomenon observed among several crustacean groups whereby general size increases with increasing depth
abyssal plain That area of the deep ocean floor lying between 4,000 and 6,000 m
acidity A measure of the concentration of hydrogen ions in a solution
aerobic Condition in which oxygen is present
ahermatypic coral A nonreef-producing coral without the symbiotic zooxanthellae in the tissues
algal ridge Coral-free ridge of encrusting coralline algae lying immediately behind the buttress zone
alien animal Animals that are not indigenous to the area
alkalinity A measure of the concentration of hydroxyl ions in a solution
allochthonous Of foreign origin; transported into the area from outside
anaerobic Condition where oxygen is absent
anchor ice Ice that forms around any convenient nucleus in the area below the permanent pack ice in Antarctica; it tends to carry organisms out of the area to be incorporated in the sea ice above
anchoveta Common name for *Engraulis ringens*
anoxic Without oxygen
aphotic Without light; that area of the oceans without light
apogee That point during the orbit of the moon around the earth when the moon is furthest from the earth
aquaculture The culture of aquatic organisms
aquatic Living or existing within or on water
asexual In reproduction, without involving sex
atoll A type of coral reef that is a modified, ring-shaped reef arising out of deep water far from continental land masses and enclosing a shallow lagoon

autochthonous Formed or occurring in the place where it is found
autotrophic Living organisms capable of producing their own energy resources
auxospore In diatoms, a reproductive cell which reestablishes the initial size of the species
baleen Horny material growing in comblike, fringed units from the upper jaws of whales of the order Mysticeti
barrier reef Reef adjacent to land masses and separated from them by a lagoon or channel of variable extent
bathyscape A free-moving, deep submersible designed to carry human observers into the deep sea
bay A partially enclosed inlet of the ocean
"bends" Name of disease occurring in humans who dive breathing compressed air; results from a rapid decrease in pressure causing nitrogen bubbles to form in the internal tissues and blood vessels
binocular vision Type of vision providing depth-of-field focus due to overlap of the field of vision of two closely set eyes
biochemistry A scientific discipline concerned with the study of the chemistry of living organisms and their products
biogeochemical cycle Cyclical movement of an element or compound through living organisms and the nonliving environment
biological interaction In ecology, a general term in which organisms have a mutual or reciprocal action or influence including predation, competition, and grazing
biological magnification Increase in concentration of chemicals in the bodies of animals with increasing trophic level
bioluminescence The production of light by living organisms
blade In a kelp plant, the flattened part of the plant that terminates the stipe
blubber The layer of lipid that serves as an insulating layer under the skin of whales and other marine mammals
blue-green algae Group of prokaryotic photosynthetic organisms of the Kingdom Monera
boiling point Temperature at which a liquid changes to a gas
bradycardia Slowing of the heart rate
buffer A chemical solution that resists or dampens changes in pH upon the addition of acids or bases
carbohydrate Group of biochemical compounds composed of carbon, hydrogen, and oxygen, often in the ratio of 1 carbon: 2 hydrogen: 1 oxygen
Carnivora An order of mammals adapted to feed on other animals
carnivore An animal that consumes other animals as food
caudal The tail or posterior end of an organism
caudal peduncle In fishes, that part of the body immediately in front of the tail
cellulose A long-chain carbohydrate composed of repeated units of the sugar glucose

centrifugal force Force tending to pull objects out from the center of rotation
Cetacea The order of mammals that contains the whales and porpoises
chemosynthetic bacteria Those bacteria able to obtain energy and therefore synthesize organic material through oxidation of reduced sulfur compounds
chitin Biochemically, a polymer of the carbohydrate glucosamine that forms the hard outer integument of crustaceans and other marine invertebrates
ciguatera Disease of humans caused by consumption of tropical fishes with toxins in the flesh and organs
cleaning behavior Special category of symbiosis in which large animals, usually fishes, permit themselves to be cleaned of various parasites by smaller fishes or invertebrates
climax In ecology, the final stage of a successional sequence that is able to persist in the absence of environmental change
clone A group of genetically identical individuals of a plant or animal species produced by asexual reproduction from a single sexually produced individual
coastal plain estuary Estuary formed by flooding of a low-lying coastal valley due to rising sea level
coccolithophore Small, unicellular, flagellated algae with an external covering of small pieces of calcium carbonate
cohesion Mutual attraction of similar molecules, which resists external forces that would break them
commensalism A symbiotic relationship between two different species in which one benefits and the other is neither benefited or harmed
compensation depth Depth at which the processes of photosynthesis and respiration are equal
compensation intensity In the water column, that point at which light intensity is equal to 1 percent of surface intensity
competition The interaction among organisms for a necessary resource that is in short supply
competitive exclusion The ecological principle that states that complete competitors cannot coexist
competitive interference Exclusion of one species by another species through interruption of its normal activities
continental shelf The shallow underwater extension of a continent; usually limited in depth to 200 m
continental slope The steeply descending bottom between the edge of the continental shelf and the abyssal plain; the ocean bottom between the depths of 200 and 4,000 m
copepod Small crustacean of the order Copepoda; the dominant planktonic herbivore
copepodite The larval stages of a copepod, which follow the nauplius and metanauplius stages and precede the adult

coral reef Massive limestone structure built up through the constructional cementing and depositional activities of anthozoans of the order Madreporaria and certain other invertebrate and algal species
corallite The cup into which a coral polyp fits
Coriolis force The deflection imparted to moving water masses due to spinning of the earth on its axis. The deflection is to the right in the Northern Hemisphere and to the left in the Southern Hemisphere.
cosmopolitan In biogeography, an organism that is distributed throughout the world in suitable habitats
crinotoxin A poisonous material secreted onto the surface of an organism
critical depth The depth at which photosynthesis for the water column is equal to respiration for the water column; the depth to which phytoplankton cells may be mixed and still spend sufficient time above the compensation depth so that respiration and photosynthesis are equal
critical tide level Points on the shore where a small change in vertical distance results in a disproportionate change in exposure time
cropper In the disturbance theory of Dayton and Hessler, the term applies to deep-sea animals that consume living and dead animals smaller than themselves
cryptic coloration Color hue and/or pattern which mimics the background
current Water movements that result in the horizontal transport of water masses
cyanellae Symbiotic blue-green algae
Cyanophyceae A class of blue-green algae
decomposer An organism that breaks down dead protoplasm, freeing simple chemical substances for use by other organisms
deep scattering layer Concentration of midwater organisms that reflect sound waves produced by sonar devices
deep sea A general term referring to that area of the ocean beyond the continental shelf and below the lighted zone
demersal Living close to the bottom of the sea
density The mass per unit volume of a substance (physics); the number of individuals per unit area (biology)
deposit feeder An animal that feeds by consuming particles on or in the substrate
desiccation The process of losing water
diatom Microscopic autotrophic organism of the algae class Bacillariophyceae characterized by being enclosed in a two-part siliceous capsule
diel Daily or once every 24 hours
dinoflagellate Microscopic organism of the order Dinoflagellata possessing two locomotory flagellae
disturbance theory A hypothesis for explaining the high diversity in the deep sea that suggests that extreme predation pressure prevents any species

from becoming numerous enough to eliminate others by competition, therefore perpetuating a large number of species and high diversity

diurnal Occurring daily or relating to daytime

diurnal tide Tide with a single high and low each day

divergence Surface horizontal flow of water away from a central area, permitting water to upwell

diversity See species diversity

drag Resistance to movement through a medium

echolocation See sonar

ecotone An area of intergradation between two biological communities or associations

ectothermic Cold-blooded; unable to regulate body temperature

electrical conductivity Relative ability of a material to allow the passage of electricity

electron A negatively charged elementary particle in all matter

element A substance composed of atoms having the same atomic number

elutriation Means of separating interstitial fauna from the sand through constant stirring with water and subsequent filtering on a fine mesh

endothermic Warm-blooded; able to regulate internal body temperature

endozoite An animal living symbiotically inside another animal

energy The capacity to do work

epicercal tail A caudal fin of fishes with the dorsal lobe larger than the ventral lobe

epifauna Benthic organisms that live on or move over the substrate surface

epiphytic Plants that are not parasitic, but that are attached to other plants

epizoite An animal that is not parasitic, but that lives attached to another animal

equilibrium species A species that has a life history characterized by long life, long development time to reach maturity, low death rates, and few reproductive periods per year

estuary A partially enclosed coastal embayment where fresh water and sea water meet and mix

euphotic The lighted uppermost zone of the world's ocean, where photosynthesis occurs

euryhaline Able to tolerate wide fluctuations in salinity

extinction coefficient Ratio between the intensity of light at a given depth with intensity at the surface

fin A flattened appendage of an aquatic animal used in locomotion or maneuvering in the water

fishery The place or act of taking fishes or other sea organisms for human use

fjord Estuaries occurring in deep, drowned valleys, originally cut by glacial action; characterized by the presence of a shallow sill at the mouth, restricting water interchange between ocean water and deeper fjord waters

flipper The pectoral or pelvic appendage of a marine mammal or reptile
flushing time Time interval required for a given mass of fresh water to be discharged from the estuary
form resistance Condition where drag is proportional to the cross-sectional area of the object in contact with the water
frictional resistance Condition where drag is proportional to the amount of surface area in contact with the water
fringing reef A coral reef that forms immediately adjacent to a land mass
fugitive species See opportunistic species
gametophyte In algae, the haploid plant that produces gametes
gas bladder Structure on dorsal side of body cavity of bony fishes that contains gas and is used by the fishes to regulate buoyancy
geomorphology Discipline dealing with the form and configuration of the surface of the earth
gorgonians Anthozoans of the subclass Octocorallia that have spiculate internal skeletons; commonly called sea fans or sea whips
gravity The attraction of terrestrial bodies toward the center of the earth (= the earth's gravitational attraction)
grazer An animal that feeds upon plants or other sessile animals
gross photosynthesis The total amount of photosynthesis before subtracting losses due to respiration
guyot A submerged, isolated, flat-topped mountain
gyre Circular motion of water in the major ocean basins
hadal Parts of the ocean below 6000 m
halophyte A plant that is adapted to grow in soils with high salt concentrations
harem In mammals, a group of females controlled by a single male during the breeding season
heat Energy which moves from a higher to a lower temperature system
heat capacity Amount of energy required to raise 1 g of material 1°C
heat of vaporization Amount of energy required to evaporate a unit mass of a substance at constant temperature and pressure
herbivore An animal that consumes plants
hermatypic coral A reef-building coral with symbiotic zooxanthellae in the tissues
heterotrophic Organisms that require an external source of food
holdfast In kelp, that part of the plant that fixes it to the substrate
holoepipelagic Nektonic animals that spend their entire lives in the open ocean
holoplankton Plankton organisms that spend their entire lives in the plankton
homoiothermic See endothermic
host The major provisioning partner of a symbiotic or commensal relationship
hydrodynamic The physical features of water motion
hydrogen bond Weak bond formed between a hydrogen atom in one molecule and an electronegative atom of the same or another molecule

hydrographic The physical and chemical features of the oceans

infauna Benthic organisms that live in or burrow through the bottom sediment

infralittoral fringe In the Stephenson Universal Zonation scheme, the lowest zone on the shore bounded above by the upper limit of laminarians and below by the lowermost tide level

inquilinism Subcategory of commensalism in which the animal lives in the home or digestive tract of another without being parasitic

insolation Being exposed directly to the sun's rays

interstitial The space between adjacent particles in a soft bottom

interzonal fauna Mesopelagic animals that migrate vertically into other zones

ion A positively or negatively charged atom or molecule produced through loss or gain of one or more electrons

isohaline Lines that join points of equal salinity concentration

isotherm Lines adjoining points of equal temperature

keel A median longitudinal ventral ridge in certain nektonic fishes

keystone species A species that is disproportionately important in the maintenance of community integrity, and without which drastic alterations of the community would occur

krill Colloquial name for crustaceans of the order Euphausiacea

lagoon (a) Type of estuary built up through the cutting off of inshore waters by the buildup of sand-bars parallel to the shore; (b) a shallow stretch of water separated from the open ocean by a coral reef or island

latent heat of fusion Increase in heat content that accompanies converting a unit mass of a substance from a solid to a liquid state

law of thc minimum Law stating that of those essential substances that are necessary to the survival of an organism, the one that is present in the smallest quantities or is reduced to a minimum first will limit the growth and survival of the organism, even if all others are plentiful

lecithotrophic larvae Planktonic larvae that do not feed in the plankton

leeward The side of an island or reef protected from winds and waves

light Electromagnetic radiation with wavelengths between 400 and 770 μm

lithothamnion ridge Synonym for algal ridge

mackerel Common name for several species of economically important fishes of the family Scombridae

macrofauna A general term referring to benthic organisms more than 1 mm in size

macroplankton Plankton with a size range from 0.2 to 2 mm

Madreporaria The order of the class Anthozoa that contains the corals

mangal A variety of tropical inshore communities dominated by several species of shrubs or trees that have the ability to grow in salt water

mangrove Common name for any of several species of inshore tropical trees or shrubs that dominate the mangal associations

mariculture The culture of marine organisms

maximum sustainable yield The largest number of fishes that can be harvested year after year without diminishing the stock
megaplankton Plankton with a size range above 2 mm
meiofauna Benthic organisms within the size range of 1 to 0.1 mm; often used synonymously with interstitial fauna
melon The large lipid-filled body in the head of whales that serves as an acoustical lens
meroepipelagic Open-ocean nekton that spend part of their lives in the open ocean and part in other areas
meroplankton Planktonic organisms that spend only part of their life cycles in the plankton
mesopsammon Organisms living between sand grains in fresh water
metabolism A general term referring to all physical and chemical processes that occur in a living organism
metabolite A biochemical produced by an organism and secreted into the surrounding medium
metamorphosis The process of structural change in an animal changing from larva to adult
microfauna Benthic organisms below the size of 0.1 mm
microplankton Plankton with a size range from 20 to 0.2 μm
midlittoral zone In the Stephenson Scheme of Zonation, that area between the upper limits of laminarians and the upper limits of barnacles
migration Periodic movement of animals from one place to another
mimic An organism that assumes the shape, pattern, or behavior of a different species, usually for protection
Minimata disease Crippling neurological disorder caused by mercury poisoning
mixed tide Tides consisting of a mixture of diurnal and semidiurnal tides
multivore See omnivore
mutualism That form of symbiosis in which two species associate to their mutual benefit
myoglobin An iron-containing protein compound in muscle, which serves as an oxygen carrier and reserve oxygen supply
nanoplankton Plankton with a size range from 2 to 20 μm
nauplius First larval form of most crustaceans; characterized by possessing three pairs of appendages
negative estuary Estuary in which evaporation is high and fresh water input low, so that salt water enters at the surface, evaporates, becomes hypersaline, and sinks, forming an outflowing bottom current
nekton Pelagic animals that are powerful enough swimmers to move at will in the water column
net photosynthesis The amount of photosynthesis in excess of respiration
net plankton Organisms caught in a standard zooplankton net (above 0.2 mm)

net production Amount of total production left after losses from respiration (or that amount left to support other trophic levels)
neutral In chemistry, having equal numbers of OH^- and H^+ ions; a pH of 7
nonpelagic development Development without a planktonic period
nucleus The central, dense, positively charged part of an atom
omnivore An organism that consumes both plant and animal material (synonym = multivore)
ooze A deep-sea sediment composed, at least partially, of the skeletal remains of certain pelagic organisms
opportunistic species A species that has a life history characterized by short life span, short development time to reach maturity, high death rate, and many reproductive periods per year
osmoconformer An organism unable to regulate its internal fluid and salt balance, and therefore one with a varying internal salt concentration
osmoregulation Physiological activity within an organism that serves to maintain the internal salt and fluid balance within a narrow acceptable range
osmosis The movement of water across a semipermeable membrane separating two solutions of differing solute concentrations, movement being from the more dilute solution to the more concentrated
oxygen minimum zone An area in the ocean, usually between 500–1,000 m, where oxygen values approach zero
parallel bottom community Concept that similar sediment types at similar depths around the world contain similar communities of organisms in which the dominant animals are closely similar taxonomically and ecologically
parasitism Association in which one species lives on, or in, another, drawing nourishment from that species at the expense of, or to the detriment of, the other
pectoral Pertaining to the chest region of the body
pelvic Pertaining to the hip region of the body
perigee That point during the orbit of the moon around the earth when the moon is closest to the earth
Petersen grab A quantitative grab designed by C. G. Joh. Petersen to take a fixed area of soft bottom, usually a tenth or a half of a square meter
pheromone A chemical secreted by a species that influences the behavior of others of the same species
photic The lighted area of the ocean; the epipelagic zone
photophore A complex organ of certain deep-sea animals that produces light and regulates its emission and use
photosynthesis Process occurring in green plants, whereby they use the energy of sunlight to synthesize energy-rich organic compounds from carbon dioxide and water
phototaxis Movement of an organism in response to light

physoclist In bony fishes, those fishes with no open duct between the gas bladder and the digestive tract
physostome In bony fishes, those fishes with an open duct between gas bladder and digestive tract
phytoplankton Planktonic plants
Pinnipedia An order of mammals containing the seals, sea lions, and walruses
plankton Those organisms free-floating or drifting in the open water of the oceans having their lateral and vertical movements determined by water motion
plankton net Fine-mesh conical nets dragged in the water to collect plankton
planktotrophic larvae Larvae that feed in the plankton
planula The sexually produced larvae of many cnidarians, including corals
pneumatocyst The gas-filled floats on certain kelps
poikilothermic See ectothermic
polar The Arctic or Antarctic zones
pollution Degradation of the natural environment
polyp A sessile or sedentary cnidarian individual with a cylindrical body and, usually, tentacles surrounding a mouth at the free end
positive estuary Estuary with substantial fresh water input and reduced evaporation to the extent that fresh water flows out at the surface and salt water flows in at the bottom; also known as a salt wedge estuary
preadaptation Possession by a species or group of species of features that favor easy occupation of a new habitat
predator An animal that feeds upon another animal
protein Complex, large molecular weight biochemical compounds composed of amino acids joined together
psammon A general term referring to organisms living interstitially
pycnocline That area in the water column where the highest rate of change in density occurs for a given change in depth
red tide The name given to massive blooms of dinoflagellates in which the concentrations of the plants are such that the water becomes a red-brown in color, toxins often being secreted in sufficient quantities to affect other marine organisms
refractive index For a medium, the ratio of the speed of light in a vacuum to that in the medium
rete mirabile A large network of anastomosing small blood vessels serving a number of functions in marine vertebrates
rhodopsin A light-sensitive pigment found in the rods of the eye
rod The sensory body of the retina that is most responsive to low light levels
salinity The total amount of dissolved material (salts) in sea water
salt marsh Communities of emergent vegetation rooted in soils that are alternately inundated and drained by tidal action
Scleractinia Synonym for Madreporaria

sea grass Collective name for marine flowering plants of the families Potamogetonaceae and Hydrocharitaceae
sea mount See guyot
seasonal succession The change of dominant species in the phytoplankton with succeeding seasons of the year
sedentary In reference to organisms, those which remain for long periods in one place or have limited movement
semidiurnal tide A tide with two highs and two lows per lunar day
sere One of the stages in a successional sequence leading to a climax community
sessile An organism that is fixed in place
silicoflagellate Flagellated photosynthetic organism of the class Chrysophyta with an internal skeleton of silicon dioxide
Sirenia The order of the class Mammalia that contains the sea cows (dugongs and manatees)
slough See lagoon
sonar A method employing sound waves by which objects may be located in the water column
species diversity In ecology, a numerical measure combining the number of species in an area with their relative abundance
species richness The number of species in a given area
spermaceti organ The melon of the sperm whale
spermatophore A special packet that contains sperm
sporophyte In algae, the diploid plant that produces spores
stability-time hypothesis Theory seeking to explain the occurrence of high diversity among communities of the marine environment. The hypothesis states that high diversity occurs because highly stable environmental conditions have persisted over long periods of time allowing numerous, highly specialized species to evolve.
standing crop The biomass of an organism or group of organisms present per unit volume or per unit area at a given point in time
stenohaline Able to tolerate only a narrow range of salinity changes
stipe That part of a kelp plant between blade and holdfast
subsidence theory Darwin's theory as to the origin of coral atolls whereby atolls begin as fringing reefs around volcanic islands. Subsequent subsidence of the island, coupled with upward reef growth, then produces first a barrier reef and finally, with the disappearance of the volcano, an atoll.
succession In ecology, the gradual and predictable replacement of communities in a given area brought about by the modification of the environment by organisms
supralittoral fringe In the Stephenson Universal Zonation scheme, the highest zone on the shore bounded below by the upper limit of barnacles and above by the upper limit of *Littorina*

surface tension With respect to water, the attraction of the molecules of water at the surface (air-water interface) so that the surface resists penetration by small objects
surge channel Deep channels in the windward face of a coral reef through which water moves in and out of the reef
suspension feeding Feeding by filtering particles out of the water column
swim bladder See gas bladder
symbiont Name given to the partners in a mutualistic relationship
symbiosis The interrelationship between two different species
tectonic estuary Estuary formed when the sea reinvades the land due to subsidence of the land as a result of earthquake activity
terrigenous Ocean sediments derived from the land
territory An area occupied and protected by a species against others of its own species, usually for breeding or feeding purposes
thalassopsammon The interstitial organisms in marine waters
thermal pollution Degradation of the environment through increases in temperature
thermocline That portion of the water column where temperature changes most rapidly with each unit change in depth
thigmotaxis Behavioral or movement response triggered by contact with a solid body
tide The periodic rise and fall in the surface water of the oceans due to gravitational attraction of the sun and moon and the rotation of the earth
tidepool Area in the rocky intertidal that retains some volume of water at low tide
toxin Any of a variety of chemical substances produced by organisms that are, in turn, poisonous to other organisms
trenches The narrow, steep-sided depressions in the ocean floor usually lying between 6,000 and 10,000 m in depth
trophic group amensalism Exclusion of one group of organisms by modification of the environment by another
trophic level An ecological system of classification of organisms according to their means of obtaining nutrition, the basic levels being autotrophs, the second herbivores, and the succeeding levels being carnivores
tsunami A very long period wave generated by an underwater earthquake, landslide, or volcanic eruption; also known as a tidal wave
tuna The common name for any of several large pelagic fishes of the family Thunnidae
turbidity Condition of reduced visibility in water due to the presence of suspended particles
ultra-abyssal See hadal
ultraplankton Plankton less than 2 μm in size
vernal The spring of the year

vertical migration Diurnal vertical movement of pelagic organisms in the water column toward the surface at night and down to depth during the day
viscosity That property of a liquid to resist movement through it
vitamin An organic compound essential to certain metabolic processes in organisms
wave With respect to the oceans, a disturbance that moves through the water but does not cause particles of the water to advance with it
whalebone See baleen
windward The side of an island or reef that faces into the prevailing wind
zoochlorellae Symbiotic algae that are green in color
zooplankton Planktonic animals
zooxanthellae Dinoflagellates symbiotic in the tissues of various marine invertebrates; symbionts that are brownish in color

Additional References

CHAPTER TWO

Cushing, D. H. 1959. On the nature of production in the sea. Fish. Invest. Lond. Series II. 22(6):1–40.

Hardy, A. C. 1953. Some problems of pelagic life, pp. 101–121. *In:* Essays in marine biology (Richard Elmhurst Memorial Lectures). Oliver and Boyd, Edinburgh. 144 pp.

Hardy, A. C., and E. R. Gunther. 1935. The plankton of the south Georgia whaling grounds and adjacent waters, 1926–27. Discovery Repts. 11:511–538.

Hart, T. J. 1942. Phytoplankton periodicity in Antarctic surface waters. Discovery Repts. 21:263–348.

Harvey, H. W., L. H. N. Cooper, M. V. Lebour, and F. S. Russell. 1935. Plankton production and its control. J. Mar. Biol. Assoc. 20:407–441.

Mare, M. F. 1940. Plankton production off Plymouth and the mouth of the English Channel in 1939. J. Mar. Biol. Assoc. 24:461–482.

Margalef, R. 1963. Succession in marine populations, pp. 137–188. *In:* Vira, R. (ed.). Advancing frontiers of plant sciences, Vol. 2.

McAllister, C. D. 1969. Aspects of estimating zooplankton production from phytoplankton production. J. Fish. Res. Bd. Can. 26:199–220.

McLaren, I. A. 1963. Effects of temperature on growth of zooplankton and the adaptive value of vertical migration. J. Fish. Res. Bd. Can. 20:685–727.

Riley, G. A. 1946. Factors controlling phytoplankton populations on Georges Bank. J. Mar. Res. 6:54–73.

CHAPTER THREE

Brooks, W. S. 1917. Notes on some Falkland Island birds. Bull. Mus. Comp. Zool. 61:135–160.

Denton, E., and N. Marshall. 1958. The buoyancy of bathypelagic fishes without a gas-filled swim bladder. J. Mar. Biol. Assoc. 37(3):753–767.

Elsner, R. 1969. Cardiovascular adjustments to diving, pp. 117–145. *In:* Andersen, H. T. (ed.). The biology of marine mammals. Academic Press, N.Y.

Kellogg, W. N. 1958. Echo ranging in the porpoise. Science 128:982–988.

Norris, K. S., J. H. Prescott, P. V. Asa-Dorian, and P. Perkins. 1961. An experimental demonstration of echo locating behavior in the porpoise, *Tursiops truncatus* (Montagu). Biol. Bull. 120:163–176.

Norris, K. S. 1969. The echolocation of marine mammals, pp. 391–423. *In:* Andersen, H. T. (ed.). The biology of marine mammals. Academic Press, N.Y.

Scholander, P. F. 1940. Experimental investigations on the respiratory function in diving mammals and birds. Hvalradets Skrifter Norske Videnskaps-Akad., Oslo. 22:1.

CHAPTER FOUR

Ballard, R. D. 1977. Notes on a major oceanographic find. Oceanus 20(3):35–44.

Blackburn, M. 1977. Studies on pelagic animal biomasses, pp. 283–299. *In:* Andersen, N. R., and B. J. Zahuranec (eds.). Oceanic sound scattering prediction. Plenum, N.Y.

Childress, J., and M. Nygaard. 1973. The chemical composition of midwater fishes as a function of depth of occurrence off southern California. Deep-Sea Res. 20:1093–1111.

Childress, J., and M. Nygaard. 1974. Chemical composition and buoyancy of midwater crustaceans as a function of depth of occurrence off southern California. Mar. Biol. 27:225–238.

Clarke, T. S. 1973. Some aspects of the ecology of lantern fishes (Myctophidae) in the Pacific Ocean near Hawaii. Fish. Bull. U.S. 71:401–434.

Clarke, T. S. 1974. Some aspects of the ecology of stomiatoid fishes in the Pacific Ocean near Hawaii. Fish. Bull. U.S. 72:337–351.

Dayton, P., and R. H. Hessler. 1972. The role of biological disturbance in maintaining diversity in the deep sea. Deep-Sea Res. 19:199–208.

Hedgpeth, J. 1957. Classification of marine environments, chpt. 2, pp. 18–27. *In:* The treatise on marine ecology and paleoecology, Vol. I, Ecology. Memoir 67, Geol. Soc. of Amer.

Hopkins, T. L., and R. C. Baird. 1977. Aspects of the feeding ecology of oceanic midwater fishes, pp. 325–360. *In:* Andersen, N. R., and B. J. Zahuranec (eds.). Oceanic sound scattering prediction. Plenum Press, N.Y.

Mauchline, J. 1977. Estimating production of midwater organisms, pp. 177–215. *In:* Anderson, N. R., and B. J. Zahuranec (eds.). Oceanic sound scattering prediction. Plenum, N.Y.

Menzies, R. J. 1964. Improved techniques for benthic trawling at depths greater than 2000 meters. Biol. Antarctic Seas. Antarctic Res. Ser., Amer. Geophys. Union 1:93–109.

Rokop, F. 1974. Reproductive patterns in the deep-sea benthos. Science 186:743–745.

Sanders, H. L. 1968. Marine benthic diversity, a comparative study. Amer. Natur. 102:243–282.

Sanders, H. L., and R. R. Hessler. 1969. Ecology of the deep-sea benthos. Science 163:1419–1424.

Siebenaller, J., and G. N. Somero. 1978. Pressure adaptive differences in lactate dehydrogenases of congeneric fishes living at different depths. Science 201(4352):255–257.

Young, R. E. 1975. Function of the dimorphic eyes in the midwater squid *Histioteuthis dofleini*. Pac. Sci. 29(2):211–218.

Zimmerman, A. M., and S. B. Zimmerman. 1972. Commentary on high pressure effects on cellular systems, pp. 140–147. *In:* Brauer, R. W. (ed.). Barobiology and the experimental biology of the deep sea. North Carolina Sea Grant Program, Univ. of North Carolina. 428 pp.

CHAPTER FIVE

Birkeland, C. 1974. Interactions between a sea pen and seven of its predators. Ecolog. Monogr. 44:211–232.

Crisp, D. J., and P. S. Meadows. 1962. The chemical bases of gregariousness in cirripedes. Roy. Soc. of Lond., Proceedings, Series B. 150:500–520.

Davis, N., and G. Van Blaricom. 1978. Spatial and temporal heterogeneity in a sand bottom epifaunal community of invertebrates in shallow water. Limnol. Oceanogr. 23(3):417–427.

Dayton, P. K., G. A. Robilliard, R. T. Paine, and L. B. Dayton. 1974. Biological accommodation in the benthic community at McMudro Sound, Antarctica. Ecolog. Monogr. 44(1):105–128.

den Hartog, C. 1977. Structure, function and classification in sea grass communities, pp. 89–121. *In:* McRoy, C. P., and C. Helfferich (eds.). Seagrass ecosystems. Dekker, N.Y.

Fenchel, T. 1977. Aspects of the decomposition of seagrasses, pp. 123–146. *In:* McRoy, E. P., and C. Helfferich (eds.). Seagrass ecosystems. Dekker, N.Y.

Kastendiek, J. 1976. Behavior of the sea pansy, *Renilla kollikeri* Pfeffer (Coelenterata, Pennatulacea) and its influence on the distribution and biological interaction of the species. Biol. Bull. 151:518–537.

Levinton, J. S. 1977. Ecology of shallow water deposit feeding communities, Quisset Harbor, Massachusetts, pp. 191–227. *In:* Coull, B. (ed.). Ecology of marine benthos. Univ. of South Carolina Press, Columbia.

MacArthur, R. 1960. On the relative abundance of species. Amer. Nat. 94 (874):25–34.

McRoy, C. P., and C. McMillan. 1977. Production ecology and physiology of sea grasses, pp. 53–88. *In:* McRoy, C. P., and C. Helfferich (eds.). Seagrass ecosystems. Dekker, N.Y.

North, W. 1971. Growth of individual fronds of the mature giant kelp, *Macrocystis,* pp. 123–168. *In:* North, W. (ed.). The biology of giant kelp beds *(Macrocystis)* in California. Beihefte zur Nova Hedwigia Heft 32, Verlag von J. Cramer. 600 pp, 166 figs.

Petersen, C. G. Joh. 1918. The sea bottom and its production of fish food. A survey of the work done in connection with the valuation of the Danish waters from 1883–1917. Rap. Danish Biol. Stat., Vol. 25. 62 pp, 10 pls., 1 chart.

Petersen, C. G. Joh. 1924. A brief survey of the animal communities in Danish waters. Amer. Jour. Sci., Ser. 5. 7(41):343–354.

Peterson, C. H. 1977. Competitive organization of the soft bottom macrobenthic communities of southern California lagoons Mar. Biol. 43:343–359.

Rasmussen, E. 1973. Systematics and ecology of the Isefjord marine fauna (Denmark). Ophelia 11(1–2):1–495.

Rhoads, D. C., and D. K. Young. 1970. The influence of deposit feeding organisms on sediment stability and community structure. J. Mar. Res. 28(2):150–178.

Rosenthal, R. J., W. D. Clarke, and P. K. Dayton. 1974. Ecology and natural history of a stand of giant kelp, *Macrocystis pyrifera*, off Del Mar, California. Fish. Bull. 72(3):670–684.

Scagel, R. F. 1947. An investigation on marine plants near Hardy Bay, B.C. Provincial Dept. Fisheries, Victoria, B.C., Canada. pp. 1–70.

Thorson, G. 1955. Modern aspects of marine level bottom animal communities. J. Mar. Res. 14:387–397.

Thorson, G. 1966. Some factors influencing the recruitment and establishment of marine benthic communities. Neth. J. Sea. Res. 3(2):267–293.

Vance, R. 1973. Reproductive strategies in marine benthic invertebrates. Amer. Nat. 107(955):339–352.

Virnstein, R. W. 1977. The importance of predation by crabs and fishes on benthic infauna in Chesapeake Bay. Ecol. 58:1199–1217.

Wilson, D. P. 1952. The influence of the nature of the substratum on the metamorphosis of the larvae of marine animals, especially the larvae of *Ophelia bicornis* Savigny. Annales de l'Institute Oceanographique Monaco 27:49–156.

Woodin, S. A. 1976. Adult-larval interactions in dense infaunal assemblages: Patterns of abundance. J. Mar. Res. 34:25–41.

CHAPTER SIX

Boyle, P. R. 1969. The survival of osmotic stress by *Sypharochiton pelliserpentis*. Biol. Bull. 136:154–165.

Chia, F. S. 1973. Sand dollar: A weight belt for the juvenile. Science 181:73–74.

Connell, J. H. 1961. The influence of interspecific competition and other factors on the distribution of the barnacle *Chthalamus stellatus*. Ecology 42:710–723.

Connell, J. H. 1970. A predator-prey system in the marine intertidal region. I, *Balanus glandula* and several predatory species of *Thais*. Ecol. Monogr. 40:49–78.

Davis, P. S. 1969. Physiological ecology of *Patella*. III, Desiccation effects. Mar. Biol. Assoc. U.K. 49:291–304.

Dayton, P. K. 1971. Competition, disturbance and community organization: The provision of and subsequent utilization of space in a rocky intertidal community. Ecol. Monogr. 41:351–389.

Dayton, P. 1973. Dispersion, dispersal and persistence of the annual intertidal alga *Postelsia palmaeformis* Ruprecht. Ecology 54(2):433–438.

Dayton, P. K. 1975. Experimental evaluation of ecological dominance in a rocky intertidal algal community. Ecol. Monogr. 45:137–159.

Doty, M. S. 1946. Critical tide factors that are correlated with the vertical distribution of marine algae and other organisms along the Pacific coast. Ecology 27:315–328.

Foster, B. A. 1971. On the determinants of the upper unit of intertidal distribution of barnacles (Crustacea, Cirripedia). J. Anim. Ecol. 40:33–48.

Francis, L. 1973. Intraspecific aggression and its effect on the distribution of *Anthopleura elegantissima* and some related anemones. Biol. Bull. 144:73–92.

Frank, P. W. 1965. The biodemography of an intertidal snail population. Ecology 46:831–844.

Haven, S. B. 1971. Niche differences in the intertidal limpets *Acmaea scabra* and *Acmaea digitalis* (Gastropoda) in central California. The Veliger 13:231–248.

Jones, N. S. 1948. Observations and experiments on the biology of *Patella vulgata* at Port

St. Mary, Isle of Man. Liverpool Biological Society Proceedings and Transactions 56:60–77.
Jones, N. S., and J. M. Kain. 1967. Subtidal algal colonization following the removal of *Echinus*. Helo. Wiss. Meeresunt. 15:460–466.
Kanwisher, J. 1957. Freezing and drying in intertidal algae. Biol. Bull. 113:275–285.
Lubchenco, J. 1978. Plant species diversity in a marine intertidal community: Importance of herbivore food preference and algal competitive abilities. Amer. Nat. 112(983):23–39.
Lubchenco, J., and B. Menge. 1978. Community development and persistence in a low rocky intertidal zone. Ecol. Monogr. 59:67–94.
MacGinitie, G. E., and N. MacGinitie. 1949. Natural history of marine animals. McGraw-Hill, N.Y. 473 pp.
Menge, B. A. 1976. Organization of the New England rocky intertidal community: Role of predation, competition and environmental heterogeneity. Ecol. Monogr. 46(4):355–393.
Paine, R. T. 1966. Food web complexity and species diversity. Amer. Nat. 100:65–75.
Segal, E., and P. Dehnel. 1962. Osmotic behavior in an intertidal limpet, *Acmaea limatula*. Biol. Bull. 122:417–430.
Stephenson, T. A., and A. Stephenson. 1949. The universal features of zonation between tide-marks on rocky coasts. J. Ecol. 37:289–305.
Stimson, J. 1970. Territorial behavior in the owl limpet, *Lottia gigantea*. Ecol. 51:113–118.
Timko, P. 1976. Sand dollars as suspension feeders: A new description of feeding in *Dendraster excentricus*. Biol. Bull. 151(1):247–259.
Vadas, R. L. 1968. The ecology of *Agarum* and the kelp community. Ph.D. dissertation, Univ. of Wash.

CHAPTER SEVEN

Coull, B. C., et al. 1977. Quantitative estimates of the meiofauna from the deep sea off North Carolina, U.S.A. Mar. Biol. 39:233–240.
Fenchel, T., B-O. Jansson, and W. vun Thun. 1967. Vertical and horizontal distribution of the metazoan microfauna and of some physical factors on a sandy beach in the northern part of the Oresund. Ophelia 4:227–243.
Fenchel, T. 1968. The ecology of marine microbenthos. II, The food of marine benthic ciliates. Ophelia 5:73–121.
Gerlach, S. A. 1965. Uber die Fauna in der Gezeitenzone von Spitzbergen. Bot. Bothab. Acta Univ. Gothab. 3:3–23.
Gerlach, S. A. 1971. On the importance of marine meiofauna for benthos communities. Oecologia 6:176–190.
McIntyre, A. D. 1969. Ecology of marine meiobenthos. Biol. Rev. 44:245–290.
Muus, B. J. 1967. The fauna of Danish estuaries and lagoons. Meddr. Danm. Fisk.-og Havunders 5:1–316.
Muus, K. 1966. Notes on the biology of *Protohydra leuckarti* Greef (Hydroidea, Protohydridae). Ophelia 3:141–150.
Tietjen, J. H., and J. J. Lee. 1977. Feeding behavior of marine nematodes, pp. 21–35. *In:* Coull, B. C. (ed.). Ecology of marine benthos. Univ. of South Carolina Press, Columbia.

Uhlig, G. 1968. Quantitative methods in the study of interstitial fauna. Trans. Amer. Microsc. Soc. 87(2):226–232.

CHAPTER EIGHT

Atwater, B. F., et al. 1979. History, land forms and vegetation of the estuary's tidal marshes, pp. 347–386. *In:* Conomos, J. (ed.). San Francisco Bay, the urbanized estuary. Pacific Division AAAS, San Francisco. 493 pp.

Barnes, R. S. K. 1967. The osmotic behaviour of a number of grapsoid crabs with respect to their differential penetration of an estuarine system. J. Exp. Biol. 47:535–551.

Darnell, R. 1961. Trophic spectrum of an estuarine community based on studies of Lake Ponchartrain, Louisiana. Ecol. 45(3):553–568.

Deevey, G. B. 1960. The zooplankton of the surface waters of the Delaware Bay region. Bull. Bing. Ocean. Coll. 17(2):54–86.

Drinnan, R. E. 1957. The winter feeding of the oyster catcher *(Haematopus ostralegus)* on the edible cockle *(Cardium edule)*. J. Anim. Ecol. 26:441–469.

Good, R. E. 1965. Salt marsh vegetation, Cape May, New Jersey. N.J. Acad. Sci. Bull. 10(1):1–11.

Goss-Custard, J. D. 1969. The winter feeding ecology of the redshank *Tringa totanus*. Ibis 111:338–356.

Haines, E. B. 1979. Interactions between Georgia salt marshes and coastal waters: A changing paradigm. pp.35–46. *In:* Ecological Processes in Coastal and Marine Systems, ed. by R. J. Livingston. Plenum Press, N.Y.

Hartley, P. H. T. 1940. The saltash tuck-net fishery and the ecology of some estuarine fishes. J. Mar. Biol. Assoc. 24:1–68.

Kinne, O. 1963. The effects of temperature and salinity on marine and brackish water animals. Oceanogr. Mar. Biol. Ann. Rev. 1:301–340.

Lackey, J. 1967. The microbiota of estuaries and their roles, pp. 291–302. *In:* Lauff, G. H. (ed.). Estuaries. Publ. no. 83. AAAS. Washington, D.C. 757 pp.

Muus, B. J. 1963. Some Danish Hydrobiidae with a description of a new species *Hydrobia neglecta*. Proc. Malac. Soc. London 35:131–138.

Odum, E. P. and A. A. de la Cruz. 1967. Particulate detritus in a Georgia salt marsh-estuarine ecosystem, pp. 383–388. *In:* Lauff, G. H. (ed.). Estuaries. Publ. No. 83. AAAS, Washington, D.C. 757 pp.

Peterson, C. H. 1979. Predation, competitive exclusion and diversity in the soft bottom benthic communities of estuaries and lagoons, pp. 233–264. *In:* Livingston, R. J. (ed.). Ecological processes in coastal and marine systems. Plenum, N.Y.

Prater, A. J. 1972. Ecology of Morecambe Bay. III, The food and feeding habits of knot *(Calidris canutus)* in Morecambe Bay. J. Applied Ecol. 9:179–194.

Remane, A. and C. Schlieper. 1971. Biology of brackish water, 2nd ed. Stuttgart, Schweizerbart'sche.

Teal, J. 1962. Energy flow in the salt marsh ecosystem of Georgia. Ecol. 43:614–624.

VanEngel, W. A. 1958. The blue crab and its fishery in Chesapeake Bay. Commercial Fisheries Review 20(16):6–17.

Zedler, J., T. Winfield, and D. Mauriello. 1978. Primary productivity in a southern California estuary. Coastal Zone 3:649–662.

Zobell, C. E., and C. B. Feltham. 1942. The bacterial flora of a marine mud flat as an ecological factor. Ecol. 23:69–78.

CHAPTER NINE

Bakus, G. 1964. The effects of fish-grazing on invertebrate evolution in shallow tropical waters. Occ. paper no. 27, Allan Hancock Foundation. 29 pp.

Bakus, G. 1973. The biology and ecology of tropical holothurians, pp. 325–367. *In:* Jones, O. A., and R. Endean (eds.). Biology and geology of coral reefs. Vol. II, Biology 1. Academic Press, N.Y.

Banner, A. H. 1976. Ciguatera: A disease from coral reef fish, pp. 177–213. *In:* Jones, O. A., and R. Endean (eds.). Biology and geology of coral reefs. Vol. III, Biology 2. Academic Press, N.Y.

Chesher, R. H. 1969. Destruction of Pacific corals by the sea star *Acanthaster planci.* Science 165:280–283.

Connell, J. 1973. Population ecology of reef building coral, pp. 205–245. *In:* Jones, O. A., and R. Endean (eds.). Biology and geology of coral reefs, Vol. II, Biology 1. Academic Press, N.Y.

Crossland, C. 1952. Madreporaria, Hydrocorallinae, Heliopora, and Tubipora. Sci. Repts. Great Barrier Reef Exped. 1928–29. 6:85–257.

Dana, T. F., W. A. Newman, and E. W. Fager. 1972. *Acanthaster* aggregations: interpreted as primarily responses to natural phenomena. Pac. Sci. 26:355–372.

Endean, R. 1973. Population explosions of *Acanthaster planci* and associated destruction of hermatypic corals in the Indo-West Pacific region, pp. 389–438. *In:* Jones, O. A., and R. Endean (eds.). Biology and geology of coral reefs. Vol. II, Biology 1. Academic Press, N.Y.

Endean, R. 1977. *Acanthaster planci* infestations of reefs of the Great Barrier Reef. Proc. 3rd Inter. Coral Reef Symp. 1:185–191.

Frankel, E. 1978. Evidence from the Great Barrier Reef of ancient *Acanthaster* aggregations. Atoll Res. Bull. 220:75–93.

Franzisket, L. 1969. Riffcorallen können autotroph leben. Naturwissenschaften 56:144.

Glynn P. 1973. Aspects of the ecology of coral reefs in the western Atlantic region, pp. 271–324. *In:* Jones, O. A., and R. Endean (eds.). Biology and geology of coral reefs. Vol. II, Biology 1. Academic Press, N.Y.

Glynn, P., G. M. Wellington, and C. Birkeland. 1979. Coral reef growth in the Galapagos: Limitations by sea urchins. Science 203:47–48.

Goldman, B., and F. Talbot. 1976. Aspects of the ecology of coral reef fishes, pp. 125–154. *In:* Jones, O. A., and R. Endean (eds.). Biology and geology of coral reefs. Vol. III, Biology 2. Academic Press, N.Y.

Goreau, T. F. 1961. On the relation of calcification to primary production in reef building organisms, pp. 269–285. *In:* Lenhoff, H. M., and W. F. Loomis (eds.). The biology of Hydra, etc. Univ. of Miami Press.

Goreau, T. F., and J. W. Wells. 1967. The shallow-water Scleractinia of Jamaica: Revised list of species and their vertical distribution range. Bull. Mar. Sci. 17(2):442–453.

Hobson, E. S. 1968. Predatory behavior of some shore fishes in the Gulf of California. Bur. Sports Fish. and Wildlife (U.S.), Res. Rept. 73. 92 pp.

Johannes, R. E., N. T. Kuenzel, and S. L. Coles. 1970. The role of zooplankton in the nutrition of some scleractinian corals. Limnol. and Oceanogr. 15:579–586.

Kohn, A., and P. Helfrich. 1957. Primary productivity of a Hawaiian coral reef. Limnol. and Oceanogr. 2(3):241–251.

Ladd, H. S., et al. 1953. Drilling on Eniwetok Atoll, Marshall Islands. Bull. Amer. Assoc. Petrol. Geol. 37:2257–2280.

Lang, J. 1973. Interspecific aggression by scleractinian corals, 2, Why the race is not only to the swift. Bull. Mar. Sci. 23(2):260–279.

Lobel, P. S. 1980. Herbivory by damsel fishes and their role in coral reef community ecology. Bull. Mar. Sci. 30:273–289.

Loya, Y. 1976. Recolonization of Red Sea corals affected by natural catastrophes and man-made perturbations. Ecol. 57(2):278–289.

McMichael, D. F. 1974. Growth rate, population size and mantle coloration in the small giant clam *Tridacna maxima* (Roding) at One Tree Island, Capricorn Group, Queensland. Proc. 2nd Inter. Coral Reef Symp. 1:241–254.

Motoda, S. 1940. The environment and life of massive reef coral, *Goniastrea aspera* Verrill inhabiting the reef flat in Palao. Palao Tropical Biol. Sta. Studies 2:61–104.

Muscatine, L., and E. Cernichiari. 1969. Assimilation of photosynthetic products of zooxanthellae by a reef coral. Biol. Bull. 137:506–523.

Odum, H. T., and E. P. Odum. 1955. Trophic structure and productivity of a windward coral reef community on Eniwetok Atoll. Ecol. Monogr. 25:291–320.

Pearson, R. G. 1975. Coral reefs, unpredictable climatic factors and *Acanthaster*. Crown-of-Thorns starfish seminar proceedings, pp. 131–134. Aust. Govt. Pub. Ser., Canberra.

Porter, J. 1972. Predation by *Acanthaster* and its effect on coral species diversity. Amer. Nat. 106(950):487–492.

Porter, J. 1974. Zooplankton feeding by the Caribbean reef-building coral *Montastrea cavernosa*. Proc. 2nd Inter. Coral Reef Symp. 1:111–125.

Sale, P. F. 1977. Maintenance of high diversity in coral reef fish communities. Amer. Naturalist 111:337–359.

Sammarco, P. W., J. S. Levinton, and J. C. Ogden. 1974. Grazing and control of coral reef community structure by *Diadema antillarum* Philippi (Echinodermata: Echinoidea): A preliminary study. J. Mar. Res. 32(1):47–53.

Tschirley, F. H. 1969. Defoliation in Vietnam. Science 163:779–786.

Vaughn, T. W. 1915. The geologic significance of the growth rate of the Floridian and Bahamian shoal-water corals. Jour. Wash. Acad. Science 5:591–600.

Walsh, G. E. 1974. Mangroves, a review, pp. 51–174. *In:* Reimold, R. J., and W. H. Queen (eds.). Ecology of Halophytes. Academic Press, N.Y.

Wells, J. W. 1954. Recent corals of the Marshall Islands, Bikini and nearby atolls. U.S. Geol. Survey Paper 260–I:385–486.

West, R. C. 1977. Tidal salt marsh and mangal formations of middle and south America, pp. 193–213. *In:* Chapman, V. J. (ed.). Wet coastal ecosystems. Elsevier, N.Y.

CHAPTER TEN

Arnold, D. C. 1957. Further studies on the behavior of the fish *Carapus acus*. Pubbl. Staz. Zool., Naples 23:91.

Beers, C. D. 1961. The obligate commensal ciliates of *Strongylocentrotus drobachiensis:* occurrence and division in urchins of diverse ages; survival in sea water in relation to infectivity. Biol. Bull. 121:69–81.
Davenport, D. 1950. Studies in the physiology of commensalism. I, The polynoid genus *Arctonoe.* Biol. Bull. 98(2):81–93.
Gorlick, D. L., P. D. Atkins, and G. S. Losey. 1978. Cleaning stations as water holes, garbage dumps and sites for evolution of reciprocal altruism. Amer. Nat. 112(984):341–353.
Greene, R. W. 1970. Symbiosis in sacoglossan opisthobranchs: Symbiosis with algal chloroplasts; translocation of photosynthetic products from chloroplast to host tissue. Malacologia 10(2):357–380.
McLaughlin, J. J. A., and P. Zahl. 1966. Endozoic algae, pp. 257–297. *In:* Henry, S. M. (ed.). Symbiosis. Vol. I, Associations of microorganisms, plants and marine organisms. Academic Press.
Nielsen, C. 1964. Studies on Danish Entoprocta. Ophelia 1(1):1–76.
Rosewater, J. 1965. The family Tridacnidae in the Indo Pacific. Indo-Pacific Mollusca 1(6):347–396.
Schlichter, D. 1976. Macromolecular mimicry: Substances released by sea anemones and their role in the protection of anemone fish, pp. 438–441. *In:* Mackie, G. O. (ed.). Coelenterate ecology and behavior. Plenum Press, N.Y.
Schoenberg, D. A. and R. K. Trench. 1976. Specificity of symboioses between marine, cnidarians and zoo xauthellae, pp. 423–432. In: Mackie, G. O. (ed). Coelenterate ecology and behavior. Plenum Press, N.Y.

CHAPTER ELEVEN

Banner, A. H. 1974. Kaneohe Bay, Hawaii: Urban pollution and a coral reef ecosystem, pp. 685–702. *In:* Proceedings of the Second International Coral Reef Symposium, Vol. 2. Great Barrier Reef Committee, Brisbane.
Banus, M. D., and S. E. Kolehmainen. 1976. Rooting and growth of red mangrove seedlings from thermally stressed trees, pp. 46–53. *In:* Esch, G. W., and R. W. McFarlane (eds.). Thermal ecology, II. Washington Technical Information Center of the Energy Research and Development Administration.
Bardach, J. E. 1968. Aquaculture. Science 161:1098–1106.
Bascom, W. (ed.). 1978. Coastal water research project annual report. 253 pp.
Carlton, J. 1975. Introduced intertidal invertebrates, pp. 17–25. *In:* Smith, R. I., and J. Carlton (eds.). Light's manual: Intertidal invertebrates of the central California coast, 3rd ed. Univ. of Calif. Press, Berkeley.
Culley, M. 1971. The pilchard, biology and exploitation. III, The California sardine, Pergamon Press, pp. 143–176.
Cushing, D. H. 1969. Upwelling and fish production. FAO Fish. Tech. Paper 84:1–40.
FAO. 1978. 1977 yearbook of fishery statistics, Vol. 44. FAO, Rome.
Goldberg, E. D. 1974. Marine pollution: Action and reaction times. Oceanus 18(1):6–18.
Goldberg, E. D. 1976. The health of the oceans. UNESCO Press, Paris. 172 pp.
Hess, W. 1978. The AMOCO CADIZ oil spill. A preliminary scientific report. NOAA/EPA. 354 pp.

Marr, J. C. 1960. The causes of major variations in the catch of the Pacific Sardine, *Sardinops caerulea*. Spec. Sci. Rept., U.S. Fish and Wildlife Ser (208):108–125.
McVay, S. 1966. The last of the great whales. Sci. Amer. 215:13–21.
Nelson-Smith, A. 1973. Oil pollution and marine ecology. Plenum, N.Y. 260 pp.
Schaefer, M. B. 1965. The potential harvest of the sea. Trans. Amer. Fish. Soc. 94:123–128.
Soutar, A., and J. D. Isaacs. 1969. History of fish populations inferred from fish scales in anaerobic sediments off California. Calif. Mar. Res. Comm., CalCOFI Rept. 13:63–70.

INDEX

Boldface type indicates the page on which the organism is illustrated; "p" indicates a color plate; "t" indicates a table.

85 9 8 7 6 5